Ergebnisse der Mathematik und ihrer Grenzgebiete

3. Folge · Band 4

A Series of Modern Surveys in Mathematics

Editorial Board

E. Bombieri, Princeton S. Feferman, Stanford
N. H. Kuiper, Bures-sur-Yvette P. Lax, New York
R. Remmert (Managing Editor), Münster
W. Schmid, Cambridge, Mass. J-P. Serre, Paris
J. Tits, Paris

W. Barth C. Peters A. Van de Ven

Compact Complex Surfaces

Springer-Verlag
Berlin Heidelberg New York Tokyo 1984

W. Barth
Mathematisches Institut
Universität Erlangen-Nürnberg
Bismarckstraße $1^1/_2$, D-8520 Erlangen

C. Peters
A. Van de Ven
Mathematisch Instituut, Rijksuniversiteit Leiden
Wassenaarseweg 80, NL-Leiden

AMS Subject Classification (1980):
14C30, 14Dxx, 14Jxx, 14J15, 32Gxx, 32J15, 32J25, 53C15, 53C55

ISBN 3-540-12172-2 Springer-Verlag Berlin Heidelberg New York Tokyo
ISBN 0-387-12172-2 Springer-Verlag New York Heidelberg Berlin Tokyo

Library of Congress Cataloging in Publication Data
Barth, Wolf.
Compact complex surfaces.
(Ergebnisse der Mathematik und ihrer Grenzgebiete; 3. Folge, Bd. 4)
Bibliography: p.
Includes index.
1. Surfaces, Algebraic. 2. Complex manifolds. I. Ven, A. Van de (Antonius) II. Peters, C. A. M. (Chris A. M.) III. Title. IV. Series. QA571.B37 1984 516.3′5 84-1421
ISBN 0-387-12172-2 (U.S.)

Printed in Germany

Typesetting, printing and bookbinding: Universitätsdruckerei H. Stürtz AG, Würzburg.
2141/3140-543210

Preface

> Par une belle matinée du mois de mai,
> une élégante amazone parcourait, sur
> une superbe jument alezane, les
> allées fleuries du Bois de Boulogne.
>
> (A. Camus, La Peste)

Early versions of parts of this work date back to the mid-sixties, when the third author started to write a book on surfaces. But for several reasons, in particular the appearance of Šafarevič's book, he postponed the project. It was revived about ten years later, when all three authors were in Leiden.

It is impossible to cover in one book the vast and rapidly developing theory of surfaces. Choices have to be made, with respect to content as well as to presentation. We have chosen for a complex-analytic point of view; this distinguishes our text from most of the existing treatments. Relations with the case of characteristic p are not discussed.

We hope to have succeeded in writing a readable book; a book that can be used by non-specialists. The specialist will find very little that is new to him anyhow.

As to acknowledgements, the authors certainly have to mention the Koninklijke Shellprijs, awarded to the third author in 1964. The numerous contacts with colleagues from other countries made possible by that award have had a very favourable influence on this book.

Our thanks are furthermore due to G. Angermüller, G. Barthel, G. Fischer, G. van der Geer, N. Hitchin, D. Husemoller, M. Reid, T.A. Springer, D. Zagier and S. Zucker. Each of them has read some part of the manuscript and has made valuable suggestions.

Editor and printer have done an excellent job, and the Springer-Verlag has been very generous in fulfilling all of our last-minute wishes.

We are also indebted to Mrs. W.M. Van de Ven who not only typed the better part of the book, but also helped in preparing it for the printer, and to Mrs. H. Dohrman who carefully typed many pages.

Finally the authors want to thank their wives for all their patience and endurance.

Erlangen/Leiden, February 1984

W. Barth
C. Peters
A. Van de Ven

Table of Contents

Introduction

Historical Note

This book is mainly concerned with the classification of smooth compact complex surfaces, i.e. of compact 2-dimensional complex manifolds, which in this introduction we shall always assume to be connected*).

Surface theory has its roots on the one hand in projective geometry and on the other in Riemann's theory of algebraic functions of a single variable. As to projective geometry, around the middle of the last century an extensive study was made of (smooth as well as singular) low-degree surfaces in complex-projective 3-space $\mathbb{P}_3$. The twenty-seven lines on a smooth cubic and names such as Cayley cubic and Steiner quartic remind us of that period. The extension of Riemann's work, in geometric form, will always be associated with mathematicians like Clebsch and M. Noether, whereas the topological and transcendental approach is linked to Poincaré and others, in particular Picard.

Soon attention was focused on a classification of algebraic surfaces with respect to birational equivalence. The classical geometers clearly had in mind something similar to what was known for curves: a coarse classification according to the value of some numerical invariants, and then a finer classification. At the beginning of this century Castelnuovo, Enriques and many others had succeeded in creating an impressive, essentially geometric theory of birational classification for smooth algebraic surfaces. (This was in fact a birational classification of *all* algebraic surfaces, smooth or not, since every algebraic surface is birationally equivalent to a smooth one; but a rigorous proof for this fact was given for the first time by R. Walker in 1935.) Among the main birational invariants, it was discovered, are the *irregularity* or, equivalently, the first Betti number $b_1(X)$ and the *plurigenera* $P_n(X)$ of a smooth algebraic surface X. For any $n \geq 1$ the n-th plurigenus $P_n(V)$ of a smooth algebraic variety V is defined as the dimension of the space of sections $\Gamma(V, \mathcal{K}_V^{\otimes n})$, where $\mathcal{K}_V = \bigwedge^n \mathcal{T}_V^{\vee}$ is the canonical line bundle of V. Traditionally, the first plurigenus $P_1(V)$ is denoted by $p_g(V)$, and called the *geometric genus* of V.

Given any smooth algebraic surface X, there are four possibilities:

1) all $P_n(X)$ vanish;
2) not all $P_n(X)$ vanish, but all are either 0 or 1;

*) From Chap. II on the meaning of the word "surface" in a given chapter is defined at the very beginning of that chapter. In Chap. I there is no danger of confusion.

3) $P_n(X) \sim n$;

4) $P_n(X) \sim n^2$.

Nowadays this fact is expressed by saying that the *Kodaira dimension* kod(X) of X is either $-\infty$, 0, 1 or 2. (For a precise definition of this concept, due to Iitaka, we refer to Chap. I, Sect. 7.) For curves the corresponding classification is the division into the rational curve $\mathbb{P}_1$ (Kodaira dimension $-\infty$), elliptic curves (Kodaira dimension 0), and curves of higher genus (Kodaira dimension 1).

It was known at the time which surfaces are in class 1), namely those surfaces which are birationally equivalent to the product of $\mathbb{P}_1$ and another curve. This includes in particular the rational surfaces, i.e. those which are birationally equivalent to $\mathbb{P}_2$. A key stone of the proof was Castelnuovo's criterion: a smooth surface X is rational if and only if its first Betti number $b_1(X)$ and its bigenus $P_2(X)$ vanish.

As to class 2), the classical geometers knew that there is a subdivision into four types, distinguished by the values of the first Betti number and the plurigenera, namely into surfaces, birationally equivalent to respectively algebraic tori, hyperelliptic surfaces, algebraic K 3-surfaces and Enriques surfaces (the names are the modern ones). The precise classification of the first two types was known, but not that of the last two types.

It had also been established that all surfaces in class 3) are elliptic (i.e. admitting a map onto a curve such that all but a finite number of fibres are elliptic curves); however, not too much was known about their further classification.

Finally, the surfaces in class 4), which are analogous to curves of genus ≥ 2, were (and still are) called surfaces of general type. The classical geometers certainly had the right idea of how to classify them, but – contrary, say, to the case of Castelnuovo's criterion or the case of hyperelliptic surfaces – they never arrived at precise results or even precise statements. Before we explain a little bit the present state of this classification, we first have to make a few remarks of a more general nature.

Today it is standard to look at the above classification of algebraic surfaces (the *Enriques classification* for algebraic surfaces) in a slightly different way. The basic idea: first a classification according to Kodaira dimension, and then a finer classification, remains the same, but it is seen as a *biregular* classification of *minimal* smooth algebraic surfaces, i.e. surfaces, which can't be obtained from another *smooth* algebraic surface by blowing up a point. Every smooth surface X can be obtained from such a minimal surface Y by blowing up. At first sight it might seem that classifying only minimal surfaces is not very satisfactory, because one and the same surface X might be obtained by blowing up different minimal surfaces Y. However, always kod$(Y)=$kod(X), and if kod$(X)\geq 0$, then Y is determined by X up to an isomorphism. So even from the biregular point of view it is sufficient to classify minimal surfaces, at least in the case of Kodaira dimension ≥ 0. (If kod$(X)=-\infty$, then different Y's can give the same X, but this case is rather easy to handle.) Furthermore, a birational transformation between two minimal surfaces of Kodaira dimension ≥ 0 is always an isomorphism, in other words, for Kodaira dimension ≥ 0

birational classification of all surfaces amounts to biregular classification of minimal surfaces. And, most important, whereas from the birational point of view good moduli spaces never exist, they do exist for many of the finer classes in the case of minimal surfaces.

Now let us return to surfaces of general type. We consider minimal ones X with given Chern numbers $c_1^2(X)=p$, $c_2(X)=q$. It turns out that for $n\geq 5$, any n-canonical map (given by the ratios $\gamma_1:\ldots:\gamma_{N+1}$, where $\gamma_1,\ldots,\gamma_{N+1}$ is a base for $\Gamma(X,\mathscr{K}_X^{\otimes n})$) is everywhere defined on X and maps this surface birationally onto a surface X' of degree $n^2c_1^2(X)$ in $\mathbb{P}_N$, with N only depending on p, q and n. Choosing a different base for $\Gamma(X,\mathscr{K}_X^{\otimes n})$ yields a surface X', which is projectively equivalent to X in $\mathbb{P}_N$. In this way minimal surfaces of general type with fixed c_1^2 and c_2 correspond one-to-one to the points of the quotient of a Zariski-open subset in a Chow variety (or Hilbert scheme) by a projective-linear group. Of course, one wants this quotient to be a variety of moduli (coarse, at least) for the surfaces under consideration. A theorem of Gieseker (1977), based on Mumford's geometric invariant theory says that for n large it indeed is.

Perhaps it should be mentioned at this point, that the results obtained by the classical geometers, their importance and wealth notwithstanding, were in many ways built on sand, for the foundations of algebraic geometry were lacking.

The years 1910–1950 did not bring too much change as far as the classification of surfaces is concerned. In the first two of these decades we see continuing great progress in the general theory of algebraic varieties, from the geometric point of view (Severi) as well as from the transcendental point of view (Lefschetz). But for both directions, a solid base was still not available. Such a base was laid in the thirties and forties, on the one hand for geometry by van der Waerden, Zariski, Weil and on the other hand for much of the topological and transcendental approach by de Rham and Hodge.

Once the foundations were present, some of the classification questions were taken up again and also considered for other ground fields: minimal models, Castelnuovo's criterion (Zariski), Enriques surfaces (M. Artin).

Decisive progress came only after the second revolution, i.e. after sheaf theory had been developed, and applied by Serre, Hirzebruch, Grothendieck and many others to analytic and algebraic geometry.

On this base Kodaira not only extended the classical results on algebraic surfaces in an essential way, but also treated non-algebraic surfaces. For these surfaces the plurigenera and Kodaira dimension can be defined in the same way as for algebraic surfaces, and thus the Enriques classification is extended to the *Enriques-Kodaira classification* of all compact, complex surfaces.

As to the algebraic surfaces, it will hardly surprise anybody that Kodaira gave the Enriques classification the necessary precision and solid base. But he went further in many directions. For example, he did the first step towards the classification of K3-surfaces. A K3-surface is a compact complex surface X with $b_1(X)=0$ and $\mathscr{K}_X$ trivial. As we have mentioned, the classification of the algebraic ones among them (which form a minority) was already an important problem in older times. Since the fifties they have been studied intensively, the main goal being to prove a conjecture of Andreotti and Weil about their classification (compare the comments at the end of Chap. VIII). Kodaira verified part of this conjecture, namely that all K3-surfaces are complex-analytic deformations of each other. The deformation theory of complex manifolds, which Kodaira created together with Spencer, realised at least part of an old ideal of Riemann and Noether: to have a theory of moduli for curves and surfaces. Another contribution of Kodaira, of far-reaching influence and significance, was his extensive study of (algebraic and non-algebraic) elliptic surfaces, something that had definitely been lacking in the work of the Italian geometers.

Though the concept of an n-dimensional complex manifold had been known implicitly for a long time (certainly since Weyl's "Die Idee der Riemannschen Fläche"), it appeared explicitly only around 1945, in the work of Ehresmann and H. Hopf. In particular, Hopf constructed an entirely new class of compact complex surfaces (the first example of what now are called the Hopf surfaces) which are topologically very different from any algebraic surface. A Hopf surface has first Betti number 1 and second Betti number 0, whereas a smooth algebraic surface always has an even first and a strictly positive second Betti number. So, contrary to tori and K3-surfaces, a Hopf surface can never be deformed into an algebraic surface.

This example shows that there is much more to non-algebraic surfaces than deforming some algebraic ones. It was again Kodaira who started with the classification of non-algebraic surfaces in general, and he completed this task to a considerable degree. In his papers he uses Atiyah-Singer's Riemann-Roch theorem in an essential way.

As to non-algebraic deformations of algebraic surfaces, their significance for a better understanding of algebraic surfaces arises clearly from the Andreotti-Weil conjecture and Kodaira's work. This point is already obvious from the case of tori, but it gains more weight if K3-surfaces and elliptic surfaces are taken into consideration.

It would be wrong to think that with Kodaira the theory of surfaces more or less came to its end. On the contrary, the interest in surfaces has only been increasing since the days that Kodaira produced most of his results, in the late fifties and the sixties.

One of the main centres of interest has already been mentioned: the conjecture of Andreotti and Weil on the classification of K3-surfaces. After important contributions by many mathematicians, in particular Šafarevich and Piateckii-Shapiro, the most important parts of the conjecture could finally be proved, but only with the help of S.-T. Yau's differential-geometric results on Calabi's conjecture.

Another centre of attention was the classification of (minimal) surfaces of general type. We spoke already about Gieseker's theorem, saying that for each ordered pair $(p,q)\in\mathbb{Z}\times\mathbb{Z}$ there is a (sometimes empty) coarse moduli scheme for minimal surfaces of general type X with $c_1^2(X)=p$, $c_2(X)=q$. The next question is of course: when is this scheme not empty? In this direction an important result was obtained in 1976 by S.-T. Yau and Miyaoka, who independently proved an older conjecture of Van de Ven, saying that for every surface X of general type the inequality $c_1^2(X) \leq 3c_2(X)$ holds. Yau obtained the inequality as a consequence of his famous work on the Calabi conjecture. Miyaoka was very much inspired by Bogomolov, who only proved the weaker inequality $c_1^2 \leq 4c_2$, but linked the question in an exciting way to the theory of stable vector bundles.

Much research was also done on surfaces with special, mostly low values of c_1^2. For example, already many years ago Severi had raised the question, whether there exist any surfaces which are homeomorphic, but not algebraically isomorphic to $\mathbb{P}_2$. It took a long time before the final answer was given by S.-T. Yau, who proved that every complex surface which is homeomorphic to $\mathbb{P}_2$ is also algebraically isomorphic to $\mathbb{P}_2$. However, as was shown by Mumford using p-adic geometry, there does exist at least one "fake projective plane", i.e. a smooth algebraic surface with $b_1 = 0$, $b_2 = 1$, which is different from $\mathbb{P}_2$. (Such a surface must be of general type and it must have infinite fundamental group.) In spite of many efforts, a direct geometric construction (within the framework of complex algebraic geometry) of such a fake projective plane is still lacking.

As a last example of a classification problem that saw much progress in recent years we mention the case of (minimal) surfaces without non-constant meromorphic functions. Kodaira had already classified these surfaces, except for surfaces with $b_1 = 1$, $b_2 = 0$, on which there are no curves, and surfaces with $b_1 = 1$, $b_2 \geq 1$. For years no example of either class was known. In 1971 Inoue found some examples of surfaces in the first class, and three years later he showed that also the second one is not empty. Since then Inoue, Bombieri, Kato and Enoki have produced many of these surfaces.

We could go on in this way, but we only wanted to indicate the progress made possible by the introduction of sheaf-theoretical methods and the use of new results in other fields. To finish, we mention two developments which don't belong to our subject, but are closely related to it.

Firstly, the extension of the Enriques classification to characteristic p by Bombieri and Mumford. By and large the classification does not differ too much from the complex-algebraic case, but much less is known about the finer classification.

Secondly, the development by Iitaka, Ueno, Viehweg, Kawamata, Mori and others of a classification theory for higher-dimensional manifolds. Since in this case uniquely determined smooth minimal models don't exist, for the time being the classification has become a birational one. The starting point is again a classification according to Kodaira dimension and – at least in dimension 3 – already much is known about the finer division.

A central role plays Iitaka's conjecture $C_{m,n}$: if X and Y are smooth, compact, irreducible algebraic varieties, of dimension m and n respectively, and if $f: X \to Y$ is a surjective morphism, then $\mathrm{kod}(X) \geq \mathrm{kod}(Y) + \mathrm{kod}(\text{general fibre of } f)$. The conjecture has been proved for certain values of m and n, in particular for the case $m = 2$, $n = 1$. This case will be treated in Chap. III.

References

Classical results in general: [C-E].
Classical theory of the Enriques classification: [Enr1], [Enr2], [Ge].

Desingularisation: [Za 3], [Li].
Zariski's work on minimal models and the Castelnuovo criterion: [Za 1], [Za 2].
M. Artin's work on Enriques surfaces: [An 1].
Enriques classification in characteristic p: [Mu 5], [B-M 1], [B-M 2].
Classification theory for higher-dimensional varieties: [Ue 1], [Es].
The other subjects mentioned are treated further on in this book.

The Content of the Book

As has been explained in the preceding section, the classification of compact, complex surfaces amounts to the classification of minimal surfaces. This is first of all a classification according to Kodaira dimension, which for a surface can assume the values $-\infty$, 0, 1 and 2. A refinement of this very coarse classification is the Enriques-Kodaira classification, a description of which is a first purpose of this book. Some of the classes occurring in the Enriques-Kodaira classification can easily be described in detail, but the others: minimal surfaces of class VII (i.e. minimal surfaces X with $b_1(X)=0$, $\mathrm{kod}(X)=-\infty$ *)), K3-surfaces, Enriques surfaces, minimal properly elliptic surfaces and minimal surfaces of general type, require further investigation. Apart from the Enriques-Kodaira classification, this book is mainly devoted to a deeper study of some of these classes, namely K3-surfaces, Enriques surfaces and surfaces of general type. On the other hand, surfaces of class VII and properly elliptic surfaces will not be treated in detail. For elliptic surfaces a number of general properties as well as a certain classification can be found in Chap. V, but these don't give "families" and cover only a small part of what is known in this direction. Surfaces of class VII occur only by way of examples, and neither the beautiful considerations of Kodaira on Hopf surfaces nor most of the work of Inoue, Kato and Enoki on surfaces without non-constant meromorphic functions can be found in this book.

It goes without saying that in a book like the present one many auxiliary results can only be quoted. As to general theorems on complex and algebraic manifolds or spaces, it has been a difficult question for us to decide whether (special) proofs for the 2-dimensional case should be included or not. Sometimes, when a more elementary treatment is available for the 2-dimensional case, we have explained this in detail. For example, we don't refer to Hironaka for the resolution of surface singularities. The 2-dimensional case is infinitely much simpler and its direct treatment very rewarding. But at other places we have used a general theorem in spite of the fact that for surfaces an elementary approach exists. For example, in Chapter IV we derive the fundamental projectivity criterion (Theorem IV.5.2) using Grauert's general ampleness theorem, though it would have been possible to avoid this by using a method of Chow

*) Our definition of a surface of class VII is slightly different from Kodaira's, compare Chap. VI, Sect. 1.

and Kodaira. The method we use is shorter, whereas in the elementary proof there is no idea that doesn't occur already elsewhere in the book.

The content of the different chapters is as follows.

In Chapter I we collect – practically without proofs – most of the definitions and results from topology, algebra, differential geometry, analytic geometry and algebraic geometry which we shall need. Since we would like to make our book as useful as possible for non-specialists, we have thought it better to deviate a little bit from the logical order by collecting some fundamental, but more technical results on complex spaces only after we have dealt with manifolds. This concerns tools like the semi-continuity theorem, the base change theorem and the comparison theorem of Grauert.

Chapter II is devoted to (possibly non-reduced) curves on (not necessarily compact) surfaces: dualising sheaf, Picard variety, singularities and their resolution. The Riemann-Roch theorem is reduced to the smooth case and Serre duality derived from the reduced projective case. Simple curve singularities are classified. Analytic intersection numbers are defined for divisors and shown to be the same as the topological ones. In the last section the foundations are laid for the vanishing theorems of Ramanujam and Mumford, which will be proved in Chap. IV.

The first part of Chap. III deals with surface singularities, their resolution and the converse of this process, the blowing down of exceptional curves. The results are applied to study bimeromorphic maps and minimal models. Rational double points and their relations with simple curve singularities are treated with care. The second part of Chap. III is devoted to (proper) curve fibrations of surfaces over curves. The main achievement here is a proof of Iitaka's conjecture $C_{2,1}$ about the Kodaira dimension of such fibrations. We base it on properties of the period map for stable curves, the Satake compactification and the Torelli theorem for curves.

Chapter IV is not very homogeneous. We have collected in this chapter several general theorems about surfaces which will play an important role later on in the book. The first sections deal with special features of the transcendental theory (differential forms) on compact surfaces. The main point is that for a compact surface the Fröhlicher spectral sequence always degenerates. Combining the consequences of this fact with the topological index theorem we find, following Kodaira, relations between topological and analytic invariants which are crucial in handling non-algebraic surfaces. We also prove the important signature theorem (known as algebraic index theorem in the case of algebraic surfaces). From the other subjects treated in this chapter we mention projectivity criteria (with an application to almost-complex surfaces without any complex structure) and the vanishing theorems of Ramanujam and Mumford.

As to Chap. V (Examples), we first of all have included this chapter as a preparation for the next one, where many of the examples occur as classes in the Enriques-Kodaira classification. Secondly we have thought that the inclusion of such a list might again be helpful to non-specialists.

In Chap. VI we present the Enriques-Kodaira classification. Our treatment is based on Castelnuovo's criterion (proved in this chapter), Iitaka's conjecture $C_{2,1}$ (proved in Chap. III) and the fundamental results of Chap. IV. At the end we apply classification to deformations of surfaces. We are aware of the possibility to avoid $C_{2,1}$, but we prefer to prove $C_{2,1}$ independently of classification, because this seems to be the most promising approach in higher dimensions.

Chapter VII is about surfaces of general type. Following closely Bombieri we base our extensive treatment of pluricanonical maps on Ramanujam's vanishing theorem. Since we don't know any simple proof for Gieseker's theorem, we have chosen to save space by referring to the original paper. In dealing with inequalities for Chern numbers, we present a simplified version of Miyaoka's proof for the inequality $c_1^2 \leq 3c_2$, as well as the more standard proofs for Noether's inequality and the inequalities $c_1^2 > 0$, $c_2 > 0$ for minimal surfaces of general type. (We don't treat Bogomolov's work on the stability of the tangent bundle.) In the last part of this chapter we describe various methods of constructing surfaces, and we list the results about Chern numbers and Gieseker schemes which these methods have yielded.

Chapter VIII deals with K3-surfaces and Enriques surfaces. We fully prove some of the main results which have been obtained during these last years: the Torelli theorem for marked K3-surfaces, the surjectivity of the period map for K3-surfaces, and the bijectivity of the period map for Enriques surfaces. Following Beauville in [Be3] we also give a short proof, based on work of Gauduchon, for Siu's result that every K3-surface is kählerian.

Standard Notations

$\mathbb{Z}$: ring of integers
$\mathbb{Z}_m$: ring of integers mod m
$\mathbb{F}_q$: field with q elements
$\mathbb{N}$: set of strictly positive integers
$\mathbb{Q}$: field of rational numbers
$\mathbb{R}$: field of real numbers
$\mathbb{R}^n$: numerical real vector space of dimension n
$\mathbb{1}_n$: unit matrix of dimension n
S^n: unit sphere in $\mathbb{R}^{n+1}$
$\mathbb{C}$: field of complex numbers
$\mathbb{C}^n$: numerical complex vector space of dimension n
$\mathbb{C}^*$: multiplicative group of complex numbers $\neq 0$
$\mathbb{P}(V)$: projective space of the (real or complex) vector space V, i.e., the space of lines in V
$\mathbb{P}_n = \mathbb{P}(\mathbb{C}^{n+1})$: n-dimensional numerical complex projective space
$\mathcal{O}(1) = \mathcal{O}_{\mathbb{P}_n}(1)$: the hyperplane bundle on $\mathbb{P}_n$
$\mathfrak{e}(z) = e^{2\pi i z}$

I. Preliminaries

Topology and Algebra

1. Notations and Basic Facts

We shall use the standard notations for some algebraic structures which frequently occur. These notations are listed on p. 9.

Let X be a topological space and $\mathscr{S}$ a sheaf of groups on X. We write $\mathscr{S}_x$ for the stalk of $\mathscr{S}$ at $x \in X$. And we shall denote by $H^i(X, \mathscr{S})$ or $H^i(\mathscr{S})$ the i-th cohomology group of X with coefficients in $\mathscr{S}$. For $H^0(X, \mathscr{S})$, the group of sections, we shall also write $\Gamma(X, \mathscr{S})$ or $\Gamma(\mathscr{S})$. If $\mathscr{S}$ is a sheaf of real or complex vector spaces, then the $H^i(X, \mathscr{S})$'s are also real or complex vector spaces, and if $\dim H^i(X, \mathscr{S})$ is finite, then $h^i(X, \mathscr{S}) = h^i(\mathscr{S})$ will denote this dimension. If G is an abelian group and G_X the corresponding constant sheaf on X, then we shall write $H^i(X, G)$ for $H^i(X, G_X)$, and $H^i_c(X, G)$ for the i-th cohomology group of G_X with compact supports.

If G is a ring, then $H^*(X, G) = \sum_{i \geq 0} H^i(X, G)$ is made into a graded ring by the cupproduct. If G has a unit element, then so has $H^*(X, G)$.

We shall often denote the cupproduct $a \cup b$ of two classes a and b by ab, or $a \cdot b$ or even (a, b). We do this mostly in the case that a and b are of complementary dimension on a connected, compact, oriented manifold, so $ab \in G$.

If X and Y are topological spaces, $\mathscr{S}$ a sheaf of groups on X, and $f: X \to Y$ a continuous map, then we shall write $f_{*i}(\mathscr{S})$ or simply $f_{*i}\mathscr{S}$ for the i-th direct image of $\mathscr{S}$ by f, but mostly just $f_*(\mathscr{S})$ or $f_*\mathscr{S}$ for the 0-th direct image. On the other hand, if $\mathscr{T}$ is a sheaf on Y, then $f^{-1}(\mathscr{T})$ will be the inverse image on X (the notation f^* will be reserved for the analytic inverse image, see Sect. 8).

For every continuous map $f: X \to Y$ and every sheaf $\mathscr{S}$ on X there is the Leray spectral sequence ([Go], Chapitre II, Théorème 17.1) with $E_2^{p,q} = H^p(f_{*q}(\mathscr{S})) \Rightarrow H^{p+q}(\mathscr{S})$. The beginning of this spectral sequence leads to the exact sequence

$$0 \to H^1(f_*(\mathscr{S})) \to H^1(\mathscr{S}) \to H^0(f_{*1}(\mathscr{S})) \to H^2(f_*(\mathscr{S})).$$

When we speak of a (mostly complex) manifold, we shall always assume it to be paracompact.

If X is an oriented, connected n-dimensional manifold, then Poincaré duality yields a canonical isomorphism

$$\mathscr{P}_X\colon H^i_c(X,G) \xrightarrow{\sim} H_{n-i}(X,G)$$

(in particular, $H^n_c(X,G) \simeq H_0(X,G) \simeq G$). This duality induces a product on $H_*(X,G)$, which for $G=\mathbb{Z}$ is nothing but the intersection product ([Dd], p. 336). We shall frequently switch between cohomology with compact supports and homology in this case, without further notice. An element of $H^n_c(X,\mathbb{Z})$ or $H_0(X,\mathbb{Z})$ will be seen as an integer.

If Y is a second connected, oriented manifold of dimension m and $f\colon X \to Y$ a continuous map, then we shall denote by

$$f^!\colon H_i(Y,\mathbb{Z}) \to H_{n-m+i}(X,\mathbb{Z})$$

the homomorphism $\mathscr{P}_X f^* \mathscr{P}_Y^{-1}$, whereas the homomorphism $\mathscr{P}_Y^{-1} f_* \mathscr{P}_X$ will be denoted by $f_!$.

(1.1) **Lemma** (Projection formula). *Let X and Y be connected, oriented manifolds, and $f\colon X \to Y$ a proper continuous map. Then*

$$f_!(x \cdot f^*(y)) = f_!(x) \cdot y \tag{1}$$

for all $x \in H^*_c(X,\mathbb{Z})$, $y \in H^*_c(Y,\mathbb{Z})$.

For a proof we refer to [Dd], p. 314.

(1.2) **Corollary.** *Let X and Y be compact, connected oriented manifolds of the same dimension. If $f\colon X \to Y$ is a continuous map of degree different from 0, then $f^*\colon H^*(Y,\mathbb{R}) \to H^*(X,\mathbb{R})$ is an injective ring homomorphism.*

(1.3) *Remark.* If in particular X and Y are complex manifolds and f a surjective holomorphic map, then the degree $\deg(f)$ of f is strictly positive and the conclusion of the Corollary applies.

For certain cases, which play an important role in this book, we shall consider in Chap. II both intersection theory and Poincaré duality more in detail.

When X is a compact oriented n-dimensional manifold, we write $T^i(X)$ for the torsion subgroup of $H^i(X,\mathbb{Z})$ and put $H^i(X,\mathbb{Z})_f = H^i(X,\mathbb{Z})/T^i(X)$. Then we have

$$\operatorname{rank}(H^i(X,\mathbb{Z})_f) = \dim_{\mathbb{R}} H^i(X,\mathbb{R}) = \dim_{\mathbb{C}} H^i(X,\mathbb{C}) = b_i(X),$$

the i-th Betti $b_i(X)$ number of X. Duality implies that $b_i(X) = b_{n-i}(X)$ and $T^i(X) \simeq T^{n-i+1}(X)$ (see [Dd], p. 167).

As for the i-th homotopy group of an arcwise connected space X, we use the standard notation $\pi_i(X)$ (or $\pi_i(X,p)$ if the base point $p \in X$ plays a role), and in particular we write $\pi_1(X)$ for the fundamental group of X.

2. Some Properties of Bilinear Forms

By a lattice $(L,\langle\ ,\ \rangle)$ we shall mean a finitely generated free $\mathbb{Z}$-module L, endowed with an integral bilinear form $\langle\ ,\ \rangle$ which is either symmetric or skew-symmetric. In the first case we speak of a *euclidean lattice*, in the second case of a *symplectic lattice*. Frequently, when confusion is unlikely, we speak of the lattice L instead of $(L,\langle\ ,\ \rangle)$.

If $\{e_1,\ldots,e_n\}$ is a basis for L, then it is easily seen that the determinant of the matrix $(\langle e_i,e_j\rangle)$ is determined uniquely, independent of the choice of basis. This number $d(L)$ is called the *discriminant* of the lattice. The lattice $(L,\langle\ ,\ \rangle)$ is non-degenerate if $\langle\ ,\ \rangle$ is non-degenerate, i.e. $d(L)\neq 0$, and $(L,\langle\ ,\ \rangle)$ is unimodular if $d(L)=\pm 1$.

Let $(L,\langle\ ,\ \rangle)$ be a lattice and $L^\vee=\mathrm{Hom}_{\mathbb{Z}}(L,\mathbb{Z})$ the dual of L. The *correlation morphism* $\phi\colon L\to L^\vee$ is defined by

$$\phi(x)=\langle\ ,x\rangle.$$

(2.1) **Lemma.** *If* $(L,\langle\ ,\ \rangle)$ *is a non-degenerate lattice, then*
(i) *the index of* $\phi(L)$ *in* $L^\vee$ *is* $|d(L)|$,
(ii) *if* M *is a submodule of* L *with* $\mathrm{rank}(M)=\mathrm{rank}(L)$, *then*

$$(L\colon M)^2=d(M)\,d(L)^{-1}.$$

The second assertion follows immediately if you write a basis for M in terms of a basis for L and observe that the determinant of the resulting matrix up to sign is $(L\colon M)$. Applying the same remark to L and $L^\vee$ yields (i).

The next lemma is a simple exercise in linear algebra over $\mathbb{Q}$.

(2.2) **Lemma.** *If* L *is a non-degenerate lattice, and* M *a submodule of* L, *then*

$$\mathrm{rank}\,M+\mathrm{rank}\,M^\perp=\mathrm{rank}\,L.\quad\square$$

A symplectic form can be non-degenerate only if the rank n of L is even.

(2.3) **Lemma.** *If* $(L,\langle\ ,\ \rangle$ *is a non-degenerate, symplectic lattice of rank* $2g$, *equipped with a non-degenerate symplectic form, then there exists a basis* $\{a_1,\ldots,a_g,b_1,\ldots,b_g\}$ *for* L *and natural numbers* $\{d_1,\ldots,d_g\}$ *with* $d_i|d_{i+1}$ $(i=1,\ldots,g-1)$ *such that the matrix of* $\langle\ ,\ \rangle$ *has the form*

$$\begin{pmatrix}0 & -\Delta\\ \Delta & 0\end{pmatrix}\quad\textit{with}\quad \Delta=\begin{pmatrix}d_1 & & & \\ & d_2 & & \\ & & \ddots & \\ & & & d_g\end{pmatrix}.$$

For a proof we refer to [Bou2], Chap. 9, 5.1.

A basis $\{a_1,\ldots,b_g\}$ as in the previous lemma is called *canonical*. For a unimodular symplectic form all d_i's are necessarily 1 and the matrix in Lemma 2.3 is:

$$\begin{pmatrix} 0 & -\mathbb{1}_g \\ \mathbb{1}_g & 0 \end{pmatrix} \quad \textit{(the standard symplectic form)}$$

The submodules of L spanned by $a_1, \ldots, a_g$ and by $b_1, \ldots, b_g$ are maximal isotropic submodules of L.

A sublattice M of a lattice L is called *primitive*, if M is a primitive submodule of L, i.e. if L/M is torsion free. If M is primitive, every basis of M can be complemented to a basis of L. In particular this holds if M is a maximal isotropic sublattice of L, since $M = M^{\perp}$ implies that M is primitive. As an application we have:

(2.4) **Lemma.** *If $(L, \langle\, , \rangle)$ is a unimodular, symplectic lattice and $\{a_1, \ldots, a_g\}$ a basis for a maximal isotropic subspace of L, then one can always complement $\{a_1, \ldots, a_g\}$ to a canonical basis.*

Proof. Since $\langle\ ,\ \rangle$ is unimodular, the correlation morphism is an isomorphism. So, if we complement $\{a_1, \ldots, a_g\}$ to a basis of L, every element of the corresponding dual basis for $L^{\vee}$ can be viewed as an element of L. In particular, there are elements $b'_1, \ldots, b'_g \in L$ such that $\langle a_i, b'_j \rangle = \delta_{i,j}$. If $\langle b'_i, b'_j \rangle = t_{i,j}$, we set $b_j = b'_j - \sum_{i=1}^{j-1} t_{i,j} a_i$ $(j \geq 2)$ and it is easily checked that $\{a_1, \ldots, b_g\}$ yields a canonical basis. □

As we have just remarked, if $M \subset L$ is primitive, every basis of M can be extended to a basis of L and in particular $L^{\vee} \to M^{\vee}$ is surjective. So we can extend every $m^* \in M^{\vee}$ to all of L and restrict the result to $M^{\perp}$, thus obtaining an element of $(M^{\perp})^{\vee}$. If we take different extensions of m^* to L, the difference of the two extensions is an element of $L^{\vee}$, which vanishes identically on M. If in addition L is unimodular, every element of $L^{\vee}$ is of the form $\langle -, x \rangle$ for some $x \in L$ and we see that the difference of the two extensions is of the form $\langle -, y \rangle$ for $y \in M^{\perp}$. If $\langle\, , \rangle | M$ is non-degenerate then so ist $\langle\, , \rangle | M^{\perp}$ and the correlation morphisms are embeddings. Moreover, we can consider M and $M^{\perp}$ as submodules of their respective dual modules. In this way we obtain a homomorphism $M^{\vee} \to (M^{\perp})^{\vee}/M^{\perp}$, and since it maps M to 0, we get a homomorphism

(2) $$\psi\colon M^{\vee}/M \to (M^{\perp})^{\vee}/M^{\perp}$$

We have:

(2.5) **Lemma.** *If L is a unimodular lattice and M a primitive sublattice such that $\langle\ ,\ \rangle | M$ is non-degenerate, then the homomorphism* (2) *is an isomorphism.*

Proof. Like any orthogonal complement $M^{\perp}$ is primitive, and since M is primitive, we have $(M^{\perp})^{\perp} = M$. So we can construct a homomorphism $(M^{\perp})^{\vee}/M^{\perp} \to M^{\vee}/M$, which is an inverse for (2). □

(2.6) **Corollary.** *Let L be a unimodular lattice and $M \subset L$ a primitive sublattice, then $|d(M)| = |d(M^{\perp})|$. If moreover M is unimodular, then $L = M \oplus M^{\perp}$.*

Proof. If $d(M)=0$, clearly $d(M^\perp)=0$, so we may assume that M is non-degenerate and in that case the previous lemma applies. Since $(M^\vee : M)=|d(M)|$ by Lemma 2.1, the first statement follows. The second one is an immediate consequence: if $|d(M)|=|d(M^\perp)|=1$, then $|d(M\oplus M^\perp)|=1$ and $M\oplus M^\perp=L$. □

From now on we only consider euclidean lattices $(L,\langle\ ,\ \rangle)$.

Associated to the lattice L we have a *quadratic form* Q defined by $Q(x) =\langle x,x\rangle$. If it takes on even values only, the lattice is called *even*, otherwise it is called *odd*. If $Q(x)>0$ (resp. ≥ 0) for all $x\in L\setminus\{0\}$ it is called *positive definite* (resp. *positive semi-definite*) and similar for *negative (semi-)definite*. A lattice is *definite* if it is either positive or negative definite. A non-degenerate lattice is called *indefinite* if it is not definite.

(2.7) **Examples.** (i) The module $\mathbb{Z}\cdot e$ with $\langle e,e\rangle=\pm 1$ is an odd, unimodular definite lattice. We denote it by $\pm\mathbb{1}$.
(ii) The *hyperbolic plane* H. As a $\mathbb{Z}$-module it is $\mathbb{Z}^2$ and if e_1, e_2 is the standard basis, the matrix $(\langle e_i,e_j\rangle)$ is just $\begin{pmatrix}0&1\\1&0\end{pmatrix}$. It is even, unimodular and indefinite.
(iii) The *root-lattice* E_8. As a $\mathbb{Z}$-module, $E_8=\mathbb{Z}^8$ and on the canonical basis the matrix $(\langle e_i,e_j\rangle)$ is the Cartan matrix of the root system E_8, that is, $2Q(E_8)$ in the notation of Lemma 2.12, or explicitely:

$$\begin{pmatrix}
2 & 0 & -1 & 0 & 0 & 0 & 0 & 0\\
0 & 2 & 0 & -1 & 0 & 0 & 0 & 0\\
-1 & 0 & 2 & -1 & 0 & 0 & 0 & 0\\
0 & -1 & -1 & 2 & -1 & 0 & 0 & 0\\
0 & 0 & 0 & -1 & 2 & -1 & 0 & 0\\
0 & 0 & 0 & 0 & -1 & 2 & -1 & 0\\
0 & 0 & 0 & 0 & 0 & -1 & 2 & -1\\
0 & 0 & 0 & 0 & 0 & 0 & -1 & 2
\end{pmatrix}.$$

The lattice E_8 is even, unimodular and positive definite. Changing all signs yields $-E_8$, a negative definite lattice.

We recall that any quadratic form can be diagonalised over $\mathbb{R}$, that the numbers τ^+, resp. τ^- of positive, resp. negative entries in every diagonalisation over $\mathbb{R}$ are the same, that $\tau^+ + \tau^-$ is called the rank, and $\tau^+ - \tau^-$ the index. Examples of odd unimodular lattices of rank n and index τ are the orthogonal direct sums $\overset{a}{\oplus}\mathbb{1}\overset{b}{\oplus}-\mathbb{1}$, where $a=\frac{1}{2}(n+\tau)$, $b=\frac{1}{2}(n-\tau)$. Examples of even unimodular lattices of rank n and index τ ($\tau\equiv 0 \bmod 8$ by [M-H] Ch. II, Th. 5.1) are the orthogonal direct sums $\overset{a}{\oplus}H\overset{b}{\oplus}\pm E_8$, where $a=\frac{1}{2}(n-\tau)$, $b=\pm\tau/8$. The following theorem whose proof can be found e.g. in [M-H], Chap. II (Theor. 5.3) or [Se6], Chap. V (§2.2–2.3), states that these examples give all possible indefinite unimodular lattices and all definite ones of low rank.

(2.8) **Theorem.** *Any indefinite unimodular lattice is up to isometry determined by its rank, index and parity (i.e. whether it is odd or even). The same holds for definite unimodular lattices of rank at most* 8.

If an even unimodular lattice L contains k hyperbolic planes, any even lattice of rank $\leq k$ can be realised as a primitive sublattice of L. In fact, we have:

(2.9) **Theorem.** *Let L be an even unimodular lattice containing a sublattice isometric to $\overset{k}{\bigoplus} H$ and let Γ be any even lattice.*
(i) If rank $\Gamma \leq k$, *then there exists a primitive embedding i: $\Gamma \to L$, i.e. i is a lattice monomorphism and $i(\Gamma)$ is primitive.*
(ii) If rank $\Gamma \leq k-1$, *and i, j: $\Gamma \to L$ are primitive embeddings, there exists an isometry ϕ of L such that $j = \phi\, i$.*

The proof of (i) is straightforward. Let $\{e_j, f_j\}$ be the standard basis for the j-th copy of H. If $\{c_1, \ldots, c_t\}$ $(t \leq k)$ is a basis of Γ we put

$$i(c_s) = e_s + \tfrac{1}{2}\langle c_s, c_s\rangle f_s + \textstyle\sum_{r<s} \langle c_r, c_s\rangle f_r$$

and it is easily checked that i is an isometry. Since the matrix $(\langle i(c_s), f_r\rangle)$ is the identity, $i(\Gamma)$ must be primitive. For the proof of (ii) we refer to [J], [Pi-S], § 6, appendix or [L-P], § 2.

Finally we state two lemmas, which will turn out to be useful in dealing with intersection forms of curve configurations on surfaces. For a proof of Lemma 2.10 we refer to [Bou3], Chap. 5, § 3.5 and for a short proof of Lemma 2.12, which uses the previous one, we refer to [Dem].

(2.10) **Lemma.** *Let Q be a symmetric bilinear form on $V = \mathbb{R}^n$ or $\mathbb{Q}^n$, given by the matrix $(q_{i,j})$. Suppose*
(i) $Q \geq 0$,
(ii) $q_{i,j} \leq 0$ *for* $i \neq j$,
(iii) there is no partition $I \cup J$ of $\{1, \ldots, n\}$ with $I \neq \emptyset$, $J \neq \emptyset$ such that for $i \in I$, $j \in J$ one has $q_{i,j} = 0$.
Then either $Q > 0$, or its annihilator is 1-dimensional and spanned by a vector $(z_1, \ldots, z_n) \in V$ with $z_i > 0$ for all $i = 1, \ldots, n$.

If instead of (i), (ii), (iii) we assume that Q satisfies (i)′, (ii), (iii) with
(i)′ the annihilator N of Q contains $z = (z_1, \ldots, z_n)$ with $z_i > 0$ for all $i = 1, \ldots, n$, then $Q \geq 0$, $\dim N = 1$ *and N is spanned by z.*

(2.11) **Corollary.** *Let the situation be as in Lemma* 2.10 *and suppose that in addition to* (i), (ii), (iii) *we know that $Q > 0$. Then there exists $x = (x_1, \ldots, x_n) \in \mathbb{Q}^n$ with $x_i > 0$ such that $Q(x, e_i) > 0$ $(i = 1, \ldots, n)$.*

Proof. Let $\lambda \in \mathbb{R}$ be the smallest positive eigenvalue of Q. The form $Q_0 = Q - \lambda\, \mathbb{1}_n$ satisfies (i), (ii), and (iii), so its annihilator is spanned by $y = (y_1, \ldots, y_n)$ with $y_i > 0$. Hence for all $i \in \{1, \ldots, n\}$ we have

$$Q(y, e_i) = \lambda\, y_i > 0.$$

Now by continuity we may find $x \in \mathbb{Q}^n$ close to y such that this inequality holds for x as well. □

(2.12) **Lemma.** *Let Γ be a connected graph with vertices $v_1, \ldots, v_n$ such that v_i and v_j are connected by at most one edge $(i \neq j)$. Let $Q(\Gamma)$ be the quadratic form with matrix $(q_{i,j})$ where $q_{i,i}=1$ $(i=1,\ldots,n)$ and (for $i \neq j$) $q_{i,j}=-\frac{1}{2}$ (resp. 0) if v_i and v_j are connected (resp. are not connected). Then $Q(\Gamma)$ satisfies the conditions (ii) and (iii) of Lemma* 2.11. *If $Q(\Gamma) \geq 0$ (i.e. if the condition (i) is also satisfied) only the following series of graphs occur:*

(i) If $Q(\Gamma) > 0$

A_n

$(n \geq 1$ *vertices*)

D_n

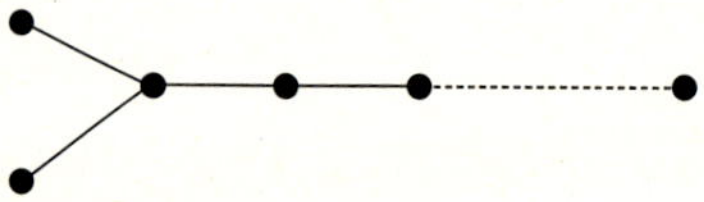

$(n \geq 4$ *vertices*)

E_n

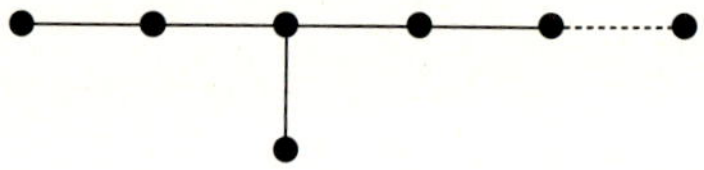

$(n = 6, 7$ *or* $8)$

(ii) If $Q(\Gamma)$ has non-trivial annihilator, the graph Γ is one out of the following list (the weights on the vertices denote the coefficients of a vector spanning the 1-dimensional annihilator):

$\tilde{A}_n$

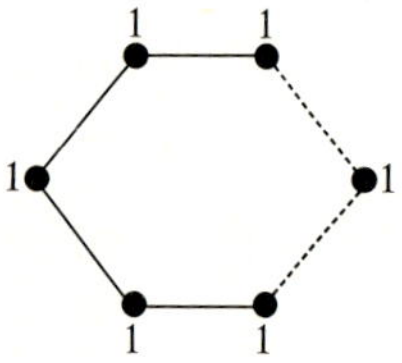

$(n+1$ *vertices*, $n \geq 2)$

$\tilde{D}_n$

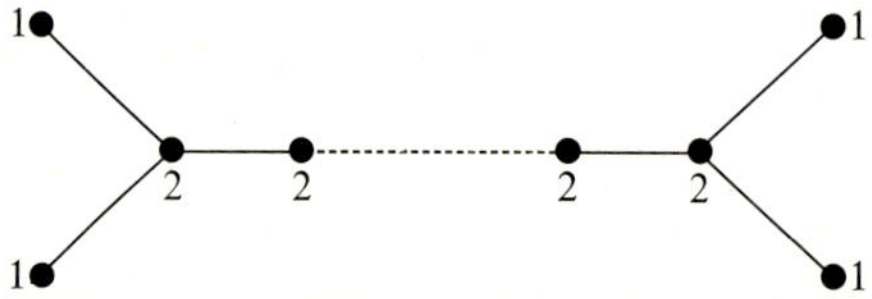

$(n+1$ *vertices*, $n \geq 4)$

$\tilde{E}_6$

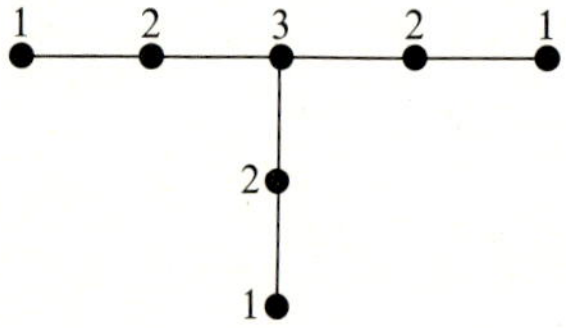

$\tilde{E}_7$

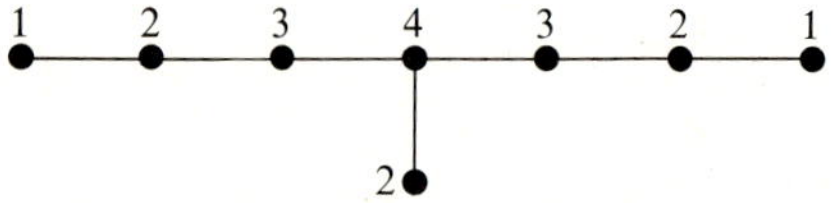

$\tilde{E}_8$

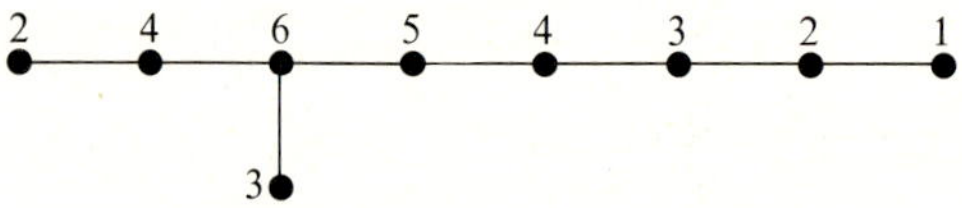

□

3. Vector Bundles, Characteristic Classes and the Index Theorem

Vector bundles, real or complex, will be denoted by $\mathscr{V}, \mathscr{W}, \ldots$, and the fibre of $\mathscr{V}$ over the point x in the base by $\mathscr{V}(x)$. We use of course the standard notations $\mathscr{V} \oplus \mathscr{W}$ for the direct sum of $\mathscr{V}$ and $\mathscr{W}$, $\mathscr{V}^\vee$ for the dual bundle of $\mathscr{V}$, $S^n\mathscr{V}$ for the n-th symmetric product of $\mathscr{V}$, etc.

If $\mathscr{V}$ is a complex vector bundle of rank d, then $\mathbb{P}(\mathscr{V})$ will denote the associated projective bundle, i.e. if ρ: $GL(d,\mathbb{C}) \to PGL(d,\mathbb{C})$ is the canonical epimorphism, and if $\mathscr{V}$ is given by coordinate transformations $\{g_{ij}\}$, then $\mathbb{P}(\mathscr{V})$ is given by the coordinate transformations $\{\rho(g_{ij})\}$. So $\mathbb{P}(\mathscr{V})$ is *not* the associated bundle of $\mathscr{V}^\vee$ as in [Ha]. If $n=2$, then $\mathbb{P}(\mathscr{V})$ and $\mathbb{P}(\mathscr{V}^\vee)$ are isomorphic, since in that case $\mathscr{V}^\vee \simeq \mathscr{V} \otimes \Lambda^2 \mathscr{V}^\vee$.

Let $\mathscr{V}$ be a real r-vector bundle on the paracompact topological space X. Then $w_i(\mathscr{V})$, $i=0,\ldots,r$ will be its i-th Stiefel-Whitney class, and $w(\mathscr{V}) = \sum_{i=0}^{r} w_i(\mathscr{V}) \in H^*(X,\mathbb{Z}_2)$ its total Stiefel-Whitney class. Similarly we have the Pontrjagin classes $p_i(\mathscr{V}) \in H^{4i}(X,\mathbb{Z})$ and the total Pontrjagin class $p(\mathscr{V}) = p_0(\mathscr{V}) + \ldots \in H^*(X,\mathbb{Z})$ if $\mathscr{V}$ is a real r-bundle, as well as the Chern classes $c_i(\mathscr{V}) \in H^{2i}(X,\mathbb{Z})$ and the total Chern class $c(\mathscr{V}) = \sum_{i=0}^{r} c_i(\mathscr{V}) \in H^*(X,\mathbb{Z})$ if $\mathscr{V}$ is a complex r-bundle.

Several times we shall use the fact ([Hir4], p. 73) that if we consider a complex vector bundle $\mathscr{V}$ as a real bundle $\mathscr{V}_{\mathbb{R}}$, then $w_i(\mathscr{V}_{\mathbb{R}})=0$ for i odd, whereas $w_{2i}(\mathscr{V}_{\mathbb{R}}) = c_i(\mathscr{V}) \pmod 2$.

Let again $\mathscr{V}$ be any real r-bundle. If we formally put $w(\mathscr{V}) = \prod_{i=1}^{r}(1+\delta_i)$, then any symmetric power series in $\delta_1,\ldots,\delta_r$ with coefficients in $\mathbb{Z}_2$, yields a polynomial in $w_1,\ldots,w_r$ and hence an element in $H^*(X,\mathbb{Z}_2)$. The same can be done, starting from a power series with rational coefficients, with the Pontrjagin classes and with the Chern classes in the case of a complex vector bundle, yielding in both cases an element $e \in H^*(X,\mathbb{Q})$. The component of e which is of dimension i, will be denoted by $t_i(e)$. Thus, once an isomorphism $H^n(X,\mathbb{Q}) \simeq \mathbb{Q}$ is given, $t_n(e)$ becomes a rational number.

Now let $\mathscr{V}$ be an oriented real r-bundle on X. Then we obtain the L-class $L(\mathscr{V}) \in H^*(X,\mathbb{Q})$ by starting from

$$p(\mathscr{V}) = \prod_{i=1}^{r}(1+\delta_i),$$

and taking the symmetric power series

$$\prod_{i=1}^{r} \frac{\sqrt{\delta_i}}{\tanh\sqrt{\delta_i}}.$$

So

$$L(\mathscr{V}) = 1 + \tfrac{1}{3}p_1(\mathscr{V}) + \tfrac{1}{45}(7p_2(\mathscr{V}) - p_1^2(\mathscr{V})) + \ldots .$$

Similarly, if $\mathscr{V}$ is a complex r-bundle, and $c(\mathscr{V}) = \prod_{i=1}^{r}(1+\delta_i)$, then the Todd class $\mathrm{Todd}(\mathscr{V})$ is obtained from $\prod_{i=1}^{r} \frac{\delta_i}{1-e^{-\delta_i}}$, whereas the Chern character

$\mathrm{ch}(\mathscr{V})$ is derived from $\sum_{i=1}^{r} e^{\delta_i}$. The beginning of $\mathrm{Todd}(\mathscr{V})$ is well known:

$$\mathrm{Todd}(\mathscr{V})=1+\tfrac{1}{2}c_1(\mathscr{V})+\tfrac{1}{12}(c_1^2(\mathscr{V})+c_2(\mathscr{V}))+\tfrac{1}{24}c_1(\mathscr{V})c_2(\mathscr{V})+\dots$$

(so, in our notation, we have for example $t_3(\mathrm{Todd}(\mathscr{V}))=0$ and $t_6(\mathrm{Todd}(\mathscr{V}))=\frac{1}{24}c_1(\mathscr{V})c_2(\mathscr{V})$).

If X is a differentiable manifold, then we denote by $\mathscr{T}_X$ its tangent bundle. Suppose now that X is compact, connected and oriented. Then $L(\mathscr{T}_X)$ is well-defined, and we set $L(X)=L(\mathscr{T}_X)$.

For any compact, connected, oriented (not necessarily differentiable) manifold X we define the index $\tau(X)$ in the following way. If $\dim X \not\equiv 0(4)$, we set $\tau(X)=0$. If $\dim X=4m$, the cupproduct form defines on $H^{2m}(X,\mathbb{R})$ a non-degenerate quadratic form $Q(X)$, and we set $\tau(X)=\tau(Q(X))$, i.e. $\tau(X)=b^+(X)-b^-(X)$, where $b^+(X)$ (resp. $b^-(X)$) is the number of positive (resp. negative) eigenvalues of Q (observe that $b_{2m}(X)=b^+(X)+b^-(X)$).

(3.1) **Theorem** (Topological index theorem = Index theorem of Thom-Hirzebruch). *Let X be a compact, connected, oriented differentiable manifold of dimension $4m$. Then*

$$\tau(X)=t_{4m}(L(X)).$$

In particular, if $m=1$, then $\tau(X)=\frac{1}{3}p_1(X)$, and if in addition X carries an almost-complex structure, then $\tau(X)=\frac{1}{3}p_1(X)=\frac{1}{3}(c_1^2(X)-2c_2(X))$ by [Hir4], p. 65.

For details we refer to [Hir4], p. 86.

Complex Manifolds

4. Basic Concepts and Facts

A 1-dimensional complex manifold will be called a *smooth curve* or a *Riemann surface* and a 2-dimensional one a *smooth surface*.

If X is an n-dimensional*) complex manifold, then we shall denote by

- $\mathscr{T}_X$: the (holomorphic) tangent bundle of X;
- $c_i(X)$: the i-th Chern class of X, that is $c_i(\mathscr{T}_X)$ (so $c_n(X)$ is equal to the Euler number $e(X)$ if X is compact);
- $\mathscr{O}_X$: the structure sheaf of X;
- $\mathscr{O}_X^*$: the sheaf of non-vanishing holomorphic function germs on X (with multiplication as group law);
- Ω_X^i: the sheaf of germs of holomorphic i-forms on X, i.e. the sheaf of sections in the bundle $\bigwedge^i \mathscr{T}_X^\vee$ $(i \geq 1)$;
- $\mathscr{K}_X$: the canonical line bundle on X, i.e. the holomorphic 1-vector bundle $\bigwedge^n \mathscr{T}_X^\vee$;
- $\mathscr{N}_{Y/X}$: the normal bundle in X of the complex submanifold Y.

*) Complex manifolds are always pure-dimensional (see Sect. 8).

We recall that a sheaf of $\mathcal{O}_X$-modules $\mathcal{S}$ is *coherent*, if locally there always is some exact sequence of sheaves of $\mathcal{O}_X$-modules

$$\mathcal{O}_X^p \to \mathcal{O}_X^q \to \mathcal{S} \to 0.$$

If X is compact and $\mathcal{S}$ coherent, then the complex vector spaces $H^i(X,\mathcal{S})$ are finite-dimensional of dimension $h^i(X,\mathcal{S})=h^i(\mathcal{S})$, and they vanish for $i>n$. Hence the Euler characteristic $\chi(X,\mathcal{S})=\sum_{i=0}^{n}(-1)^i h^i(\mathcal{S})$ is well-defined. We put $\chi(X,\mathcal{O}_X)=\chi(X)$. The integer $p_a(X)=(-1)^n(\chi(X)-1)$ is called the *arithmetic (al) genus* of X.

Setting for a moment $\Omega_X^0=\mathcal{O}_X$, we define

$$h^{p,q}(X)=h^q(\Omega_X^p)$$
$$q(X)=h^{0,1}(X)$$
$$p_g(X)=h^{0,n}(X) \quad \textit{(geometric genus)}.$$

5. Holomorphic Vector Bundles, Serre Duality and the Riemann-Roch Theorem

The sheaf of sections of a holomorphic r-vector bundle on the complex manifold X is locally free, i.e. locally isomorphic to $\mathcal{O}_X^r$; conversely every locally free sheaf $\mathcal{S}$ of $\mathcal{O}_X$-modules is the sheaf of sections of a holomorphic r-vector bundle, which is determined up to isomorphism. Because of this equivalence we shall sometimes use for a holomorphic bundle and its sheaf of sections the same symbol. But tradition does not always permit to do so; thus it prescribes that the sheaf of sections of $\bigwedge^i \mathcal{T}_X^\vee$ is denoted by Ω_X^i (Sect. 4).

Let $\mathcal{V}$ be a holomorphic d-vector bundle on the complex manifold X, let $Y=\mathbb{P}(\mathcal{V})$ and p: $Y\to X$ the projection. Then $p^*(\mathcal{V})$ has a canonical holomorphic subbundle $\mathcal{L}^\vee$ of rank 1, the tautological line bundle of $\mathcal{V}$. The fibre $\mathcal{L}^\vee(y)$ is nothing but the 1-dimensional subspace of $p^*(\mathcal{V})(y)\cong\mathcal{V}(p(y))$ which is represented by y. The quotient bundle is isomorphic to $\mathcal{L}^\vee\otimes\mathcal{W}$, where $\mathcal{W}$ is the bundle along the fibres of Y (see [Hir4], §13). So on Y there is an exact sequence of holomorphic vector bundles

$$0\to\mathcal{L}^\vee\to p^*(\mathcal{V})\to\mathcal{L}^\vee\otimes\mathcal{W}\to 0. \tag{3}$$

By eliminating the Chern classes of $\mathcal{W}$ from $p^*(c(\mathcal{V}))=c(\mathcal{L}^\vee)\cdot c(\mathcal{L}^\vee\otimes\mathcal{W})$ we find

$$c_1^d(\mathcal{L})+p^*(c_1(\mathcal{V}))\cdot c_1^{d-1}(\mathcal{L})+\ldots+p^*(c_d(\mathcal{V}))=0.$$

In Chap. VII we shall need the following result from [Ha1], p. 68.

(5.1) **Theorem.** *Let X be a complex manifold, and $\mathcal{V}$ a holomorphic vector bundle on X. Let $Y=\mathbb{P}(\mathcal{V}^\vee)$, p: $Y\to X$ the projection and $\mathcal{L}^\vee$ the tautological line bundle on Y. Then for every coherent sheaf $\mathcal{S}$ on X and for every $n\geq 1$ there are natural isomorphisms of $\mathcal{O}_X$-modules*

$$p_*(p^*(\mathcal{S})) \xrightarrow{\sim} \mathcal{S}$$
$$p_*(\mathcal{L}^{\otimes n}\otimes p^*(\mathcal{S}))\xrightarrow{\sim} S^n\mathcal{V}\otimes\mathcal{S}.$$

Furthermore, $p_{*i}(\mathscr{L}^{\otimes n}\otimes p^*(\mathscr{S}))=0$ *for all* $i\geq 1$. *Hence by the Leray spectral sequence there are isomorphisms*

$$H^i(Y,p^*(\mathscr{S})) \xrightarrow{\sim} H^i(X,\mathscr{S})$$
$$H^i(Y,\mathscr{L}^{\otimes n}\otimes p^*(\mathscr{S}))\xrightarrow{\sim} H^i(X,S^n\mathscr{V}\otimes\mathscr{S})$$

for all $i\geq 0$.

A result that we shall use frequently is Serre's duality theorem for manifolds (compare [Se1], p. 17–20, and Sect. 12 of this chapter.)

(5.2) **Theorem** (Serre's duality theorem for manifolds in its simplest form). *Let* X *be a compact, connected complex manifold of dimension* n *and* $\mathscr{V}$ *a holomorphic vector bundle on* X. *Then*

$$h^i(\mathscr{V})=h^{n-i}(\mathscr{V}^\vee\otimes\mathscr{K}_X).$$

As a special case we find $p_g(X)=h^{0,n}(X)=h^{n,0}(X)$.

In Sect. 11 we shall give a more precise formulation and in Chap. II, Sect. 6 we shall give a version for singular *curves*.

An equally fundamental corner-stone of this book is the Riemann-Roch theorem. We don't need it in full generality; the following part of it will be sufficient in our case (compare [Hir4], p. 155 and p. 188).

(5.3) **Theorem** (Hirzebruch-Atiyah-Singer Riemann-Roch theorem). *Let* $\mathscr{V}$ *be a holomorphic vector bundle on the compact, connected* n*-dimensional complex manifold* X. *Then*

$$\chi(X,\mathscr{V})=t_{2n}(\mathrm{Todd}(X)\cdot\mathrm{ch}(\mathscr{V})).$$

In particular, when we take for $\mathscr{V}$ the trivial line bundle, we obtain $\chi(X)=\chi(\mathscr{O}_X)=t_{2n}(\mathrm{Todd}(X))$. The right hand side of this equation is a polynomial with rational coefficients in the Chern classes of X; it is called the *Todd genus* of X and denoted by $T(X)$.

(5.4) **Theorem** (Todd-Hirzebruch formula). *If* X *is any compact, connected complex manifold, then*

$$\chi(X)=T(X).$$

For $n=1$ this gives that $q(X)=g(X)$, where $g(X)$ is the topological genus of X.

For $n=2$ we find *Noether's formula*

(4) $$1-q(X)+p_g(X)=\tfrac{1}{12}(c_1^2(X)+c_2(X)).$$

Applying Theorem 5.3 to a line bundle $\mathscr{L}$ on a compact, connected smooth curve X, we find

$$h^0(X,\mathscr{L})-h^1(X,\mathscr{L})=c_1(\mathscr{L})-g(X)+1.$$

The integer $c_1(\mathscr{L})$ is called the degree $\deg(\mathscr{L})$ of $\mathscr{L}$, so we can also write

(5) $$h^0(X,\mathscr{L})-h^1(X,\mathscr{L})=\deg(\mathscr{L})-p(X)+1,$$

which is the classical formula for curves.

Furthermore we find:

for dim $X=2$, rank $\mathscr{V}=1$

(6) $$h^0(X,\mathscr{V})-h^1(X,\mathscr{V})+h^2(X,\mathscr{V})=\tfrac{1}{2}c_1(\mathscr{V})(c_1(\mathscr{V})+c_1(X))+T(X),$$

and, combining with Serre duality:

(7) $$h^0(X,\mathscr{V})-h^1(X,\mathscr{V})+h^0(X,\mathscr{K}_X\otimes\mathscr{V}^\vee)=\tfrac{1}{2}c_1(\mathscr{V})(c_1(\mathscr{V})+c_1(X))+T(X);$$

for dim $X=2$, rank $\mathscr{V}=2$

(8) $$h^0(\mathscr{V})-h^1(\mathscr{V})+h^2(\mathscr{V})=\tfrac{1}{2}(c_1^2(\mathscr{V})-2c_2(\mathscr{V}))+\tfrac{1}{2}c_1(\mathscr{V})c_1(X)+2T(X);$$

for dim $X=1$, rank $\mathscr{V}=n$

(9) $$h^0(\mathscr{V})-h^1(\mathscr{V})=c_1(\mathscr{V})-n(g-1);$$

for dim $X=n$, rank $\mathscr{V}=1$

(10) $$h^0(\mathscr{V})-h^1(\mathscr{V})+\dots+(-1)^n h^n(\mathscr{V})$$
$$=\frac{c_1^n(\mathscr{V})}{n!}+\text{terms containing strictly lower powers of } c_1.$$

6. Line Bundles and Divisors

Let X be a complex manifold of dimension n. With respect to the tensor product, the set of holomorphic line bundles on X forms a group, which is naturally isomorphic to $H^1(X,\mathcal{O}_X^*)$. This group is by definition the *Picard group* $\mathrm{Pic}(X)$.

The exponential sequence of sheaves

$$0\to\mathbb{Z}_X\xrightarrow{i}\mathcal{O}_X\xrightarrow{j}\mathcal{O}_X^*\to 0$$

(where i denotes the inclusion and j attaches $\mathrm{e}(g)$ to any germ g) gives rise to the *exponential cohomology sequence*

(11) $$0\to H^1(\mathbb{Z}_X)\to H^1(\mathcal{O}_X)\to H^1(\mathcal{O}_X^*)\xrightarrow{\delta} H^2(\mathbb{Z}_X)\to H^2(\mathcal{O}_X).$$

We put kernel$(\delta)=\mathrm{Pic}^0(X)$, so $\mathrm{Pic}(X)/\mathrm{Pic}^0(X)$ is isomorphic to a subgroup of $H^2(X,\mathbb{Z})$, the *Néron-Severi group* of X.

(6.1) **Proposition.** *For any holomorphic line bundle* $\mathscr{L}\in H^1(\mathcal{O}_X^*)$ *we have that* $\delta(\mathscr{L})=c_1(\mathscr{L})$.

For a proof we refer to [Hir4], Theorem 4.3.1.

A hypersurface $H\subset X$ is any (non-empty) closed subset of X with the following property: every point $p\in H$ has a connected open neighbourhood U, such that $H\cap U$ is the zero set of a non-constant holomorphic function on U. A hypersurface is irreducible if it is not the union of two other hypersurfaces.

A *divisor* on X is a formal sum $D=\sum_{i=1}^{\infty} d_i D_i$, where $d_i\in\mathbb{Z}$ and $\{D_i\}_{i\in\mathbb{N}}$ a locally finite sequence of irreducible hypersurfaces on X (locally finite means that every point has a neighbourhood which meets only finitely many D_i's). The divisor D is called *effective* or *positive* (notation: $D>0$) if all d_i are non-negative and not all zero; and D is called *non-negative* if it is effective or the zero divisor. Let X and Y be connected complex manifolds and $f\colon X\to Y$ a holomorphic map. If D is any divisor on Y and $f(X)\not\subset\operatorname{supp}(D)$, then $f^*(D)$ is defined in the obvious way by lifting the local equation of its irreducible components.

If $\mathscr{L}$ is a line bundle on X, then every meromorphic section m, different from the zero section, determines a divisor (m) on X, namely its zero divisor minus its polar divisor. Conversely, if any divisor D is given, then there is (up to isomorphism) exactly one $\mathscr{L}$ with a meromorphic section m, such that $(m)=D$. This $\mathscr{L}$ will be denoted $\mathscr{O}_X(D)$. Two divisors D and E are linearly equivalent if and only if $\mathscr{O}_X(D)\cong\mathscr{O}_X(E)$.

The set $|D|$ of non-negative divisors which are linearly equivalent to a given divisor D, is naturally isomorphic to $\mathbb{P}(\Gamma(\mathscr{O}_X(D))$ ($\mathbb{P}(0)=\emptyset$).

Canonical divisors on X will be denoted by K_X or K, so $\mathscr{K}_X=\mathscr{O}_X(K_X)$.

Now let Y be a $(n-1)$-dimensional complex submanifold of X; it is a divisor on X in an obvious way. If we denote by $|$ the analytic restriction, we have ([Hir4], Theorem 4.8.1)

(6.2) **Proposition.** $\mathscr{O}_X(Y)|\,Y\cong\mathscr{N}_{Y/X}$.

If we combine this fact with the isomorphism

$$\Lambda^n\mathscr{T}_X^\vee\,|\,Y\cong\Lambda^{n-1}\mathscr{T}_Y^\vee\otimes\mathscr{N}_{Y/X}^\vee,$$

which follows from the exact sequence

$$0\to\mathscr{T}_Y\to\mathscr{T}_X|\,Y\to\mathscr{N}_{Y/X}\to 0$$

of holomorphic vector bundles on Y, we obtain

(6.3) **Theorem** (Adjunction formula). *If Y is a complex submanifold of codimension* 1 *of the complex manifold X, then*

$$\mathscr{K}_Y=\mathscr{K}_X\otimes\mathscr{O}_X(Y)|\,Y.$$

If Y is compact, then $c_1(\mathscr{O}_X(Y))$ comes from a uniquely determined element of $H_c^2(X,\mathbb{Z})$, and we have like in [Hir4], Theorem 4.9.1:

(6.4) **Proposition.** *If Y is compact, then the element of $H_{2n-2}(X,\mathbb{Z})$, determined by Y is Poincaré dual to $c_1(\mathscr{O}_X(Y))\in H_c^2(X,\mathbb{Z})$.*

7. Algebraic Dimension and Kodaira Dimension

In this section all manifolds are connected.

As to the algebraic dimension, the main result (compare [Rem1], p. 279) is

(7.1) **Theorem.** *The field of meromorphic functions on a compact connected complex manifold is a finitely generated algebraic function field with a transcendency degree over* $\mathbb{C}$, *that does not exceed the dimension of* X.

This transcendency degree is called the *algebraic dimension* of X; it will be denoted by $\mathrm{a}(X)$.

It follows from GAGA (see Sect. 19) that if X is algebraic, then every meromorphic function is rational, and the field of meromorphic functions is nothing but the field of rational functions on X.

As to the Kodaira dimension, let X be any compact complex manifold. Since there is a pairing

$$\Gamma(X, \mathscr{K}_X^{\otimes m_1}) \otimes \Gamma(X, \mathscr{K}_X^{\otimes m_2}) \to \Gamma(X, \mathscr{K}_X^{\otimes (m_1+m_2)})$$

we can make the direct sum $\mathbb{C} \oplus \sum_{m \geq 1} \Gamma(X, \mathscr{K}_X^{\otimes m})$ into a commutative ring $R(X)$ with unit element. This ring is called the *canonical ring* of X. It can be proved that $R(X)$ has a finite degree of transcendency, say $\mathrm{tr}(R(X))$, over $\mathbb{C}$. Thus we can define the *Kodaira dimension* $\mathrm{kod}(X)$ of X as follows:

$$\mathrm{kod}(X) = \begin{cases} -\infty & \text{if } R(X) \cong \mathbb{C} \\ \mathrm{tr}(R(X)) - 1 & \text{otherwise.} \end{cases}$$

We have always that $\mathrm{kod}(X) \leq \mathrm{a}(X) \leq \dim X$.

Let $P_m(X) = h^0(\mathscr{K}_X^{\otimes m})$, $m \geq 1$. This number is called the *m-th plurigenus* of X (so $P_1(X) = p_g(X)$). The Kodaira dimension yields precise information about the behaviour of $P_m(X)$ for $m \to \infty$.

(7.2) **Theorem.** *Let X be a compact connected complex manifold. Then:*

$\mathrm{kod}(X) = -\infty \Leftrightarrow P_m(X) = 0$ *for all* $m \geq 1$
$\mathrm{kod}(X) = 0 \Leftrightarrow P_m(X) = 0$ *or* 1 *for* $m \geq 1$, *but not always* 0
$\mathrm{kod}(X) = k (1 \leq k \leq \dim X) \Leftrightarrow$ *there exist an integer k and strictly positive constants α, β such that*

$$\alpha m^k < P_m(X) < \beta m^k \quad \text{for } m \text{ large enough.}$$

So for $k \geq 1$ we have that P_m grows like m^k.

For this result we refer to [Ue1], p. 86.

Furthermore we shall need ([Ue1], p. 69 and 73).

(7.3) **Theorem.** *If X_1 and X_2 are connected compact complex manifolds, then* $\mathrm{kod}(X_1 \times X_2) = \mathrm{kod}(X_1) + \mathrm{kod}(X_2)$.

(7.4) **Theorem.** *Let X and Y be compact, connected complex manifolds of the same dimension. If there exists a generically finite holomorphic map from X onto Y, then $P_n(X) \geq P_n(Y)$ for $n \geq 1$, hence* $\mathrm{kod}(X) \geq \mathrm{kod}(Y)$. *If the map is an unramified covering, then* $\mathrm{kod}(X) = \mathrm{kod}(Y)$.

General Analytic Geometry

8. Complex Spaces

Let $B \subset \mathbb{C}^n$ be an open ball. A closed analytic subspace $Y = (|Y|, \mathcal{O}_Y)$ of B is any local-ringed space which can be obtained in the following way. Let $\{f_i\}_{i \in I}$ be a set of holomorphic functions on B, and $\mathcal{I}$ the subsheaf of $\mathcal{O}_B$, generated (as $\mathcal{O}_B$-module) by the f_i's. Now let $|Y| = \{z \in B, f_i(z) = 0, i \in I\}$ and $\mathcal{O}_Y = \mathcal{O}_B/\mathcal{I}$. The sheaf $\mathcal{I}$ is the ideal sheaf of Y in B.

A *complex space* is a local-ringed space $X = (|X|, \mathcal{O}_X)$, for which $|X|$ is a Hausdorff space with countable base, and which is everywhere locally isomorphic to a closed analytic subspace of an open ball in some $\mathbb{C}^n$.

The structure sheaf $\mathcal{O}_X$ of a complex space $X = (|X|, \mathcal{O}_X)$ is a sheaf of $\mathbb{C}$-algebras. If $Y = (|Y|, \mathcal{O}_Y)$ is any other complex space, then an *analytic* or *holomorphic* map from X into Y is a morphism $f = (|f|, \tilde{f})$ of local-ringed spaces, such that $\tilde{f}\colon \mathcal{O}_Y \to |f|_* \mathcal{O}_X$ is a morphism of $\mathbb{C}$-algebras.

Thus the complex spaces form a category, in which direct products and fibred products exist.

On any complex space *coherent sheaves* are defined in exactly the same way as on complex manifolds, i.e. as sheaves $\mathcal{S}$ of $\mathcal{O}_X$-modules, for which there exists locally an exact sequence

$$\mathcal{O}_X^r \xrightarrow{h_1} \mathcal{O}_X^s \xrightarrow{h_2} \mathcal{S} \to 0.$$

If $f\colon X \to Y$ is a holomorphic map and $\mathcal{S}$ a coherent sheaf on Y, then the *analytic inverse image* or *analytic pull-back* $f^*(\mathcal{S}) = |f|^{-1}(\mathcal{S}) \otimes_{\mathcal{O}_Y} \mathcal{O}_X$ is a coherent sheaf on X.

Let $X = (|X|, \mathcal{O}_X)$ be an analytic space and $\mathcal{I} \subset \mathcal{O}_X$ a coherent $\mathcal{O}_X$-sheaf of ideals. If $|Y| = \operatorname{supp}(\mathcal{O}_X/\mathcal{I})$, then $(|Y|, \mathcal{O}_Y)$ with $\mathcal{O}_Y = \mathcal{O}_X/\mathcal{I}$ is called a complex subspace of X. The exact sequence

$$0 \to \mathcal{I} \to \mathcal{O}_X \to \mathcal{O}_Y \to 0$$

is called the *structure sequence* of Y (in X). If there is any danger of confusion we shall write $\mathcal{I}_Y$ or $\mathcal{I}_{Y,X}$ instead of $\mathcal{I}$. If X is smooth and Y locally given by one equation, then $\mathcal{I}_{Y/X} \cong \mathcal{O}_X(-Y)$, and if in addition Y is smooth, then $\mathcal{I}/\mathcal{I}^2 \cong \mathcal{N}_{Y/X}^\vee$, the conormal bundle of Y in X.

A *closed embedding* of a complex space Y into a complex space X is an isomorphism from Y onto a complex subspace of X.

If $X = (|X|, \mathcal{O}_X)$ is a complex space and $A \subset X$ a closed subset, then A is called a closed analytic subset if there exists a complex subspace Y of X with $|Y| = A$.

Let again $X = (|X|, \mathcal{O}_X)$ be any complex space, and let $\mathcal{R}$ be the coherent sheaf on X, associated to the presheaf which attaches to an open subset $E \subset |X|$ the $\mathcal{O}_X$-module $\{f \in \mathcal{O}_E, f^k = 0 \text{ for some } k \in \mathbb{N}\}$. Then $X_{\text{red}} = \{|X|, \mathcal{O}_X/\mathcal{R}\}$ is again a complex space and the canonical morphism $X_{\text{red}} \to X$ is analytic. This space X_{red} is called the *reduction* of X, and a space X is called *reduced* if $X = X_{\text{red}}$.

A reduced space X is *irreducible* if $|X| \neq |X_1| \cup |X_2|$, where X_1, X_2 are closed subspaces of X, with $|X_1|$, $|X_2| \neq |X|$. Every reduced space uniquely decomposes into a locally finite union of irreducible subspaces, the *irreducible components.*

Now let X be reduced. A point $x \in |X|$ is *regular* or *smooth* if locally around x the space X is isomorphic to $(B, \mathcal{O}_B)$ for an open ball B in some $\mathbb{C}^N$. A point $x \in |X|$ is *singular* if it is not smooth; the singular points of X form a proper closed analytic subset of X.

A complex space is a complex manifold if and only if it is reduced and all its points are smooth.

Again, let X be reduced. The *dimension* $\dim_x X$ at $x \in |X|$ is the Krull dimension of the local ring $\mathcal{O}_{X,x}$. If x is regular, then the dimension is the N above.

If $\dim_x X$ is the same for all $x \in |X|$, then we call this number the dimension of X. This is in particular the case if X is irreducible. The dimension of a reduced complex space is defined as the maximum of the dimensions of its irreducible components, if this maximum exists. If all irreducible components of X have the same dimension d, then we call X of *pure* dimension d.

We shall say that the complex space X has dimension d if $\dim X_{\mathrm{red}}$ is defined and equal to d. A *curve* (Riemann surface) is a complex space X with X_{red} of pure dimension 1 such that all irreducible components of X_{red} have dimension 1.

Let X be reduced. A point $x \in |X|$ is called *normal* if its local ring $\mathcal{O}_{X,x}$ is integrally closed in its ring of quotients. A complex space is normal if all its points are normal. The non-normal points always form a thin closed analytic subset, i.e. a closed analytic subset that does not contain any irreducible components of X. The singular locus of a normal complex space has codimension ≥ 2.

A regular point of X is normal, but the converse does not hold if $\dim_x X \geq 2$.

Every reduced complex space X has a *normalisation*, defined uniquely up to analytic isomorphisms, i.e. there exists a normal complex space X_{norm} and a finite, surjective analytic map $\nu_X\colon X_{\mathrm{norm}} \to X$, such that ν_X maps $X_{\mathrm{norm}} \setminus \nu_X^{-1}(N)$ isomorphically onto $X \setminus N$, where N is the set of non-normal points on X.

Given any holomorphic map $f\colon X \to Y$ between reduced complex spaces such that no irreducible component of X is mapped into the locus of non-normal points on Y, then there exists a unique lifting $g\colon X_{\mathrm{norm}} \to Y_{\mathrm{norm}}$ such that the diagram

$$\begin{array}{ccc} X_{\mathrm{norm}} & \xrightarrow{\;g\;} & Y_{\mathrm{norm}} \\ \downarrow{\scriptstyle \nu_X} & & \downarrow{\scriptstyle \nu_Y} \\ X & \xrightarrow{\;f\;} & Y \end{array}$$

is commutative.

Now we state a number of fundamental theorems which will be used in this book.

(8.1) **Theorem** (Stein factorisation). *Let X, Y be reduced complex spaces and $f\colon X \to Y$ a proper analytic map. Then there exists a reduced complex space Z and* *holomorphic maps $g\colon X\to Z$, $h\colon Z\to Y$, with $f=hg$ such that g is surjective and connected (i.e. all fibres are connected). If X and Y are normal, then Z can be taken to be normal too.*

A proof can be found in [F], p. 70.

(8.2) **Theorem** (Grauert's direct image theorem). *Let X, Y be complex spaces, $f\colon X\to Y$ a proper analytic map. Then for every coherent sheaf $\mathscr{S}$ on X the direct image sheaves $f_{*i}\mathscr{S}$, $i=0,1,2,\dots$ are also coherent.*

For this result we refer to [Gr1], [F-K] and [Ki-V]. As an immediate consequence we have the following two facts.

(8.3) **Theorem** (Finiteness theorem of Cartan-Serre). *Let X be a compact complex space and $\mathscr{S}$ a coherent sheaf on X. Then $H^i(X,\mathscr{S})$ has finite dimension for all $i\geq 0$ and $H^i(X,\mathscr{S})=0$ for $i>\dim X$.*

(8.4) **Theorem** (Remmert's mapping theorem). *Let X, Y be reduced complex spaces and $f\colon X\to Y$ a proper analytic map. If $A\subset X$ is a closed analytic subset, then so is $|f|(A)\subset Y$.*

Let X and Y be complex spaces, $f\colon X\to Y$ a surjective holomorphic map and $y\in|Y|$. The *analytic fibre* over y, denoted by X_y, is defined as the fibre product $y\times_Y X$. It can be described more explicitly in the following way. Let $\mathscr{I}$ be the ideal sheaf of y in $\mathscr{O}_Y$. Then X_y is the complex subspace $(|f|^{-1}(y), \mathscr{O}_X/f^*(\mathscr{I}))$ of X. Furthermore, for any coherent sheaf $\mathscr{S}$ on X we denote by $\mathscr{S}_y$ the analytic restriction of $\mathscr{S}$ to X_y (i.e. the analytic inverse image with respect to the embedding of X_y in X).

The next two theorems are e.g. proved in [B-S], Chap. III, §3.

(8.5) **Theorem.** *Let X, Y be reduced complex spaces and $f\colon X\to Y$ a proper holomorphic map. If $\mathscr{S}$ is any coherent sheaf on X, which is flat over y (i.e. $\mathscr{S}_x$ is a flat $\mathscr{O}_{|f|(x)}$-module for all $x\in X$), we have*

(i) *the Euler characteristic $\chi(X_y,\mathscr{S}_y)$ is locally constant;*
(ii) (Grauert's semi-continuity theorem) *$h^q(X_y,\mathscr{S}_y)$ is an upper-semi-continuous function of y for all $q\geq 0$;*
(iii) *if $h^q(X_y,\mathscr{S}_y)$ is constant, then $f_{*q}\mathscr{S}$ is locally free;*
(iv) (Grauert's base change theorem) *if $h^q(X_y,\mathscr{S}_y)$ is constant, then the "base-change map" $(f_{*q}\mathscr{S})_y/\mathscr{I}_y\cdot(f_{*q}\mathscr{S})_y\to H^q(X_y,\mathscr{S}_y)$ is bijective.*

Let X, Y be reduced complex spaces, $f\colon X\to Y$ a proper analytic map and $\mathscr{S}$ a coherent sheaf on X. If $y\in Y$, then $(f_{*q}\mathscr{S})_y$ is a finite $\mathscr{O}_{Y,y}$-module by Theorem 8.2; we put

$$\varprojlim_k (f_{*q}\mathscr{S})_y/\mathscr{I}_y^k(f_{*q}\mathscr{S})_y=(f_{*q}(\mathscr{S}))_y^{\wedge},$$

where $\mathscr{I}_y$ is the maximal ideal in $\mathscr{O}_{Y,y}$. There is a natural homomorphism

$$h^q\colon (f_{*q}(\mathscr{S}))_y^{\wedge}\to\varprojlim_k H^q(X_y,\mathscr{S}/(f^*(\mathscr{I}^k))\mathscr{S}).$$

(8.6) **Theorem** (Grauert's comparison theorem). *The homomorphism h^q is an isomorphism.*

A more elementary auxiliary result, that we shall use several times is

(8.7) **Theorem** (Levi's extension theorem). *Let X be an irreducible complex space and $A \subset X$ an analytic subset of codimension at least 2. Then every meromorphic function on $X \setminus A$ extends uniquely to a meromorphic function on X. If X is a complex manifold and the function holomorphic, then the extension is also holomorphic.*

The last part is frequently called Riemann's extension theorem.
For a proof of Levi's theorem we refer to [F], p. 185.
For the proof of the following theorem we refer to [Lj].

(8.8) **Theorem.** *Let X be a complex space and A an analytic subset of X. Then there exists a triangulation of $|X|$ in which A appears as the support of a subcomplex. In particular, there exists an open neighbourhood E of A such that E and $\bar{E}$ can be retracted onto A.*

Let X be a complex space and A an irreducible, compact analytic subset of X. Then it is possible to attach to A an element $a \in H_i(|X|, \mathbb{Z})$, with $i = 2 \dim A$. This can be done for example by the Borel-Haefliger method ([B-Ha]). In the case that A has a desingularisation (e.g., by Hironaka ([Hik1]), when A is a projective-algebraic variety), you can just take for a the image of the fundamental class of the desingularisation in $H_i(|X|, \mathbb{Z})$. For curves on surfaces this becomes an elementary approach, as we shall see later (Chap. II, Sect. 10).

Finally we shall need

(8.9) **Proposition** (Properness criterion). *Let $f\colon X \to Y$ be a holomorphic map of complex spaces, $y \in Y$, and $A \subset X$ a connected component of the fibre $f^{-1}(y)$. If A is compact, then there is an open neighbourhood $U \subset X$ of A such that $f|U$ is proper.*

Proof. Let $V \subset X$ be some open neighbourhood of A such that $\bar{V}$ is compact in X. Since $f^{-1}(y)$ is closed, we may take V so small that $\bar{V} \cap f^{-1}(y) = A$. Then $f(\partial V) \subset Y$ is compact with $y \notin f(\partial V)$. So $U = V \setminus f^{-1}(f(\partial V))$ is an open neighbourhood of A. Whenever $K \subset f(U)$ is compact, then $(f|U)^{-1}(K) = U \cap f^{-1}(K)$ is closed in $\bar{V}$, hence compact.

9. The σ-Process

Let $(z_1, \ldots, z_n)$ be the standard coordinates in $\mathbb{C}^n$, $n \geq 2$, and let $(\xi_1 : \ldots : \xi_n)$ be the standard homogeneous coordinates in $\mathbb{P}_{n-1}$. We take a neighbourhood U of $a = (a_1, \ldots, a_n)$ in $\mathbb{C}^n$, and consider on the product $U \times \mathbb{P}_{n-1}$ the subset $\bar{U}$ given by the equations $(z_i - a_i)\xi_j - (z_j - a_j)\xi_i = 0$, $i, j = 1, \ldots, n$. Application of the jacobian criterion shows that $\overline{U}$ is an n-dimensional complex submanifold of $U\times\mathbb{P}_{n-1}$; the projection $p\colon \overline{U}\to U$ maps $\overline{U}\backslash p^{-1}(a)$ biregularly onto $U\backslash a$, whereas $p^{-1}(a)$ is an $(n-1)$-dimensional submanifold of $\overline{U}$, isomorphic to $\mathbb{P}_{n-1}$.

Sometimes we shall refer to this procedure as the *σ-process*, applied to U in a, or we shall say that we have applied a *monoidal transformation* to U with centre a. But most frequently we shall say that $\overline{U}$ is obtained from U by *blowing up* (in) a.

Using local coordinates you can blow up any point x_0 of an n-dimensional complex manifold ($n\geq 2$); up to an isomorphism the result is independent of the coordinates used. The resulting complex manifold will be denoted by $\overline{X}(x_0)$ or simply $\overline{X}$, and often we shall loosely speak of the blowing-up $p\colon \overline{X}\to X$ (at some given point). The divisor $E=p^{-1}(x_0)$, which is isomorphic to $\mathbb{P}_{n-1}$, will be called the *exceptional divisor*.

In the following theorem we collect some properties of the σ-process, which will be used at one place or another in this book.

(9.1) **Theorem.** *Let X be a complex manifold of dimension ≥ 2 and $p\colon \overline{X}\to X$ the blowing-up of X at some point. Then* $\mathcal{N}_{E/\overline{X}}\cong\mathcal{O}_{\mathbb{P}_{n-1}}(-1)$, and

(i) *p induces an isomorphism between the fields of meromorphic functions on X and $\overline{X}$. In particular, if X (and hence $\overline{X}$) is compact, then $a(\overline{X})=a(X)$.*

(ii) *$p_*(\mathcal{O}_{\overline{X}})=\mathcal{O}_X$ and $p_{*i}(\mathcal{O}_{\overline{X}})=0$ for $i\geq 1$.*

(iii) *$p^*\colon H^i(X,\mathcal{O}_X)\to H^i(\overline{X},\mathcal{O}_{\overline{X}})$ is an isomorphism for all $i\geq 0$.*

(iv) *$p^*\colon H^i(X,\mathbb{Z})\to H^i(\overline{X},\mathbb{Z})$ is bijective for $i=1$ and injective for $i=2$. Furthermore,*

$$H^2(\overline{X},\mathbb{Z})\cong p^*(H^2(X,\mathbb{Z}))\oplus\mathbb{Z}\{e\},\qquad \textit{where } e=c_1(\mathcal{O}_{\overline{X}}(E)).$$

(v) *For every $a\in H^2(X,\mathbb{Z})$ we have $p_!p^*(a)=a$.*

(vi) *$p^*\colon H^1(X,\mathcal{O}_X^*)\to H^1(\overline{X},\mathcal{O}_{\overline{X}}^*)$ is injective, and thus $Pic(\overline{X})$ is isomorphic to the product of $Pic(X)$ and the infinite cyclic subgroup, generated by $\mathcal{O}_{\overline{X}}(E)$,*

(vii) *$\mathcal{K}_{\overline{X}}=p^*(\mathcal{K}_X)\otimes\mathcal{O}_{\overline{X}}((\dim X-1)E)$.*

(viii) *p induces an isomorphism $p^*\colon \Gamma(X,\mathcal{K}_X^{\otimes m})\to\Gamma(\overline{X},\mathcal{K}_{\overline{X}}^{\otimes m})$ for all $m\geq 1$, so if X is compact, $P_m(\overline{X})=P_m(X)$ for $m\geq 1$ and $\mathrm{kod}(\overline{X})=\mathrm{kod}(X)$.*

In Chap. II we shall consider blowing-up of points more in detail, but then only for the case that X is a surface.

As to the proof of all these facts, the isomorphism $\mathcal{N}_{E/\overline{X}}\cong\mathcal{O}_{\mathbb{P}_{n-1}}(-1)$ is elementary, whereas (i) and (viii) are direct consequences of Levi's extension theory for meromorphic functions (Theorem 8.7). The Leray spectral sequence combined with (ii) yields (iii). For (iv) we refer to [Ae], p. 269, (v) is a direct consequence of (iv) and Lemma 1.1. Then (vi) is a consequence of (iv), (v) and the exponential sequence. Property (vii) is again elementary.

So it remains to prove (ii). The first statement, namely that p induces an isomorphism $\mathcal{O}_X\xrightarrow{\sim}p_*\mathcal{O}_{\overline{X}}$ follows from (i). As to the vanishing of $p_{*i}\mathcal{O}_{\overline{X}}$, $i\geq 1$, it is clear that $p_{*i}\mathcal{O}_{\overline{X}}=0$ outside of x_0 (in fact, you use that $H^i(\mathcal{O}_B)=0$ for all $i\geq 1$ and any open ball B). So the only thing left is to prove that the finite-dimensional complex vector space $(p_{*i}\mathcal{O}_{\overline{X}})_{x_0}=0$ for $i\geq 1$. By Krull's theorem

([G-R3], p. 211) it is sufficient to show that $(p_{*i}\mathcal{O}_X)_{x_0}/\mathcal{I}^k_{x_0}(p_{*i}\mathcal{O}_X)_{x_0}=0$ for all $i\geq 1$. Grauert's comparison theorem shows that we are ready as soon as we know that $H^i(E,\mathcal{O}_X/\mathcal{I}^k_{E,X})=0$ for all k and all $i\geq 1$. Using the exact sequence

$$0\to\mathcal{I}^k_E/\mathcal{I}^{k+1}_E\to\mathcal{O}_X/\mathcal{I}^{k+1}_E\to\mathcal{O}_X/\mathcal{I}^k_E\to 0$$

and the isomorphism $\mathcal{I}^k_E/\mathcal{I}^{k+1}_E\cong\mathcal{N}^{\otimes(-k)}_{E/X}\cong\mathcal{O}_{\mathbb{P}_{n-1}}(k)$, this follows by induction.

10. Deformations of Complex Manifolds

A *family of compact, complex manifolds* $\mathcal{X}=(X,p,S)$ consists of

(i) a pair of connected complex spaces: X and S;
(ii) a surjective, proper holomorphic map $p\colon X\to S$ which is flat (i.e. $\mathcal{O}_{X,x}$ is a flat $\mathcal{O}_{S,p(x)}$-module for all $x\in X$) and whose fibres X_s are all connected manifolds.

The space S is the *base (space)* of the family, or the family is a *family over* S or parametrised by S.

If X and S are smooth, then flatness implies that p is everywhere of maximal rank and the family $\mathcal{X}$ is called a *smooth* family. A theorem of Ehresmann (compare [M-K], p. 19) says that a smooth family is differentiably locally trivial; in particular all of its fibres are diffeomorphic.

The simplest examples are given by trivial families, i.e. products $X=V\times S$, where V is a connected compact manifold and S some connected complex space, or by locally trivial families, i.e. holomorphic fibre bundles. (It is by no means true that a smooth family is automatically a fibre bundle. Consider for example the family of elliptic curves in $\mathbb{P}_2$ given by

$$X=\{((x_0:x_1:x_2),\lambda)\in\mathbb{P}_2\times\mathbb{C};\, x_0^3+x_1^3+x_2^3=3\lambda x_0x_1x_2\}.$$

The family is smooth over $\{\lambda\in\mathbb{C};\,\lambda^3\neq+1\}$ and the j-invariant of X_λ is $\lambda^3(\lambda+8)^3(\lambda-1)^{-3}$, so not all fibres are isomorphic.)

Let $\mathcal{X}=(X,p,S)$ be any family, S' a complex space and $f\colon S'\to S$ a holomorphic map. Then the fibred product $X\times_S S'$ is in a natural way a family over S', called the *pull-back* of $\mathcal{X}$ by f.

If $\mathcal{X}_i=(X_i,p_i,S_i)$, $i=1,2$, is a pair of families, then a morphism from $\mathcal{X}_1$ into $\mathcal{X}_2$ consists of a pair of analytic maps, $g\colon X_1\to X_2$ and $f\colon S_1\to S_2$, such that $p_2g=fp_1$. There is always a morphism from the pull-back into the original family.

(10.1) **Theorem** (Local-triviality theorem of Grauert-Fischer). *A smooth family of compact complex manifolds is locally trivial if and only if all fibres are analytically isomorphic.*

This theorem is proved in [F-G].

Let V be a connected compact complex manifold. A *deformation* of V, parametrised by (or over) the complex space S consists of a connected complex space S, a base point $s_0\in S$, a family $\mathcal{X}=(X,p,S)$ and an isomorphism from V onto the fibre X_{s_0}. (Confusingly, the fibres of p are also called deformations of

V.) Morphisms between deformations are defined as "base point preserving" morphisms between the families in question which are compatible with the isomorphisms from V onto the fibres over the base points.

In the sequel we shall consider only germs of deformations, tacitly identifying two deformations over S, both with base point s_0, if they coincide in a neighbourhood of this point. Thus if S is smooth in s_0, we can assume all deformations over S (with base point s_0) to be smooth (if the base of a family is smooth, then so is the family).

Let again V be a fixed manifold. A deformation $\mathscr{X}=(X,p,S)$ of V with base point s_0 is called *(locally) complete* if (locally) every deformation $\mathscr{X}'=(X',p',S')$ of V with base point s'_0 is obtained as the pull-back from $\mathscr{X}$ by a suitable analytic map $f\colon S'\to S$ with $f(s'_0)=s_0$. If in addition f is always uniquely determined by $\mathscr{X}'$, the deformation $\mathscr{X}$ is called *(locally) universal*. As soon as it exists, the universal deformation of V is unique up to isomorphisms. Although it exists in many cases such as for curves and tori, it does not exist always, as the following example shows. You take

$$X=\{((y_0:y_1),(x_0:x_1:x_2),t)\in\mathbb{P}_1\times\mathbb{P}_2\times\mathbb{C};\, y_0^2x_1-y_1^2x_0-t\,y_0y_1x_2=0\}$$

and consider it as a family over $\mathbb{C}$ via projection onto the t-factor. For $t\neq 0$ the fibre is isomorphic to $\mathbb{P}_1\times\mathbb{P}_1$, but $X_0=\Sigma_2$, the Hirzebruch surface (see V, Sect. 4) which is not isomorphic to $\mathbb{P}_1\times\mathbb{P}_1$. If we take $X=X'$ and if $f(t)=t(1+\varepsilon(t))$ $(\varepsilon(0)=0)$ is a holomorphic function on a neighbourhood U of $0\in\mathbb{C}$, we can lift f to a holomorphic map g defined for $t\in U$ by setting

$$g((y_0:y_1),(x_0:x_1:x_2),t)=(((1+\varepsilon(t))\,y_0:y_1),((1+\varepsilon(t))^2\,x_0:x_1:x_2),f(t)).$$

Since for $\varepsilon(t)$ you can take any holomorphic function with $\varepsilon(0)=0$ we see that $\{X_t\}_{t\in U}$ can never be universal at 0.

Exactly because of such examples the notion of versal deformation has been introduced. A deformation $\mathscr{X}=(X,p,S)$ of V with base point s_0 is called *versal* if it has the following property: for every $\mathscr{X}'$ as above there is an f as above which need not be unique, but whose derivative at s'_0 is uniquely determined. The preceding example gives a versal deformation of Σ_2. Versal deformations always exist:

(10.2) **Theorem** (Kuranishi's theorem). *Every compact complex manifold has a versal deformation.*

A proof of this basic result can be found in [Dy].

Furthermore there are the following facts, for which we refer to [Wav] and [Dy].

(10.3) *The versal deformation in Theorem 10.2 is complete for any of its fibres, and versal for any of its fibres as soon as $h^1(X_s,\mathscr{T}_{X_s})$ is constant.*

(10.4) *If $H^2(V,\mathscr{T}_V)=0$, then V admits a smooth versal deformation.*

(10.5) *If $H^0(V,\mathscr{T}_V)=0$ then V has a universal deformation.*

(10.6) *If a universal deformation exists, then every versal deformation is isomorphic to it.*

Remark. In the case 10.5, i.e. if $H^0(V, \mathcal{T}_V)=0$, then the universal deformation $\mathcal{X} =(X, p, S)$ has an additional property: any deformation $\mathcal{X}'=(X', p', S')$ of X not only determines the inducing map from S' into S, but also the map from X' into X, i.e. the whole morphism from $\mathcal{X}'$ into $\mathcal{X}$ (see [L-P], p. 170). This does not hold for any universal deformation.

Now let $\mathcal{X}=(X, p, S)$ be a smooth deformation of V with base point s_0. From the exact cohomology sequence of

$$0 \to \mathcal{T}_{p^{-1}(s_0)} \to \mathcal{T}_X | p^{-1}(s_0) \to \mathcal{N}_{p^{-1}(s_0)/X} \to 0$$

we obtain a morphism

$$\Gamma(p^{-1}(s_0), \mathcal{N}_{p^{-1}(s_0)/X}) \to H^1(p^{-1}(s_0), \mathcal{T}_{p^{-1}(s_0)}).$$

Since $\mathcal{N}_{p^{-1}(s_0)/X}$ is trivial, the first of these spaces is naturally isomorphic to $\mathcal{T}_S(s_0)$, and since the second can be identified with $H^1(V, \mathcal{T}_V)$ because of the *given* isomorphism from V onto $p^{-1}(s_0)$, we obtain a homomorphism

$$\rho_{\mathcal{X}}\colon \mathcal{T}_S(s_0) \to H^1(V, \mathcal{T}_V)$$

which is called the *Kodaira-Spencer map*. Using the Zariski tangent space (to S at s_0) it can be defined for any family, smooth or not, but we don't need this.

A classical result of Kodaira and Spencer (see [K-S 1]) is

(10.7) **Theorem.** *A smooth family is complete if and only if the Kodaira-Spencer map is surjective.*

If $\mathcal{X}'$ is the pull-back of S by an analytic map $f\colon S' \to S$, then there is a commutative diagram

$$\begin{array}{ccc} \mathcal{T}_{S'}(s_0') & \xrightarrow{\rho_{\mathcal{X}'}} & \\ \downarrow df & & H^1(V, \mathcal{T}_V). \\ \mathcal{T}_S(s_0) & \xrightarrow{\rho_{\mathcal{X}}} & \end{array}$$

Consequently, if $\mathcal{X}$ is complete and $\rho_{\mathcal{X}}$ bijective, then $\mathcal{X}$ is versal.

Next, suppose we are given a complete smooth deformation $\mathcal{X}$ of V. The Kodaira-Spencer map is surjective, and we may select a smooth subspace S' of the base space of $\mathcal{X}$, such that $\rho_{\mathcal{X}} | \mathcal{T}_S(s_0)$ is an isomorphism. The above diagram shows that this isomorphism coincides with the Kodaira-Spencer map of the restricted family $\mathcal{X}|S'$. By the preceding remark, this family is versal. Consequently: *if V admits a smooth complete deformation, then it admits a versal smooth deformation.*

Differential Geometry of Complex Manifolds

11. De Rham Cohomology

Let X be a compact, connected, oriented n-dimensional differentiable manifold. We write $\mathscr{D}_X^p$ for the sheaf of real-valued C^∞-forms of rank p on X, and $H_{\mathrm{DR}}^p(X)$ for the p-th de Rham group, i.e.

$$H_{\mathrm{DR}}^p(X) = \{\xi \in \Gamma(\mathscr{D}_X^p),\ d\xi = 0\}/d\Gamma(\mathscr{D}_X^{p-1}).$$

The p-th homology group $H_p(X, \mathbb{Z})$ for the complex of singular chains is the same as the one for the subcomplex of differentiable chains. So Stokes' theorem implies that there is a well-determined homomorphism

$$H_{\mathrm{DR}}^p(X) \to \mathrm{Hom}_{\mathbb{Z}}(H_p(X, \mathbb{Z}), \mathbb{R}),$$

defined by sending the class of a closed p-form ξ to the functional

$$c \mapsto \int_\gamma \xi,$$

where the differentiable chain γ represents the class $c \in H_p(X, \mathbb{Z})$. De Rham's theorem asserts (see [S-T1], p. 147) that in fact this is an isomorphism. The right hand side is of course nothing but the p-th singular cohomology group $H^p(X, \mathbb{R})$. So we obtain the de Rham isomorphism

$$H_{\mathrm{DR}}^p(X) \xrightarrow{\sim} H^p(X, \mathbb{R}). \tag{12}$$

The Kronecker pairing

$$H^p(X, \mathbb{R}) \times H_p(X, \mathbb{R}) \to \mathbb{R}$$

translates into integration

$$(x, c) \mapsto \int_\gamma \xi,$$

where ξ represents x and γ represents c.

If we compare the perfect pairing

$$H^p(X, \mathbb{R}) \times H^{n-p}(X, \mathbb{R}) \to H^n(X, \mathbb{R}) \cong \mathbb{R}$$

given by the cupproduct with the Kronecker pairing, we obtain an isomorphism between $H_p(X, \mathbb{R})$ and $H^{n-p}(X, \mathbb{R})$ which is nothing but Poincaré duality $\mathscr{P} = \mathscr{P}_X$. In other words, we have

$$x \cdot y = \int_{\mathscr{P}(x)} y, \tag{13}$$

where $x \in H^p(X, \mathbb{R})$ and $y \in H^{n-p}(X, \mathbb{R})$.

If the singular cohomology over $\mathbb{R}$ is naturally identified with the sheaf cohomology over $\mathbb{R}$ (for example by way of a triangulation) then the isomorphism between de Rham cohomology and sheaf cohomology we thus obtain is nothing else but the isomorphism, obtained from the resolution

$$0 \to \mathbb{R}_X \to \mathscr{D}_X^0 \to \ldots \to \mathscr{D}_X^n \to 0.$$

If X is a *non-compact* oriented manifold, everything remains the same up to and including the de Rham isomorphism (12), but then the fact that the Poincaré duality is an isomorphism between the cohomology group $H^p_c(X, \mathbb{R})$ *(compact supports)* and $H_{n-p}(X, \mathbb{R})$ has to be taken into account (see [Rh]).

12. Dolbeault Cohomology

Let X be a connected compact complex manifold of dimension n. We denote by $\mathscr{D}^{p,q}_X$ the sheaf of $\mathbb{C}$-valued C^∞-forms of type (p, q) on X, and by $H^{p,q}(X)$ the Dolbeault (cohomology) group

$$H^{p,q}(X) = \{\alpha \in \Gamma(\mathscr{D}^{p,q}_X),\ \bar{\partial}\alpha = 0\} / \bar{\partial}\Gamma(\mathscr{D}^{p,q-1}_X).$$

Then we have Dolbeault's isomorphism ([Hir4], p. 119)

$$H^{p,q}(X) \cong H^q(X, \Omega^p_X), \quad \text{and} \quad \dim H^{p,q}(X) = h^{p,q}(X).$$

The complexified de Rham-complex $\{\Gamma(\mathscr{D}^{\cdot}_X \otimes \mathbb{C}), d\}$ is the single complex, associated to the double complex $\{\Gamma(\mathscr{D}^{\cdot,\cdot}_X, \partial, \bar{\partial})\}$. Its cohomology groups are $H^k_{\mathrm{DR}}(X) \otimes \mathbb{C} = H^k(X, \mathbb{R}) \otimes \mathbb{C} = H^k(X, \mathbb{C})$. Hence the spectral sequence for a double complex reads:

$$E^{p,q}_1 = H^{p,q}(X) \Rightarrow H^{p+q}(X, \mathbb{C}).$$

This spectral sequence is called the *Fröhlicher spectral sequence*, and the resulting filtration on $H^{p+q}(X, \mathbb{C})$ is called the *Hodge filtration*. Explicitly, we have

$$H^k(X, \mathbb{C}) = F^0(H^k) \supset F^1(H^k) \supset \ldots \supset F^k(H^k) \supset 0,$$

where

(14) $$F^p(H^k) = \{[\alpha],\ \alpha \in \bigoplus_{\substack{p'+q'=k \\ p' \geq p}} \Gamma(\mathscr{D}^{p',q'}_X),\ d\alpha = 0\}.$$

Now the isomorphism of Dolbeault can be generalised to the case of holomorphic vector bundles $\mathscr{V}$, where it says

$$H^p(X, \mathscr{V}) \cong \{\alpha \in \Gamma(\mathscr{V} \otimes \mathscr{D}^{0,q}),\ (1 \otimes \bar{\partial})\alpha = 0\} (1 \otimes \bar{\partial}) \Gamma(\mathscr{V} \otimes \mathscr{D}^{0,q-1}).$$

In Chap. II we shall need the following explicit description of a special case of Serre duality. As recalled in Theorem 5.2 we have $h^p(\mathscr{V}) = h^{n-p}(\mathscr{V}^\vee \otimes \mathscr{K}_X)$ for any holomorphic vector bundle $\mathscr{V}$.

Then a duality between $H^i(\mathscr{V})$ and $H^{n-i}(\mathscr{V}^\vee \otimes \mathscr{K}_X)$ can be obtained in the following way. Considering $H^p(\mathscr{V})$ and $H^{n-p}(\mathscr{V}^\vee \otimes \mathscr{K}_X)$ as Dolbeault groups there is a pairing

(15) $$H^p(\mathscr{V}) \otimes H^{n-p}(\mathscr{V}^\vee \otimes \mathscr{K}_X) \to H^n(\mathscr{K}_X) \cong \mathbb{C}$$

and we have

(12.1) **Proposition.** *The pairing* (15) *of Dolbeault groups is a perfect pairing.*

For this result we refer to Serre's original paper [Se1], p. 17–20.

13. Kähler Manifolds

A hermitian metric g on a complex manifold X is called a *Kähler metric* if the associated real differential form of type $(1,1)$ is closed. Forms arising in this way are called *Kähler forms.*

A complex manifold which can be provided with at least one Kähler metric is called a Kähler manifold. Since every submanifold of a Kähler manifold is again a Kähler manifold, and since the Fubini metric on $\mathbb{P}_n$ is kählerian, we have that every projective algebraic manifold is a Kähler manifold. The existence of non-algebraic tori (on which the standard flat metric is a Kähler metric) shows that, even for compact manifolds, the converse does not hold (compare Theorem 19.4.)

A cohomology class $h \in H^2(X, \mathbb{R})$ is called a Kähler class if it can be represented by a Kähler form.

Now let X be a compact connected Kähler manifold and Y a compact, connected complex manifold of dimension k. Suppose there is a holomorphic map $f\colon Y \to X$ such that $f(Y)$ is a k-dimensional complex subvariety of X. Then if ω is any Kähler form on X, we have

$$\int_Y f^*(\omega^k) > 0.$$

So if $h \in H^2(X, \mathbb{R})$ is the cohomology class of ω, then $f^*(h^k) > 0$. Applying Lemma 1.1 we see that $h^k f_!(1) > 0$, with $f_!(1)$ dual to the homology class of $f(Y)$.

By resolution of singularities ([Hik2]) we thus obtain

(13.1) **Lemma.** *If X is a compact Kähler manifold, h a Kähler class on X and y dual to the homology class of any k-dimensional closed analytic subvariety on X, then $h^k y > 0$.*

Remark. Hironaka's theorem can be avoided at this point by using a suitable integration theory for forms on arbitrary analytic subsets of a complex manifold.

For the proof of the following important fact we refer to [Wei1] and [Del].

(13.2) **Theorem.** *Let X be a compact Kähler manifold. Then*

(i) *the Fröhlicher spectral sequence degenerates at E_1-level;*

(ii) *there is a direct sum decomposition*

$$H^k(X, \mathbb{C}) = \bigoplus_{p+q=k} {}'H^{p,q}, \qquad \text{with } {}'H^{p,q} = F^p(H^k) \cap \overline{F^q(H^k)}$$

("formal Hodge decomposition");

(iii) *a class in $H^k(X, \mathbb{C})$ belongs to ${}'H^{p,q}$ if and only if it can be represented by a d-closed form of type (p,q);*

(iv) *a d-closed form is d-exact if and only if it is $\bar{\partial}$-exact.*

Properties (iii) and (iv) imply that there is a well-defined injective homomorphism from ${}'H^{p,q}$ into $H^{p,q}$. This map is an isomorphism since both spaces have dimension $h^{p,q}$ (for $'H^{p,q}$ this is a direct consequence of the degeneracy of the Fröhlicher spectral sequence). If we identify $'H^{p,q}$ and $H^{p,q}$ in this way we obtain:

(13.3) **Corollary.** *For any compact Kähler manifold there is a direct sum decomposition*

$$H^k(X,\mathbb{C})=\bigoplus_{p+q=k} H^{p,q}(X).$$

This decomposition is the famous *Hodge decomposition.*

Furthermore we clearly have

(13.4) **Corollary.** *If X is a compact Kähler manifold, then*

(i) $h^{p,q}(X)=h^{q,p}(X)$

(ii) $b_k(X)=\sum_{p+q=k} h^{p,q}(X).$

Now let X be a connected compact Kähler manifold. To X there are associated two complex tori, both of dimension $g=h^{1,0}(X)$, namely the Albanese torus $\mathrm{Alb}(X)$ and the Picard torus $\mathrm{Pic}^0(X)$. We define them in a rather primitive way which however is the best one for our purposes.

Firstly, the Albanese torus. Let $\omega_1,\dots,\omega_g$ be a base for $\Gamma(X,\Omega_X^1)$. It follows from Corollary 13.3 that $\omega_1,\dots,\omega_g,\ \bar\omega_1,\dots,\bar\omega_g$ form a base of $H^1(X,\mathbb{C})$. Furthermore, let $h_1,\dots,h_{2g}$ be a base for $H_1(X,\mathbb{Z})$ mod torsion. We consider the vectors

$$v_j=\Big(\int_{h_j}\omega_1,\dots,\int_{h_j}\omega_g\Big)\in\mathbb{C}^g,\qquad j=1,\dots,2g,$$

and claim that they are independent over $\mathbb{R}$. Indeed, if they were not, then the $2g$ vectors

$$w_j=\int_{h_j}\omega_1+\bar\omega_1,\dots,\int_{h_j}\omega_g+\bar\omega_g,\quad \sqrt{-1}\int_{h_j}\omega_1-\bar\omega_1,\dots,\sqrt{-1}\int_{h_j}\omega_g-\bar\omega_g$$

would be dependent in $\mathbb{R}^{2g}$. Because of (12) there would be real numbers $\lambda_1,\dots,\lambda_g,\ \mu_1,\dots,\mu_g$, not all 0, such that

$$\lambda_1(\omega_1+\bar\omega_1)+\dots+\lambda_g(\omega_g+\bar\omega_g)+\sqrt{-1}\,\mu_1(\omega_1-\bar\omega_1)+\dots+\sqrt{-1}\,\mu_g(\omega_g-\bar\omega_g)$$

would be cohomologous to 0. But this would mean that $\omega_1,\dots,\omega_g,\ \bar\omega_1,\dots,\bar\omega_g$ would be $\mathbb{C}$-dependent, which is not the case.

So, from the real point of view, the vectors $v_1,\dots,v_{2g}$ span a lattice in $\mathbb{C}^g$ and thus determine a complex torus. If we replace the h_j's or the ω_k's by another base, then, up to an isomorphism, we obtain the same torus. This torus is $\mathrm{Alb}(X)$.

There is an obvious holomorphic map $f\colon X\to\mathrm{Alb}(X)$, determined up to a translation of $\mathrm{Alb}(X)$. Given any holomorphic map g from X into any complex torus T, then if we choose the proper origin on T, g is the composition of f and a homomorphism from $\mathrm{Alb}(X)$ into T. In fact, $\mathrm{Alb}(X)$ can be characterised by this property (see [Bl]) which in particular implies that $f(X)$ is never contained in a translate of a proper subtorus and also that $X=\mathrm{Alb}(X)$ if X is a torus itself.

Remark. The Albanese torus has been defined by Blanchard for any compact, connected complex manifold. Observe in any case that as soon as we have a Hodge decomposition in dimension 1, then the preceding construction works also in the non-kählerian case (but there exist compact complex manifolds without such a Hodge decomposition).

The identity component of $\mathrm{Pic}(X)$ is the *Picard torus* $\mathrm{Pic}^0(X)$. We have

$$\mathrm{Pic}^0(X) = H^1(X, \mathcal{O}_X)/i(H^1(X, \mathbb{Z})),$$

where i comes from the exponential cohomology sequence

$$H^1(X, \mathbb{Z}) \xrightarrow{i} H^1(X, \mathcal{O}_X) \to H^1(X, \mathcal{O}_X^*).$$

The point is that if X is a compact Kähler manifold, then $i(H^1(X, \mathbb{Z}))$ is always a lattice in $H^1(X, \mathcal{O}_X)$, and so $\mathrm{Pic}^0(X)$ a complex torus (of dimension $h^{1,0}(X)$). In general this is not the case, even if X is compact.

To prove this property of $\mathrm{Pic}^0(X)$ for a compact Kähler manifold, we observe that there is a factorisation of i:

$$\begin{array}{ccc} H^1(X, \mathbb{Z}) & \xrightarrow{\quad i \quad} & H^1(X, \mathcal{O}_X) \\ & {\scriptstyle j}\searrow \quad \nearrow{\scriptstyle p} & \\ & H^1(X, \mathbb{C}). & \end{array}$$

Now the Hodge decomposition (Corollary 13.3):

$$H^1(X, \mathbb{C}) \cong H^{1,0}(X) \oplus H^{0,1}(X)$$

is such, that the projection onto the first factor becomes the p of the diagram. We shall not prove this here, but we shall prove a similar statement for dimension 2 in Chap. IV (Theorem 2.12). Once this is known, our claim follows immediately: the image of $H^1(X, \mathbb{Z})$ in $H^1(X, \mathbb{C})$ is a lattice in $H^1(X, \mathbb{R})$, and its projection into $H^{1,0}(X)$ remains a lattice, since p maps $H^1(X, \mathbb{R})$ isomorphically onto $H^{1,0}(X)$, if we consider also this last space as a real vector space.

The functorial properties of $\mathrm{Pic}^0(X)$ which we shall use, are obvious from the definition.

14. Weight-1 Hodge Structures

Let $H_{\mathbb{C}}$ be a complex vector space of dimension $2g$. By definition, a *Hodge structure of weight* 1 on $H_{\mathbb{C}}$ consists of

i) a $\mathbb{Z}$-submodule $H_{\mathbb{Z}} \subset H_{\mathbb{C}}$ of rank $2g$ spanning $H_{\mathbb{C}}$ over $\mathbb{C}$ ($H_{\mathbb{Z}}$ is called the integral lattice);

ii) a direct sum decomposition $H_{\mathbb{C}} = H^{1,0} \oplus H^{0,1}$ with $H^{1,0} = \overline{H^{0,1}}$.

Here and in the sequel the bar denotes complex conjugation in $H_{\mathbb{C}} = H_{\mathbb{Z}} \otimes \mathbb{C}$ induced by conjugation on $\mathbb{C}$. Its fixed points form the real subvectorspace $H_{\mathbb{R}} = H_{\mathbb{Z}} \otimes \mathbb{R}$ of $H_{\mathbb{C}}$.

(14.1) **Example.** Let X be a compact Kähler manifold. Then $H=H^1(X,\mathbb{C})$ carries a weight-1 Hodge structure given by the Hodge-decomposition (Corollary 13.3), and thc integral lattice is $H^1(X,\mathbb{Z})$.

Denote by $q\colon H_{\mathbb{C}}^{\vee}\to(H^{1,0})^{\vee}$ the projection dual to the inclusion $H^{1,0}\subset H_{\mathbb{C}}$. The kernel of q is $(H^{0,1})^{\vee}$. By property ii) we have $H_{\mathbb{R}}^{\vee}\cap(H^{0,1})^{\vee}=0$, so q induces an $\mathbb{R}$-isomorphism $H_{\mathbb{R}}^{\vee}\to(H^{1,0})^{\vee}$ and the image $q(H_{\mathbb{Z}}^{\vee})$ in $(H^{1,0})^{\vee}$ is a lattice of maximal rank. The torus $\operatorname{Alb} H_{\mathbb{C}}=(H^{1,0})^{\vee}/q(H_{\mathbb{Z}}^{\vee})$ is called the Albanese torus of the given Hodge structure.

In the preceding example the map q is nothing but the map $H_1(X,\mathbb{C})\to(H^{1,0})^{\vee}$ sending the class of the 1-cycle γ to the functional $\omega\to\int_{\gamma}\omega$, $[\gamma]\in H_1(X,\mathbb{C})$. It follows that the Albanese torus of the Hodge structure on $H^1(X,\mathbb{C})$ coincides with the Albanese torus $\operatorname{Alb}(X)$ defined in Sect. 13.

For tori this has the following useful consequence:

(14.2) **Theorem** (Torelli theorem for tori). *Let T, T' be two tori of the same dimension and let $\varphi\colon H^1(T,\mathbb{Z})\to H^1(T',\mathbb{Z})$ be an isomorphism such that its $\mathbb{C}$-linear extension maps $H^{1,0}(T)$ isomorphically onto $H^{1,0}(T')$. Then φ is induced by an isomorphism from T' onto T.*

Proof. The universal property of the Albanese map implies that any torus is isomorphic to its Albanese torus. So there are isomorphisms $\alpha\colon T\xrightarrow{\sim}\operatorname{Alb}(T)$ and $\alpha'\colon T'\xrightarrow{\sim}\operatorname{Alb}(T')$. But $\operatorname{Alb}(T)=\operatorname{Alb} H^1(T,\mathbb{C})$ and similarly for T'. So φ induces $\varphi^*\colon\operatorname{Alb}(T')\xrightarrow{\sim}\operatorname{Alb}(T)$ and $\alpha^{-1}\circ\varphi^*\circ\alpha'\colon T'\xrightarrow{\sim}T$ induces φ by construction. □

Next, we consider *polarised Hodge structures of weight* 1. Suppose we are given a Hodge structure of weight 1 on the vector space $H_{\mathbb{C}}$ of dimension $2g$. Given an $\mathbb{R}$-bilinear form Q on $H_{\mathbb{R}}$ we denote its $\mathbb{C}$-bilinear extension to $H_{\mathbb{C}}$ also by Q. We say that a skew form Q on $H_{\mathbb{R}}$ *polarises the Hodge structure* if

(i) $Q(c,c')=0$ for all $c,c'\in H^{1,0}$;

(ii) $\sqrt{-1}\,Q(c,\bar{c})>0$ for all $c\in H^{1,0}$, $c\neq 0$.

We call Q *integral* if it assumes integral values on $H_{\mathbb{Z}}$. In this case (by Lemma 2.3) $H_{\mathbb{Z}}$ admits a *canonical basis* in which Q is given by a matrix

$$\begin{pmatrix}0 & -\Delta\\ \Delta & 0\end{pmatrix};\qquad \Delta=\begin{pmatrix}\delta_1 & 0 & \dots & 0\\ 0 & \ddots & & \vdots\\ \vdots & & \ddots & 0\\ 0 & \dots & 0 & \delta_g\end{pmatrix}.$$

We call Q a *polarisation of type* Δ, and if all δ_i are 1 we call it a *principal polarisation.*

(14.3) **Example.** Let X be a compact, connected Riemann surface. The weight-1 Hodge structure on $H^1(X,\mathbb{C})$ considered in Example 14.1 is obviously polarised by the cupproduct pairing

$$(x,y)=\int_X \xi\wedge\eta$$

(here x, y are the classes represented by the forms ξ, η). The cupproduct pairing is integral, even unimodular by the Poincaré duality theorem, so defines a principal polarisation. A canonical basis consists of classes $\alpha_1, \dots, \alpha_g$, $\beta_1, \dots, \beta_g \in H^1(X, \mathbb{Z})$ with $(\alpha_i, \alpha_j) = (\beta_i, \beta_j) = 0$ and $(\alpha_i, \beta_j) = \delta_{i,j}$. Let $a_1, \dots, a_g$, $b_1, \dots, b_g \in H_1(X, \mathbb{Z})$ be the Poincaré-duals (see (13)), i.e. $a_i = \mathscr{P}_X \alpha_i$, $b_i = \mathscr{P}_X \beta_i$. Then

$$a_i \cdot a_j = b_i \cdot b_j = 0, \qquad a_i \cdot b_j = \delta_{i,j},$$

so this is a canonical basis for $H_1(X, \mathbb{Z})$ in the classical sense.

Weight-1 Hodge structures on a fixed space $H_{\mathbb{C}}$ of dimension $2g$, with fixed integral lattice $H_{\mathbb{Z}}$ and integral polarisation Q of type Δ, can be classified in the following way. Let $\alpha_1, \dots, \alpha_g$, $\beta_1, \dots, \beta_g$ be a canonical basis for $H_{\mathbb{Z}}$ and $\varphi_1, \dots, \varphi_g$ a basis for $H^{1,0}$. We may express the φ_i in terms of the α_i and β_i

$$(\varphi_1, \dots, \varphi_g) = (\alpha_1, \dots, \beta_g)\, \Omega,$$

where Ω is a complex $(2g \times g)$-matrix. It is called the *period matrix* of the polarised Hodge structure with respect to the given bases. If we set

$$\Omega = \begin{pmatrix} \Omega_1 \\ \Omega_2 \end{pmatrix}$$

with $g \times g$-matrices Ω_i, properties i) and ii) are equivalent with the matrix conditions:

$$\Omega_1^t \,\Delta\, \Omega_2 \qquad \text{symmetric,}$$
$$\sqrt{-1}\,(\Omega_1^t \,\Delta\, \bar{\Omega}_2 - \Omega_2^t \,\Delta\, \bar{\Omega}_1) > 0.$$

The second condition implies in particular that Ω_1 is invertible. We therefore may substitute

$$(\varphi_1', \dots, \varphi_g') = (\varphi_1, \dots, \varphi_g)\, \Omega_1^{-1} \Delta^{-1}$$

to obtain a new period matrix

$$\Omega' = \begin{pmatrix} \Delta^{-1} \\ Z \end{pmatrix}.$$

Now the matrix conditions above become

$$(16) \qquad Z = Z^t, \qquad \operatorname{Im} Z > 0$$

(Riemann period conditions). A basis $\varphi_1', \dots, \varphi_g'$ as above and the corresponding period matrix Ω' are called *normalised* with respect to the canonical basis $\alpha_1, \dots, \beta_g \in H_{\mathbb{Z}}$. The set of complex $g \times g$-matrices Z satisfying (16) is called the *Siegel upper half space* $\mathfrak{H}_g$.

For later use we need another description of the part Z of a normalised period matrix for a principal polarisation. Consider the images $p\alpha_1, \dots$, $p\beta_g \in H^{0,1}$ under the projection $p\colon H_{\mathbb{C}} \to H^{0,1}$. Since $p\varphi = 0$ for all $\varphi \in H^{1,0}$, the equation $(\varphi_1, \dots, \varphi_g) = (\alpha_1, \dots, \alpha_g) + (\beta_1, \dots, \beta_g) Z$ defining Z in $H^{0,1}$ becomes

$$(p\alpha_1, \dots, p\alpha_g) = -(p\beta_1, \dots, p\beta_g)\, Z.$$

So $p\beta_1, \dots, p\beta_g$ form a $\mathbb{C}$-basis of $H^{0,1}$ and $-Z$ is the matrix of coefficients of the vectors $p\alpha_1, \dots, p\alpha_g$ with respect to this basis.

If we have a *principal polarisation*, then two canonical bases of $H_{\mathbb{Z}}$ are related by

$$(17) \qquad (\alpha_1, \ldots, \beta_g) = (\alpha_1', \ldots, \beta_g') \cdot \sigma, \qquad \sigma = \begin{pmatrix} A & B \\ C & D \end{pmatrix} \in Sp(g, \mathbb{Z}).$$

The new period matrix is $\begin{pmatrix} \mathbb{1} \\ Z' \end{pmatrix}$ with

$$(18) \qquad Z' = (DZ + C)(BZ + A)^{-1}.$$

By way of σ: $Z \to Z'$ the group $Sp(g, \mathbb{Z})$ operates on $\mathfrak{H}_g$. Since $\pm \mathbb{1}$ operates trivially, there is an induced action of the *modular group* $\Gamma_g = Sp(g, \mathbb{Z})/\{\pm 1\}$. The *period domain* for principally polarised Hodge structures on a $2g$-dimensional vector space is the quotient $D_g = \mathfrak{H}_g / \Gamma_g$. The group Γ_g acts properly and discontinuously, so this period domain carries the structure of a normal analytic variety. (It is even quasi-projective by [B-B].)

Example. As above let $H = H^1(X, \mathbb{C})$, where X is a connected compact Riemann surface. If $a_1, \ldots, b_g \in H_1(X, \mathbb{Z})$ is a canonical homology basis and $\varphi_1, \ldots, \varphi_g$ is a basis of the space of holomorphic differentials on X, the classical period matrix is

$$\begin{pmatrix} \int_{a_i} \varphi_k \\ \int_{b_i} \varphi_k \end{pmatrix}.$$

This matrix is equivalent with our Ω. In fact, let $\mathscr{P} = \mathscr{P}_X$ and $a_1 = \mathscr{P} \alpha_1, \ldots, b_g = \mathscr{P} \beta_g$. These classes form a canonical cohomology basis with

$$\int_{a_i} \alpha_j = \int_{b_i} \beta_j = 0, \qquad \int_{b_i} \alpha_j = -\int_{a_i} \beta_j = \delta_{ij}.$$

So the basis $-\beta_1, \ldots, -\beta_g, \alpha_1, \ldots, \alpha_g$, which is also a canonical cohomology basis, is Kronecker dual to $a_1, \ldots, b_g$. Our period matrix Ω for this basis therefore coincides with the classical period matrix, up to the action of the symplectic group. (Instead of the original normalised period matrix $\begin{pmatrix} \mathbb{1} \\ Z \end{pmatrix}$ we have $\begin{pmatrix} \mathbb{1} \\ -Z^{-1} \end{pmatrix}$.)

15. Yau's Results on Kähler-Einstein Metrics

Let X be a Kähler manifold, and ω_g the closed real (1, 1)-form belonging to a Kähler metric g on X. We can associate to g a second closed real (1, 1)-form, the Ricci form Ric(g) of g (see [Sem P.], p. 79). The metric g is called a *Kähler-Einstein metric* if there exists a $\lambda \in \mathbb{R}$, such that $\mathrm{Ric}(g) = \lambda \, \omega_g$.

Around 1955, Calabi formulated several conjectures concerning the existence of such metrics, a (highly non-trivial) part of which was proved by S.-T. Yau in 1976. Yau's results have important implications for the topology of algebraic manifolds. We shall formulate here only those special cases of Yau's results which will be needed in this book. For more details we refer to [Y2], [Y3] and [Sem P.].

(15.1) **Theorem.** *Let X be a compact complex manifold, of dimension at least 2, with $c_1(X)=0$. If an element of $H^2(X,\mathbb{R})$ can be represented by a Kähler form, then it can be represented by exactly one Kähler form ω_g such that* $\mathrm{Ric}(g)=0$.

(15.2) **Theorem.** *Let X be a compact complex manifold of dimension at least 2, such that $-c_1(X)$ can be represented by a Kähler form. Then X admits a Kähler-Einstein metric.*

On the other hand there are the following facts, which were known before. For the first theorem we refer to [K-N], vol. II, p. 17 and for the second to [S-T2], p. 359.

(15.3) **Theorem.** *Let X be an n-dimensional compact, connected Kähler manifold with constant holomorphic sectional curvature. Then the universal covering space of X is either $\mathbb{P}_n$ (with the Fubini metric), or $\mathbb{C}^n$ (with the standard flat metric) or the unit ball in $\mathbb{C}^n$ (with the Bergmann metric).*

(15.4) **Theorem.** *If on the 2-dimensional compact, connected complex manifold X with $c_1^2(X)=3c_2(X)$ there exists a Kähler-Einstein metric, then the holomorphic sectional curvature is constant.*

(15.5) **Corollary.** *Let the 2-dimensional compact connected complex manifold X be endowed with a Kähler-Einstein metric. If $c_1^2(X)=3c_2(X)>0$ and if X is different from $\mathbb{P}_2$, then the universal covering of X is the unit ball in $\mathbb{C}^2$.*

Proof. If the universal covering of X is $\mathbb{P}_2$, then X is isomorphic to $\mathbb{P}_2$, since every automorphism of $\mathbb{P}_2$ has a fixed point. On the other hand, if the universal covering of X is $\mathbb{C}^2$, and hence the metric on X flat, by Theorems (15.4) and (15.3) we must have $e(X)=c_2(X)=0$. □

Coverings

16. Ramification

Apart from the meaning: "system of subsets, covering a whole set", we shall use the word "covering" in two (related) ways.

Firstly in the sense of *analytic covering space*. This is a triple (X, Y, π) (also to be denoted by $\pi\colon X \to Y$) where X and Y are connected complex spaces and $\pi\colon X \to Y$ a surjective holomorphic map such that all points $y \in Y$ have a connected neighbourhood V_y, with the property that $\pi^{-1}(V_y)$ consists of the union of disjoint open subsets of X, each of which is mapped isomorphically onto V_y by π. Mostly X and Y will be complex manifolds.

If we use the corresponding topological concept we shall speak of a "topological covering". Given any topological covering (X, Y, π) such that Y is a complex manifold, then, up to an isomorphism, there is one structure of complex manifolds on Y such that (X, Y, π) becomes an analytic covering.

Secondly, we shall use the word "covering" for triples (X, Y, π) (also to be denoted by $\pi: X \to Y$) where X and Y are connected *normal* complex spaces and π a *finite*, surjective proper holomorphic map.

In this last case there exists a proper analytic subset of X, outside of which π is a topological covering. Indeed, on $X' = X \setminus \pi^{-1}(\pi(\operatorname{Sing} X) \cup \operatorname{Sing} Y)$ the map π is a covering between manifolds, so it is a topological covering outside of $\pi^{-1}(\pi(S))$ where $S = \{x \in X';\ \operatorname{rank}(d\pi)_x \leq \dim X - 1\}$. By definition the degree of π is the degree of $\pi | X' \setminus \pi^{-1}(\pi(S))$. Properness of π implies that for every $x \in X$ there exists at least one connected open neighbourhood V of $\pi(x) \in X$ such that $\pi^{-1}(V)$ is a union of disjoint connected open neighbourhoods U_i of x_i $(i = 1, \ldots, n)$, where $\pi^{-1}\pi(x) = \{x = x_1, x_2, \ldots, x_n\}$. If $V' \subset V$ is another connected neighbourhood of $\pi(x)$, we claim that $\pi^{-1}(V') \cap U_1 = U_1'$ is connected. Indeed, if U_1'' is any connected component of U_1', its image must be V' by Theorem 8.4, so $x \in U_1''$ and $U_1'' = U_1'$. It follows that the degree of $\pi | \pi^{-1}(V') \cap U_1$ is the same for all connected neighbourhoods $V' \subset V$ of x. This degree is called the *local degree* e_x of π at x or the *branching order of* π at x. If $e_x \geq 2$ we say that π is *ramified* at x, and x is called a *ramification point*. The images of ramification points are called *branch points*.

Now let us assume that both X and Y are manifolds. Then the set of ramification points is the zero divisor R of the canonical section in $\operatorname{Hom}(p^*(\mathcal{K}_Y), \mathcal{K}_X)$, i.e.

$$\mathcal{K}_X = \pi^*(\mathcal{K}_Y) \otimes \mathcal{O}_X(R). \tag{19}$$

The divisor D is called the *ramification divisor* of π. Formula (19), together with the specification of R, given by Lemma 16.1 below, is called the *Hurwitz-formula*.

We observe that the properness of π implies that $\pi: X \setminus \pi^{-1}(\pi(R)) \to Y \setminus \pi(R)$ is a covering in the first sense; in particular, a covering in the second sense with $R = 0$ is one in the first sense too.

To emphasise the difference we shall frequently call a covering in the first sense an *unbranched, unramified, or étale covering* and one in the second sense a *branched* or *ramified covering*, as soon as $R \neq 0$.

(16.1) **Lemma.** *If* $R = \sum r_j R_j$, *where* R *is the ramification divisor of some branched covering and the* R_j*'s its irreducible components, then* $r_j = e_j - 1$, *where* e_j *is the branching order at any point* $x \in R_j$ *which is smooth on* R_{red}, *and for which* $y = \pi(x)$ *is smooth on* $B_j = \pi(R_j)$.

Proof. Let $(t_1, \ldots, t_n)$ be local coordinates on Y, centered at y, such that B_j is given by $t_1 = 0$. If $s = 0$ is a local equation for R_j at x, then we have $\pi^*(t_1) = \varepsilon \cdot s^{e_j}$, where ε does not vanish around x, and in fact can be taken to be 1 if s is suitably chosen. If we set $\omega = dt_1 \wedge \ldots \wedge dt_n$, then $\pi^*(\omega) = s^{e_j - 1}\, ds \wedge d\pi^*(t_2) \wedge \ldots \wedge d\pi^*(t_n)$. This not only shows that $(s, \pi^*(t_2), \ldots, \pi^*(t_n))$ is a local coordinate system at x (so $e_j=e_x$), but also that the zero divisor of $\pi^*(\omega)$ is $(e_j-1)R_j$. Hence $r_j=e_j-1$. □

(16.2) **Lemma.** *Let X and Y be compact connected complex manifolds and $f: X\to Y$ a covering of degree d. If $\mathcal{L}$ is a line bundle on Y with $f^*\mathcal{L}=\mathcal{O}_X$, then $\mathcal{L}^{\otimes d}=\mathcal{O}_Y$.*

Proof. Since $f_*\mathcal{O}_X$ is locally free of rank d (compare [Se3]) this is an immediate consequence of $f_*\mathcal{O}_X=f_*f^*\mathcal{L}=\mathcal{L}\otimes f_*\mathcal{O}_X$. □

17. Cyclic Coverings

Let Y be a connected complex manifold and B a divisor on Y which is either effective or zero. Suppose we have a line bundle $\mathcal{L}$ on Y such that

$$\mathcal{O}_Y(B)=\mathcal{L}^{\otimes n},$$

and a section $s\in\Gamma(Y,\mathcal{O}_Y(B))$ vanishing exactly along B (if $B=0$, we take for s the constant function 1). We denote by L the total space of $\mathcal{L}$ and we let $p: L\to Y$ be the bundle projection. If $t\in\Gamma(L, p^*\mathcal{L})$ is the tautological section, then the zero divisor of p^*s-t^n defines an analytic subspace X in L.

If $B\neq 0$ and reduced, X is an irreducible normal analytic subspace of L, and $\pi=p|X$ exhibits X as an n-fold ramified covering of Y with branch-locus B. We call (X,Y,π) (or X, or π) the n-cyclic covering of Y branched along B, determined by $\mathcal{L}$.

On the other hand, if $B=0$, we must take n minimal (i.e. $\mathcal{L}$ is exactly of order n in $\mathrm{Pic}(Y)$) in order to obtain a connected manifold X. In this case (X,Y,π) (or X, or π) is called the n-cyclic unramified covering of Y determined by the torsion bundle $\mathcal{L}$.

If $\mathrm{Pic}(Y)$ has no torsion, then B uniquely determines $\mathcal{L}$ and we may speak of *the* n-cyclic covering of Y, branched along B.

It is clear from the above description that X has at most singularities over singular points of B. In particular if B is reduced and smooth, then also X is smooth.

(17.1) **Lemma.** *Let π: $X\to Y$ be the n-cyclic covering of Y branched along a smooth divisor B and determined by $\mathcal{L}$, where $\mathcal{L}^{\otimes n}=\mathcal{O}_Y(B)$. Let B_1 be the reduced divisor $\pi^{-1}(B)$ on X. Then*

(i) $\mathcal{O}_X(B_1)=\pi^*\mathcal{L}$

(ii) $\pi^*B=nB_1$ *(in particular n is the branching order along B_1)*

(iii) $\mathcal{K}_X=\pi^*(\mathcal{K}_Y\otimes\mathcal{L}^{n-1})$.

Proof. If we embed Y as the zero-section in L, then the section $t\in\Gamma(L,p^*\mathcal{L})$ has divisor Y, so $\mathcal{O}_L(Y)=p^*\mathcal{L}$. By construction Y and $X\subset L$ intersect transversally in B_1, so $\mathcal{O}_X(B_1)=\mathcal{O}_L(Y)|X=\pi^*\mathcal{L}$. The identity $\pi^*B=nB_1$ follows from the equation $p^*s-t^n=0$ for X in L. The formula for $\mathcal{K}_X$ is an application of Lemma 16.1. □

(17.2) **Lemma.** *Let* $\pi: X \to Y$ *be as in Lemma* 17.1. *Then* $\pi_* \mathcal{O}_X \simeq \bigoplus_{j=0}^{n-1} \mathcal{L}^{-j}$.

Proof. For an open set $V \subset Y$, any holomorphic function f on $p^{-1}(V)$ has a unique power series expansion $f = \sum_{k=0}^{\infty} a_k t^k$, $a_k \in \Gamma(V, \mathcal{L}^{-k})$. Every function on $\pi^{-1}(V) \subset p^{-1}(V)$ is the restriction of such an f. Using the equation $t^n = \pi^* s$, we obtain a unique expansion $\sum_{k=0}^{n-1} b_k t^k$, $b_k \in \Gamma(V, \mathcal{L}^{-k})$ for holomorphic functions on $\pi^{-1}(V)$.

18. Covering Tricks

The "*unbranched covering trick*" is nothing but the following remark.

(18.1) **Proposition** (Unbranched covering trick). *Let X be a connected complex manifold.*

(i) *If* $b_1(X) \neq 0$, *then X admits unbranched coverings of any order;*
(ii) *If* $H_1(X, \mathbb{Z})$ *contains k-torsion, then X has an unbranched covering of order k.*

Proof. As to (i), if $b_1(X) \neq 0$, then $H_1(X, \mathbb{Z})$ is infinite and therefore admits quotient groups of any order. Consequently, also $\pi_1(X)$ has such quotients, and so there exist unbranched coverings of X of any order.

The proof of (ii) is similar. □

The "*branched covering trick*", though trivial from an algebraic point of view, requires a little bit more care.

(18.2) **Theorem** (Branched covering trick). *Be given a holomorphic* $\mathbb{P}_1$*-bundle over the irreducible, normal complex space X, with total space B and projection* $p: B \to X$. *If S is any irreducible divisor on B, meeting a general fibre in n points, then there exists a normal complex space Y, a generically finite surjective map* $f: Y \to X$, *and n effective divisors* $S_1, \ldots, S_n$ *on* $B' = B \times_Y X$, *all meeting a general fibre in one point, such that* $g^*(S) = S_1 + \ldots + S_n$, *where* $g: B' \to B$ *is the projection.*

Proof. For $n = 1$ there is nothing to prove. If $n \geq 2$ we consider the normalisation $\bar{S}$ of S and observe that, if we put $B_1 = \bar{S} \times_X B$ and denote by $g_1: B_1 \to B$ the natural projection, we have in a canonical way that $g^*(S) = S_1 + S'$, where S_1 (resp. S') meets a general fibre of $B_1 \to \bar{S}$ in one (resp. $(n-1)$) point(s). Applying the same procedure to $\bar{S}$, and so on, we finally obtain the desired result. □

As a consequence we have (compare [Mi3], Lemma 11):

(18.3) **Theorem.** *Let X be a compact, connected complex manifold and* $\mathcal{L}$ *a holomorphic line bundle on X with* $h^0(\mathcal{L}^{\otimes n}) \geq 2$ *for some* $n \geq 1$. *Then there exists a compact complex manifold Y and a generically finite-to-one map* $f: Y \to X$, *such that* $h^0(f^*(\mathcal{L})) \geq 2$.

Projective-Algebraic Varieties

19. GAGA Theorems and Projectivity Criteria

In the same way as we defined complex-analytic spaces, but using Zariski-topology and affine charts instead of "open ball-charts", we can define algebraic varieties. We shall only need projective-algebraic varieties, i.e. closed algebraic subvarieties of projective spaces $\mathbb{P}_n(\mathbb{C})$, or at most quasi-projective varieties, that is, locally closed algebraic subvarieties of a $\mathbb{P}_n$.

First of all a word about the results of Serre's classical GAGA-paper [Se4], of which we shall need only a modest part.

To every (projective-)algebraic variety X we can attach an analytic space X^{an} in an obvious way such that there is a morphism $m\colon X^{\mathrm{an}} \to X$ of ringed spaces. If $\mathcal{S}$ is any coherent sheaf on X in the algebraic sense, we can make $m^{-1}(\mathcal{S})$ to a coherent sheaf $\mathcal{S}^{\mathrm{an}}$ on X^{an} in the analytic sense by tensoring with $\mathcal{O}_{X^{\mathrm{an}}}$. One has that $m^*\colon H^i(X,\mathcal{S}) \xrightarrow{\sim} H^i(X^{\mathrm{an}},\mathcal{S}^{\mathrm{an}})$ for all $i \geq 0$.

(19.1) **Theorem.** *Let X be a projective-algebraic variety and $\tilde{\mathcal{S}}$ a coherent sheaf on X^{an}. Then there is one coherent sheaf $\mathcal{S}$ on X with $\mathcal{S}^{\mathrm{an}} = \tilde{\mathcal{S}}$. If $\tilde{\mathcal{S}}$ is locally free, then so is $\mathcal{S}$.*

Roughly we can say: every analytic coherent sheaf on X is algebraic. In particular, every analytic vector bundle on X is algebraic.

An easy consequence of Theorem 19.1 is

(19.2) **Theorem** (Chow's theorem). *Every closed analytic subspace of $\mathbb{P}_n$ is an algebraic subvariety.*

Let X be an analytic space and $\mathcal{L}$ a holomorphic line bundle on X. If for every point $x \in X$ there is at least one section $s \in \Gamma(X,\mathcal{L})$ with $s(x) \neq 0$, then (upon choosing a base) $\Gamma(X,\mathcal{L})$ yields a holomorphic map into $\mathbb{P}_N$, where $N = \dim \Gamma(X,\mathcal{L}) - 1$. If this map is an isomorphism from X onto its image, then we say that $\mathcal{L}$ is *very ample*. A line bundle $\mathcal{M}$ is called *ample* if there exists an $n \geq 1$ such that $\mathcal{M}^{\otimes n}$ is very ample.

The following classical criterion for a line bundle to be very ample will play an important role in this book (compare [Gr2]).

(19.3) **Theorem** (Grauert's criterion). *Let X be a compact complex space and $\mathcal{L}$ a holomorphic line bundle on X. Then $\mathcal{L}$ is very ample if and only if the following holds: given any irreducible analytic subset Y of strictly positive dimension on X, there exists an $n = n(Y)$, such that $\mathcal{L}^{\otimes n}|Y$ has a section which has at least one zero, but does not vanish identically.*

A projectivity criterion of quite different nature is Kodaira's criterion.

Let X be a complex manifold. A Kähler metric is called a *Hodge metric* if the associated element of $H^2(X,\mathbb{R})$ is integral, i.e. in the image of $H^2(X,\mathbb{Z})$. A

Kähler manifold is called a *Hodge manifold* if it admits at least one Hodge metric, and a famous result of Kodaira ([Ko1], Theorem 4) says that the converse is also true:

(19.4) **Theorem** (Kodaira's criterion for Hodge manifolds). *A compact complex manifold is (isomorphic to) a projective-algebraic manifold if and only if it is a Hodge manifold.*

Both Theorems 19.3 and 19.4 are trivial in one direction (as to 19.4, a projective space $\mathbb{P}_n$ obviously has a Hodge metric, and a submanifold of a Hodge manifold is again a Hodge manifold).

Finally, one more fact we shall need:

(19.5) **Proposition.** *The normalisation of a projective-algebraic space is again a projective-algebraic space.*

This is, for example, an easy consequence of Grauert's criterion.

20. Theorems of Bertini and Lefschetz

Again we shall need only a limited part of the classical theorems in question.

But first of all we recall a well-known elementary result which is sometimes also called Bertini's theorem, at least in the case that $\dim Y=1$.

(20.1) **Theorem.** *Let $f: X\to Y$ be a proper, surjective holomorphic map between the complex manifolds X and Y. Then the set of points on X where f is not of maximal rank, is a proper analytic subset of X, the image of which is a proper analytic subset on Y.*

Now the special case of Bertini's theorem which we need (see [G-H], p. 137 and [FL], p. 33).

(20.2) **Theorem** (Bertini's theorem). *Let X be a connected n-dimensional compact complex manifold and $f: X\to\mathbb{P}_N$ a holomorphic map such that* $\dim f(X)\geq 2$. *Then for a general hyperplane $H\subset\mathbb{P}_N$ (i.e. for all hyperplanes outside of a proper algebraic subset of $\mathbb{P}_N^\vee$) the (analytic) inverse image of H is a connected $(n-1)$-dimensional complex submanifold on X.*

So, apart from the case that f is constant, we have either the case of the theorem or the case $\dim f(X)=1$. In this last case there is a Stein factorisation $f=h\circ g$ where g is a connected map from X onto a smooth curve C, and $h: C\to\mathbb{P}_N$ finite. By Theorem 20.1, the general fibre of g is smooth, hence the general fibre of f consists of the disjoint union of finitely many submanifolds on X.

In this last case, g is called a *pencil* (rational if $C\cong\mathbb{P}_1$, irrational otherwise), and f is "*composed with*" this pencil (compare IV, Sect. 1).

(20.3) **Corollary.** *Let X be a connected n-dimensional algebraic submanifold of $\mathbb{P}_N$, with $n \geq 2$. Then a general hyperplane H in $\mathbb{P}_N$ intersects X transversally along a smooth irreducible divisor.*

Now let X be any compact, connected complex manifold and $\mathscr{L}$ a line boundle on X. Let $B \subset X$ be the base point set of $\mathscr{L}$, i.e. the analytic set of those points $x \in X$, for which $\gamma(x)=0$ for all $\gamma \in \Gamma(X, \mathscr{L})$. If $B=\emptyset$, then we have exactly the situation described above if we take for f the map in a projective space, obtained by taking a base of $\Gamma(X, \mathscr{L})$. If $\dim B \leq \dim X-2$, we obtain a similar result by considering the desingularisation of the graph of the meromorphic map obtained from $\Gamma(X, \mathscr{L})$, after choosing a base.

Either the image in $\mathbb{P}_N$ has dimension at least 2 and the inverse image of a general hyperplane section (i.e. the zero set of a general member of $\Gamma(X, \mathscr{L})$) on X is irreducible, with singularities at most in B, or f is composed with a "pencil with base points". But we shall have to face this situation for surfaces only, where the desingularisation is simple, so we shall return to this point in Chap. IV, Sect. 1.

(20.4) **Theorem** (Lefschetz theorem for hyperplane sections). *Let X be an n-dimensional submanifold of $\mathbb{P}_N$, $n \geq 2$, and let $H \subset \mathbb{P}_N$ be a hyperplane, such that $H \cap X$ is again a complex manifold. Then the inclusion homomorphisms*

$$H_i(X \cap H, \mathbb{Z}) \rightarrow H_i(X, \mathbb{Z})$$
$$\pi_i(X \cap H, \mathbb{Z}) \rightarrow \pi_i(X, \mathbb{Z})$$

are isomorphisms for $0 \leq i \leq n-2$.

For a proof we refer to [Mil2].

II. Curves on Surfaces

Unless stated otherwise, X denotes in this chapter a 2-dimensional complex manifold, not necessarily compact or connected.

Embedded Curves

1. Some Standard Exact Sequences

A *curve on* X is a 1-dimensional subspace of X locally defined by one equation, so there is a natural $1-1$ correspondence between curves and effective divisors on X.

If C is a curve on X, $f\colon C\to X$ the embedding and D a divisor on X, then we denote by $\mathcal{O}_C(D)$ the analytic restriction $f^*(\mathcal{O}_X(D))$. In particular $\mathcal{O}_C(C) = \mathcal{O}_X(C)|C$. The structure sequence (I, Sect. 8)

(1) $$0\to\mathcal{I}_C\to\mathcal{O}_X\to\mathcal{O}_C\to 0$$

yields, in combination with $\mathcal{I}_C\cong\mathcal{O}_X(-C)$, the sequence

(2) $$0\to\mathcal{O}_X\to\mathcal{O}_X(C)\to\mathcal{O}_C(C)\to 0.$$

If C is smooth, then $\mathcal{O}_C(C)$ is the normal bundle of C in X (see Proposition I.6.2). In case C is singular, we still shall call the line bundle $\mathcal{O}_C(C)$ the normal bundle of C in X. If $\{u_i\in\Gamma(\mathcal{O}_X|U_i)\}_{i\in I}$ is a collection of local equations for C, then $\mathcal{O}_C(C)$ is the line bundle defined by the cocycle $(u_i/u_j)|C$. Of course, there is a tangential sequence

(3) $$0\to\mathcal{T}_C\to\mathcal{T}_X|C\to\mathcal{N}_{C/X}\to 0$$

only when C is smooth.

Let $C=A+B$ be the sum of two effective divisors A, B on X.

The inclusions $\mathcal{I}_C\subset\mathcal{I}_B\subset\mathcal{O}_X$ induce a diagram

$$\begin{array}{cccccccccc}
 & & & & 0 & & 0 & & & \\
 & & & & \downarrow & & \downarrow & & & \\
 & & 0 & \longrightarrow & \mathcal{O}_X(-C) & \longrightarrow & \mathcal{O}_X(-B) & \longrightarrow & \mathcal{O}_A(-B) & \longrightarrow 0 \\
 & & & & \downarrow & & \downarrow & & & \\
 & & & & \mathcal{O}_X & = & \mathcal{O}_X & & & \\
 & & & & \downarrow & & \downarrow & & & \\
0 & \longrightarrow & \mathcal{I}_B/\mathcal{I}_C & \longrightarrow & \mathcal{O}_C & \longrightarrow & \mathcal{O}_B & \longrightarrow & 0. & \\
 & & & & \downarrow & & \downarrow & & & \\
 & & & & 0 & & 0 & & &
\end{array}$$

The first row is the structure sequence for A tensored with $\mathcal{O}_X(-B)$. This shows that $\mathcal{I}_B/\mathcal{I}_C$ is isomorphic with $\mathcal{O}_A(-B)$, a line bundle on A. The resulting exact sequence

$$0 \to \mathcal{O}_A(-B) \to \mathcal{O}_C \xrightarrow{\text{restr}} \mathcal{O}_B \to 0 \tag{4}$$

will be called a *decomposition sequence* for $C=A+B$. The special case $2C=C+C$ shows that $\mathcal{I}_C/\mathcal{I}_C^2 \simeq \mathcal{O}_C(-C)$ (compare I, Sect. 8). Also in the non-smooth case this sheaf is called the co-normal bundle of C in X.

Now let $C \subset X$ be *reduced*. If $\nu\colon \tilde{C} \to C$ is the normalisation of C (I, Sect. 8), then $\tilde{C}$ is a (non-singular) Riemann surface. For $x \in C$ singular, $\nu^{-1}(x)$ consists of finitely many pointa corresponding to the different branches of C through x. On C there is the *normalisation sequence*

$$0 \to \mathcal{O}_C \to \nu_* \mathcal{O}_{\tilde{C}} \to S \to 0 \tag{5}$$

with S concentrated at the singularities of C.

For a smooth curve C, we have the adjunction formula (I.6.3)

$$\mathcal{K}_C = \mathcal{K}_X \otimes \mathcal{O}_C(C).$$

If C is singular (and possibly non-reduced) we can formally define a canonical bundle for C by setting

$$\omega_C = \mathcal{K}_X \otimes \mathcal{O}_C(C).$$

This line bundle is mostly called the *dualising sheaf* on C. According to our definition it depends on the embedding of C in X, although general theory shows that this is actually not the case. Tensoring (2) with $\mathcal{K}_X$ yields the *residue sequence*

$$0 \to \mathcal{K}_X \to \mathcal{K}_X \otimes \mathcal{O}_X(C) \xrightarrow{r} \omega_C \to 0. \tag{6}$$

For *reduced* C, we give here an explicit description of r. The relation with the residue theorem will be discussed in Sect. 4.

Let (u, v) be local coordinates on X such that u is not constant on any open subset of C. If $f=0$ is a local equation for C, the partial derivative f_v cannot vanish on any open subset of C, since

$$f_u \nu^*(du) + f_v \nu^*(dv) = \nu^*(df) = 0.$$

If for a local section $\varphi = h\, du \wedge dv/f$ in $\mathcal{K}_X \otimes \mathcal{O}_X(C)$ we define

$$r'(\varphi) = \nu^*(h\, du/f_v),$$

then $r'(\varphi)$ is a meromorphic differential on an open piece of $\tilde{C}$. It is obvious that $r'(\varphi)$ does not depend on the choice of f. But it also is independent of the choice of the local coordinates (u, v). Namely, let (w, z) be other coordinates, with $\delta = u_w v_z - u_z v_w$. Then

$$\begin{aligned}
\nu^*(f_z\, du) &= \nu^*(f_z u_w\, dw + f_z u_z\, dz) \\
&= \nu^*(f_z u_w - f_w u_z)\, \nu^*(dw), \\
f_z u_w - f_w u_z &= f_u u_z u_w + f_v v_z u_w - f_u u_w u_z - f_v v_w u_z = f_v \cdot \delta, \\
\nu^*(\delta\, dw/f_z) &= \nu^*(du/f_v).
\end{aligned}$$

So r' is a globally defined morphism of $\mathscr{K}_X \otimes \mathcal{O}_X(C)$ into the sheaf of meromorphic 1-forms on $\tilde{C}$. Its kernel consists of those $\varphi = h\,du \wedge dv/f$, for which $v^*(h) = 0$, i.e. $\ker(r') = \operatorname{im}(\mathscr{K}_X)$. So ω_C equals $v^*(\operatorname{im}(r'))$, i.e. ω_C is the sheaf (on C) of meromorphic differentials on $\tilde{C}$ of the form $r'(\varphi)$ and r can be identified with r'.

2. The Picard-Group of an Embedded Curve

If C is a smooth curve, then on C there is the exponential sequence (I, Sect. 6)

$$0 \to \mathbb{Z}_C \to \mathcal{O}_C \xrightarrow[\mathfrak{e}]{} \mathcal{O}_C^* \to 0. \tag{7}$$

This sequence can be generalised to the case of any curve C, reduced or not, on a surface X, as we shall now explain. Let $\mathcal{O}_C^* \subset \mathcal{O}_C$ be the subsheaf of those germs f in $\mathcal{O}_C$ for which $f|C_{\mathrm{red}}$ has no zero. Such an f can locally be represented as $f = \tilde{f}|C$, where $\tilde{f}$ is holomorphic without zeros on an open set in X. For $f \in \mathcal{O}_C$ we put $\mathfrak{e}(f) = e^{2\pi\sqrt{-1}f}$, where the *exponential* e^f is defined in the following way. If locally f is represented as $\tilde{f}|C$, with $\tilde{f}$ holomorphic in a neighbourhood on X of the point on C in question, and if we put $e^f = e^{\tilde{f}}|C$, then this e^f depends on f only, for if $g \in \Gamma(X, \mathscr{I}_C)$, then

$$e^g - 1 = \sum_{m \geq 1} g^m/m! \in \Gamma(X, \mathscr{I}_C)$$

by the closedness property ([G-R3], IV.4.1) of the ideal $\mathscr{I}_C$. To prove exactness in the middle of (7), let $f = \tilde{f}|C$ with $e^f = 1$, i.e. $e^{\tilde{f}} - 1 \in \mathscr{I}_C$. After subtracting the (locally) constant function $f|C_{\mathrm{red}} \in \mathbb{Z} \cdot 2\pi\sqrt{-1}$, we may assume $\tilde{f} \in \mathscr{I}_{C_{\mathrm{red}}}$, say $\tilde{f} \in \mathscr{I}_{C_i}^n$ but $\tilde{f} \notin \mathscr{I}_{C_i}^{n+1}$, for some irreducible component $C_i \subset C$. Then also

$$e^{\tilde{f}} - 1 = \sum_{m \geq 1} \tilde{f}^m/m! \in \mathscr{I}_{C_i}^n \setminus \mathscr{I}_{C_i}^{n+1},$$

so n is just the multiplicity of C_i in C, and $\tilde{f} \in \mathscr{I}_C$.
Finally, $\mathfrak{e}$ is surjective, because it is surjective on every open set $U \subset X$.

Exactly like in the smooth case, the exponential cohomology sequence

$$H^1(C, \mathbb{Z}) \to H^1(\mathcal{O}_C) \to H^1(\mathcal{O}_C^*) \to H^2(C, \mathbb{Z})$$

describes the group $\operatorname{Pic}(C) = H^1(\mathcal{O}_C^*)$ of line bundles on C.

(2.1) **Proposition.** *If C is compact, the inclusion $\mathbb{Z} \to \mathcal{O}_C$ identifies $H^1(C, \mathbb{Z})$ with a discrete subgroup of $H^1(\mathcal{O}_C)$. So $\operatorname{Pic}^0(C) = H^1(\mathcal{O}_C)/H^1(C, \mathbb{Z})$ carries the structure of a (Hausdorff) abelian complex Lie group.*

Proof. The diagram

$$\begin{array}{ccc} H^1(C, \mathbb{Z}) & \longrightarrow & H^1(\mathcal{O}_C) \\ \Big\Vert & & \Big\downarrow \text{restriction} \\ H^1(C_{\mathrm{red}}, \mathbb{Z}) & \longrightarrow & H^1(\mathcal{O}_{C_{\mathrm{red}}}) \end{array}$$

shows that the assertion holds for C if it holds for C_{red}. So we may assume C reduced.
Since $\Gamma(\mathcal{O}_C)$ and $\Gamma(\mathcal{O}_C^*)$ are spaces of locally constant functions, the map $\mathbf{e}$: $\Gamma(\mathcal{O}_C)\to\Gamma(\mathcal{O}_C^*)$ is surjective and the injectivity of $H^1(C,\mathbb{Z})\to H^1(\mathcal{O}_C)$ follows from the exponential cohomology sequence on C.
To prove that the image of $H^1(C,\mathbb{Z})$ is discrete in $H^1(\mathcal{O}_C)$ we consider the normalisation sequence (5) for C. From it we obtain a diagram of sheaves

$$\begin{array}{ccccccccc} 0 & \longrightarrow & \mathbb{Z}_C & \longrightarrow & \nu_*\mathbb{Z}_{\tilde{C}} & \longrightarrow & \Sigma & \longrightarrow & 0 \\ & & \downarrow & & \downarrow & & \downarrow & & \\ 0 & \longrightarrow & \mathcal{O}_C & \longrightarrow & \nu_*(\mathcal{O}_{\tilde{C}}) & \longrightarrow & S & \longrightarrow & 0 \end{array}$$

and a diagram in cohomology

$$\begin{array}{ccccc} \Gamma(\Sigma) & \longrightarrow & H^1(C,\mathbb{Z}) & \xrightarrow{\nu^*} & H^1(\tilde{C},\mathbb{Z}) \\ \downarrow & & \downarrow & & \downarrow \\ \Gamma(S) & \longrightarrow & H^1(\mathcal{O}_C) & \xrightarrow{\nu^*} & H^1(\mathcal{O}_{\tilde{C}}). \end{array}$$

Since $\tilde{C}$ is non-singular, the right hand vertical arrow has a discrete image. It suffices to prove the same for the left hand vertical arrow. But this can be done locally. If $x\in C$ is a singular point, $\nu^{-1}(x)=\{x_1,\ldots,x_r\}$, $U\subset C$ a small neighbourhood of x, and $\nu^{-1}(U)=U_1\cup\ldots\cup U_r\subset\tilde{C}$, then we consider the diagram

$$\begin{array}{ccccccccc} 0 & \longrightarrow & \Gamma(U,\mathbb{Z}) & \xrightarrow{\nu^*} & \bigoplus\Gamma(U_i,\mathbb{Z}) & \longrightarrow & \Sigma_x & \longrightarrow & 0 \\ & & \downarrow & & \downarrow & & \downarrow & & \\ 0 & \longrightarrow & \Gamma(\mathcal{O}_U) & \longrightarrow & \bigoplus\Gamma(\mathcal{O}_{U_i}) & \longrightarrow & S_x & \longrightarrow & 0. \end{array}$$

If $e_i\in\Gamma(U_i,\mathbb{Z})$ denotes the constant 1, then $e_1,\ldots,e_r$ generate $\bigoplus\Gamma(U_i,\mathbb{Z})$. The images of these elements in $\bigoplus\Gamma(\mathcal{O}_{U_i})$ are linearly independent over $\mathbb{C}$ and span over $\mathbb{R}$ a vector space V of dimension r. The intersection $V_0=V\cap\Gamma(\mathcal{O}_U)$ is the one-dimensional real vector space spanned by $1\in\Gamma(\mathcal{O}_U)$. The image of Σ_x in S_x is a group of rank $r-1$ spanning over $\mathbb{R}$ the $(r-1)$-dimensional vector space V/V_0. So the image of Σ_x is a lattice in $V/V_0\subset S_x$, hence discrete in S_x. □

If $C=C_1\cup\ldots\cup C_r\subset X$ is compact and reduced, then ν^* induces an isomorphism $H^2(C,\mathbb{Z})\to\bigoplus H^2(\tilde{C}_i,\mathbb{Z})=r\mathbb{Z}$ (after distinguishing in $H^2(\tilde{C}_i,\mathbb{Z})$ the generator corresponding to the canonical orientation of $\tilde{C}_i$). If $\mathcal{L}\in\mathrm{Pic}(C)$, its image $c_1(\mathcal{L})\in H^2(C,\mathbb{Z})$ can be viewed as an r-tuple $(\ell_1,\ldots,\ell_r)$ of integers. The *degree* of $\mathcal{L}$ is $\deg(\mathcal{L})=\ell_1+\ldots+\ell_r$. If $C=n_1C_1+\ldots+n_rC_r$ is not reduced, we put $\deg(\mathcal{L})=n_1\ell_1+\ldots+n_r\ell_r$ with $\ell_i=\deg(\mathcal{L}|(C_i)_{\mathrm{red}})$. For every locally free $\mathcal{O}_C$-sheaf $\mathcal{F}$ we define $\deg(\mathcal{F})$ as $\deg(\det\mathcal{F})$.

3. Riemann-Roch for an Embedded Curve

(3.1) **Theorem** (Riemann-Roch theorem for an embedded curve). *Let $C \subset X$ be a compact curve and $\mathscr{F}$ an $\mathcal{O}_C$-sheaf, locally free of rank r. Then*

$$\chi(\mathscr{F}) = \deg(\mathscr{F}) + r \cdot \chi(\mathcal{O}_C).$$

If C is smooth, this formula already appeared as formula I (9). If C is reduced, it can be proved as follows. Consider the normalisation $\nu\colon \tilde{C} \to C$ and put $\tilde{\mathscr{F}} = \nu^* \mathscr{F}$. Then

$$\deg(\tilde{\mathscr{F}}) = \deg(\mathscr{F}), \quad \chi(\tilde{\mathscr{F}}) = \chi(\nu_* \tilde{\mathscr{F}}), \quad \text{and} \quad \chi(\mathcal{O}_{\tilde{C}}) = \chi(\nu_* \mathcal{O}_{\tilde{C}})$$

by the Leray spectral sequence (see I, Sect. 1). So Riemann-Roch on $\tilde{C}$ implies

$$\deg(\mathscr{F}) = \chi(\tilde{\mathscr{F}}) - r\,\chi(\mathcal{O}_{\tilde{C}}) = \chi(\nu_* \tilde{\mathscr{F}}) - r\,\chi(\nu_* \mathcal{O}_{\tilde{C}}).$$

Since $\mathscr{F}$ is locally free, $\nu_* \tilde{\mathscr{F}} / \mathscr{F} \cong r(\nu_* \mathcal{O}_{\tilde{C}} / \mathcal{O}_C)$ and

$$\begin{aligned} \chi(\mathscr{F}) &= \chi(\nu_* \tilde{\mathscr{F}}) - \chi(\nu_* \tilde{\mathscr{F}} / \mathscr{F}) \\ &= \deg(\mathscr{F}) + r\,\chi(\nu_* \mathcal{O}_{\tilde{C}}) - r\,\chi(\nu_* \mathcal{O}_{\tilde{C}} / \mathcal{O}_C) \\ &= \deg(\mathscr{F}) + r\,\chi(\mathcal{O}_C), \end{aligned}$$

If C is not reduced, the formula can be proved using a decomposition $C = A + B$ and the corresponding sequences

$$\begin{array}{ccccccccc} 0 & \longrightarrow & \mathcal{O}_A(-B) & \longrightarrow & \mathcal{O}_C & \longrightarrow & \mathcal{O}_B & \longrightarrow & 0 \\ 0 & \longrightarrow & \mathscr{F} | A \otimes \mathcal{O}_A(-B) & \longrightarrow & \mathscr{F} & \longrightarrow & \mathscr{F} | B & \longrightarrow & 0. \end{array}$$

From these sequences it follows that

$$\begin{aligned} \chi(\mathcal{O}_C) &= \chi(\mathcal{O}_B) + \chi(\mathcal{O}_A(-B)) \\ \chi(\mathscr{F}) &= \chi(\mathscr{F} | B) + \chi(\mathscr{F} | A \otimes \mathcal{O}_A(-B)). \end{aligned}$$

Riemann-Roch on A and B implies

$$\begin{aligned} \chi(\mathscr{F} | B) &= \deg(\mathscr{F} | B) + r\,\chi(\mathcal{O}_B) \\ \chi(\mathscr{F} | A \otimes \mathcal{O}_A(-B)) &= \deg(\mathscr{F} | A \otimes \mathcal{O}_A(-B)) + r\,\chi(\mathcal{O}_A) \\ &= \deg(\mathscr{F} | A) + r \deg(\mathcal{O}_A(-B)) + r\,\chi(\mathcal{O}_A) \\ &= \deg(\mathscr{F} | A) + r\,\chi(\mathcal{O}_A(-B)). \end{aligned}$$

Adding these equations and using $\deg(\mathscr{F}) = \deg(\mathscr{F} | A) + \deg(\mathscr{F} | B)$ we obtain Riemann-Roch for $\mathscr{F}$ on C. $\square$

4. The Residue Theorem

Let X be a complex manifold and $H \subset X$ a non-singular complex hypersurface. We shall say that a C^∞-differentiable complex-valued q-form φ on $X \setminus H$ has *ordinary poles along H*, if in local coordinates $(z_1, \ldots, z_n)$, with $z_1 = 0$ a local equation for H, it can be written

$$\varphi=\varphi_1\wedge\frac{dz_1}{z_1}+\varphi_2, \tag{8}$$

where φ_1 and φ_2 extend over H as C^∞-forms. This notion is independent of the local coordinates chosen. (If the system $(z_1,\ldots,z_n)$ overlaps with $(w_1,\ldots,w_n)$, then $f=z_1/w_1$ extends holomorphically over H, so

$$\frac{df}{f}=\frac{dz_1}{z_1}-\frac{dw_1}{w_1},$$

$$\varphi=\varphi_1\wedge\frac{dw_1}{w_1}+\varphi_1\wedge\frac{df}{f}+\varphi_2.)$$

We denote by $\mathscr{D}_X^q(H)$ the sheaf on X of such q-forms (more precisely, the direct image on X of that sheaf).

If in the system $(z_1,\ldots,z_n)$ the form $\varphi\in\mathscr{D}_X^q(H)$ has two representations (8), say

$$\varphi=\varphi_1\wedge\frac{dz_1}{z_1}+\varphi_2=\psi_1\wedge\frac{dz_1}{z_1}+\psi_2,$$

then we may write

$$\varphi_1-\psi_1=\omega\wedge dz_1+\sigma,\qquad \sigma=\sum_{i_k\geq 2}\sigma_{i_2\ldots i_q}dz_{i_2}\wedge\ldots\wedge dz_{i_q}.$$

So

$$\varphi_1|H-\psi_1|H=\sigma|H\qquad\text{with } \sigma\wedge dz_1=z_1(\psi_2-\varphi_2).$$

This shows that $\sigma|H$ vanishes and $\varphi_1|H\in\mathscr{D}_H^{q-1}$ is independent of the representation (8) for φ. Since we saw already that $\varphi_1|H$ is independent of the coordinate system, there is a well-defined residue map

$$\operatorname{res}\colon\ \mathscr{D}_X^q(H)\to\mathscr{D}_H^{q-1},$$

defined locally for a form (8) by

$$\operatorname{res}(\varphi)=\varphi_1|H.$$

(4.1) **Theorem** (Residue theorem). *The residue map has the following properties:*

(i) $\operatorname{res}(g\,\varphi)=g\operatorname{res}(\varphi)$ *for any C^∞-function* g;
(ii) $\partial(\operatorname{res}\varphi)=\operatorname{res}(\partial\varphi)$, $\bar\partial(\operatorname{res}\varphi)=\operatorname{res}(\bar\partial\varphi)$;
(iii) *for every real submanifold $M\subset X$ of dimension $q+1$ intersecting H transversally in an oriented submanifold S such that* $\operatorname{supp}(\varphi)\cap S$ *is compact, the residue formula*

$$2\pi\sqrt{-1}\int_S\operatorname{res}(\varphi)=\lim_{\varepsilon\to 0}\int_{M\cap\{\delta=\varepsilon\}}\varphi,$$

holds, with δ being some tubular neighbourhood function for H on a neighbourhood of $S\cap\operatorname{supp}(\varphi)$.

(By this we mean that δ is a real-valued C^∞-function vanishing on H together with its derivatives $\frac{\partial\delta}{\partial x_1}$ and $\frac{\partial\delta}{\partial x_2}$, where $z_1=x_1+\sqrt{-1}\,x_2$ is a local equation for H, and the matrix

$$\begin{pmatrix} \partial^2 \delta/\partial x_1^2 & \partial^2 \delta/\partial x_1\,\partial x_2 \\ \partial^2 \delta/\partial x_1\,\partial x_2 & \partial^2 \delta/\partial x_2^2 \end{pmatrix}$$

is positive definite on H. For small ε, the set $M \cap \{\delta = \varepsilon\}$ is a real manifold of dimension q and it inherits an orientation from S in a natural way.)

Proof. First we prove (i), (ii), (iii) locally where φ is of the form (8) and $\operatorname{res}(\varphi)$ is given by $\varphi_1 | H$. Assertions (i) and (ii) are obvious. To prove (iii) we put $z_1 = x_1 + \sqrt{-1}\,x_2$ and consider only those φ for which $\operatorname{supp}(\varphi)$ is compact and so small, that $x_1|M, x_2|M$ can be extended to a system $x_1, \ldots, x_{q+1}$ of local real coordinates on M in a neighbourhood of $M \cap \operatorname{supp}(\varphi)$. Then (iii) becomes a statement in $\mathbb{R}^{q+1}$, namely

$$2\pi\sqrt{-1} \int_{x_1 = x_2 = 0} \varphi_1 = \lim_{\varepsilon \to 0} \int_{\delta(x) = \varepsilon} \varphi.$$

But this is just a version with $q-1$ parameters of the usual residue formula in one complex variable:

$$2\pi\sqrt{-1}\, h(0) = \lim_{\varepsilon \to 0} \int_{\delta(x_1, x_2) = \varepsilon} h(x_1, x_2) \frac{dx_1 + \sqrt{-1}\, dx_2}{x_1 + \sqrt{-1}\, x_2},$$

if $h(x_1, x_2)$ is a C^∞-function near $0 \in \mathbb{R}^2$ and

$$\delta(0) = \frac{\partial \delta}{\partial x_1}(0) = \frac{\partial \delta}{\partial x_2}(0) = 0, \qquad \begin{pmatrix} \partial^2 \delta/\partial x_1^2 & \partial^2 \delta/\partial x_1\,\partial x_2 \\ \partial^2 \delta/\partial x_1\,\partial x_2 & \partial^2 \delta/\partial x_2^2 \end{pmatrix} > 0.$$

By a partition of unity argument, the global residue formula is reduced to the local one. □

To conclude, a word about the case of a smooth curve C on a surface X. Let (u, v) be local coordinates on X with $v = 0$ a local equation for C. If $\varphi \in \Gamma(\mathcal{D}_X^2(C))$ is holomorphic, it can be written locally as $\varphi = h\, du \wedge dv/v$ with h holomorphic. So φ is nothing but a section in $\Gamma(\mathcal{K}_X \otimes \mathcal{O}_X(C))$. Also the map $\varphi \to \operatorname{res}(\varphi) = h\, du | C$ is the same as the map $r\colon \mathcal{K}_X \otimes \mathcal{O}_X(C) \to \omega_C$ from Sect. 1.

5. The Trace Map

Let $D \subset X$ be some curve, not necessarily reduced, and ω_D its dualising sheaf, defined in Sect. 1. In this section we define the *trace map*

$$\operatorname{tr}_D\colon H_c^1(\omega_D) \to \mathbb{C}.$$

First we recall how to describe $H_c^n(\mathcal{K}_Y)$, where Y is a complex *manifold* of dimension n, in terms of differential forms. Since the sheaves $\mathcal{D}_Y^{n,q}$ are fine, we have $H_c^i(\mathcal{D}_Y^{n,q}) = 0$ for $i \geq 1$, and Dolbeault's $\bar{\partial}$-resolution of $\mathcal{K}_Y$ gives us an isomorphism

$$H_c^n(\mathcal{K}_Y) = \Gamma_c(\mathcal{D}_Y^{n,n}) / \bar{\partial}\, \Gamma_c(\mathcal{D}_Y^{n,n-1}).$$

So each class $[\varphi]\in H^n_c(\mathscr{K}_Y)$ is represented by some (n,n)-form φ with compact support. If we modify φ by $\bar\partial\chi=d\chi$, $\chi\in\Gamma_c(\mathscr{D}_Y^{n,n-1})$, then $\int_Y\varphi$ does not change. So integration over Y defines a $\mathbb{C}$-linear map

$$\mathrm{Tr}_Y\colon H^n_c(\mathscr{K}_Y)\to\mathbb{C}.$$

If Y is compact connected, then $H^n_c(\mathscr{K}_Y)=H^{2n}(Y)$ is of dimension one and Tr_Y is an isomorphism (I, Sect. 11).

This definition of Tr_Y does not work for singular Y. Therefore, for an arbitrary $D\subset X$ we proceed as follows. Let $\delta\colon H^1_c(\omega_D)\to H^2_c(\mathscr{K}_X)$ be the map obtained from the sequence (6) and let $\mathrm{tr}_D=\mathrm{Tr}_X\circ\delta$. (This tr_D a priori depends on the embedding, but this will not matter in the sequel.) Of course, we need that for non-singular D this tr_D is essentially the same as Tr_D (Proposition 5.1 below). So we look for an explicit description of δ in the case of *non-singular D*. Recall from Sect. 4 that $\mathscr{D}^q_X(D)$ is the sheaf of q-forms with poles along D. For $q=0,1,2$ we put (in accordance with I, Sect. 12)

$$\mathscr{D}_X^{2,q}(D)=\text{subsheaf of } \mathscr{D}_X^{2+q}(D) \text{ of forms of type } (2,q)$$
$$\mathscr{D}_D^{1,q}=\text{sheaf of } C^\infty\text{-forms on } D \text{ of type } (1,q)$$
$$\mathscr{K}^q=\text{kernel }\{\text{res}\colon \mathscr{D}_X^{2,q}(D)\to\mathscr{D}_D^{1,q}\}.$$

Since res is C^∞-linear, $\mathscr{K}^q$ is a fine sheaf. Furthermore, $\mathscr{K}^0\subset\mathscr{D}_X^{2,0}(D)$ is the sheaf of forms $f\,du\wedge dv/u$, where u is a local holomorphic equation of D and the C^∞-function f vanishes on D. If such a form is $\bar\partial$-closed, then f/u is holomorphic, which implies

$$\mathscr{K}^0\cap\text{kernel }\{\bar\partial\colon\mathscr{D}_X^{2,0}(D)\to\mathscr{D}_X^{2,1}(D)\}=\mathscr{K}_X.$$

So the double complex

$$0\to\mathscr{K}^q\to\mathscr{D}_X^{2,q}(D)\xrightarrow{\text{res}}\mathscr{D}_D^{1,q}\to 0,\qquad q=0,1,2,$$

with vertical arrows induced by $\bar\partial$ is a fine resolution of the sequence (6). If now $[\varphi]\in H^1_c(\omega_D)$ is represented by the form $\varphi\in\Gamma_c(\mathscr{D}_D^{1,1})$, then $\delta[\varphi]=[\bar\partial\psi]$ where $\psi\in\Gamma_c(\mathscr{D}_X^{2,1}(D))$ is some pre-image of φ under res and

$$\bar\partial\psi\in\Gamma_c(\mathscr{K}^2)=\Gamma_c(\mathscr{D}_X^{2,2}(D)).$$

(5.1) **Proposition.** *For every class $[\varphi]\in H^1_c(\omega_D)$ one has*

$$2\pi\sqrt{-1}\int_D[\varphi]=\int_X\delta[\varphi].$$

Proof. Let ψ be as above and let τ be a tubular neighbourhood function for D, valid in a neighbourhood of the support of ψ. The residue formula together with Stokes' theorem gives

$$2\pi\sqrt{-1}\int_D\varphi=\lim_{\varepsilon\to0}\int_{\tau=\varepsilon}\psi=\lim_{\varepsilon\to0}\int_{\tau\geq\varepsilon}\bar\partial\psi.$$

Now $\mathscr{K}^2$ contains $\mathscr{D}_X^{2,2}$ as subsheaf and if $\Phi\in\Gamma_c(\mathscr{D}_X^{2,2})$ is a smooth form representing in $H^{2,2}_c(X)$ the class $\delta[\varphi]=[\bar\partial\psi]$, then $\Phi=\bar\partial\psi+\bar\partial\chi$ with some

$\chi \in \Gamma_c(\mathscr{K}^1)$. And if τ is adapted to Φ and χ also, then

$$\int_X \delta[\varphi] = \int_X \Phi = \lim_{\varepsilon \to 0} \int_{\tau \geq \varepsilon} \Phi$$
$$= \lim_{\varepsilon \to 0} \left(\int_{\tau \geq \varepsilon} \bar{\partial}\psi + \int_{\tau \geq \varepsilon} \bar{\partial}\chi \right) = 2\pi\sqrt{-1} \int_D \varphi,$$

because by the residue formula

$$\lim_{\varepsilon \to 0} \int_{\tau \geq \varepsilon} \bar{\partial}\chi = 2\pi\sqrt{-1} \int_D \operatorname{res}(\chi) = 0. \quad \square$$

6. Serre Duality on an Embedded Curve

In this section we treat in an elementary way Serre-Grothendieck duality for a (not necessarily reduced) compact curve C on a (not necessarily compact) surface X.

(6.1) **Theorem** (Duality theorem for an embedded curve). *For every compact curve C, embedded in a smooth surface X, there is an epimorphism*

$$\operatorname{tr}: H^1(\omega_C) \to \mathbb{C},$$

such that the cupproduct pairing

$$H^1(\mathscr{F}) \otimes H^0(\mathscr{F}^\vee \otimes \omega_C) \to H^1(\omega_C) \xrightarrow{\operatorname{tr}} \mathbb{C}$$

(defined for every $\mathscr{O}_C$-sheaf $\mathscr{F}$) is perfect as soon as $\mathscr{F}$ is locally free.

If C is smooth, then this is nothing but the Serre-duality of Proposition I.12.1. If C is reduced and projective, then the result is again due to Serre ([Se5], Chap. IV). For general projective schemes it is due to Grothendieck (see for example [A-K]), whereas for complex spaces it can be found in [R-R-V]. An elementary treatment, however, does not seem to be available in the literature, so we venture to include one here.

We start by observing that every reduced compact curve C is projective. To see this we remark that there is a line bundle $\mathscr{L}$ on C with a section, vanishing exactly in $p_1 \cup \ldots \cup p_k$, where $p_1, \ldots, p_k$ are smooth points on C, one on every irreducible component. By Grauert's criterion (Theorem I.19.3) the existence of such an $\mathscr{L}$ ensures the projectivity of C.

Next we recall that Serre proved his theorem with ω_C being the sheaf of *rational* Rosenlicht-differentials (see below); it follows from GAGA (I, Sect. 19) that ω_C can also be taken to be the sheaf of *meromorphic* Rosenlicht-differentials as soon as it is clear that this sheaf is coherent.

From here on we proceed as follows:
a) we prove that our sheaf ω_C is the same as the sheaf of meromorphic Rosenlicht-differentials. This implies in particular that this last sheaf is coherent and consequently we may indeed apply Serre's result to any reduced C, using our ω_C;

b) we define a trace map for non-reduced C:
c) we prove duality on a non-reduced C by decomposition.

a) Rosenlicht-differentials

Let C be a reduced curve with normalisation $\nu: \tilde{C} \to C$. *A meromorphic differential* on C is nothing but a section in $\nu_* \mathscr{M}_{\tilde{C}}$ with $\mathscr{M}_{\tilde{C}}$ the sheaf of meromorphic 1-forms on $\tilde{C}$. Such a meromorphic σ on C is called a *Rosenlicht-differential* if for all $x \in C$, $g \in \mathcal{O}_{C,x}$

$$\sum_{x_k \in \nu^{-1}(x)} \operatorname{res}(x_k, g\sigma) = 0,$$

where res denotes the ordinary residue on the Riemann surface $\tilde{C}$. A Rosenlicht-differential is holomorphic in all regular points of C.

Let C be embedded in a non-singular surface X. In Sect. 1 we defined ω_C and showed how to view it as subsheaf of the sheaf of meromorphic differentials on C.

(6.2) **Proposition.** *$\omega_C \subset \nu_* \mathscr{M}_{\tilde{C}}$ is the sheaf of Rosenlicht-differentials.*

Proof. Let $x \in C$ be a singular point, (u, v) local coordinates centered at x, and $f = 0$ a local equation for C at x such that $f_v = \partial f / \partial v$ does not vanish identically on any branch of C at x. We put

$$B_\rho = \{|u|^2 + |v|^2 \leq \rho\}, \quad M_\rho = \partial B_\rho, \quad S_\rho = C \cap M_\rho.$$

If ρ is small, B_ρ contains no singularity of C and no zero of $f_v | C$ but for x itself. Also, M_ρ is a compact 3-sphere intersecting C transversally in S_ρ, a union of smooth circles. For every meromorphic function t on C, holomorphic on $(C \cap B_\rho) \setminus \{x\}$, we have

$$\text{(9)} \qquad \sum_{x_k \mapsto x} \operatorname{res}(x_k, t\, du/f_v) = (2\pi\sqrt{-1})^{-1} \int_{S_\rho} t\, du/f_v$$

by the ordinary residue theorem. If t is holomorphic on a neighbourhood in X of $C \setminus \{x\}$, the Residue Theorem 4.1 states:

$$\text{(10)} \qquad \int_{S_\rho} t\, du/f_v = (2\pi\sqrt{-1})^{-1} \lim_{\varepsilon \to 0} \int_{M_\rho \cap \{|f| = \varepsilon\}} t\, du \wedge dv/f.$$

Now we take some $\sigma \in \omega_{C,x}$, which by definition may be written $\sigma = h\, du/f_v$, $h \in \mathcal{O}_{C,x}$. Since h and any other $g \in \mathcal{O}_{C,x}$ is the restriction of a function holomorphic on some B_ρ, ρ small, formula (9) and (10) for $t = gh$ show that

$$\sum_{x_k \to x} \operatorname{res}(x_k, g\sigma) = (2\pi\sqrt{-1})^{-2} \lim_{\varepsilon \to 0} \int_{M_\rho \cap \{|f| = \varepsilon\}} gh\, du \wedge dv/f.$$

Since the 2-form $gh\, du \wedge dv/f$ is holomorphic on $X \setminus C$, it is closed on $M_\rho \setminus C$ and by Stokes' theorem

$$\int_{M_\rho \cap \{|f| = \varepsilon\}} gh\, du \wedge dv/f = \int_{M_\rho \cap \{|f| \geq \varepsilon\}} d(gh\, du \wedge dv/f) = 0.$$

This shows that σ is a Rosenlicht-differential.

Conversely, let σ be some Rosenlicht-differential on C near x. It can be written as $\sigma = s\,du/f_v$ with s meromorphic on C. We show that s is holomorphic. To begin with, for large n, $u^n s$ will be holomorphic, hence the restriction of some $h_n \in \Gamma(B_\rho, \mathcal{O}_X)$. By assumption $\sum \operatorname{res}(x_k, g u^{n-1} \sigma) = 0$ for all holomorphic g. So by formulas (9) and (10) for $t = g h_n/u = g u^{n-1} s$

$$\lim_{\varepsilon \to 0} \int_{M_\rho \cap \{|f| = \varepsilon\}} g \frac{h_n}{u} du \wedge dv/f = 0.$$

But the integrand is a closed 2-form on $M_\rho \cap \{f \neq 0\} \cap \{u \neq 0\}$. So

$$\lim_{\varepsilon \to 0} \int_{M_\rho \cap \{|u| = \varepsilon\}} g \frac{h_n}{u} du \wedge dv/f = 0.$$

Now we consider the curve $D = \{u = 0\}$ and apply the residue theorem to $M_\rho \cap D \subset M_\rho$. Then it follows that

$$\operatorname{res}\left(x, \frac{g h_n}{f} dv \,\middle|\, D\right) = 0$$

for all holomorphic g. But this means h_n/f is holomorphic when restricted to D, i.e. there are holomorphic functions h_{n-1} and a on some B_ρ with $h_n = u h_{n-1} + af$. In other words, $h_n | C$ is divisible by u and we might as well have started with h_{n-1}. Repeating this procedure $(n-1)$ times we find that s itself was already holomorphic. $\square$

b) The trace map

We consider the trace map $\operatorname{tr}_C \colon H^1(\omega_C) \to \mathbb{C}$ as defined in Sect. 5.

(6.3) **Proposition.** *If C is an irreducible (hence reduced) compact curve, then $h^1(\omega_C) = 1$ and $\operatorname{tr}_C \neq 0$. So up to a non-zero constant, our tr_C coincides with the algebraic trace map.*

Proof. By duality on the reduced curve C, we have $h^1(\omega_C) = h^0(\mathcal{O}_C) = 1$. If we denote by $S \subset C$ the set of singular points and by D the smooth curve $C \setminus S$, then there is a commutative diagram

$$\begin{array}{ccccc} H^1_c(\omega_D) & \longrightarrow & H^1(\omega_C) & \longrightarrow & H^1(S, \omega_C | S) \\ \downarrow \delta & & \downarrow \delta & & \\ H^2_c(\mathcal{K}_X | X \setminus S) & \longrightarrow & H^2_c(\mathcal{K}_X) & & \end{array}$$

with $H^1(S, \omega_C | S) = 0$ since $\dim S = 0$. So the generator $\xi \in H^1(\omega_C)$ is the image of some $[\eta] \in H^1_c(\omega_D)$ and

$$\operatorname{tr}_C(\xi) = \int_X \delta[\eta] = 2\pi\sqrt{-1} \int_D \eta$$

by Proposition 5.1.

Now let $\nu \colon \tilde{C} \to C$ be the normalisation. By Proposition 6.2 $\nu_* \omega_{\tilde{C}} \subset \omega_C$. So there is a factorisation

$$H^1_c(\omega_D) \to H^1(\omega_{\tilde{C}}) = H^1(\nu_* \omega_{\tilde{C}}) \to H^1(\omega_C)$$

showing that $[\eta]\in H^1(\omega_{\tilde{C}})$ is non-trivial. But this implies

$$\int_D \eta = \int_{\tilde{C}} \eta \neq 0. \qquad \square$$

Next we consider a decomposition $C=A+B$ with two curves A, B on X.

(6.4) **Proposition.** *There is a canonical $\mathcal{O}_C$-map $a\colon \omega_A \to \omega_C$, and* $\operatorname{tr}_C H^1(a)=\operatorname{tr}_A$, *where $H^1(a)\colon H^1(\omega_A)\to H^1(\omega_C)$ is induced by a.*

Proof. The decomposition sequence combines with (6) to a diagram

$$\begin{array}{ccccccccc}
 & & 0 & & 0 & & & & \\
 & & \downarrow & & \downarrow & & & & \\
 & & \mathcal{K}_X & = & \mathcal{K}_X & & & & \\
 & & \downarrow & & \downarrow & & & & \\
0 & \longrightarrow & \mathcal{K}_X\otimes\mathcal{O}_X(A) & \longrightarrow & \mathcal{K}_X\otimes\mathcal{O}_X(C) & \longrightarrow & \omega_B\otimes\mathcal{O}_B(A) & \longrightarrow & 0 \\
 & & \downarrow & & \downarrow & & & & \\
0 & \longrightarrow & \omega_A & \xrightarrow[a]{} & \omega_C & \longrightarrow & \omega_B\otimes\mathcal{O}_B(A) & \longrightarrow & 0. \\
 & & \downarrow & & \downarrow & & & & \\
 & & 0 & & 0 & & & &
\end{array}$$

c) Duality on non-reduced C

The duality theorem can also be phrased as follows:

For every locally free $\mathcal{O}_C$-sheaf $\mathcal{F}$ the map

(11) $$\operatorname{Hom}_{\mathcal{O}_C}(\mathcal{F},\omega_C)\to H^1(\mathcal{F})^\vee,$$

sending an $\mathcal{O}_C$-morphism $h\colon \mathcal{F}\to\omega_C$ to the linear form

$$H^1(\mathcal{F})\longrightarrow H^1(\omega_C)\xrightarrow{\operatorname{tr}}\mathbb{C},$$

is an isomorphism.

We show that this statement is true for a compact curve $C=A+B$ if it is true on A and B.

Step 1: injectivity of (11).

Consider the commutative pairing of decomposition sequences

$$\begin{array}{ccccccccc}
0 & \longrightarrow & \mathcal{O}_A(-B) & \longrightarrow & \mathcal{O}_C & \xrightarrow{\text{restr}} & \mathcal{O}_B & \longrightarrow & 0 \\
 & & \otimes & & \otimes & & \otimes & & \\
0 & \longleftarrow & \mathcal{O}_A & \xleftarrow{\text{restr}} & \mathcal{O}_C & \longleftarrow & \mathcal{O}_B(-A) & \longleftarrow & 0 \\
 & & \downarrow & & \downarrow & & \downarrow & & \\
 & & \mathcal{O}_A(-B) & \longrightarrow & \mathcal{O}_C & \longleftarrow & \mathcal{O}_B(-A). & &
\end{array}$$

Tensoring the first row with $\mathscr{K}_X \otimes \mathcal{O}_C(C)$ we obtain

$$\begin{array}{ccccccccc}
0 & \longrightarrow & \omega_A & \xrightarrow{a} & \omega_C & \longrightarrow & \omega_B \otimes \mathcal{O}_B(A) & \longrightarrow & 0 \\
 & & \otimes & & \otimes & & \otimes & & \\
0 & \longleftarrow & \mathcal{O}_A & \longleftarrow & \mathcal{O}_C & \longleftarrow & \mathcal{O}_B(-A) & \longleftarrow & 0 \\
 & & \downarrow & & \downarrow & & \downarrow & & \\
 & & \omega_A & \xrightarrow{a} & \omega_C & \xleftarrow{b} & \omega_B. & &
\end{array}$$

Applying $\operatorname{Hom}(\mathscr{F}, -)$ to the first row and $H^1(\mathscr{F} \otimes -)$ to the second one, we obtain a pairing of exact sequences

$$\begin{array}{ccccccc}
0 & \longrightarrow & \operatorname{Hom}(\mathscr{F}, \omega_A) & \longrightarrow & \operatorname{Hom}(\mathscr{F}, \omega_C) & \longrightarrow & \operatorname{Hom}(\mathscr{F}, \omega_B \otimes \mathcal{O}_B(A)) \\
 & & \otimes & & \otimes & & \otimes \\
0 & \longleftarrow & H^1(\mathscr{F}|A) & \longleftarrow & H^1(\mathscr{F}) & \longleftarrow & H^1(\mathscr{F} \otimes \mathcal{O}_B(-A)) \\
 & & \downarrow & & \downarrow & & \downarrow \\
 & & H^1(\omega_A) & \xrightarrow{H^1(a)} & H^1(\omega_C) & \xleftarrow{H^1(b)} & H^1(\omega_B).
\end{array}$$

The left hand pairing followed by tr_A is the perfect duality pairing on A and similarly for the right hand pairing. Proposition 5.1 thus gives the diagram

$$\begin{array}{ccccccc}
0 & \longrightarrow & \operatorname{Hom}(\mathscr{F}, \omega_A) & \longrightarrow & \operatorname{Hom}(\mathscr{F}, \omega_C) & \longrightarrow & \operatorname{Hom}(\mathscr{F}, \omega_B \otimes \mathcal{O}_B(A)) \\
 & & \| & & \downarrow & & \| \\
0 & \longrightarrow & H^1(\mathscr{F}|A)^\vee & \longrightarrow & H^1(\mathscr{F})^\vee & \longrightarrow & H^1(\mathscr{F} \otimes \mathcal{O}_B(-A))^\vee.
\end{array}$$

It follows that the morphism $\operatorname{Hom}(\mathscr{F}, \omega_C) \to H^1(\mathscr{F})^\vee$ obtained from the column in the middle is injective.

Step 2: equality $h^0(\mathscr{F}^\vee \otimes \omega_C) = h^1(\mathscr{F})$.

We know already $h^0(\mathscr{F}^\vee \otimes \omega_C) \leq h^1(\mathscr{F})$ and $h^0(\mathscr{F}) \leq h^1(\mathscr{F}^\vee \otimes \omega_C)$. So it suffices to prove $\chi(\mathscr{F}^\vee \otimes \omega_C) = -\chi(\mathscr{F})$. But

$$\chi(\mathscr{F}) = \chi(\mathscr{F}|B) + \chi(\mathscr{F} \otimes \mathcal{O}_A(-B))$$
$$\chi(\mathscr{F}^\vee \otimes \omega_C) = \chi(\mathscr{F}^\vee \otimes \omega_B) + \chi(\mathscr{F}^\vee \otimes \mathcal{O}_A(B) \otimes \omega_A),$$

and duality on A and B implies $\chi(\mathscr{F}^\vee \otimes \omega_C) = -\chi(\mathscr{F})$. □

7. The σ-Process

Let X be a surface and $x \in X$. In I, Sect. 9 we have recalled what it means to blow up X at x, or, in other words, to apply the σ-process (σ-transformation) to X at x. Thus a new surface $\bar{X}$ is constructed, together with a regular map σ: $\bar{X} \to X$, such that $E = \sigma^{-1}(x)$ is a smooth rational curve, the exceptional curve, whereas σ maps $\bar{X}\setminus E$ biregularly onto $X\setminus x$. If (u, v) are local coordinates on X, centred at x, and $(z_0 : z_1)$ homogeneous coordinates on $\mathbb{P}_1$, then locally $\bar{X}$ can be described as

$$\{(y,(z_0:z_1))\in X\times\mathbb{P}_1,\, u(y)\,z_0 = v(y)\,z_1\},$$

with $\sigma(y,(z_0:z_1)) = y$.

On $\bar{X}$, and near E, we can take as local coordinates outside of $z_0=0$ the functions v, $\frac{z_1}{z_0}$, and outside of $z_1=0$ the functions u, $\frac{z_0}{z_1}$. So u, v vanish along E to the first order. With respect to the covering $E=E_0\cup E_1$, $E_i=\{z_i\neq 0\}$, the normal bundle $\mathcal{N}_{E/X}$ is defined by the cocycle $\frac{v}{u}=\frac{z_0}{z_1}\Big|E_0\cap E_1$. This implies $\mathcal{N}_{E/X} = \mathcal{O}_{\mathbb{P}_1}(-1)$, the line bundle on E of degree -1.

Now let C be a curve on X, with a point x of multiplicity $\mu_x(C)=\mu$. Let $f=\sum_{p\ge\mu} f_p(u,v)=0$ be a local equation for C at x, with f_p homogeneous of degree p. Then

$$\sigma^* f_p = f_p\left(u, \frac{z_1}{z_0}u\right) = u^p f_p\left(1, \frac{z_1}{z_0}\right)$$

where $z_0\neq 0$ and similarly where $z_1\neq 0$. This proves that the curve $f^{-1}(C)$ contains E with multiplicity μ. So $\sigma^{-1}(C)=\mu E+\bar{C}$ with $\bar{C}$ the closure in $\bar{X}$ of $\sigma^{-1}(C\setminus\{x\})$. The curve $\sigma^{-1}(C)$ is called the *total transform* of C, and $\bar{C}$ its *proper transform*. By linearity this definition extends to divisors. If $\tau=\sigma_k\sigma_{k-1}\dots\sigma_1$ is a succession of finitely many σ-processes, the proper transform under τ is the composition of the proper transforms under the σ_i. The intersections x_i of E with $\bar{C}$ are just the zeros (a^i, b^i) *of* $f_\mu(z_0, z_1)$ and thus correspond bijectively with the tangents $T_i=\{u\,b^i = v\,a^i\}$ of C at x. Let $C_i\subset C$ be the union of branches of C at x tangent to T_i, and $\mu_i=\mu(C_i, x)$. Then $\mu=\sum\mu_i$, because there is a factorisation $f=\prod f^i$, with $f^i=0$ a local equation for C_i. For the multiplicity $\bar{\mu}_i$ of $\bar{C}_i$ at x_i one has the estimate $\bar{\mu}_i\le\mu_i$. Indeed, if we assume $(a^i, b^i)=(1,0)$, then a local equation for $\bar{C}_i$ is $\sigma^* f^i/u^{\mu_i} = f^i_{\mu_i}\left(1,\frac{z_1}{z_0}\right)+\dots$, vanishing to an order $\le\mu_i$. So blowing up does not increase multiplicities of curves in the sense that

$$\mu\ge\sum\bar{\mu}_i.$$

An important application of the σ-process is the desingularisation of curves.

(7.1) **Theorem.** *Let X be a (smooth) surface and $C\subset X$ an embedded reduced curve. Then there is a non-singular surface Y and a proper map τ: $Y\to X$ consisting of a succession of σ-transformations (locally with respect to X, finitely many), such that the proper transform of C in Y is smooth.*

Proof. We may restrict our attention to one singularity x of the curve C. We have to show that the σ-process with centre x transforms C into a "less singular" curve $\bar{C}$. By the estimate above for the multiplicities, this is clear as soon as C has at least two different tangents at x, or if C has only one tangent and if the multiplicity of $\bar{C}$ at the corresponding point is smaller than $\mu_x(C)$. If none of these two favorable cases occurs, we can choose coordinates (u, v) such that the line $v=0$ is tangent to the singularity, but not a branch of the curve. Then the singularity has the equation $f(u, v)=0$ with

$$f=a v^\mu+\sum_{l+m\geq\mu+1} a_{lm} u^l v^m,$$

and the proper transform $\bar{C}$ has equation $\bar{f}(u, z)=0$, $z=\frac{z_0}{z_1}$, with

$$\bar{f}=a z^\mu+\sum_{l+m\geq\mu+1} a_{lm} u^{l+m-\mu} z^m.$$

Since the multiplicity of $\bar{C}$ is μ again, the coefficients a_{lm} vanish for $l+2m<2\mu$. Since $\bar{C}$ has one tangent only, $a_{lm}=0$ for $l+2m=2\mu$ too. This shows that the curve $z=0$, the proper transform of the line $v=0$, is tangent to the singularity of $\bar{C}$. By assumption $a_{l0}\neq 0$ for some l. The minimum l_0 of these l is the order to which $f(u, 0)$, the restriction of f to the tangent $v=0$, vanishes at the origin. Then $\bar{f}(u, 0)$, the restriction to the tangent $z=0$ of $\bar{f}$, vanishes to the order $l_0-\mu$. We can repeat the σ-process arbitrarily often, but this order of vanishing can decrease only a finite number of times. So after finitely many σ-transformations we shall be in one of the favorable situations described above, and the multiplicity of the proper transform of C will be $<\mu$. Now the assertion follows by induction on μ. □

Sometimes information on the total transform of C is needed.

(7.2) **Theorem.** *Let X be a (smooth) surface and C a reduced curve on X. Then there is a map $\tau\colon Y\to X$ as in Theorem 7.1 above such that the reduction of the total transform of C has only ordinary double points as singularities.*

Proof. Let $\tau_1\colon Y_1\to X$ be a succession of σ-processes with non-singular proper transform $C_1\subset Y_1$ of C. The reduction of the total transform $\tau_1^{-1}(C)$ is $C_1\cup D$ with D the union of all exceptional curves for τ_1. Since the exceptional curve of a σ-process intersects transversally the proper transform of every non-singular curve, it suffices to take $\tau=\tau_1\circ\tau_2$, with $\tau_2\colon Y_2\to Y_1$ resolving the singularities of $C_1\cup D$. In fact, a singularity y of the reduction of $\tau^{-1}(C)$ is an ordinary double point unless at y the proper transform of $C_1\cup D$ meets two exceptional curves E' and E'' for τ_2. Blowing down E' and E'' would create at $\tau_2(y)$ a one-branch singularity for $C_1\cup D$, a contradiction. □

8. Simple Singularities of Curves

In this section by a curve we always mean a curve on a smooth surface. We give the *A-D-E* classification of simple curve-singularities. A singularity of a reduced curve is called *simple*, if it is itself a double or triple point, and if when resolving the singularity to a collection of ordinary nodes according to Theorem 7.2, after each blowing up, the (reduced) total transform of our curve again has double or triple points only. We classify simple singularities by classifying all *double points*, all *triple points with two or three different tangents*, and *simple triple points with one tangent*.

The singularity under consideration will always be at the centre (0,0) of a coordinate system (x, y) and will have equation $f(x, y)=0$. By $\mathfrak{m}\subset\mathcal{O}_{(0,0)}$ we denote the maximal ideal.

Classification of double points. These are the singularities with multiplicity $\mu=2$. The residue $f_2=f \bmod \mathfrak{m}^3$ is a non-zero homogeneous quadratic polynomial, which after a linear transformation can be put into the form

$$f_2=x^2+y^2 \quad \text{or} \quad f_2=x^2.$$

In the first case we can write $f=x^2\varphi_1+y^2\varphi_2$ with $\varphi_1(0,0)\neq 0$, $\varphi_2(0,0)\neq 0$, and we can introduce $x\sqrt{\varphi_1}$ and $y\sqrt{\varphi_2}$ as new coordinates. Then f is in *normal form*

$$f=x^2+y^2 \quad \text{(ordinary double point, node, type } A_1).$$

In the second case, we consider the *Milnor number* $n=\dim_{\mathbb{C}}\mathcal{O}_{(0,0)}/(f_x,f_y)$. It is independent of the choice of coordinates and finite, because we assumed our curve to be reduced. (Indeed, if $n=\infty$, the analytic set $f_x=f_y=0$ will contain a curve passing through (0,0) and f will vanish on this curve.) We write

$$f(x,y)=x^2 e(x,y)+x\varphi(y)+\psi(y) \tag{12}$$

with $e(0,0)\neq 0$ and φ, ψ vanishing at $y=0$ to the orders $k\geq 2$, $l\geq 3$ respectively. Since up to a unit $\varphi=y\cdot\varphi_y$, we have the inclusion

$$(f_x,f_y)=(2xe+x^2e_x+\varphi,\ x^2e_y+x\varphi_y+\psi_y)\subset(x,\varphi_y,\psi_y),$$

from which we conclude

$$\min\{k,l\}\leq n+1. \tag{13}$$

Replacing x by $x\sqrt{e}+\varphi/2\sqrt{e}$ we find

$$f=x^2-\varphi^2/4e+\psi.$$

Next we may expand

$$-1/4e=c(y)+x\varphi_1(y)+x^2 g(x,y)$$

and put

$$e'=1+\varphi^2 g, \quad \varphi'=\varphi^2\varphi_1, \quad \psi'=\psi+\varphi^2 c.$$

Then f appears again in the original form (12), but now k, the vanishing order of φ, has at least doubled. When repeating this procedure arbitrarily often, the finiteness condition (13) prevents l from going to ∞. So it is no loss of generality to assume $k\geq l$ and to write $x\varphi(y)+\psi(y)=y^l e''(x,y)$ with $e''(0,0)\neq 0$. After introducing new coordinates $x\sqrt{e}$ and $y\sqrt[l]{e''}$, the equation appears in *normal form* $(l=n+1)$

$$f=x^2+y^{n+1}, \quad n\geq 2 \quad (\text{type } A_n).$$

Classification of triple points with two or three different tangents. These are the singularities of multiplicity $\mu=3$ where the cubic homogeneous polynomial $f_3 = f \bmod \mathfrak{m}^4$ has at least two different roots. One of them must be simple and corresponds to a non-singular branch which we can normalise to $y=0$. Since we exclude the possibility of a non-reduced branch, we have $f=h(x,y)\cdot y$, where $h=0$ is one of the double points discussed above. Since the quadratic polynomial $h_2 = h \bmod \mathfrak{m}^3$ does not contain the factor y, by a linear transformation, leaving invariant the axis $y=0$, we can put it into the form $h_2=x^2+y^2$ or $h_2=x^2$. Now we apply the normalising procedure above to the double point $h=0$. Noticing that in this procedure we did not change the axis $y=0$, we find that the triple point can be given by some equation

$$y(x^2+y^{n-2})=0, \qquad n\geq 4 \quad (\text{type } D_n).$$

Classification of simple triple points with one tangent. Now $f_3 = f \bmod \mathfrak{m}^4$ has one root only, say $x=0$. Then there is an expansion

$$f(x,y)=x^3\, e(x,y)+x^2 y^2 \varphi_1(y)+x y^3 \varphi_2(y)+y^4 \varphi_3(y)$$

with $e(0,0)\neq 0$. Putting $x=u\,y$ and dividing by y^3 we obtain the equation

$$\bar{f}(x,y)=u^3\, \bar{e}(u,y)+u^2 y\, \varphi_1(y)+u\, y\, \varphi_2(y)+y\, \varphi_3(y) \tag{14}$$

for the proper transform of our curve after blowing up the origin. This transform is at worst a double point, so one of the following must hold:

$$E_6\colon\ \varphi_3(0)\neq 0$$
$$E_7\colon\ \varphi_3(0)=0, \qquad \varphi_2(0)\neq 0$$
$$E_8\colon\ \varphi_3(0)=\varphi_2(0)=0, \qquad \varphi_3'(0)\neq 0.$$

The case E_6. Replacing y by $y\sqrt[4]{\varphi_3}+x\,\varphi_2/4\sqrt[4]{\varphi_3}^{\,3}$ we bring f into the form (with new e and φ_1)

$$f(x,y)=x^3\, e(x,y)+x^2 y^2 \varphi_1(y)+y^4.$$

Substituting $\xi = x\sqrt[3]{e}+y^2\varphi_1/3\sqrt[3]{e^2}$ we change this into

$$\begin{aligned} f(x,y) &= \xi^3+y^4(1-x\,\varphi_1^2/3e-y^2\varphi_1^3/27e^2)\\ &= \xi^3+y^4 e'(\xi,y)\end{aligned}$$

with $e'(0,0)\neq 0$. Finally, replacing y by $y\sqrt[4]{e'}$, we arrive at the *normal form*

$$f=x^3+y^4 \qquad (\text{type } E_6).$$

The case E_7. By Eq. (14), our singularity is an ordinary double point of the proper transform. So the original triple point was reducible. Taking for axis $x=0$ one of the non-singular branches, the equation is put into the form

$$f(x,y)=x(x^2 e(x,y)+x\,y^2\varphi_1(y)+y^3\varphi_2(y)).$$

Since $\varphi_2(0)\neq 0$, we may replace y by

$$\eta = y\sqrt[3]{\varphi_2}+x\,\varphi_1/3\sqrt[3]{\varphi_2}^{\,2}$$

to find

$$f(x,\eta)=x(x^2 e(x,\eta)+\eta^3-x^2 y\varphi_1^2/3\varphi_2-x^3\varphi_1^3/27\varphi_2^2)$$
$$=x(x^2 e'(x,\eta)+\eta^3).$$

Since $e'(0,0)\neq 0$ we can replace x by $x\sqrt{e'}$ and divide by a unit to arrive at the normal form

$$f=x(x^2+y^3)\qquad (\text{type } E_7).$$

The case E_8. We write

$$f(x,y)=x^3 e(x,y)+x^2 y^2\varphi_1(y)+x y^4\varphi_2(y)+y^5\varphi_3(y)$$

with $\varphi_3(0)\neq 0$. Substituting as above $\xi=x\sqrt[3]{e}+y^2\varphi_1/3\sqrt[3]{e^2}$, we obtain

$$f(\xi,y)=\xi^3+y^4(x\varphi_2(y)-x\varphi_1^2/3e-y^2\varphi_1^3/27e^2)+y^5\varphi_3.$$

Expanding the bracket as power series in ξ and y we find (with new functions e and φ_i)

$$f(\xi,y)=\xi^3 e(\xi,y)+\xi^2 y^4\varphi_1(y)+\xi y^4\varphi_2(y)+y^5\varphi_3(y),$$

where still $\varphi_3(0)\neq 0$. Replacing y by $y\sqrt[5]{\varphi_3}+\xi\varphi_2/5\sqrt[5]{\varphi_3}^4$ we can change this into (again with new e and φ_1)

$$f(\xi,y)=\xi^3 e(\xi,y)+\xi^2 y^3\varphi_1(y)+y^5.$$

With $\zeta=\xi^3\sqrt[3]{e}+y^3\varphi_1/3\sqrt[3]{e^2}$ this becomes

$$f(\zeta,y)=\zeta^3+y^5(1-\xi y^2\varphi_1/3e-y^4\varphi_1^3/27e^2)$$
$$=\zeta^3+y^5 e'(\zeta,y)$$

with $e'(0,0)\neq 0$. Passing from y to $y\sqrt[5]{e'}$ we obtain the normal form

$$f=x^3+y^5\qquad (\text{type } E_8).$$

Having completed these classifications we can easily prove the basic

(8.1) **Theorem.** *The simple singularities of curves are exactly the double points with equations*

$$A_n\colon\ x^2+y^{n+1}=0\qquad n\geq 1$$

and the triple points with equations

$$D_n\colon\ y(x^2+y^{n-2})=0\qquad n\geq 4$$
$$E_6\colon\ x^3+y^4=0$$
$$E_7\colon\ x(x^2+y^3)=0$$
$$E_8\colon\ x^3+y^5=0.$$

Proof. By the classification above, it is clear that all simple singularities must occur in this list. To prove that the *A-D-E* singularities are all simple, it suffices to compute their total transforms under blowing up the origin and to observe that the new singularities are again of A-D-E type. In fact we have:

singularity	A_2	$A_n, n \geq 3$	D_4	D_5	$D_n, n \geq 6$	E_6	E_7	E_8
transform	A_3	D_{n+1}	$3A_1$	A_1, A_3	A_1, D_{n-2}	A_5	D_6	E_7

□

Intersection Theory

9. Intersection Multiplicities

Let x be an isolated intersection of two reduced curves C and D on a (non-singular) surface X. Let $f, g \in \mathcal{O}_{X,x}$ be local equations for C, D respectively. We define the following numbers:

$$i_1 = \dim_{\mathbb{C}} \mathcal{O}_X/(f_x, g_x)$$

$$i_2 = \sum_{k=1}^{r} \operatorname{ord}_{x_k}(g_k),$$

where $v\colon \tilde{C} \to C$ is the normalisation, $g_1, \ldots, g_r$ are the germs of $g \circ v$ in the points $x_1, \ldots, x_r \in v^{-1}(x)$, and $\operatorname{ord}_{x_k}(g_k)$ denotes their order of vanishing.

(9.1) **Proposition.** *The numbers i_1 and i_2 are equal.*

The number $i_1 = i_2 = i_x(C, D)$ is called the *intersection multiplicity* of C and D in x.

Proof of Proposition 9.1. Let $\mathcal{I} \subset \mathcal{O}_C$ be the ideal generated by the function $g|C$. Then $i_1 = \dim_{\mathbb{C}}(\mathcal{O}_C/\mathcal{I})_x$. Consider the exact diagram on C:

$$\begin{array}{ccccccccc}
 & & 0 & & 0 & & & & \\
 & & \downarrow & & \downarrow & & & & \\
0 & \longrightarrow & \mathcal{I} & \longrightarrow & v_* v^* \mathcal{I} & \longrightarrow & \mathcal{C}_1 & \longrightarrow & 0 \\
 & & \downarrow & & \downarrow & & \downarrow & & \\
0 & \longrightarrow & \mathcal{O}_C & \longrightarrow & v_* \mathcal{O}_{\tilde{C}} & \longrightarrow & \mathcal{C}_2 & \longrightarrow & 0 \\
 & & \downarrow & & \downarrow & & & & \\
 & & \mathcal{O}_C/\mathcal{I} & \xrightarrow{q} & v_*(\mathcal{O}_{\tilde{C}}/v^* \mathcal{I}) & & & & \\
 & & \downarrow & & \downarrow & & & & \\
 & & 0 & & 0. & & & &
\end{array}$$

Since $\mathscr{I}$ is locally isomorphic with $\mathcal{O}_C$, the two quotients $\mathscr{C}_1$ and $\mathscr{C}_2$ will be isomorphic too. Both have support in x, so $\ker p$ and $\operatorname{coker} p$ are sheaves of the same length supported in x. Now the familiar snake lemma shows $\ker p = \ker q$ and $\operatorname{coker} p = \operatorname{coker} q$. Hence

$$i_1 = \dim_{\mathbb{C}} \mathcal{O}_C/\mathscr{I} = \dim_{\mathbb{C}} v_*(\mathcal{O}_{\tilde{C}}/v^*\mathscr{I})$$
$$= \sum_1^r \dim_{\mathbb{C}} \mathcal{O}_{\tilde{C}}/(g_k) = i_2. \quad \square$$

The preceding descriptions of $i_x(C,D)$ allow to draw the following conclusions:

1) $i_x(C,D)=1$ if and only if f and g span $m_x \subset \mathcal{O}_{X,x}$, i.e. if C and D are non-singular and do not touch at x.

2) $i_x(C,D)$ is symmetric in C and D. Since $\operatorname{ord}_{x_k}(g_k h_k) = \operatorname{ord}_{x_k}(g_k) + \operatorname{ord}_{x_k}(h_k)$, the intersection number depends bilinearly on C and D. So by linearity the definition can be extended to divisors.

3) Let $\sigma\colon \bar{X} \to X$ be the σ-process with centre x. Then i_2 does not change if one replaces C by $\bar{C}$, the proper transform, and D by $v^{-1}(D)$, the total transform. So one has

$$i_x(C,D) = \sum_{y \in \bar{C} \cap \sigma^{-1}(x)} i_y(\bar{C}, \sigma^{-1}(D)).$$

10. Intersection Numbers

Let X be a connected surface. Then $H_c^4(X,\mathbb{Z}) \cong H_0(X,\mathbb{Z}) \cong \mathbb{Z}$, in a canonical way, and we have the cupproduct pairing

$$\begin{array}{ccccc} H_c^2(X,\mathbb{Z}) & \times & H^2(X,\mathbb{Z}) & \to & H_c^4(X,\mathbb{Z}) \cong \mathbb{Z} \\ \xi & \times & \eta & \to & (\xi,\eta) \end{array}$$

We shall denote (ξ,η) sometimes by $\xi\eta$.

Definition. If D is a divisor on X with compact support, E some other divisor and $\mathscr{M}$ a line bundle on X, we put

$$(D,E) = DE = (c_1(\mathcal{O}_X(D)), c_1(\mathcal{O}_X(E)))$$
$$(D,\mathscr{M}) = (c_1(\mathcal{O}_X(D)), c_1(\mathscr{M})).$$

(Recall that $c_1(\mathcal{O}_X(D)) \in H_c^2(X,\mathbb{Z})$, cf. Proposition I.6.4.) If X is compact we set

$$(\mathscr{L},\mathscr{M}) = (c_1(\mathscr{L}), c_1(\mathscr{M}))$$

for any two line bundles $\mathscr{L},\mathscr{M}$ on X.
All the integers thus defined are called *intersection numbers.*

The intersection product has the following properties.

(i) *It is bilinear with respect to the tensor product operation on line bundles and with respect to addition of divisors.*

(ii) *It is symmetric.*

(iii) *If* $\pi\colon Y\to X$ *is a proper map from another surface* Y *onto* X, *then the product formula* I(1) *implies*

$$(\pi^*(\mathscr{L}), \pi^*(\mathscr{M})) = (\deg(\pi))(\mathscr{L}, \mathscr{M}).$$

(iv) *If* C *is any compact curve, then*

$$(C, \mathscr{M}) = \deg(\mathscr{M}\,|\,C). \tag{15}$$

Proof. If C is a smooth compact curve, then the homology class of C is dual to $c_1(\mathscr{O}_X(C))$ (Proposition I.6.4). Therefore, if $\mathscr{M}$ is any line bundle on X, we have (15) again by the product formula.
If C is singular, but still reduced and irreducible, and if $v\colon \tilde{C}\to C$ is the desingularisation (Theorem 7.1) then $v^*\colon H^2(C,\mathbb{Z})\to H^2(\tilde{C},\mathbb{Z})$ is an isomorphism, and consequently (15) holds in this case too. The general case follows by linearity. □

(v) *If* D_1 *and* D_2 *are two compact divisors, then* $D_1 D_2$ *is the intersection product of their homology classes.* To see this it is sufficient to show that for any compact divisor D the class $c_1(\mathscr{O}_X(D))$ is dual to the homology class of D. By linearity this can be reduced to the case of an irreducible D. We take a succession of σ-processes $\tau\colon \bar{X}\to X$ such that the proper transform $\bar{D}\subset\bar{X}$ of D is smooth. Using Proposition I.6.4 and linearity of c_1 we see that $c_1(\tau^*(\mathscr{O}_X(D))$ is dual to $\bar{D}+$(sum of exceptional curves). Applying Lemma I.1.1, we find the result we want.

(vi) *If the compact divisors* D, E *have no common component (i.e. if they intersect in a finite set of points), then*

$$DE = \sum_{x\in D\cap E} i_x(D, E).$$

To prove this, by linearity we may assume D and E to be effective, even irreducible. Let $v\colon \tilde{D}\to D$ be the normalisation, then by (iv) above

$$DE = \deg \mathscr{O}_D(E) = \deg v^*(E).$$

But the canonical section in $\mathscr{O}_{\tilde{D}}(v^*(E))$ vanishes exactly in the points $y\in\tilde{D}$ with $x=v(y)\in D\cap E$ and there to the order $\operatorname{ord}_y(v^*(f_x))$, with $f_x\in\mathscr{O}_{X,x}$ a local equation for E. So

$$\deg v^*(E) = \sum_{x\in D\cap E}\ \sum_{y\in v^{-1}(x)} \operatorname{ord}_y(v^*(f_x))$$

and this proves the assertion, because

$$\sum_{y\in v^{-1}(x)} \operatorname{ord}_y(v^*(f_x)) = i_x(D, E).$$

11. The Arithmetical Genus of an Embedded Curve

Let C be a compact irreducible curve. The genus $g(\tilde{C})$ of its normalisation $\tilde{C}$ is called the *geometric(al) genus* of C. If C is embedded in a non-singular surface

X, then the *arithmetic(al) genus* of C (compare I, Sect. 4) is

$$g(C)=1-\chi(\mathcal{O}_C)=1+\chi(\omega_C)$$

(the second equality is due to duality). This $g(C)$ has the following properties:
a) *If C is irreducible non-singular, then arithmetic and geometric genus are the same. If $C=\bigcup_1^k C_i$ is non-singular with C_i connected, then $g(C)=\sum_1^k g(C_i)-(k-1)$. If $C\subset X$ is reduced with ν: $\tilde{C}\to C$ its normalisation, then the normalisation sequence* (5) *shows that $\chi(\mathcal{O}_C)=\chi(\nu_*\mathcal{O}_{\tilde{C}})-h^0(\nu_*\mathcal{O}_{\tilde{C}}/\mathcal{O}_C)$. Hence*

$$g(C)=g(\tilde{C})+\delta(C)$$

with

$$\delta=\sum_{x\in C}\dim_{\mathbb{C}}(\nu_*\mathcal{O}_{\tilde{C}}/\mathcal{O}_C).$$

b) *For every $C\subset X$ we have the adjunction formula or genus formula*

(16) $$g(C)=1+\tfrac{1}{2}\deg(\mathcal{K}_X\otimes\mathcal{O}_X(C)|C).$$

To prove this, we start from

$$\deg(\omega_C)=\deg(\mathcal{K}_X\otimes\mathcal{O}_X(C)|C).$$

Riemann-Roch plus duality show

$$\deg(\omega_C)=\chi(\omega_C)-\chi(\mathcal{O}_C)=2\chi(\omega_C),$$

yielding (16).

(If C is smooth then this formula is indeed equivalent to the "old" adjunction formula of Theorem I.6.3.) By this formula we can also define the arithmetical genus for divisors C. If $C=A+B$, then

(17) $$g(A+B)=g(A)+g(B)+AB-1.$$

c) *If C is reduced and connected, then $g(C)\geq 0$, and $g(C)=0$ implies that C is a tree of non-singular rational curves.* Here $C=\sum R_i$ is a tree if

(i) $R_iR_j\leq 1$ for $i\neq j$,
(ii) there is no cycle $R_{i_1},\ldots,R_{i_n}\subset C$, $n\geq 3$, with $R_{i_j}R_{i_{j+1}}\neq 0$ for $j=1,\ldots,n-1$ and $R_{i_1}R_{i_n}\neq 0$,
(iii) three different curves never have a point in common.

Proof of c). Since $h^0(\mathcal{O}_C)=1$ for every reduced connected C, it follows that $g(C)=1-h^0(\mathcal{O}_C)+h^1(\mathcal{O}_C)\geq 0$. Conversely, if $g(C)=0$, then the assertion can be proved by induction on the number of irreducible components: if $C=A+B$, then $AB\geq 1$, so $g(C)\geq g(A)+g(B)$, and $g(C)$ vanishes only if $g(A)=g(B)=0$ and $AB=1$. □

12. 1-Connected Divisors

We shall call an effective divisor C *connected*, if supp(C) is connected. For compact reduced C, connectedness implies $h^0(\mathcal{O}_C)=1$, but this is no longer true if C is not reduced. The following lemma deals with this situation. For $\mathscr{L}=\mathcal{O}_C$, it is due to Ramanujam ([Ram], Lemma 3).

(12.1) **Lemma.** *Let C be a compact curve on the non-singular surface X and $\mathscr{L}$ a line bundle on C, the restriction of which to any irreducible component of C has degree 0. If $h\in H^0(\mathscr{L})$, and $C=C_1+C_2$, with $C_1\leq C$ a maximal divisor satisfying $h|C_1\equiv 0$, then*

$$C_1 C_2\leq 0.$$

Proof. By assumption $h\in H^0(\mathscr{I}_{C_1}\cdot\mathscr{L})=H^0(\mathcal{O}_{C_2}(-C_1)\otimes\mathscr{L})$. The map $h\colon \mathcal{O}_{C_2}\to\mathcal{O}_{C_2}(-C_1)\otimes\mathscr{L}$ is injective and its cokernel $\mathscr{Q}$ has finite support. The exact sequence

$$(18)\qquad 0\to\mathcal{O}_{C_2}\to\mathcal{O}_{C_2}(-C_1)\otimes\mathscr{L}\to\mathscr{Q}\to 0$$

and Riemann-Roch on C_2 then show

$$\begin{aligned}-C_2C_1&=\deg(\mathcal{O}_{C_2}(-C_1)\otimes\mathscr{L})\\&=\chi(\mathcal{O}_{C_2}(-C_1)\otimes\mathscr{L})-\chi(\mathcal{O}_{C_2})=h^0(\mathscr{Q})\geq 0. \quad\square\end{aligned}$$

Definition. A compact effective divisor C on a surface is called *m-connected*, if $C_1C_2\geq m$ for each effective decomposition $C=C_1+C_2$.

Of course, every 1-connected ("numerically connected") divisor C has connected support, but the converse is false. (Take for example an irreducible curve D with negative self-intersection. Then for $n\geq 2$ the curve $C=nD$ is connected but not 1-connected.)

(12.2) **Lemma.** *Let C be 1-connected and $\mathscr{L}$ a line bundle on C as in Lemma* 12.1 *above. Then $h^0(\mathscr{L})\leq 1$ and $h^0(\mathscr{L})=1$ if and only if $\mathscr{L}=\mathcal{O}_C$.*

Proof. For $h\in H^0(\mathscr{L})$ define $C=C_1+C_2$ as above. If $h\neq 0$, then $C_2\neq 0$, and so $C_1=0$ by 1-connectedness. But now the exact sequence (18) and Riemann-Roch show $h^0(\mathscr{Q})=0$, i.e. $\mathscr{Q}=0$, and $h\colon\mathcal{O}_C\to\mathscr{L}$ is an isomorphism. $\square$

As a special case we note

(12.3) **Corollary.** *If C is a 1-connected effective divisor, then $h^0(\mathcal{O}_C)=1$.*

Later we shall need the following auxiliary result.

(12.4) **Lemma.** *Let D be a 1-connected effective divisor on X and $E\not\subset\operatorname{supp}(D)$ an irreducible curve on X with $DE=1$. Then $h^0(\mathcal{O}_D(E))=1$, unless the component R of D intersecting E is non-singular rational.* (Notice that because of $DE=1$ the curve E intersects only one component R of D, which is necessarily reduced.)

Proof. If we write $D=C+R$, then $E\cap\operatorname{supp}(C)=\emptyset$ and we have two decomposition sequences

$$\begin{array}{ccccccccc} 0 & \longrightarrow & \mathcal{O}_C(E-R) & \longrightarrow & \mathcal{O}_D(E) & \longrightarrow & \mathcal{O}_R(E) & \longrightarrow & 0 \\ & & \| & & & & & & \\ 0 & \longrightarrow & \mathcal{O}_C(-R) & \longrightarrow & \mathcal{O}_D & \longrightarrow & \mathcal{O}_R & \longrightarrow & 0. \end{array}$$

Since D is 1-connected, by Corollary 12.3 the restriction $H^0(\mathcal{O}_D) \to H^0(\mathcal{O}_R)$ is injective, so $h^0(\mathcal{O}_C(E-R)) = h^0(\mathcal{O}_C(-R)) = 0$. And if $h^0(\mathcal{O}_D(E)) > 1$, the line bundle $\mathcal{O}_R(E)$ of degree 1 will have two independent sections, defining an isomorphism from R onto $\mathbb{P}_1$. □

The content of the Sects. 1–7 is absolutely standard. In algebraic formulation, most of it is contained in Serre's book [Se5], in particular chapitre I. Serre duality for reduced projective curves is formulated and proved there for the first time. Duality on projective schemes is due to Grothendieck, see the lecture notes [A-K]. There is also an analytic version of this duality, see e.g. [R-R]. A simple treatment of the curve case in analytic language however does not seem available. Of course, every compact analytic curve is projective, but the easiest way to prove this is using duality.

It should be mentioned that the "fundamental class" in analytic and algebraic duality differs by a power of $2\pi\sqrt{-1}$. If X is a compact algebraic manifold of dimension n, then under the identification

$$H^n_{\mathrm{alg}}(X, \Omega^n) = H^n_{\mathrm{an}}(X, \Omega^n)$$

we have

$$\text{algebraic fund. class} = (2\pi\sqrt{-1})^n \text{ analytic fund. class.}$$

For our purposes, such differences are not important.

The relation between simple singularities and simple Lie groups is one of the most beautiful discoveries in mathematics. It is impossible to attribute it to a single author. Any list of references will contain the names V.I. Arnold, P. Du Val, F. Klein, and it will lead back to the classification of platonic solids. Section 8 contains the classification of simple singularities of curves in the most elementary way we know of. Dynkin diagrams will appear in III, Sect. 7.

Sections 9, 10, 11 are again standard knowledge, although references in the analytic case are scarce. Section 12 is essentially due to Ramanujam (see [Ram]).

III. Mappings of Surfaces

In this chapter a surface is a reduced 2-dimensional complex space, unless specified otherwise. In this respect we draw in particular attention to the convention valid for Sects. 8–18.

Bimeromorphic Geometry

1. Bimeromorphic Maps

Let X, Y be irreducible surfaces. A proper holomorphic surjective map $\pi\colon X \to Y$ is called *bimeromorphic* if there are proper analytic subsets $T \subset X$ and $S \subset Y$ such that $\pi\colon X \setminus T \to Y \setminus S$ is biholomorphic. A basic example of such a map is provided by the *normalisation* $\nu\colon \tilde{X} \to X$, introduced in I, Sect. 8.

In the sequel we consider only bimeromorphic maps $\pi\colon X \to Y$ with X and Y *normal*. Then there is a discrete set $S \subset Y$ such that $\pi|\pi^{-1}(Y \setminus S)$ is biholomorphic and such that $\pi^{-1}(y)$ is a curve in X for every $y \in S$ (see [G-R1]). The points $y \in S$ are called *fundamental points* for π and the curves $\pi^{-1}(y)$ *exceptional curves* for π. It is a consequence of the Riemann and Levi extension theorems on the normal surface Y (Theorem I.8.7) that π^* and π_* give isomorphisms between the rings of holomorphic functions and the fields of meromorphic functions. This implies in particular that the exceptional curves $\pi^{-1}(y)$ are *connected.*

A special type of bimeromorphic map – the σ-process – appeared already before (I, Sect. 9 and II, Sect. 7). Like in that special case we define for a bimeromorphic map $\pi\colon X \to Y$ between smooth surfaces X and Y the *total transform* of any divisor D on Y to be the divisor $\pi^*(D)$ on X, whereas the *proper transform* $\bar{D}$ of D is the divisor obtained from $\pi^*(D)$ by removing all components which are exceptional for π.

If X and Y are normal projective surfaces, then by Chow's theorem (Theorem I.19.2) a bimeromorphic map is nothing but a birational map.

Let X, Y be an ordered pair of normal surfaces. A *bimeromorphic correspondence* between X and Y is a triple (Z, π_1, π_2), where Z is a third irreducible and normal surface and $\pi_1\colon Z \to X$, $\pi_2\colon Z \to Y$ are bimeromorphic maps. Such a correspondence induces a *bimeromorphic transformation* $\tau\colon X \to Y$ given by $\tau = \pi_2 \circ \pi_1^{-1}$. Given a bimeromorphic transformation (Z_1, π_1, π_2) between X_1

and X_2 and a bimeromorphic transformation (Z_2, π_3, π_4) between X_2 and X_3, there exists a diagram

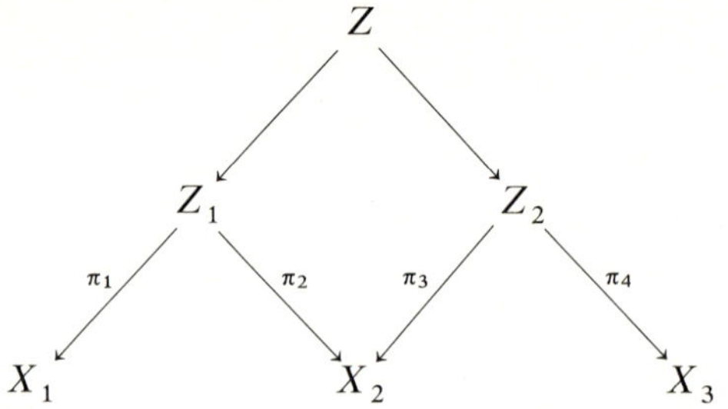

where Z is a properly chosen component of the normalisation of $Z_1 \times_{X_2} Z_2$. This diagram provides a bimeromorphic transformation from X_1 onto X_3. So, if we call two normal surfaces *bimeromorphically equivalent* as soon as there is at least one bimeromorphic equivalence between them, we see that this equivalence is indeed an equivalence relation.

Remark. When two surfaces are bimeromorphically equivalent, then their fields of meromorphic functions are isomorphic. The converse does not hold in general, but it is true in the case of algebraic surfaces, for which bimeromorphic equivalence is the same as birational equivalence.

2. Exceptional Curves

A compact, reduced, connected curve C on a *nonsingular* surface X is called *exceptional*, if there is a bimeromorphic map $\pi: X \to Y$ such that C is exceptional for π, i.e., if there is an open neighbourhood U of C in X, a point $y \in Y$, and a neighbourhood V of y in Y, such that π maps $U \setminus C$ biholomorphically onto $V \setminus \{y\}$, whereas $\pi(C) = y$. We shall express this situation also by saying that C is *contracted* to y. Since by agreement we assume $y \in Y$ to be a normal point, this singularity is uniquely determined by the embedding $C \subset X$ up to biholomorphic equivalence. (Compare the proof of Proposition 8.5.) The following characterisation is due to Grauert ([Gr2], p. 367) and its earlier algebraic version to Mumford (see [Mu1]).

(2.1) **Theorem** (Grauert's criterion). *A reduced, compact connected curve C with irreducible components C_i on a smooth surface is exceptional if and only if the intersection matrix $(C_i C_j)$ is negative definite.*

The following examples of exceptional curves are the most important ones:

i) Exceptional curves of the first kind. These are nonsingular rational curves with self-intersection -1. Frequently we shall call such curves (-1)-*curves.* A very useful characterisation of (-1)-curves is given by

(2.2) **Proposition.** *An irreducible curve $C \subset X$ is a (-1)-curve if and only if*

$$C^2 < 0 \quad \textit{and} \quad (\mathcal{K}_X, C) < 0.$$

Proof. If C is a (-1)-curve, then by definition $C^2=-1$, hence $(\mathscr{K}_X, C)=-1$ by the adjunction formula. Conversely, if the inequalities of the proposition hold, then $\deg(\omega_C)<0$, hence C is nonsingular rational (II, Sect. 11), and $C^2=-1$, again by the adjunction formula.

(2.3) **Proposition.** *Let X be a smooth compact connected surface with* $\mathrm{kod}(X)\geq 0$, *and D an effective divisor on X such that $(\mathscr{K}_X, D)<0$. Then D contains a (-1)-curve.*

Proof. It is sufficient to show that if D is an irreducible curve with $(\mathscr{K}_X, D)<0$, then D is a (-1)-curve. But by assumption, for some $n\geq 1$ there is a non-negative n-canonical divisor $K=\sum c_i C_i$, $c_i\geq 0$. Since $KD<0$, the curve D must be one of the C_i's, say $D=C_0$. Hence $D(K-c_0 D)\geq 0$ and $D^2<0$. The assertion thus follows from Proposition 2.2. □

ii) Hirzebruch-Jung strings. These are unions $C=\bigcup_1^r C_i$ of smooth rational curves C_i such that

$$\begin{array}{ll} C_i^2\leq -2 & \text{for all } i, \\ C_i C_j=1 & \text{if } |i-j|=1, \\ C_i C_j=0 & \text{if } |i-j|\geq 2. \end{array}$$

If $e_i=C_i^2$, then this configuration is visualised by the dual graph

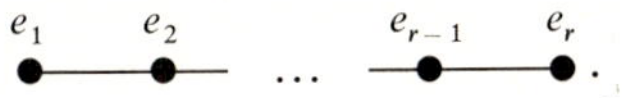

The intersection matrix

$$\begin{matrix} e_1 & 1 & 0 & \cdots \\ 1 & e_2 & 1 & \ddots \\ 0 & 1 & e_3 & \ddots \\ \vdots & \ddots & \ddots & \ddots \end{matrix}$$

is negative definite. Concrete examples of such curves are easily constructed: let U_i be a small neighbourhood of the zero-section in $\mathcal{O}_{\mathbb{P}_1}(e_i)$, the line bundle over $C_i=\mathbb{P}_1$ of degree e_i. Then patch the U_i together like this:

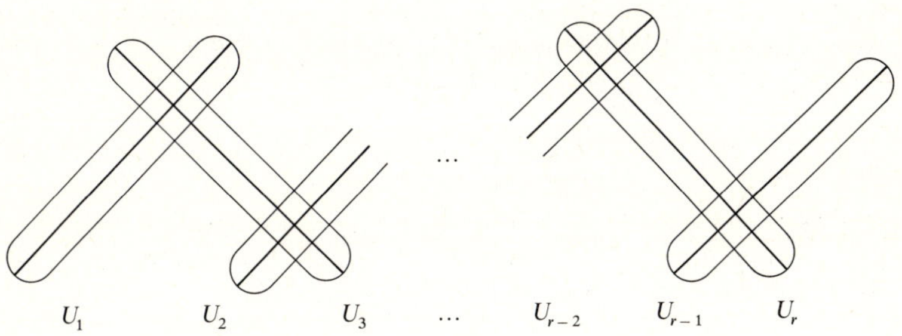

The simplest Hirzebruch-Jung string is a smooth rational curve with selfintersection -2. We shall call such a curve a (-2)-*curve*.

iii) A-D-E curves. These are the exceptional curves $C=\bigcup C_i$ for which all irreducible components are (-2)-curves. Because of

$$(C_i+C_j)^2=2(C_i C_j-2)<0,$$

for all $i \neq j$ one has $C_i C_j \leq 1$, i.e., two such curves can intersect in at most one point and then transversally. The intersection form of C being negative definite, it must be one of the forms described by Dynkin diagrams A_n with $n \geq 1$, D_n with $n \geq 4$, or E_6, E_7, or E_8 (I, Sect. 2). So these Dynkin diagrams are the dual graphs of our curves. Notice that the curves A_n are Hirzebruch-Jung strings.

A-D-E-curves can be recognised by the following criterion.

(2.4) **Proposition.** *Let $C \subset X$ be an exceptional curve with $(\mathscr{K}_X, C_i)=0$ for each irreducible component C_i of C. Then C is an A-D-E curve.*

Proof. From $(\mathscr{K}_X, C_i)=0$ and $C_i^2<0$ it follows that $g(C_i) \leq 0$. Hence each C_i is nonsingular rational and $C_i^2=-2$ by the adjunction formula. □

Grauert has shown in [Gr2] (p. 357) that an exceptional curve $C \subset X$ possesses arbitrarily small strictly pseudo-convex neighbourhoods $U \subset X$ with the following property: For every locally free $\mathcal{O}_U$-sheaf $\mathscr{S}$ there is some $k_0>0$ such that for all $k \geq k_0$

$$\text{restr:}\ H^i(U, \mathscr{S}) \rightarrow H^i(\mathscr{S} | k C) \quad (i \geq 1)$$

is injective. C being of dimension one, this shows in particular $H^2(U, \mathscr{S})=0$. Using $H^2(U, \mathscr{S} \otimes \mathcal{O}_U(-k C))=0$, one even finds that

(1) $$\text{restr:}\ H^1(U, \mathscr{S}) \rightarrow H^1(\mathscr{S} | k C)$$

is bijective for $k \gg 0$.

3. Rational Singularities

Let $y \in Y$ be the singularity obtained by contracting the exceptional curve $C \subset X$ via $\pi: X \rightarrow Y$. The singularity is called *rational* if $\pi_{*1} \mathcal{O}_X$ vanishes. By Grauert's result (1) and the cohomology sequences of the "decomposition series"

$$0 \rightarrow \mathcal{O}_C(-k C) \rightarrow \mathcal{O}_{(k+1)C} \rightarrow \mathcal{O}_{kC} \rightarrow 0.$$

$k \geq 1$, we see that this is the case if and only if $h^1(\mathcal{O}_{kC})=0$ for all $k \geq 1$. From this criterion we can now deduce that the examples i)–iii) of the preceding section are rational.

(3.1) **Proposition.** *A (-1)-curve or a Hirzebruch-Jung string gives rise to a rational singularity.*

Proof. In view of the decomposition sequences above it is sufficient to show that

$$h^1(\mathcal{O}_C(-k C))=h^0(\omega_C \otimes \mathcal{O}_C(k C))=0 \quad \text{for all } k \geq 0.$$

But if C is a (-1)-curve, then $\mathcal{O}_C(C)=\mathcal{O}_{\mathbb{P}_1}(-1)$ and $h^1(\mathcal{O}_C(-kC))=h^1(\mathcal{O}_{\mathbb{P}_1}(k))$ obviously vanishes.

If $C=\bigcup C_i$ is a Hirzebruch-Jung string, then

$$\begin{aligned}&\deg(\omega_C\otimes\mathcal{O}_C(kC)|C_i)\\&=\deg(\omega_{C_i})+C_i(C-C_i)+kCC_i\\&=-2-e_i+(k+1)CC_i\\&=\begin{cases}k-1+ke_i & \text{if } C_i \text{ is an end of the string,}\\ 2k+ke_i & \text{if } C_i \text{ is not an end.}\end{cases}\end{aligned}$$

This degree is always ≤ 0 and strictly negative if C_i is an end curve. So $\omega_C\otimes\mathcal{O}_C(kC)$ admits the trivial section only. □

To deal with the curves of type D_n, E_6, E_7, E_8, we use (compare [An2], Proposition 1)

(3.2) **Theorem** (Artin's criterion). *The exceptional curve $C=\bigcup C_i$ contracts to a rational singularity, if and only if each divisor $Z=\sum r_i C_i$, $r_i\geq 0$, has arithmetical genus $g(Z)\leq 0$.*

Proof. Firstly, let the singularity be rational. For each divisor Z there is some $k>0$ and an epimorphism $\mathcal{O}_{kC}\to\mathcal{O}_Z$. Since $\operatorname{supp}(C)$ is of dimension one, $H^1(\mathcal{O}_Z)=0$ as soon as $H^1(\mathcal{O}_{kC})$ vanishes. But this last group always vanishes for rational singularities.

Conversely, assume that $g(Z)\leq 0$ for all Z. Putting in particular $Z=C_i$, we find that all C_i are nonsingular rational (II, Sect. 11). Now we use induction on $r=\sum r_i$ to show $h^1(\mathcal{O}_Z)=0$. So let $r\geq 2$ and $h^1(\mathcal{O}_{Z'})=0$ for all $Z'=\sum r_i' C_i$ with $\sum r_i'=r-1$. Let C_0 be a component of Z, and $Z_0=Z-C_0$. Because of the decomposition sequence

$$0\to\mathcal{O}_{C_0}(-Z_0)\to\mathcal{O}_Z\to\mathcal{O}_{Z_0}\to 0,$$

it suffices to prove $h^1(\mathcal{O}_{C_0}(-Z_0))=0$, which is equivalent with $Z_0C_0\leq 1$, that is

$$ZC_0\leq 1+C_0^2.$$

As we may choose C_0 arbitrarily, we are ready unless $ZC_i\geq 2+C_i^2$ for each C_i. But if this were the case, then

$$\begin{aligned}\deg(\omega_Z)&=(\mathcal{O}_Z(Z)\otimes\mathcal{K}_X, Z)\\&=\sum r_i\{ZC_i+\deg(\mathcal{K}_X|C_i)\}\\&=\sum r_i\{ZC_i-2-C_i^2\}\geq 0,\end{aligned}$$

and we would have the contradiction $g(Z)=1+\frac{1}{2}\deg(\omega_Z)>0$. □

(3.3) **Proposition.** *The A-D-E curves contract to rational singularities.*

Proof. Since

$$(\mathscr{K}_X, C_i) = -2 - C_i^2 = 0,$$

for any effective divisor $Z = \sum r_i C_i$ we have

$$\deg(\omega_Z) = Z^2 = (\sum r_i C_i)^2 < 0,$$

because C is exceptional. So $g(Z) = 1 + \frac{1}{2}\deg(\omega_Z) \leq 0$. □

Rational singularities are easy to deal with, because they admit arbitrarily small neighbourhoods $U \subset X$ with $H^1(U, \mathcal{O}_U^*) = H^2(U, \mathbb{Z})$, and

$$H^2(U, \mathbb{Z}) = H^2(C, \mathbb{Z}) = \bigoplus H^2(C_i, \mathbb{Z})$$

is the free group with one generator for each component C_i. Since for an exceptional *A-D-E* curve $C = \bigcup C_i \subset X$ we have $(\mathscr{K}_X, C_i) = 0$ for all i, this implies

(3.4) **Proposition.** *Any A-D-E curve has a neighbourhood U with* $\mathscr{K}_U = \mathcal{O}_U$.

An effective divisor D on U is the divisor (f) of a holomorphic function if and only if $DC_i = 0$ for all i. Any effective D is of the form $D = Z + \sum U_j$ where $Z = \sum r_i C_i$ has support on C and the U_j intersect C in finitely many points at most. Since $((f), C_i) = 0$ for all i and $U_j C_i \geq 0$, we see that Z satisfies

(2) $$ZC_i \leq 0 \quad \text{for all } i.$$

Conversely, if Z satisfies (2), then (perhaps after shrinking U) one may pick U_j's such that $C_i(Z + \sum U_j) = 0$ for all i. Thus we have

(3.5) **Proposition.** *The divisors* $Z = \sum r_i C_i, r_i \geq 0$, *satisfying* (2) *are exactly the parts contained in C of divisors of holomorphic functions defined on some neighbourhood of C.*

Given two divisors $Z = \sum r_i C_i$ and $Z' = \sum r_i' C_i$ which are the parts in C of the principal divisors of two holomorphic functions f, f', then for general $\alpha, \alpha' \in \mathbb{C}$ the part of $(\alpha f + \alpha' f')$ in C is $\sum \min\{r_i, r_i'\} C_i$, which therefore also satisfies (2). So there is a minimal effective divisor Z satisfying (2). In [An2] Artin has called this Z the *fundamental cycle* of the singularity.

(3.6) **Proposition.** *Let* $\mathfrak{m} \subset \mathcal{O}_{y,Y}$ *denote the maximal ideal. Then for all* $k \geq 1$

$$\pi^*(\mathfrak{m}^k) = \mathcal{O}_X(-kZ) \subset \mathcal{O}_X.$$

Proof. If suffices to show $\pi^*(\mathfrak{m}) = \mathcal{O}_X(-Z)$. Now if $g \in \mathfrak{m}$, then $g \circ \pi$ vanishes on C and $(g \circ \pi) \geq Z$ by construction of the fundamental cycle. This shows $\pi^*(\mathfrak{m}) \subset \mathcal{O}_X(-Z)$.

To prove the converse, let $x \in C$. As pointed out above, there is on a neighbourhood of C a holomorphic function f such that $(f) = Z + \sum U_j$. The U_j's may be chosen not to contain x. So in a neighbourhood of x, the sheaf $\mathcal{O}_X(-Z)$ is generated by f. But $\pi_* f$ is holomorphic on a neighbourhood of $y \in Y$. So $f = \pi^*(\pi_* f) \in \pi^*(\mathfrak{m})$. □

The fundamental cycles for the different types are the following (Lemma I.2.12):

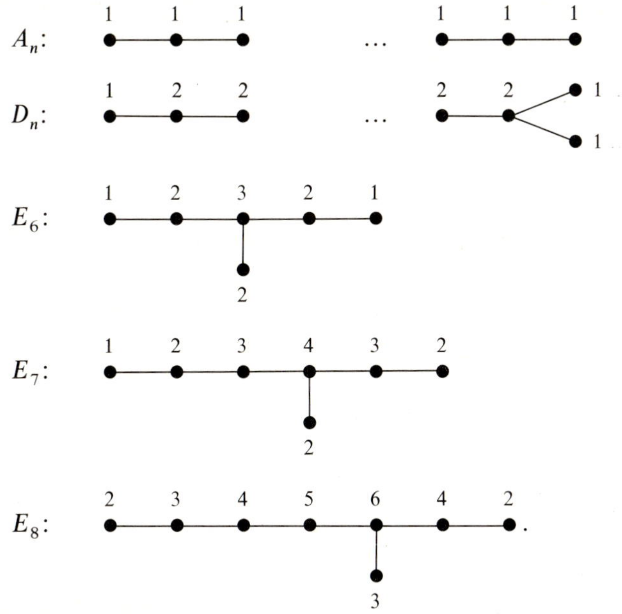

As an immediate consequence we have

(3.7) **Proposition.** *For the fundamental cycle Z of an exceptional A-D-E curve we have*

$$Z^2 = -2.$$

The fundamental cycles can also be used for constructing functions, which make *A-D-E* singularities locally double coverings of the plane.

(3.8) **Lemma.** *Let $C \subset X$ be an exceptional A-D-E curve and $y \in Y$ the singularity arising from contracting it. Then locally Y can be realised as a double covering of a nonsingular surface.*

Proof. We map a neighbourhood of y into $\mathbb{C}^2$ by constructing two functions f, g on a neighbourhood of C. Since the singularity is rational, it suffices to give effective divisors (f), (g) satisfying $((f), C_i) = ((g), C_i) = 0$ for all irreducible components $C_i \subset C$. We want f and g to satisfy the following conditions:

(i) $\{f = g = 0\}$ is the set C,

(ii) $(f) = \sum r_i C_i + R$ where R is a nonsingular curve intersecting C either in one point, at which $g|R$ vanishes to the second order, or in two different points, at which $g|R$ has simple zeros.

Property (i) implies that the map $Y \to \mathbb{C}^2$ is proper in a neighbourhood of y, and (ii) forces the degree of this covering to be 2. For finding the divisors (f) and (g), we distinguish between two cases.

a) In the case of an A_n-singularity, let R_1 and R_2 be two curves intersecting only the end curves C_1 and C_n, and both of them transversally in one point. Let S_1 and S_2 be two curves with the same properties, but disjoint from R_1 and R_2. Then we put $(f)=Z+R_1+R_2$ and $(g)=Z+S_1+S_2$.
b) In the remaining cases there is a distinguished component C_{i_0} of multiplicity 2 in Z with $ZC_{i_0}=-1$. Let R and S be two disjoint curves intersecting no C_i but C_{i_0}, and C_{i_0} transversally in one point. Then we put $(f)=Z+R$ and $(g)=Z+S$. $\square$

4. Exceptional Curves of the First Kind

The contraction of a (-1)-curve is always the converse of a σ-transform. This will follow from Theorems 4.1 and 4.2 below.

(4.1) **Theorem.** *Let X be a nonsingular surface, $E \subset X$ a (-1)-curve and $\pi\colon X \to Y$ the map contracting E. Then $y=\pi(E)$ is nonsingular on Y.*

Proof. We may assume X to be so small a neighbourhood of E that there are disjoint nonsingular curves $U, V \subset X$ intersecting E transversally in two distinct points:

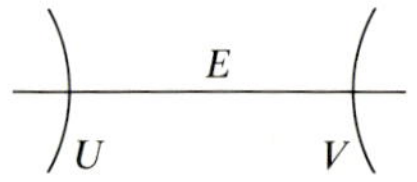

By Sect. 3, we also may assume that there are holomorphic functions u, v on X with divisors $(u)=U+E$, $(v)=V+E$. If $\rho\colon X \to \mathbb{C}^2$ is the map $x \to (u(x), v(x))$, then $\rho^{-1}(0)=\{u=v=0\}=E$. Since ρ is proper on a neighbourhood of E (see I.8.9), and since $\rho(E)=0$, there is a factorisation $\rho=\gamma\pi$ with $\gamma\colon Y \to B$ a (perhaps ramified) covering of a neighbourhood $B \subset \mathbb{C}^2$ of the origin. The curve U is mapped into the v-axis $\{u=0\} \subset \mathbb{C}^2$ and since v vanishes on U to the first order at $U \cap E$, the map $\rho|U$ is injective near this point. Since u also vanishes to the first order only along U, there is no ramification along this curve. This proves that the degree of the covering γ is one, so $Y=B$ and $\rho=\pi$. $\square$

(4.2) **Theorem** (Uniqueness of the σ-process). *Let X and Y be smooth surfaces and $\pi\colon X \to Y$ a bimeromorphic map. If $E=\pi^{-1}(y)$ is an irreducible curve, then near E, the map π is equivalent to the σ-process with centre y.*

Proof. Let u, v be local coordinates on Y centered at y. Put $U=\{u=0\}$, $V=\{v=0\} \subset Y$ and let $\bar{U}, \bar{V} \subset X$ be the proper transforms of these curves. If $\bar{U} \cap E=\{x\}$, then $i_x(\bar{U}, \pi^*(V))=\operatorname{ord}_x(\pi^*(v)|\bar{U})=\operatorname{ord}_y(v|U)=1$. So $\bar{U} \cap \bar{V}=\emptyset$ and $\bar{U}E=\bar{V}E=1$. This implies $(\pi^*(u))=\bar{U}+E$ and $(\pi^*(v))=\bar{V}+E$. So the meromorphic function $\pi^*(u/v)$ on X has no points of indeterminacy and defines a map $\lambda\colon X \to \mathbb{P}_1$ such that $\lambda|E\colon E \to \mathbb{P}_1$ is an isomorphism. Then $\pi \times \lambda\colon X \to Y \times \mathbb{P}_1$ is an isomorphism of X onto the result $\bar{Y} \subset Y \times \mathbb{P}_1$ of the σ-process with centre y. $\square$

(4.3) **Lemma** (Factorisation lemma). *Let* $\pi\colon X\to Y$ *be a bimeromorphic map with* X, Y *nonsingular surfaces. Unless* π *is an isomorphism, there is a factorisation* $\pi=\pi'\circ\sigma$, *where* $\sigma\colon X\to X'$ *is a* σ*-process.*

Proof. Take some $y\in Y$ with $C=\pi^{-1}(y)\subset X$ an exceptional curve. We may assume Y so small that there are global coordinates u, v on Y. Then $\pi^*(du\wedge dv)$ is an effective canonical divisor on X, say $K=\sum k_i C_i$, with $k_i>0$ and C_i irreducible components of C. Then $K^2<0$, and there will be some C_i with $KC_i<0$ and $C_i^2<0$. This C_i is a (-1)-curve by Proposition 2.2. By Theorem 4.2 the contraction $\sigma\colon X\to X'$ of this curve is a σ-transform. There is a factorisation $\pi=\pi'\circ\sigma$ with $\pi'\colon X'\to Y$ holomorphic outside of $x'=\sigma(C_i)$. Since however $u\circ\pi'$ and $v\circ\pi'$ extend holomorphically over x', this π' is holomorphic in x' too. □

(4.4) **Corollary** (Decomposition of bimeromorphic maps). *Let* X, Y *be nonsingular and* $\pi\colon X\to Y$ *a bimeromorphic map. Then* π *is equivalent to a succession of* σ*-transforms, which locally* (w.r. to Y) *are finite in number.*

Proof. Induction on the number of irreducible components of each exceptional curve.

A smooth surface is called *minimal*, if it does not contain any (-1)-curve. A nonsingular surface $X_{\min}$ is called a *minimal model* of the nonsingular surface X, if $X_{\min}$ is minimal itself, and if there is a bimeromorphic map (i.e., a succession of σ-transforms) from X onto $X_{\min}$.

(4.5) **Theorem.** *Every compact nonsingular surface* X *has a minimal model.*

Proof. Suppose that X contains a (-1)-curve and let X_1 be obtained from X by contracting it. If X_1 contains another (-1)-curve, the process can be repeated, and so on. This must lead to a surface without (-1)-curves after a finite number of blowing downs, since each time the second Betti number diminishes by 1 (Theorem I.9.1, (iv)). □

(4.6) **Proposition.** *If* X *is a compact connected surface with* $\operatorname{kod}(X)\geq 0$, *then all minimal models of* X *are isomorphic.*

Proof. It is sufficient to show that on a surface X with $\operatorname{kod}(X)\geq 0$, there never can be a pair E_1, E_2 of (-1)-curves with $(E_1,E_2)\geq 1$. Suppose there were such curves. By assumption, for some n there is a non-negative n-canonical divisor $D\geq 0$. Since $(\mathscr{K}_X,E_1)=(\mathscr{K}_X,E_2)=-1$, we would have $D=E_1+E_2+D_1$ with some divisor $D_1\geq 0$. Since

$$D_1 E_1=E_i(D-E_1-E_2)\leq DE_i\leq -1$$

for $i=1,2$, also D_1 would split off E_1+E_2, and so on. This is clearly impossible and the proposition follows. □

Remark. If $\operatorname{kod}(X) = -\infty$, then the statement is no longer true, for $\mathbb{P}_1 \times \mathbb{P}_1$ blown up in one point can be blown down to $\mathbb{P}_2$. (Compare VI, Sect. 6.)

5. Hirzebruch-Jung Singularities

Let $C = \sum_{i=1}^{r} C_i$ with $C_i^2 = e_i \leq -2$ as in Sect. 2, Example ii). For sufficiently small $X \supset C$ there is a (closed, but not necessarily compact) smooth curve C_0 which intersects C_1 transversally in one point, without meeting any of the other curves C_i. Similarly there is such a curve C_{r+1} intersecting C_r transversally in one point which does not intersect any of the curves $C_0, \ldots, C_{r-1}$. Thus we have a dual diagram

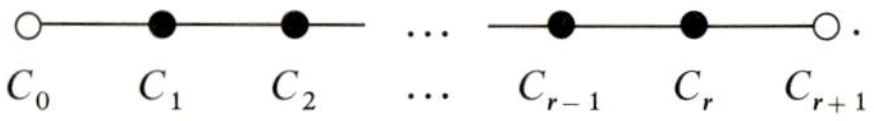

Let $n_i \in \mathbb{Z}$, $n_i \geq 0$, $i = 1, \ldots, r+1$. By Sect. 3 there is a holomorphic function φ on X with divisor $(\varphi) = \sum_0^{r+1} n_i C_i$ if and only if for $k = 1, \ldots, r$

$$n_{k-1} + e_k n_k + n_{k+1} = \sum_0^{r+1} n_i C_i C_k = 0.$$

Given n_0 and n_1, the coefficients n_k, $k = 2, \ldots, r+1$ are determined uniquely by the recursion formula

$$n_k = |e_{k-1}| \cdot n_{k-1} - n_{k-2}. \tag{3}$$

If $n_0 < n_1$ (resp. $n_0 \leq n_1$), then it follows by induction that $n_k < n_{k+1}$ (resp. $n_k \leq n_{k+1}$) for $k = 1, \ldots, r$. So if we determine integers μ_k, ν_k from (3) starting with initial data

$$\mu_0 = 0, \quad \mu_1 = 1 \Rightarrow \mu_k,$$
$$\nu_0 = 1, \quad \nu_1 = 1 \Rightarrow \nu_k,$$

then for $k \geq 1$ these integers will be positive. Therefore, we have holomorphic functions g, h on X with divisors

$$(g) = \sum_{i=0}^{r+1} \mu_i C_i, \qquad (h) = \sum_{i=0}^{r+1} \nu_i C_i.$$

Notice that the integers μ_i satisfy

$$\mu_1 = 1, \qquad \mu_2 = |e_1|,$$

$$\frac{\mu_3}{\mu_2} = |e_2| - 1/|e_1|$$

$$\frac{\mu_{k+1}}{\mu_k} = |e_k| - 1 \Big/ |e_{k-1}| - \ldots - 1 \Big/ |e_1|,$$

where the last symbol means expansion as a continued fraction. The recursion formula (3) implies

$$\text{g.c.d.}\ (\mu_{k+1}, \mu_k) = \text{g.c.d.}\ (\mu_k, \mu_{k-1}) = \ldots = \text{g.c.d.}\ (\mu_2, \mu_1) = 1.$$

It follows that μ_k and μ_{k+1} are coprime, and so they might also have been defined by the above expansion.

We put in particular $n' = \mu_{r+1}$, $q' = \mu_r$. Then the expansion

$$\frac{n'}{q'} = |e_r| - 1 \Big| \overline{|e_{r-1}| - \ldots - 1 \Big| \overline{|e_1|}}$$

shows that the self-intersections e_i are determined by the two integers n' and q'.

Finally, we define a divisor

$$(f) = \sum_{i=0}^{r+1} \lambda_i C_i$$

by integers λ_i satisfying (3) and

$$\lambda_{r+1} = 0, \qquad \lambda_r = 1.$$

These λ's are the μ's we would have obtained when starting with our index i at the other end of the string. So

$$\lambda_1 = q, \qquad \lambda_0 = n,$$

with

(4) $$\frac{n}{q} = |e_1| - 1 \Big| \overline{|e_2| - \ldots - 1 \Big| \overline{|e_r|}}.$$

By induction we obtain

(5) $$\lambda_k + (n-q)\,\mu_k = n\,\nu_k$$

(6) $$\lambda_k\,\mu_{k+1} - \lambda_{k+1}\,\mu_k = n.$$

For $k = r$, Eq. (6) implies

$$n' = n,$$

whereas Eq. (5) shows that q' is the integer determined uniquely by

$$0 < q' < n, \qquad q\,q' \equiv 1\,(n).$$

For the functions f, g and h, Eq. (5) means

$$(f g^{n-q}) = (f) + (n-q)(g) = n(h) = (h^n).$$

So $f g^{n-q}/h^n$ is a function in $\Gamma(\mathcal{O}_X^*)$, which can be squeezed into f for example. Then we have the relation

$$h^n = f g^{n-q}.$$

In other words: by

$$w = h, \qquad z_1 = f, \qquad z_2 = g,$$

X is mapped into the surface

$$W = \{(w, z_1, z_2) \in \mathbb{C}^3 : w^n = z_1 z_2^{n-q}\} \in \mathbb{C}^3.$$

(5.1) **Theorem.** *For $0 < q < n$, n and q coprime, let $C \subset X$ be the Hirzebruch-Jung string with self-intersection numbers e_i satisfying (4), and let $y \in Y$ be the singu-*

larity resulting from contracting C. Then this singularity is isomorphic with the unique singularity lying over $0 \in \mathbb{C}^3$ in the normalisation of the surface W above.

Remark. This theorem shows in particular that the singularity $y \in Y$ (hence the embedding $C \subset X$) depends on n and q only. We therefore call it *the* singularity $A_{n,q}$.

Proof of the Theorem. We denote by $\pi\colon X \to Y$ the contraction of C and by $\rho\colon X \to \mathbb{C}^2$ the map defined by $z_1 = f$, $z_2 = g$. Since the divisors (f) and (g) intersect in C only, we have $\rho^{-1}(0) = C$. By Proposition I.8.9 we may assume ρ proper. There is a factorisation $\rho = \gamma\pi$, where $\gamma\colon Y \to \mathbb{C}^2$ is proper and finite over a neighbourhood $Z \subset \mathbb{C}^2$ of the origin, so a (ramified) covering. Since γ factors through the normalisation $\tilde{W}$ of W, the assertion follows as soon as we know that $\deg(\gamma) = n$. But consider for example the curve $\pi(C_0) = \gamma^{-1}\{z_1 = 0\}$. At $C_0 \cap C_1$ the function g vanishes to the first order only, and f vanishes along C_0 to the n-th order. So γ is ramified cyclically of order n along C_0, hence $\deg(\gamma) = n$ near $y \in Y$. □

Next we describe some frequently occuring situations which give rise to $A_{n,q}$ singularities.

i) Coverings Branched over the Coordinate Axes

Let Z denote the unit dicylinder $\{(z_1, z_2) \in \mathbb{C}^2\colon |z_1| < 1,\ |z_2| < 1\}$. We consider here coverings $\gamma\colon B \to Z$, with B a connected normal surface, which are only branched over $z_1 \cdot z_2 = 0$. Put $Z^* = Z \setminus \{z_1 z_2 = 0\}$ and $B^* = \gamma^{-1}(Z^*)$. Then γ and B are determined uniquely by the nonramified covering $B^* \to Z^*$ ([St], Satz 1), i.e., by the subgroup $\Gamma = \gamma_* \pi_1(B^*) \subset \pi_1(Z^*)$. If $w_i \in \pi_1(Z^*)$ is the class of a positively oriented little loop around the z_i-axis, then $\pi_1(Z^*) = \mathbb{Z} \times \mathbb{Z}$ with generators $w_1 = (1,0)$ and $w_2 = (0,1)$.

a) First we notice that if over one axis, say $z_1 = 0$, no ramification takes place, then Γ is generated by $(0,1)$ and some element $(b,0)$, $b > 0$. But this is the case also for the covering $Z \to Z$, $z_1 \mapsto z_1^b$, $z_2 \mapsto z_2$.

b) Next we determine the subgroup Γ in case $B = Z$ is smooth and $\gamma(z_1, z_2) = (z_1^{m_1}, z_2^{m_2})$. Then obviously $\gamma_*(w_i) = m_i w_i$, so Γ is the subgroup generated by $(m_1, 0)$ and $(0, m_2)$.

c) Now we identify Γ if $B = Y$ is a neighbourhood of the Hirzebruch-Jung singularity $A_{n,q}$ mapped onto Z by $z_1 = f$, $z_2 = g$ as above. Let $\delta_k \in \pi_1\left(X \setminus \bigcup_1^{r+1} C_i\right)$ be the class of a positively oriented little loop around the curve C_k. Since f and g vanish along C_k to the order λ_k and μ_k respectively, we find $(\gamma\pi)_*(\delta_k) = \lambda_k w_1 + \mu_k w_2$. In particular, $\Gamma \subset \mathbb{Z} \times \mathbb{Z}$ contains the elements

$$(\gamma\pi)_*(\delta_0) = (n, 0),$$

$$(\gamma\pi)_*(\delta_1) = (q, 1).$$

But these elements generate in $\mathbb{Z} \times \mathbb{Z}$ a subgroup of index n, which must coincide with Γ.

d) Finally we treat the general case. Let $\Gamma=\gamma_*(\pi_1(B^*))$ be an arbitrary subgroup of finite index. We pick generators for Γ as follows: $\Gamma\cap(\mathbb{Z},0)$ is nontrivial, so there is some $(n',0)\in\Gamma$ generating this intersection, $n'>0$. The quotient $\Gamma/\mathbb{Z}\cdot(n',0)$ is isomorphic with $\mathbb{Z}$, so there is some $(q',m_2)\in\Gamma$, $0\le q'<n'$, $m_2>0$, such that $(n',0)$ and (q',m_2) generate Γ. We may assume $q'>0$, because $q'=0$ is case b), above. Let $m_1=\text{g.c.d.}(n',q')$ and $n'=n\,m_1$, $q'=q\,m_1$. Then Γ is contained in the subgroup $\Gamma_1=\mathbb{Z}(m_1,0)+\mathbb{Z}(0,m_2)$. By b) above, this corresponds to a covering $\gamma_1\colon Z_1\to Z$ with Z_1 nonsingular. The original covering factors as $\gamma=\gamma_1\gamma_2$. If we identify again $\pi_1(Z_1^*)$ with $\mathbb{Z}\times\mathbb{Z}$, then $\gamma_2\colon B\to Z_1$ corresponds to the subgroup with generators $(n,0)$ and $(q,1)$. So B carries an $A_{n,q}$ singularity over the origin. This proves

(5.2) **Theorem.** *Let $\gamma\colon Y\to Z$ be a covering with Z a nonsingular surface and Y normal. If γ is branched over a curve in Z with at worst nodes, then the singularities of Y can only be of Hirzebruch-Jung type. If in particular the ramification curve is nonsingular, then Y is nonsingular too.*

ii) Riemann Existence Domain of the Function $\sqrt[n]{z_1^a z_2^b}$

By this existence domain we mean the normalisation $\tilde{W}$ of the surface

$$W=\{(w,z_1,z_2)\in\mathbb{C}^3;\ w^n=z_1^a z_2^b\}.$$

The projection $(w,z_1,z_2)\mapsto(z_1,z_2)$ exhibits $\tilde{W}$ as an n-fold covering of $\mathbb{C}^2$, branched only over $z_1\cdot z_2=0$. So $\tilde{W}$ will be smooth, except at the points lying over $(0,0)$, where singularities of $A_{n,q}$-type appear. Explicitly, the situation is as follows:

a) If we put $d=\text{g.c.d.}(n,a,b)$, $n=\nu d$, $a=\alpha d$, and $b=\beta d$, then

$$w^n-z_1^a z_2^b=\prod_{j=1}^{d}\left(w^\nu-z_1^\alpha z_2^\beta\,\mathrm{e}\left(\frac{d}{j}\right)\right)$$

and $\tilde{W}$ decomposes into d different coverings, all of which are isomorphic with the existence domain of $\sqrt[\nu]{z_1^\alpha z_2^\beta}$.

b) Assume now that $\text{g.c.d.}(n,a,b)=1$ and let

$$d_a=\text{g.c.d.}(n,a),\qquad a=\alpha d_a,$$
$$d_b=\text{g.c.d.}(n,b),\qquad b=\beta d_b,$$
$$n=\nu d_a d_b.$$

In $\mathbb{C}^3$ we consider the surface

$$U=\{(u,y_1,y_2)\in\mathbb{C}^3: u^\nu=y_1^\alpha y_2^\beta\}.$$

It is mapped into W by

$$w=u,\qquad z_1=y_1^{d_b},\qquad z_2=y_2^{d_a}.$$

Using $\text{g.c.d.}(\alpha,d_b)=\text{g.c.d.}(\beta,d_a)=\text{g.c.d.}(d_a,d_b)=1$, we find that the map $U\to W$ is injective over $\mathbb{C}^2\setminus\{z_1z_2=0\}$. Since both the coverings $\tilde{W}$ and $\tilde{U}$ over $\mathbb{C}^2(z_1,z_2)$ are of degree n, there is an isomorphism $\tilde{U}\to\tilde{W}$. In other words:

the covering $\tilde{W}\to\mathbb{C}^2(z_1,z_2)$ factors through $\mathbb{C}^2(y_1,y_2)\to\mathbb{C}^2(z_1,z_2)$, and over $\mathbb{C}^2(y_1,y_2)$ the surface $\tilde{W}$ is the existence domain of $\sqrt[v]{y_1^\alpha y_2^\beta}$.
c) Finally, let g.c.d.$(n,a)=$g.c.d.$(n,b)=1$. We define the integer $q, 0<q<n$, by

$$a q \equiv -b(n),$$

say $aq=rn-b$ with $0<r\leq a$. Then q and n are coprime because g.c.d.$(n,b)=1$. Consider in $\mathbb{C}^3(u,z_1,z_2)$ the surface $u^n=z_1 z_2^{n-q}$. Since a and n are coprime, via $v=u^a$ it is mapped bijectively onto the surface

$$v^n=z_1^a z_2^{a(n-q)}=z_1^a z_2^b\cdot z_2^{n(a-r)}.$$

Outside of the axis $z_2=0$, the function $w=v/z_2^{a-r}$ is holomorphic and maps this last surface bijectively onto W. This shows that $\tilde{W}$ is the existence domain of $\sqrt[n]{z_1 z_2^{n-q}}$. Now the functions f, g, h from Theorem 5.1 define explicitly an isomorphism between $A_{n,q}$ and a neighbourhood of the singularity in this domain.

iii) Cyclic Quotients

In this section we denote the elements of $\mathbb{Z}_n$ by integers $k\in\mathbb{Z}$ (identified mod n). Every *linear* operation of $\mathbb{Z}_n$ on $\mathbb{C}^2$ can be given with respect to suitable coordinates (u_1,u_2) by

$$k\begin{pmatrix}u_1\\u_2\end{pmatrix}=\begin{pmatrix}\mathrm{e}(q_1 k/n)\cdot u_1\\ \mathrm{e}(q_2 k/n)\cdot u_2\end{pmatrix}$$

with integers q_1, q_2 satisfying $0\leq q_i<n$. Except for their ordering these integers q_i are determined uniquely by the operation. We call them the *weights* of the operation. If one of them vanishes, the operation is essentially 1-dimensional and the quotient smooth. From here on we exclude this possibility.

Furthermore, if $c=$g.c.d.$(n,q_1,q_2)>1$, then the action of $\mathbb{Z}_n$ can be considered as an action of $\mathbb{Z}_{n/c}$. So without loss of generality we may assume that g.c.d.$(n,q_1,q_2)=1$.

We use the following notation $(i=1,2)$:

$$d_i=\text{g.c.d.}(n,q_1),\qquad n=n_i d_i,\qquad q_i=p_i d_i$$
$$m=\text{g.c.d.}(n_1,n_2),$$
$$p_i' \text{ the integer with } p_i p_i'\equiv 1(m),\qquad 0<p_i'<m,$$
$$q \text{ the integer with } q\equiv p_1 p_2'(m),\quad 0<q<m.$$

Notice that $n=$l.c.m.(n_1,n_2) and $n_1 n_2=mn$.

(5.3) **Proposition.** *The image of* $(0,0)\in\mathbb{C}^2$ *in the quotient* $\mathbb{C}^2/\mathbb{Z}_n$ *is a singularity of type* $A_{m,q}$.

Proof. Let $\mathbb{Z}_n$ be embedded in $\mathbb{Z}_{n_1}\times\mathbb{Z}_{n_2}$ as the subgroup generated by the element $(p_1 \bmod n_1,\ p_2 \bmod n_2)$. By way of this embedding the action of $\mathbb{Z}_n$ is induced by the action

$$\begin{pmatrix}k_1 \bmod n_1\\ k_2 \bmod n_2\end{pmatrix}\begin{pmatrix}u_1\\u_2\end{pmatrix}=\begin{pmatrix}\mathrm{e}(k_1/n_1)u_1\\ \mathrm{e}(k_2/n_2)u_2\end{pmatrix}$$

of the group $\mathbb{Z}_{n_1} \times \mathbb{Z}_{n_2}$ on $\mathbb{C}^2$. The quotient map γ of this action can be identified with the covering $z_1 = u_1^{n_1}$, $z_2 = u_2^{n_2}$ of $\mathbb{C}^2$ branched over $\{z_1 \cdot z_2 = 0\}$ only. It is determined by the subgroup $\Gamma \subset \mathbb{Z} \times \mathbb{Z} = \pi_1(\mathbb{C}^2 \setminus \{z_1 \cdot z_2 = 0\})$ which is generated by $(n_1, 0)$ and $(0, n_2)$.

If we denote by γ_0 the quotient map $\mathbb{C}^2 \to \mathbb{C}^2/\mathbb{Z}_n$, then there is a factorisation $\gamma = \delta \gamma_0$, where δ displays our quotient $\mathbb{C}^2/\mathbb{Z}_n$ as a covering of $\mathbb{C}^2$ of degree $n_1 n_2/n = m$, branched over $\{z_1 \cdot z_2 = 0\}$ only. We have to determine the subgroup $\Delta \subset \mathbb{Z} \times \mathbb{Z} = \pi_1(\mathbb{C}^2 \setminus \{z_1 \cdot z_2 = 0\})$ which corresponds to δ.

Under the canonical epimorphism $\mathbb{Z} \times \mathbb{Z} \to \mathbb{Z} \times \mathbb{Z}/\Gamma = \mathbb{Z}_{n_1} \times \mathbb{Z}_{n_2}$ the group Δ is mapped onto $\mathbb{Z}_n$, embedded in $\mathbb{Z}_{n_1} \times \mathbb{Z}_{n_2}$ as above. So $\Delta \subset \mathbb{Z} \times \mathbb{Z}$ is generated by

$$(n_1, 0), \quad (0, n_2), \quad \text{and} \quad (p_1, p_2).$$

Since g.c.d. $(n_1, n_2 p_1) = m$, there are integers a and b with

$$a(n_1, 0) + b(n_2 p_1, n_2 p_2) = (m, b\, n_2 p_2) = (m, 0) + b\, p_2 (0, n_2).$$

This proves $(m, 0) \in \Delta$. Similarly one finds $(0, m) \in \Delta$. Moreover,

$$p_2'(p_1, p_2) \equiv (q, 1) \bmod (m, m),$$

hence $(q, 1) \in \Delta$. On the other hand, the elements $(m, 0)$ and $(q, 1)$ generate a subgroup of index m in $\mathbb{Z} \times \mathbb{Z}$, which must coincide with Δ. □

By a result of H. Cartan ([Car], Lemma 2, p. 98) every action of a finite group on a manifold can locally be linearised. Applying this result together with Proposition 5.3 we find

(5.4) **Theorem.** *If the finite cyclic group G acts on the smooth surface X, then the quotient X/G has only singularities of Hirzebruch-Jung type.*

6. Resolution of Surface Singularities

In this section we shall show that the preceding considerations quickly lead to a proof of a central and classical result: every surface is bimeromorphically equivalent to a nonsingular surface.

(6.1) **Theorem.** *For every normal surface X there is a bimeromorphic map $\pi\colon Y \to X$ with Y nonsingular.*

Proof. The singularities of a normal surface are isolated, so we may concentrate on one point $x \in X$ and proceed locally on X. There is a (local) embedding $X \subset \mathbb{C}^n$ and a linear projection $\gamma\colon \mathbb{C}^n \to \mathbb{C}^2$ with $x = \gamma^{-1}(\gamma(x)) \cap X$, the Remmert-Stein projection (see [G-R2], Satz IV., p. 256). This γ makes X (locally) into a covering of some domain $B \subset \mathbb{C}^2$ ramified over some curve $C \subset B$. Singularities of X will lie over singularities of B only (Theorem 5.2). By Theorem II.7.2 we can choose a bimeromorphic map $\beta\colon \bar{B} \to B$ such that $\beta^{-1}(C) \subset \bar{B}$ has normal crossings only. Forming $X \times_B \bar{B}$ and (if necessary) normalising this surface, we obtain a surface X_1 which is provided with two proper projections, a bimeromorphic one onto X and a second one onto $\overline{B}$, which is a covering with ramification locus contained in $\beta^{-1}(C)$. This shows that the only singularities of X_1 are those which appear on coverings of $\mathbb{C}^2$ ramified over the coordinate axes. By Theorem 5.2 these singularities can be resolved by Hirzebruch-Jung strings. □

A resolution of singularities $\pi: Y \to X$ is called *minimal*, if π does not contract any (-1)-curve in Y.

(6.2) **Theorem.** *Every normal surface X admits a minimal resolution of singularities which is determined uniquely by X.*

Proof. The existence of a minimal resolution is shown as in the proof of Theorem 4.5: you start with some resolution, the existence of which is assured by Theorem 6.1, then you blow down all (-1)-curves contained in the exceptional curves and repeat this process if necessary.

Now assume that there are two minimal resolutions $\mu_i: M_i \to X$, $i=1,2$. Let $M \subset M_1 \times_X M_2$ be the 2-dimensional component mapped bimeromorphically onto X. After normalising and resolving the singularities of M we obtain a nonsingular surface $\overline{M}$. The maps $\overline{M} \to M_i$, $i=1,2$, are bimeromorphic, and by Corollary 4.4 both consist of successions of σ-processes. So M_1 and M_2 are isomorphic, unless at some stage of blowing down $\overline{M}$ to M_1, say, we would meet an exceptional curve E containing two intersecting (-1)-curves C_1 and C_2. But this would imply $(C_1+C_2)^2 \geq 0$, contradicting Theorem 2.1. □

(6.3) **Theorem** (Structure theorem for bimeromorphic transformations). *Let $\tau: X \to Y$ be a bimeromorphic transformation of non-singular surfaces. Then τ is the composition of σ-processes (in both directions), i.e., there is a non-singular surface Z, two maps $\pi_1: Z \to X$, $\pi_2: Z \to Y$ which are compositions of σ-processes such that $\pi = \pi_2 \circ \pi_1^{-1}$.*

Proof. By definition, there is a normal surface Z' such that $\tau = \pi'_2 \circ (\pi'_1)^{-1}$ with $\pi'_1: Z' \to X$, $\pi'_2: Z' \to Y$ bimeromorphic. If Z is the desingularisation of Z' and π_1, π_2 the maps induced by π'_1, π'_2, then by Corollary 4.4 both maps π_1 and π_2 are compositions of σ-processes. □

Theorem I.9.1, (iii) and (viii) yield

(6.4) **Corollary.** *The plurigenera P_n and the irregularity q of smooth, compact, connected surfaces are invariant under bimeromorphic transformations.*

7. Singularities of Double Coverings, Simple Singularities of Surfaces

Let X be a normal surface and Y a smooth surface, both connected, and let $X \to Y$ be a double covering, ramified over $B \subset Y$. Singularities of X (lying over the singularities of B) can be resolved by the method used in the proof of Theorem 6.1. But sometimes a variation, the *canonical resolution*, is more economical, though in general it does not lead to a minimal resolution. It goes as follows. Let μ_y be the multiplicity of $y \in B$, and let $\sigma_1: Y_1 \to Y$ be the σ-process simultaneously applied in all singularities $y \in B$. If $E_y = \sigma_1^{-1}(y)$ is the exceptional curve over y, then $\sigma^*(B) = \bar{B} + \sum \mu_y E_y$ is the total transform of B. Let X_1 be the normalisation of $X \times_Y Y_1$. It is a double covering of Y_1, ramified over

$$B_1 = \bar{B} + \sum_{\mu_y \text{odd}} E_y.$$

Unless B_1 is nonsingular, we repeat this construction, and so on. Since the new ramification curves $B_1, B_2, \ldots$, are contained in the total transforms of B, Theorem II, 7.2 implies that after finitely many steps (locally with respect to Y) we arrive at a ramification curve B_k with at worst nodes. So all singularities of B_k have multiplicities $\mu_y = 2$, hence $B_{k+1} = \bar{B}_k$ is a nonsingular curve, and X_{k+1} is a resolution of the singularities of X.

We demonstrate this method by resolving *simple surface singularities*. These are the singularities of double coverings branched over a curve B having an *A-D-E* singularity. In II, Sect. 8, we have given explicit equations for simple curve singularities. Thus we obtain the following explicit equations for simple surface singularities:

$$\begin{array}{ll} A_n (n \geq 1): & w^2 + x^2 + y^{n+1} = 0 \\ D_n (n \geq 4): & w^2 + y(x^2 + y^{n-2}) = 0 \\ E_6: & w^2 + x^3 + y^4 = 0 \\ E_7: & w^2 + x(x^2 + y^3) = 0 \\ E_8: & w^2 + x^3 + y^5 = 0. \end{array}$$

(Here the covering is $(w, x, y) \mapsto (x, y)$ and the singularity lies in the origin.)

The result is shown in Table 1. The total transform of B is given at the moment when the canonical resolution stops, and the exceptional curve is shown in the covering lying over it. Straight lines are symbols for copies of $\mathbb{P}_1$, mapped biregularly onto their images, the little fat curves form the proper transform of B, and the symbol

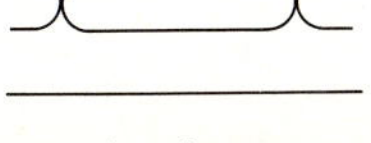

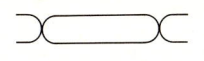

means a projective line lying over its image ramified in two points. Numbers without brackets are multiplicities and numbers within brackets denote selfintersections.

Table 1 shows that all the irreducible components of the exceptional curve in the covering surface are (-2)-curves. So the canonical resolution in these cases is minimal. By inspection one even finds:

(7.1) **Theorem.** *A simple surface singularity is resolved by an exceptional A-D-E curve of the corresponding type.*

As we shall presently show (Theorem 7.3), the converse of this theorem is also true. In other words, *A-D-E* surface singularities and simple surface singularities are one and the same thing. But first some preparations.

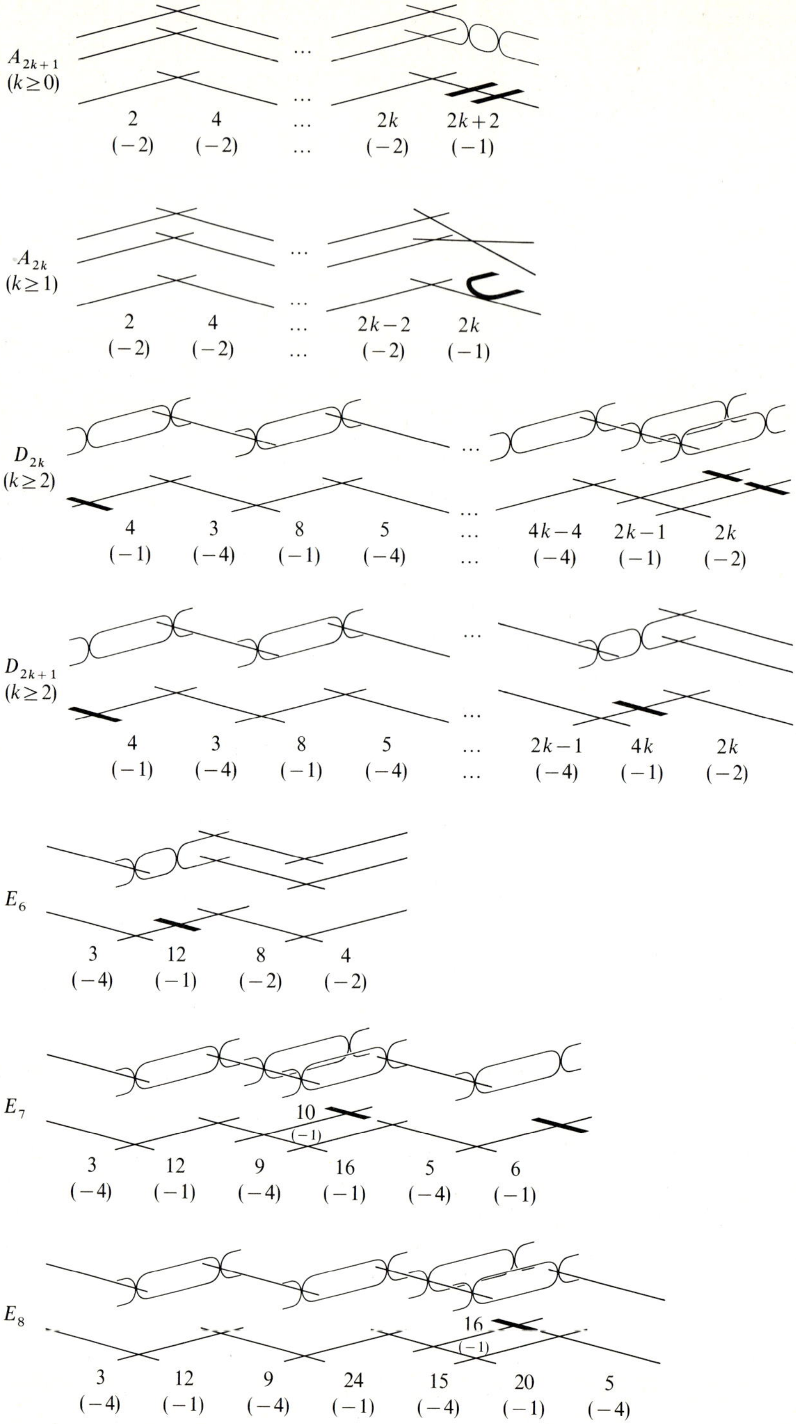
A_{2k+1} ($k \geq 0$)
2 (−2) 4 (−2) … 2k (−2) 2k+2 (−1)
A_{2k} ($k \geq 1$)
2 (−2) 4 (−2) … 2k−2 (−2) 2k (−1)
D_{2k} ($k \geq 2$)
4 (−1) 3 (−4) 8 (−1) 5 (−4) … 4k−4 (−4) 2k−1 (−1) 2k (−2)
D_{2k+1} ($k \geq 2$)
4 (−1) 3 (−4) 8 (−1) 5 (−4) … 2k−1 (−4) 4k (−1) 2k (−2)
E_6
3 (−4) 12 (−1) 8 (−2) 4 (−2)
E_7
10 (−1)
3 (−4) 12 (−1) 9 (−4) 16 (−1) 5 (−4) 6 (−1)
E_8
16 (−1)
3 (−4) 12 (−1) 9 (−4) 24 (−1) 15 (−4) 20 (−1) 5 (−4)

Table 1

(7.2) **Theorem.** *Let* $\gamma\colon X \to Y$ *be a double covering with* X *normal and* Y *nonsingular, ramified over the (reduced) curve* $B \subset Y$. *Let* $\mathscr{L}$ *be the line bundle on* Y, *satisfying* $\mathscr{L}^{\otimes 2} = \mathscr{O}_Y(B)$, *which determines the covering as in I, Sect. 17. Consider the canonical resolution diagram*

$$\begin{array}{ccc} \bar{X} & \xrightarrow{\pi} & X \\ {\scriptstyle\bar{\gamma}}\downarrow & & \downarrow{\scriptstyle\gamma} \\ \bar{Y} & \xrightarrow{\tau} & Y \end{array}$$

where τ *is a sequence of* σ*-processes and* $\bar{X}$ *is nonsingular. Then there is a divisor* $Z \geq 0$ *on* $\bar{X}$, *with* $\operatorname{supp}(Z)$ *contained in the union of the exceptional curves for* π *such that*

$$\mathscr{K}_{\bar{X}} = (\gamma\pi)^*(\mathscr{K}_Y \otimes \mathscr{L}) \otimes \mathscr{O}_{\bar{X}}(-Z). \tag{7}$$

Furthermore, $Z=0$ *if and only if the singularities of* B *(hence of* X*) are simple.*

Proof. Let $\tau = \sigma\tau'$, where $\sigma\colon Y_1 \to Y$ is the σ-transform in one singularity $y \in B$. Since the covering $X \to Y$ is obtained as inverse image of the canonical section in $\mathscr{O}_Y(B)$ under the squaring map $\mathscr{L} \to \mathscr{O}_Y(B)$, the normalised covering $X_1 \to Y_1$ is obtained from the canonical section in $\mathscr{O}_{Y_1}(B_1)$ under the squaring map $\mathscr{L}_1 \to \mathscr{O}_{Y_1}(B_1)$, where B_1 is either $\bar{B}$ or $\bar{B} + E_y$ and

$$\mathscr{L}_1 = \sigma^*(\mathscr{L}) \otimes \mathscr{O}_{Y_1}\left(-\left[\frac{\mu_y}{2}\right] E_y\right).$$

So we have

$$\mathscr{K}_{Y_1} \otimes \mathscr{L}_1 = \sigma^*(\mathscr{K}_Y \otimes \mathscr{L}) \otimes \mathscr{O}_{Y_1}\left(\left(1 - \left[\frac{\mu_y}{2}\right]\right) E_y\right).$$

Now the adjunction formula for coverings (I, Sect. 16) shows $\mathscr{K}_{\bar{X}} = \bar{\gamma}^*(\mathscr{K}_{\bar{Y}} \otimes \bar{\mathscr{L}})$, where $\bar{\mathscr{L}}$ is the line bundle on $\bar{Y}$ satisfying $\bar{\mathscr{L}}^{\otimes 2} = \mathscr{O}_{\bar{Y}}(\bar{B})$ and determining the covering $\bar{\gamma}$. By repeating the blowing up procedure above we see that $\mathscr{K}_{\bar{Y}} \otimes \bar{\mathscr{L}} = \tau^*(\mathscr{K}_Y \otimes \mathscr{L}) \otimes \mathscr{O}_Y(-\bar{Z})$ with $\bar{Z} \geq 0$ a divisor on $\bar{Y}$ contained in the union of exceptional curves. Putting $Z = (\bar{\gamma})^*(\bar{Z})$ we obtain formula (7).

Furthermore, $Z=0$ if and only if $1 - \left[\frac{\mu_y}{2}\right]$ vanishes for all singularities of all the ramification curves $B, B_1, B_2, \ldots$, i.e., if all these singularities are only double or triple points. By II, Sect. 8 this holds if all the singularities of B are simple. Conversely, if a singularity $y \in B$ is a double point, it is of type A_n, by Theorem II.8.1 hence simple. If it is a triple point with two different tangents, it is simple of type D_n. If it is a triple point with one tangent, and if $B_1 = \bar{B} + E_y$ has at most triple points, then the singularity is simple of type E_6, E_7, or E_8. □

(7.3) **Theorem.** *An exceptional A-D-E curve contracts to a simple surface singularity with the corresponding equation. In particular, the singularity (and the embedding of the exceptional curve) is determined uniquely by the corresponding dual graph.*

Proof. Let $x \in X$ be the singularity in question and $\gamma\colon X \to Y$ the double covering constructed in Lemma 3.8. By Proposition 3.4 and Theorem 7.2, the ramification curve $B \subset Y$ has a simple singularity at $y = \gamma(x)$.

Fibrations of Surfaces

In Sects. 8–18, X will always be a connected smooth surface (not necessarily compact), S a smooth connected curve and $f\colon X \to S$ a proper surjective holomorphic map. Unless stated otherwise, the map f is also assumed to be connected. Sometimes we write $X \to S$ instead of $f\colon X \to S$.

8. Generalities on Fibrations

A point $x \in X$ is a *critical point* of f, if $df = 0$ at x. The critical points form an analytic set. By Remmert's Theorem I.8.4, the *critical values* of f (i.e., the images $f(x) \in S$ of critical points $x \in X$) form in S an analytic subset of dimension 0, i.e., a discrete subset. If $s \in S$, with $\mathfrak{m}_s \subset \mathcal{O}_S$ its maximal ideal, then the fibre X_s is the curve $f^{-1}(s)$ on X with sheaf of ideals $f^*(\mathfrak{m}_s)$. This fibre is singular, if and only if s is a critical value. So almost all fibres are smooth. All smooth fibres are diffeomorphic, hence they all have the same genus.

If S is not compact, $H^1(\mathcal{O}_S) = H^2(S, \mathbb{Z}) = 0$, so the bundle $\mathcal{O}_S(s)$, $s \in S$, is trivial. Hence $\mathcal{O}_X(X_s)$ is trivial too. We have

(8.1) **Lemma.** *Unless X is compact, all line bundles $\mathcal{O}_X(X_s)$, $s \in S$, are trivial. Whether X is compact or not, the normal bundle $\mathcal{O}_{X_s}(X_s)$ of each fibre is trivial.*

(8.2) **Lemma** (Zariski's lemma). *Let $X_s = \sum n_i C_i$, $n_i > 0$, $C_i \subset X$ irreducible, be a fibre of the fibration $X \to S$. Then we have*

(8) $C_i X_s = 0$ *for all i.*

(9) *If $D = \sum m_i C_i$, $m_i \in \mathbb{Z}$, then $D^2 \leq 0$.*

(10) $D^2 = 0$ *holds in* (9) *if and only if $D = r X_s$, $r \in \mathbb{Q}$ (i.e., $pD = qX_s$ with $p, q \in \mathbb{Z}$, $p \neq 0$).*

Proof. Property (8) follows from Lemma 8.1. Consider the subvectorspace $V \subset H_2(X, \mathbb{Q})$, spanned by the classes of the curves C_i. The class of X_s belongs to the annihilator of the bilinear form on V, induced by the intersection product on X. Furthermore, let W be the abstract vector space $\bigoplus \mathbb{Q}\, C_i$. The canonical homomorphism from W onto V induces a bilinear form Q on W. We can apply Lemma I.2.10 to $-Q$: conditions (i′) and (ii) are obviously satisfied, whereas (iii) is true since X_s is connected. It follows that Q is semi-negative definite and that $\sum n_i C_i$ spans its annihilator. Thus we obtain (9) and (10). □

A singular fibre $X_s=\sum n_i C_i$ is called *multiple fibre (of multiplicity n)* if $n = \text{g.c.d.}\{n_i\} > 1$. Then $X_s = n \cdot F$ with F another effective divisor on X. Lemma 8.2 shows that F is 1-connected in the sense of Ramanujam (II, Sect. 12).

(8.3) **Lemma.** *Let $S=\Delta\subset\mathbb{C}$ be the unit disc and $X_0=n\cdot F$ a multiple fibre with multiplicity n. Then $\mathcal{O}_X(F)$ on X and $\mathcal{O}_F(F)$ are both torsion bundles of order n.*

Proof. Since $\mathcal{O}_X(nF)=\mathcal{O}_X(X_0)$ is the trivial bundle, the order of $\mathcal{O}_X(F)$ is finite and does not exceed n. If $\mathcal{O}_X(F)$ would be of order $<n$ then there would be some holomorphic function on X vanishing along X_0 of lower order than $z\circ f$, z being the coordinate function on Δ. But this is impossible, because any function on X is the pull-back of some function on Δ.

To prove that $\operatorname{ord}\mathcal{O}_F(F)$ is also n, we first shrink Δ a little bit (compare Theorem I, 8.8) so that the restriction $H^i(X,\mathbb{Z})\to H^i(F,\mathbb{Z})$ is bijective, $i=1,2$. Then we consider the exponential diagram

$$\begin{array}{ccccccc}
H^1(X,\mathbb{Z}) & \longrightarrow & H^1(\mathcal{O}_X) & \longrightarrow & H^1(\mathcal{O}_X^*) & \longrightarrow & H^2(X,\mathbb{Z}) \\
\| & & \downarrow & & \downarrow & & \| \\
H^1(F,\mathbb{Z}) & \longrightarrow & H^1(\mathcal{O}_F) & \longrightarrow & H^1(\mathcal{O}_F^*) & \longrightarrow & H^2(F,\mathbb{Z}). \\
 & & & & & & \| \\
 & & & & & & \mathbb{Z}^N
\end{array}$$

Since $\mathcal{O}_X(F)\in H^1(\mathcal{O}_X^*)$ has finite order and since $H^2(X,\mathbb{Z})$ is free of torsion, there is a pre-image $\xi\in H^1(\mathcal{O}_X)$ of $\mathcal{O}_X(F)$. If m is the order of $\mathcal{O}_F(F)$, then $m|n$, and there is some class $c\in H^1(X,\mathbb{Z})$ mapped onto $m\cdot(\xi|F)$. So $\frac{n}{m}c$ and $n\xi$ both have the same image in $H^1(\mathcal{O}_F)$. Since the map $H^1(F,\mathbb{Z})\to H^1(\mathcal{O}_F)$ is injective by Proposition II.2.1, the class $m\xi$ equals the image of c in $H^1(\mathcal{O}_X)$, i.e., $\mathcal{O}_X(F)^{\otimes m}=0$. So $m=n$. □

Let f: $X\to S$, g: $Y\to S$ be two (connected) fibrations. We call them *bimeromorphically equivalent*, if there exists a bimeromorphic correspondence between X and Y respecting the fibrations, i.e., if there is a surface Z and bimeromorphic maps ξ: $Z\to X$, η: $Z\to Y$ such that $f\circ\xi=g\circ\eta$.

(8.4) **Proposition.** *If the genus of its general fibre is strictly positive, then $X\to S$ factors over a unique nonsingular surface Y containing in none of its fibres a (-1)-curve.*

Proof. We blow down all (-1)-curves contained in fibres X_s, and repeat this until a surface Y is obtained containing no more such curves (at least locally, this stage is reached after a finite number of blowing downs, and this is enough). This Y is uniquely determined, unless some fibre X_s contains two intersecting (-1)-curves C_1, C_2. Then $(C_1+C_2)^2\geq 0$ and $(C_1+C_2)^2=0$ only if

$(C_1, C_2)=1$. By Lemma 8.2 (10), the fibre X_s containing both C_1 and C_2 must be of the form $n(C_1+C_2)$. Then for the (connected) general fibre X_t we have

$$(\mathcal{K}_X, X_t)=n(\mathcal{K}_X, C_1+C_2)=-2n.$$

So $n=1$ and the general fibre is rational. □

A fibration which has no (-1)-curves in any of its fibres is called *relatively minimal.*

(8.5) **Proposition.** *Let $f\colon X\to S$, $g\colon Y\to S$ (with also Y smooth, connected and g connected) be relatively minimal. Furthermore let $s\in S$, and $C\subset X_s$, $D\subset Y_s$ be curves with $X_s\setminus C\neq\varnothing$ and $Y_s\setminus D\neq\varnothing$. Then the existence of a fibre-preserving biholomorphic map $X\setminus C\to Y\setminus D$ implies that the fibrations f and g are biholomorphically equivalent. If the relative minimality is dropped, then the conclusion is that the fibrations are bimeromorphically equivalent.*

Proof. By Lemma 8.2 (10), the curves C and D are exceptional. Let $(X, C)\to(\bar{X}, x)$ and $(Y, D)\to(\bar{Y}, y)$ be the maps contracting them, and let $\varphi\colon \bar{X}\setminus\{x\}\to\bar{Y}\setminus\{y\}$ be the induced map. Then $\lim_{\nu\to\infty}\varphi(x_\nu)=y$ for each sequence $(x_\nu)_{\nu=1}^\infty$ in $\bar{X}\setminus\{x\}$ converging to x. So φ extends continuously and for every open neighbourhood V of y in $\bar{Y}$ there exists an open neighbourhood U of x in $\bar{X}$ with $\varphi(U\setminus x)\subset V$. We may take V so small that it is an analytic subvariety of some open ball in a $\mathbb{C}^N$. Since x is normal, the coordinate functions of φ extend holomorphically over x and so φ extends to a biholomorphic map from $\bar{X}$ onto $\bar{Y}$. The assertion now follows from Theorem 6.2.

9. The n-th Root Fibration

Let $f\colon X\to\Delta$ be a fibration over the unit disc Δ. We define its n-th root as follows. Let $\delta_n\colon\Delta\to\Delta$ be the map $z\to z^n$, $X'=\Delta\times_\Delta X$ the fibre product with respect to δ_n, X'' the normalisation of X', and $X^{(n)}$ the minimal desingularisation of X''. Then there is a diagram

(11)
$$\begin{array}{ccccccc}
 & & & \overset{\tau^{(n)}}{\frown} & & & \\
X^{(n)} & \xrightarrow{\tau''} & X'' & \xrightarrow{\tau'} & X' & \xrightarrow{\tau} & X \\
\downarrow{\scriptstyle f^{(n)}} & & \downarrow{\scriptstyle f''} & & \downarrow{\scriptstyle f'} & & \downarrow{\scriptstyle f} \\
\Delta & = & \Delta & = & \Delta & \xrightarrow{\delta_n} & \Delta
\end{array}$$

and we call $f^{(n)}\colon X^{(n)}\to\Delta$ the *n-th root fibration of f.*

Remark. The smooth surface $X^{(n)}$ does not contain any (-1)-curves over the singular points of X'', but in general the singular fibres of $f^{(n)}$ do contain (-1)-curves.

(9.1) **Proposition.** *Let X_0 be a multiple fibre of multiplicity n as in Lemma 8.3. Then X'' is nonsingular, so $X^{(n)}=X''$. Furthermore $\tau^{(n)}=\tau\tau'$: $X^{(n)}\to X$ is an unramified covering, and $f^{(n)}$: $X^{(n)}\to\Delta$ has no multiple fibre over $0\in\Delta$.*

Proof. If f is locally given by the function g^n (with g in general reducible), then a local equation for $X'\subset\Delta\times X$ is

$$z^n-g^n=(z-g)(z-\varepsilon g)\cdot\ldots\cdot(z-\varepsilon^{n-1}g)=0$$

with $\varepsilon=\mathfrak{e}\left(\frac{1}{n}\right)$. The normalisation X'' decomposes into n pieces with local equations $z=g,\ldots,z=\varepsilon^{n-1}g$. So X'' is nonsingular and the multiplicities n_i'' of the components of X_0'' are n_i/n, where n_i is the multiplicity of the corresponding component in X_0.

(9.2) **Proposition.** *Let $X_0=\sum n_i C_i$ with C_i irreducible and such that $(X_0)_{\mathrm{red}}=\sum C_i$ has only normal crossings. If n is chosen such that $n_i|n$ for all i, then $(X^{(n)})_0$ is reduced and has only normal crossings.*

Proof. Every $x\in X_0$ is the centre of a local coordinate system (z_1,z_2) with $f=z_1^a z_2^b$. Let us first assume that x is a regular point of $(X_0)_{\mathrm{red}}$. Then, say, $b=0$ and f is given by z_1^a, where a is one of the n_i's. Locally $X^{(n)}=X'$ is defined as $\{(w,z_1,z_2)\colon w^n=z_1^a\}$ with $f^{(n)}$ given by w. If we write $n=v\cdot a$, then X' decomposes into a surfaces $w^v=z_1\,\mathfrak{e}(k/a)$, $k=1,\ldots,a$. These surfaces are all nonsingular and w vanishes on X_0' to the first order.

In the general case, i.e., if $ab\neq 0$, we put $d=\text{g.c.d.}(a,b)$, $a=\alpha d$, $b=\beta d$, $n=r\alpha\beta d$. Then X', defined by $z^n=z_1^a z_2^b$, decomposes into d copies of the existence domain of $w=\sqrt[r\alpha\beta]{z_1^\alpha z_2^\beta}$. In X'' these copies become separated and f'' is given by w. Following the procedure of Sect. 5, ii), one finds that X'' is locally isomorphic with the existence domain of $w=\sqrt[r]{y_1 y_2}$ where $z_1=y_1^\beta$, $z_2=y_2^\alpha$. Now Sect. 5, ii), shows that X'' has a singularity of type $A_{r,r-1}=A_{r-1}$ and w becomes the function h from Theorem 5.1. But if $q=n-1$ in (4), all e_k equal 2 and all $v_k=1$. So w vanishes on the exceptional curve only to the first order. □

10. Stable Fibrations

Let f: $X\to S$ be a fibration with $g>0$, where $g=h^1(\mathcal{O}_{X_s})$ is the genus of the general fibre X_s.

Definition. *A fibre X_s, $s\in S$, is stable, if it has the following three properties:*

i) *X_s is reduced,*

ii) *the only singularities of X_s are nodes,*

iii) *X_s contains no (-1)-curves.*

The fibration f is stable, if all fibres X_s are stable.

Remarks. a) For any irreducible component C of X_s, the conditions i) and ii) imply that $-C^2$ equals the number of intersections of C with other components of X_s. So iii) can be replaced by

iii′) *every nonsingular rational component (of a fibre) has at least two points in common with the union of the remaining components.*

b) When Deligne and Mumford ([D-M]) first defined stability of curves, instead of iii′) they imposed the stronger condition that each nonsingular rational component should have at least three points in common with the union of the other components. In other words: X_s should not contain (-2)-curves. Of course one can achieve this by blowing down exceptional curves in X_s consisting of (-2)-curves. Since this produces singularities on X, we prefer to work with the above definition. Our stable curves usually are called semistable, but for the sake of brevity we drop the prefix "semi".

Example. The only stable fibres with $g(X_s)=1$ are of the type I_b, $b \geq 0$, where following Kodaira in [Ko2], part II we define

I_0: non-singular elliptic,
I_1: irreducible rational with one node,
I_b: cycle of b different (-2)-curves $(b \geq 2)$.

This is an easy consequence of II, (17)

$$g(C)+g(X_s-C)+C(X_s-C)-1=g(X_s)=1$$

for each irreducible component $C \subset X_s$.

The following results are crucial in the sense that they help to reduce the study of fibrations in general to the study of stable fibrations, which are much easier to handle.

(10.1) **Theorem** (Local stable reduction theorem). *Let $f: X \to \Delta$ be such that $X_0 = \sum n_i C_i$, the only singular fibre, has a reduction $(X_0)_{\text{red}} = \sum C_i$ with at worst nodes. If $n \in \mathbb{N}$ is a multiple of all n_i, then, after blowing down the (-1)-curves contained in its fibres, the n-th root fibration $f^{(n)}: X^{(n)} \to \Delta$ is a stable fibration.*

This theorem follows at once from Proposition 9.2, because blowing down a (-1)-curve in $(X^{(n)})_0$ does not affect the remaining components of $(X^{(n)})_0$.

Theorem 10.1 makes it possible to define for every fibration $f: X \to \Delta$ and arbitrarily big $n \in \mathbb{N}$ a *stable reduction*. This is a diagram

(12)
$$\begin{array}{ccc}
\bar{Y} & \longrightarrow & \bar{X} \\
\downarrow & & \downarrow \\
Y & & X \\
{\scriptstyle g}\downarrow & & \downarrow{\scriptstyle f} \\
\Delta & \xrightarrow{\delta_n} & \Delta
\end{array}
\qquad \bar{g}: \bar{Y} \to \Delta, \quad \bar{f}: \bar{X} \to \Delta$$

where
- the map $\overline{X} \to X$ is bimeromorphic and resolves $(X_0)_{\text{red}}$ to a curve $(\overline{X}_0)_{\text{red}}$ with only nodes (always possible by Theorem II.7.2),
- n is some multiple of all multiplicities of components in $\overline{X}_0$,
- $\bar{g}$ is the n-th root fibration for $\bar{f}$ (Proposition 9.2),
- $\overline{Y} \to Y$ blows down all (-1)-curves in the fibres of $\bar{g}$,
- $g: Y \to \Delta$ is stable.

Morally, X is the quotient $Y/\mathbb{Z}_n$, but unfortunately this holds only bimeromorphically:

(10.2) Proposition. *In the situation* (12), *the cyclic group* $\mathbb{Z}_n$ *operates on* Y *covering its action* $z \mapsto \mathfrak{e}(1/n) z$ *on* Δ. *The quotient* $Y/\mathbb{Z}_n$ *is bimeromorphically equivalent with* X.

Proof. If $X' = \Delta \times_\Delta \overline{X}$ with respect to $\delta: \Delta \to \Delta$, $z \mapsto z^n$, and X'' is its normalisation as in (11), then $\mathbb{Z}_n$ acts on X'' compatibly with the fibrations and such that the quotient is $\overline{X}$. This action lifts to $\overline{Y}$ and as a consequence of Proposition 8.5, $\overline{Y}/\mathbb{Z}_n$ is bimeromorphically equivalent with $\overline{X}$. The $\mathbb{Z}_n$-operation descends to Y, the relatively minimal model of $\overline{Y}$, and $Y/\mathbb{Z}_n$ is bimeromorphically equivalent with $\overline{Y}/\mathbb{Z}_n$. □

So diagram (12) can be extended as to contain $\widetilde{Y/\mathbb{Z}_n}$ in the following way ($\widetilde{Y/\mathbb{Z}_n}$ denotes the desingularisation of $Y/\mathbb{Z}_n$):

(13)
$$\begin{array}{ccccc}
\overline{Y} & \longrightarrow & & & \overline{X} \\
\downarrow & & \widetilde{Y/\mathbb{Z}_n} & & \downarrow \\
& \nearrow & & \searrow & \\
Y & \longrightarrow & Y/\mathbb{Z}_n & & X \\
\downarrow g & & \downarrow & & \downarrow f \\
\Delta & \xrightarrow{\delta_n} & \Delta & = & \Delta.
\end{array}$$

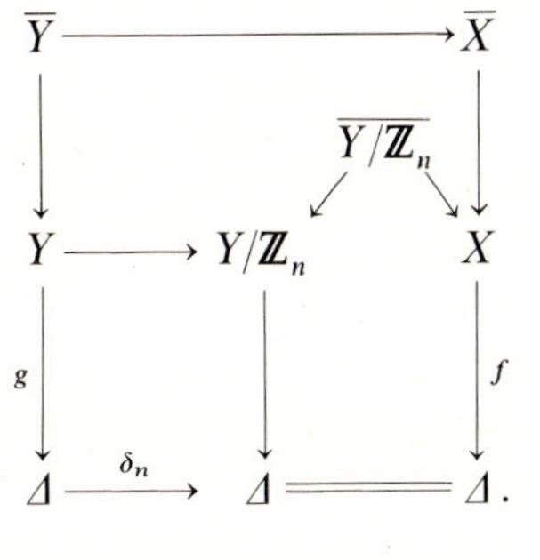

Finally, it is easy to patch local stable reductions together to obtain (compare [D-M]):

(10.3) Theorem (Stable reduction theorem). *Consider a fibration* $f: X \to S$, *with* S *compact. Let* $\overline{X} \to X$ *be a bimeromorphic map resolving all fibres* X_s *to fibres* $\overline{X}_s$ *with* $(\overline{X}_s)_{\text{red}}$ *having at worst nodes. Then there exists a cyclic covering* $\delta: T \to S$, *ramified only over the critical values* $s_1, \ldots, s_k \in S$ *of* f *and one more point* $s_0 \in S$ *such, that* $T \times_S \overline{X}$ *is bimeromorphically equivalent to a stable fibration.*

Proof. We choose some multiple $N \in \mathbb{N}$ of all multiplicities n_i occurring as multiplicities of fibre components. Then we take $s_0 \in S \setminus \{s_1, \ldots, s_k\}$ arbitrarily. If, furthermore, $l \in \mathbb{N}$ is chosen such that $N | k + l$, then there is at least one line bundle $\mathscr{L} \in \text{Pic}(S)$ with $\mathscr{L}^{\otimes N} = \mathscr{O}_S(l s_0 + s_1 + \ldots + s_k)$. Such an $\mathscr{L}$ admits an N-

valued section T, the pre-image under $\mathscr{L}\to\mathscr{L}^{\otimes N}$, $t\mapsto t^N$, of the canonical section in $\mathcal{O}_S(ls_0+s_1+\ldots+s_k)$. This T can be interpreted as a Riemann surface, lying cyclically of order N over S. Locally at $s_1,\ldots,s_k$, the map $T\to S$ is just δ_N from Sect. 9, so the assertion follows now from the local version of the theorem. □

11. Direct Image Sheaves

Every fibration $f\colon X\to S$ of the type we consider is flat (see e.g. [*F*], p. 154), so Theorem I, 8.5 on direct image sheaves applies.

(11.1) **Lemma.** *Let $f\colon X\to S$ be a fibration as before, but not necessarily connected. Then $h^i(\mathcal{O}_{X_s})$, $i=0,1$ is independent of s. ($h^0(\mathcal{O}_{X_s})$ equals the number of components of a nonsingular fibre.)*

Proof. Firstly, let f be connected. Then $h^0(\mathcal{O}_{X_s})=1$ if s is not a critical value and $h^0(\mathcal{O}_{X_s})\geq 1$ for all $s\in S$. Assume that there is a critical value $s\in S$ with $h^0(\mathcal{O}_{X_s})>1$. Ramanujam's Lemma II.12.2 shows that the divisor X_s cannot be 1-connected. So $X_s=nF$ must be a multiple fibre with F 1-connected by Zariski's Lemma 8.2. Now consider for $1\leq v\leq n-1$ the exact sequences

$$0\to\mathcal{O}_F(-vF)\to\mathcal{O}_{(v+1)F}\to\mathcal{O}_{vF}\to 0.$$

By Lemma 8.3 all the bundles $\mathcal{O}_F(-vF)$ are non-trivial of finite order. Ramanujam's lemma shows that $h^0(\mathcal{O}_F(-vF))$ vanishes. By induction on v it follows that

$$h^0(\mathcal{O}_{X_s})=h^0(\mathcal{O}_{nF})\leq h^0(\mathcal{O}_F)$$

with $h^0(\mathcal{O}_F)=1$, because F is 1-connected. So in this case we have $h^0(\mathcal{O}_{X_s})=1$ for all $s\in S$.

If f is not connected, let $f=\gamma\circ\rho$ be its Stein factorisation, so $\gamma\colon\tilde{S}\to S$ is a ramified covering and ρ is connected. Then $h^0(\mathcal{O}_{\rho^{-1}(t)})=1$ for all $t\in\tilde{S}$. If $s\in S$ is not a ramification point, then $h^0(\mathcal{O}_{X_s})=d$, with $d=\deg(\gamma)$. If however $s\in S$ is a ramification point, say $\gamma^{-1}(s)=\{t_1,\ldots,t_k\}$ with v_k the ramification order at t_k, then $v_1+\ldots+v_k=d$ and $X_s=v_1\rho^{-1}(t_1)+\ldots+v_k\rho^{-1}(t_k)$. So it suffices to show $h^0(\mathcal{O}_{v\cdot\rho^{-1}(t)})\leq v$ for every fibre of the connected fibration ρ. But this follows as usual from the exact sequences

$$0\to\mathcal{O}_{\rho^{-1}(t)}((1-v)\rho^{-1}(t))\to\mathcal{O}_{v\cdot\rho^{-1}(t)}\to\mathcal{O}_{(v-1)\cdot\rho^{-1}(t)}\to 0,$$

because $\mathcal{O}_{\rho^{-1}(t)}(\rho^{-1}(t))=\mathcal{O}_{\rho^{-1}(t)}$ by Lemma 8.1.

So we have completed the proof for $i=0$. The case $i=1$ follows from Theorem I.8.5, (i). □

From Theorem I.8.5, (iii) we now obtain

(11.2) **Corollary.** *If $f\colon X\to S$ is a not necessarily connected fibration, then the sheaves $f_*\mathcal{O}_X$ and $f_{*1}\mathcal{O}_X$ are locally free and have the base-change property.*

(11.3) **Proposition.** *Let* $g\colon Y\to T$ *be a stable reduction of* $f\colon X\to S$ *with* $S=T/\mathbb{Z}_N$ *as in Theorem 10.3. Then, if* $\delta\colon T\to S$ *is the quotient map, there is a natural injection* $\delta^*(f_{*1}\mathcal{O}_X)\to g_{*1}\mathcal{O}_Y$. *If* δ *is ramified at* $t\in T$ *cyclically of order* n, $n|N$, *then* $(\delta^*(f_{*1}\mathcal{O}_X))_t$ *is the submodule of* $\mathbb{Z}_n$*-invariants of* $(g_{*1}\mathcal{O}_Y)_t$.

Proof. Since δ is ramified at t cyclically of order n, we can identify a suitable neighbourhood of t with Δ and also a neighbourhood of s with Δ such that locally the situation (13) occurs (with $\delta|\Delta=\delta_n$). Let $f'\colon \overline{Y/\mathbb{Z}_n}\to X\to\Delta$ and $f''\colon Y/\mathbb{Z}_n\to\Delta$ be the induced maps. By Proposition 10.2, the map $\overline{Y/\mathbb{Z}_n}\to X$ is bimeromorphic, hence a composition of σ-processes (Theorem 4.4) and so by Proposition 3.1, the natural morphism $f_{*1}\mathcal{O}_X\to f'_{*1}\mathcal{O}_{\overline{Y/\mathbb{Z}_n}}$ is bijective. The natural map $f''_{*1}\mathcal{O}_{Y/\mathbb{Z}_n}\to f'_{*1}\mathcal{O}_{\overline{Y/\mathbb{Z}_n}}$ is bijective too, because $Y/\mathbb{Z}_n$ has only cyclic quotient singularities, which are of type $A_{n,q}$ by Theorem 5.4, hence rational by Proposition 3.1. The assertion now follows from [Gk2], p. 202, Cor. of Prop. 5.2.3: $H^1(\mathcal{O}_{Y/\mathbb{Z}_n})$ is the subspace of $\mathbb{Z}_n$-invariants in $H^1(\mathcal{O}_Y)$. □

(11.4) **Proposition.** *Let* $f\colon X\to S$ *be a fibration and* X_{gen} *a non-singular fibre. Then*

i) $e(X_s)\geq e(X_{\text{gen}})$ *for all fibres* X_s;
ii) *if* X *is compact, then*

$$e(X)=e(X_{\text{gen}})\cdot e(S)+\sum_{s\in S}(e(X_s)-e(X_{\text{gen}})).$$

Proof. i) Since $h^1(\mathcal{O}_{X_s})$ is independent of s by Lemma 11.1, we have to show that $e(X_s)\geq 2-2h^1(\mathcal{O}_{X_s})$. Since $e(X_s)=e((X_s)_{\text{red}})$ and $h^1(\mathcal{O}_{X_s})\geq h^1(\mathcal{O}_{(X_s)_{\text{red}}})$, it suffices to prove that $e(C)\geq 2-2h^1(\mathcal{O}_C)$ for a connected compact reduced curve C. But for such a curve we have $H^2(C,\mathbb{Z})=\mathbb{Z}^N$, where N is the number of irreducible components, whereas $b_1(C)=\operatorname{rank} H^1(C,\mathbb{Z})\leq 2h^1(\mathcal{O}_C)$ by Proposition II.2.1.

ii) Let $s_1,\ldots,s_k\in S$ be the critical values of f. Then we have

$$\begin{aligned} e(X)&=e\left(X\setminus\bigcup_{i=1}^{k}X_{s_i}\right)+e\left(\bigcup_{i=1}^{k}X_{s_i}\right)\\ &=(e(S)-k)\,e(X_{\text{gen}})+\sum_{i=1}^{k}e(X_{s_i})\\ &=e(S)\,e(X_{\text{gen}})+\sum_{i=1}^{k}(e(X_{s_i})-e(X_{\text{gen}})). \quad \square \end{aligned}$$

(11.5) *Remark.* Assertion i) can easily be sharpened. In fact, for singular X_s one always has $e(X_s)>e(X_{\text{gen}})$ unless X_s is a multiple fibre with $(X_s)_{\text{red}}$ nonsingular elliptic.

(11.6) **Corollary.** *If* X *is a compact surface and* $f\colon X\to S$ *a fibration with fibre genus* g_1 *and base genus* g_2, *then* $e(X)\geq 4(g_1-1)(g_2-1)$.

12. Relative Duality

The differential df of $f\colon X \to S$ can be viewed as an injection of sheaves $f^*(\mathcal{K}_S) \to \Omega_X^1$. Its cokernel $\Omega_{X/S}$ is called the sheaf of *relative differentials*. In general, $\Omega_{X/S}$ is very far from being locally free. Indeed, if some fibre has a multiple component, then df vanishes along this component and $\Omega_{X/S}$ contains a torsion subsheaf with one-dimensional support. However, away from the singularities of f, there is an exact sequence *of vector bundles*

$$0 \to f^*(\mathcal{K}_S) \to \Omega_X^1 \to \Omega_{X/S} \to 0$$

inducing outside of the singularities an isomorphism between $\Omega_{X/S}$ and $\mathcal{K}_X \otimes f^*(\mathcal{K}_S^{\vee})$.

Definition. The line bundle $\omega_{X/S} = \mathcal{K}_X \otimes f^*(\mathcal{K}_S^{\vee})$ on X is called the *dualising sheaf of f*.

By definition (see II, Sect. 1) and Lemma 8.1 the restriction $\omega_{X/S}|X_s$ is isomorphic with ω_{X_s} for all $s \in S$. So by duality on X_s the dimensions $h^0(\omega_{X/S}|X_s) = h^1(\mathcal{O}_{X_s})$ and $h^1(\omega_{X/S}|X_s) = h^0(\mathcal{O}_{X_s})$ are independent of $s \in S$. Because of Theorem I.8.5, (iii) and Lemma 11.1, this proves

(12.1) **Proposition.** *The sheaves $f_*\omega_{X/S}$ and $f_{*1}\omega_{X/S}$ on S are locally free and have the base-change property.*

The following result is the heart of this section.

(12.2) **Proposition.** *There is an epimorphism* Tr: $f_{*1}\omega_{X/S} \to \mathcal{O}_S$ *which on every fibre restricts to the trace morphism* tr_{X_s} *from* II, Sect. 5. *If f is connected,* Tr *is an isomorphism.*

Proof. Let X be embedded in $X \times S$ by way of $j = (\mathrm{id}_X, f)$, and let $X' = j(X)$. Then we have a diagram

$$\begin{array}{ccccc} X & \xrightarrow{\;j\;} & X \times S & \xrightarrow{\;q\;} & X \\ & {\scriptstyle f}\searrow & \downarrow{\scriptstyle p} & & \\ & & S & & \end{array}$$

where p, q are the projections.

Tensoring the structure sequence

$$0 \to \mathcal{O}_{X \times S} \to \mathcal{O}_{X \times S}(X') \to \mathcal{O}_{X'}(X') \to 0$$

with $q^*(\mathcal{K}_X)$, and using the adjunction formula I.6.3 as well as $\mathcal{K}_{X \times S} \cong p^*(\mathcal{K}_S) \otimes q^*(\mathcal{K}_X)$, we obtain

$$0 \to q^*(\mathcal{K}_X) \to q^*(\mathcal{K}_X) \otimes \mathcal{O}_{X \times S}(X') \to \omega_{X/S} \to 0. \qquad (14)$$

Now, for any sheaf $\mathscr{F}$ on $X \times S$, let $p^c_*(\mathscr{F})$ be the direct image sheaf with compact supports in X-direction (see [Bre], p. 135). Then, by Dolbeault a germ in $p^c_{*2}(q^*(\mathscr{K}_X))_{s_0}$, $s_0 \in S$, is represented by a $\bar\partial$-closed (2,2)-form Φ on a neighbourhood of $X \times \{s_0\} \subset X \times S$. This can be written as

$$\Phi = \varphi(s) + \psi(s) \wedge d\bar{s} + \psi'(s) \wedge ds.$$

Here $\varphi(s) = \Phi | X \times \{s\}$ is a (2,2)-form and $\psi(s)$, $\psi'(s)$ are forms on $X \times \{s\}$ of type (2,1), resp. (1,2). If we denote by $\bar\partial_X$ the derivative in X-direction, then

$$\frac{\partial \varphi}{\partial \bar{s}} \wedge d\bar{s} + \bar\partial_X \psi(s) \wedge d\bar{s} + \frac{\partial \psi'}{\partial \bar{s}} \wedge d\bar{s} \wedge ds = \bar\partial \Phi = 0.$$

So

$$\begin{aligned} \frac{\partial}{\partial \bar{s}} \int_{X \times \{s\}} \Phi &= \frac{\partial}{\partial \bar{s}} \int_X \varphi(s) = \int_X \frac{\partial \varphi(s)}{\partial \bar{s}} \\ &= -\int_X \bar\partial_X \psi(s) = -\int_X d\psi(s) = 0, \end{aligned}$$

because $\psi(s)$ has compact support. This shows that the function $s \mapsto \int_{X \times \{s\}} \Phi$ is holomorphic near s_0, i.e., integration over the p-fibres defines a morphism $p^c_{*2}(q^*(\mathscr{K}_X)) \to \mathscr{O}_S$.

Now we define Tr as the composition of this morphism with the morphism

$$\delta\colon f_{*1}\, \omega_{X/S} = p^c_{*1}\, \omega_{X'/S} \to p^c_{*2}(q^* \mathscr{K}_X)$$

induced by the sequence (14). Evaluated at every $s \in S$, this Tr is the trace morphism tr_{X_s} (base-change for $\omega_{X'/S}$). So the surjectivity of Tr follows from the surjectivity of all morphisms tr_{X_s}, $s \in S$. □

Next let $\mathscr{F}$ be some locally free $\mathscr{O}_X$-sheaf. The cup-product (of classes defined in a neighbourhood of f-fibres) defines a pairing of $\mathscr{O}_S$-sheaves

$$f_*(\mathscr{F}^\vee \otimes \omega_{X/S}) \otimes_{\mathscr{O}_S} f_{*1} \mathscr{F} \to f_{*1}\, \omega_{X/S} \xrightarrow{\mathrm{Tr}} \mathscr{O}_S.$$

For every $s \in S$, there is a commutative diagram

$$\begin{array}{ccccc} f_*(\mathscr{F}^\vee \otimes \omega_{X/S})/m_s f_*(\mathscr{F}^\vee \otimes \omega_{X/S}) & \otimes_{\mathscr{O}_S} & f_{*1}\mathscr{F}/m_s f_{*1}\mathscr{F} & \to & \mathbb{C} \\ \downarrow \varphi_0 & & \downarrow \varphi_1 & & \| \\ H^0(\mathscr{F}^\vee \otimes \omega_{X_s}) & \otimes_{\mathbb{C}} & H^1(\mathscr{F}|X_s) & \to & \mathbb{C} \end{array}$$

relating this pairing of sheaves with the duality pairing on the curve X_s. The induced morphism

$$f_*(\mathscr{F}^\vee \otimes \omega_{X/S}) \to (f_{*1} \mathscr{F})^\vee \tag{15}$$

is called the *relative duality morphism.*

(12.3) **Theorem** (Relative duality theorem). *If $f\colon X \to S$ is a fibration and $\mathscr{F}$ some locally free $\mathscr{O}_X$-sheaf, then the relative duality morphism* (15) *is bijective.*

Proof. We consider the restriction morphisms φ_0, φ_1 in the commutative diagram of pairings above.

Since $\dim S=1$, we have $m_s f_*(\mathscr{F}^\vee \otimes \omega_{X/S}) = f_*(\mathscr{I}_{X_s} \cdot \mathscr{F}^\vee \otimes \omega_{X/S})$, so φ_0 is injective. We denote its image by E. It is the subspace in $H^0(\mathscr{F}^\vee \otimes \omega_{X_s})$ of sections extending to a neighbourhood of X_s and its dimension is $h^0(\mathscr{F}^\vee \otimes \omega_{X_{\mathrm{gen}}})$, where X_{gen} is a general fibre of f.

Since the fibre dimension is 1, φ_1 is surjective. We denote by $(f_{*1}\mathscr{F})_{\mathrm{tors}}$ the torsion submodule of $(f_{*1}\mathscr{F})_s$ and by $T \subset H^1(\mathscr{F}|X_s)$ its image $\varphi_1(f_{*1}\mathscr{F})_{\mathrm{tors}}$. The free quotient $(f_{*1}\mathscr{F})^{\vee\vee}$ then maps onto $H^1(\mathscr{F}|X_s)/T$, so the codimension of T is at most $h^1(\mathscr{F}|X_{\mathrm{gen}})$, which equals $h^0(\mathscr{F}^\vee \otimes \omega_{X_{\mathrm{gen}}})$ by duality on X_{gen}.

Since $f_*(\mathscr{F}^\vee \otimes \omega_{X/S}) \otimes (f_{*1}\mathscr{F})_{\mathrm{tors}}$ goes to zero in $\mathcal{O}_S$, the spaces E and T annihilate each other under the duality pairing on the curve X_s. Since this pairing is perfect by Theorem II.6.1, we find $\operatorname{codim} T = \dim E$, and the composed map

$$f_*(\mathscr{F}^\vee \otimes \omega_{X/S})/m_s f_*(\mathscr{F}^\vee \otimes \omega_{X/S}) \to E \to$$
$$\to (H^1(\mathscr{F}|X_s)/T)^\vee \to ((f_{*1}\mathscr{F})^{\vee\vee}/m_s(f_{*1}\mathscr{F})^{\vee\vee})^\vee = (f_{*1}\mathscr{F})^\vee/m_s(f_{*1}\mathscr{F})^\vee$$

is an isomorphism.

This proves that the reduction of (15) modulo m_s is bijective, and so (15) is an isomorphism too. □

The Period Map of Stable Fibrations

13. Period Matrices of Stable Curves

For a nonsingular connected compact Riemann surface F we have defined in I, Sect. 14 what we mean by a canonical basis of $H^1(X, \mathbb{Z})$ and by the corresponding period matrix. Here we generalise this to the case where F is a reduced, connected, compact curve with no other singularities but ordinary double points.

Let $\nu\colon \tilde{F} \to F$ be the normalisation. We put $g = h^1(\mathcal{O}_F)$ and $\tilde{g} = h^1(\mathcal{O}_{\tilde{F}})$, the sum of the genera of the irreducible components $\tilde{F}_k \subset \tilde{F}$. For all $\alpha, \beta \in H^1(F, \mathbb{Z})$ we can define a product $(\alpha, \beta) \in \mathbb{Z}$ by mapping

$$\alpha \cup \beta \in H^2(F, \mathbb{Z}) \simeq \bigoplus_k H^2(\tilde{F}_k, \mathbb{Z}) \simeq \bigoplus_k \mathbb{Z} \to \mathbb{Z} \qquad ((\varphi_k)_k \mapsto \sum \varphi_k)$$

(both isomorphisms are canonical).

(13.1) **Proposition.** *There is a basis $\alpha_1, \ldots, \alpha_{\tilde{g}}, \beta_1, \ldots, \beta_g \in H^1(F, \mathbb{Z})$ such that*

i) *the images $p\beta_1, \ldots, p\beta_g \in H^1(\mathcal{O}_F)$ span $H^1(\mathcal{O}_F)$ over $\mathbb{C}$,*

ii) $(\alpha_i, \alpha_j) = (\beta_i, \beta_j) = 0$, $(\alpha_i, \beta_j) = \delta_{ij}$.

Proof. We choose canonical bases for all components $\tilde{F}_k$. They combine to a set $\tilde{\alpha}_1, \ldots, \tilde{\alpha}_{\tilde{g}}, \tilde{\beta}_1, \ldots, \tilde{\beta}_{\tilde{g}} \in H^1(\tilde{F}, \mathbb{Z})$ with $p\tilde{\beta}_1, \ldots, p\tilde{\beta}_{\tilde{g}}$ generating $H^1(\mathcal{O}_{\tilde{F}})$ and

$$(\tilde{\alpha}_i, \tilde{\alpha}_j) = (\tilde{\beta}_i, \tilde{\beta}_j) = 0, \quad (\tilde{\alpha}_i, \tilde{\beta}_j) = \delta_{ij}. \tag{16}$$

Now the cohomology diagram from II, Sect. 2

$$\begin{array}{ccccccccccc}
0\to H^0(F,\mathbb{Z}) & \to & H^0(\tilde F,\mathbb{Z}) & \to & H^0(\Sigma) & \to & H^1(F,\mathbb{Z}) & \xrightarrow{v^*} & H^1(\tilde F,\mathbb{Z})\to 0\\
\downarrow & & \downarrow & & \downarrow & & \downarrow p & & \downarrow p\\
0\to H^0(\mathcal{O}_F) & \to & H^0(\mathcal{O}_{\tilde F}) & \to & H^0(S) & \to & H^1(\mathcal{O}_F) & \longrightarrow & H^1(\mathcal{O}_{\tilde F})\to 0
\end{array}$$

shows that there are pre-images $\alpha_1,\ldots,\alpha_{\tilde g}$, $\beta_1,\ldots,\beta_{\tilde g}\in H^1(F,\mathbb{Z})$ for the classes $\tilde\alpha_1,\ldots,\tilde\beta_{\tilde g}$. Since v^* commutes with cup products, they have the same products (16) as their images on $\tilde F$.

Since each singularity $x_i\in F$ is ordinary, we have $\Sigma_{x_i}=\mathbb{Z}$ and $S_{x_i}=\mathbb{C}$. This implies $\operatorname{rank} H^1(F,\mathbb{Z})=g+\tilde g$. So the image of $H^0(\Sigma)$ in $H^1(F,\mathbb{Z})$ will be a direct summand of rank $g-\tilde g$. We pick a basis $\beta_{\tilde g+1},\ldots,\beta_g$ for this $\mathbb{Z}$-module. Since $v^*\beta_{\tilde g+1}=\ldots=v^*\beta_g=0$, cup products with $\beta_{\tilde g+1},\ldots,\beta_g$ annihilate all of $H^1(F,\mathbb{Z})$. This proves ii) for $\alpha_1,\ldots,\beta_g$.

Obviously the image of $H^0(\Sigma)$ in $H^0(S)$ spans this $\mathbb{C}$-vector space. This implies that $p\beta_1,\ldots,p\beta_g$ span $H^1(\mathcal{O}_F)$. □

A basis as in the lemma will be called a *canonical basis* for $H^1(F,\mathbb{Z})$. We may express $p\alpha_1,\ldots,p\alpha_{\tilde g}\in H^1(\mathcal{O}_F)$ in the basis $p\beta_1,\ldots,p\beta_g$. The result is a $g\times\tilde g$ matrix Z with

$$(\alpha_1,\ldots,\alpha_{\tilde g})=-(\beta_1,\ldots,\beta_g)Z.$$

After applying v^*, the first $\tilde g\times\tilde g$ square block of Z becomes the direct sum $\tilde Z$ of the period matrices for the components $\tilde F_k$. So the Riemann period relations I, (16) hold for this block: $\tilde Z=\tilde Z^t$, $\operatorname{Im}\tilde Z>0$.

14. Topological Monodromy of Stable Fibrations

Let $f\colon X\to\Delta$ be a fibration with only one singular fibre X_0, which is reduced and has no singularities but ordinary double points. For more details of the following discussion we refer to [Le] or [La].

The pair (X,X_0) is triangulable, so X_0 is a deformation retract of arbitrarily small neighbourhoods in X (Theorem I.8.8). The function $|f|$ can be used as a Morse-function on $X\backslash X_0$. This shows that there is a deformation retraction $\pi\colon X\to X_0$. It follows that the canonical maps

$$\text{restr}\colon H^i(X,\mathbb{Z})\to H^i(X_0,\mathbb{Z})\quad\text{and}\quad H_i(X_0,\mathbb{Z})\to H_i(X,\mathbb{Z})$$

are isomorphisms. Near a double point $x_i\in X_0$ we can find local coordinates (u,v) on X such that $f=u^2+v^2$. For every $\varepsilon>0$ we can choose π such that outside of the ball $B_i=\{|u|^2+|v|^2<2\varepsilon\}$ the map $\pi|X_s$ is a diffeomorphism for all s with $|s|<\varepsilon$. For $s\neq 0$ the circle $S_i=\{(u,v)\in B_i,\ u^2+v^2=s,\ \operatorname{Im}u=\operatorname{Im}v=0\}$ is a deformation retract of $B_i\cap X_s$ (in fact $B_i\cap X_s$ is homeomorphic to $S_i\times(0,1)$). For $s=0$ this circle becomes the point x_i. The retraction π can be chosen such that it retracts S_i to x_i and induces a map $X_s\backslash\bigcup_i S_i\to X_0\backslash\bigcup x_i$, homotopic to a diffeomorphism. This explains why the circle S_i, or rather its class $e_i\in H_1(X_s,\mathbb{Z})$

(determined up to sign) is called the *vanishing cycle*. Also it becomes clear that X_s is homeomorphic to a connected sum, obtained from the components of $\tilde{X}_0$, the normalisation of X_0, in the following way. Each time X_0 has a double point x_i with inverse images y_i, $y_i' \in \tilde{X}_0$ one must remove two discs about y_i and y_i' and close up with a cylinder (compare IV, Sect. 7):

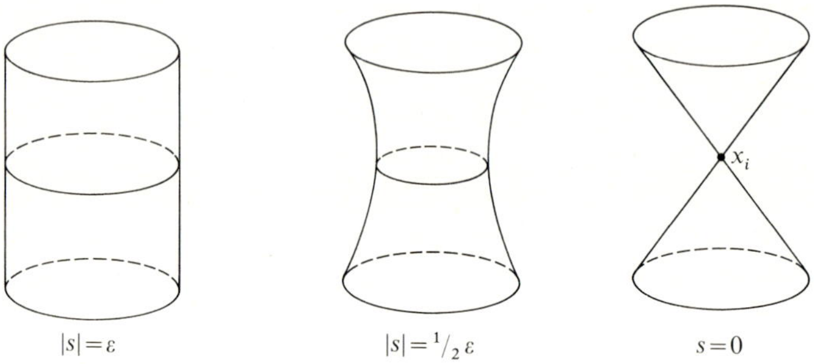

Part of X_s in B_i for various values of s

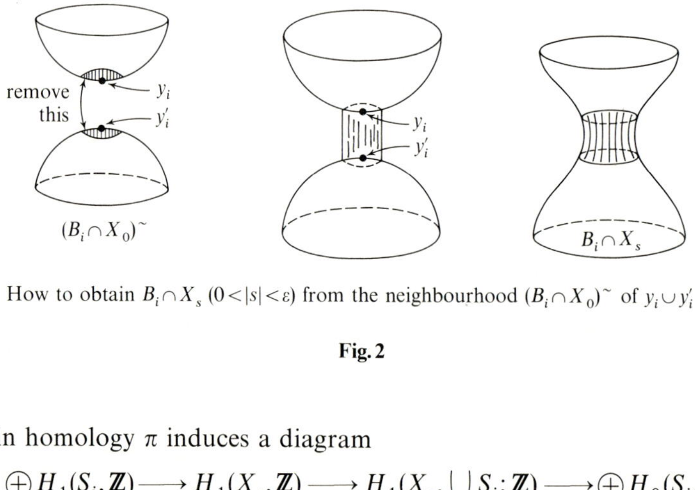

How to obtain $B_i \cap X_s$ $(0<|s|<\varepsilon)$ from the neighbourhood $(B_i \cap X_0)^{\sim}$ of $y_i \cup y_i'$ in $\tilde{X}_0$,

Fig. 2

So in homology π induces a diagram

$$\begin{array}{ccccccc}
\bigoplus_i H_1(S_i,\mathbb{Z}) & \longrightarrow & H_1(X_s,\mathbb{Z}) & \longrightarrow & H_1(X_s,\bigcup S_i;\mathbb{Z}) & \longrightarrow & \bigoplus_i H_0(S_i,\mathbb{Z}) \\
\downarrow & & \downarrow{\scriptstyle \pi_*} & & \| & & \| \\
0 & \longrightarrow & H_1(X_0,\mathbb{Z}) & \longrightarrow & H_1(X_0,\bigcup x_i;\mathbb{Z}) & \longrightarrow & \bigoplus_i H_0(x_i,\mathbb{Z}).
\end{array}$$

If we denote by $V \subset H_1(X_s,\mathbb{Z})$ the subspace spanned by the vanishing cycles, then the sequence

(17) $$0 \to V \to H_1(X_s,\mathbb{Z}) \xrightarrow{\pi_*} H_1(X_0,\mathbb{Z}) \to 0$$

is exact.

Differentiably, f is locally trivial over Δ^*. Circling the origin $0 \in \Delta$ once in counter-clockwise direction we obtain an automorphism

$$T\colon H_1(X_s, \mathbb{Z}) \to H_1(X_s, \mathbb{Z}),$$

the *Picard-Lefschetz monodromy*. It is determined by the vanishing cycles e_i by way of the classical formula (compare [P-Si], [Le], [La])

$$T(a) = a - \sum_i (a.e_i) e_i, \quad a \in H_1(X_s, \mathbb{Z}).$$

On $H^1(X_s, \mathbb{Z}) = \mathrm{Hom}_{\mathbb{Z}}(H_1(X_s, \mathbb{Z}), \mathbb{Z})$ the monodromy T operates canonically by

$$(T\alpha)(b) = \alpha(T^{-1} b), \quad \alpha \in H^1, \; b \in H_1.$$

T respects the intersection pairing on H_1 and this implies that T commutes with $P\colon H^1 \to H_1$, the inverse of Poincaré duality.

(14.1) **Theorem** (Local invariant cycle theorem). *The image of*

$$\mathrm{restr}\colon \; H^1(X, \mathbb{Z}) \to H^1(X_s, \mathbb{Z})$$

is exactly the subgroup of classes invariant under T.

Proof. The sequence (17) dualises to

$$0 \to H^1(X, \mathbb{Z}) \xrightarrow{\mathrm{restr}} H^1(X_s, \mathbb{Z}) \to \mathrm{Hom}(V, \mathbb{Z}), \tag{18}$$

and so the image of restr consists exactly of the classes $\alpha \in H^1(X_s, \mathbb{Z})$ vanishing on V.

Now let $\alpha = Pa$. From $TP = PT$ it follows that

$$\begin{aligned} T\alpha = TPa &= P(a - \textstyle\sum (a.e_i) e_i) \\ &= \alpha - P(\textstyle\sum \alpha(e_i) e_i). \end{aligned}$$

So $T\alpha = \alpha$ if and only if $\sum \alpha(e_i) e_i$ vanishes.

Let v_k be a basis for V and $e_i = \sum c_{ik} v_k$. Then the two integral matrices c_{ik} and $(\sum_i c_{ik} c_{il})$ are of maximal rank. So

$$\sum_i \alpha(e_i) e_i = \sum c_{ik} c_{il} \alpha(v_k) v_l$$

vanishes if and only if $\sum_k c_{ik} c_{il} \alpha(v_k) = 0$ for all l, i.e., $\alpha(v_k) = 0$ for all k, in other words, if α vanishes on V. □

15. Monodromy of the Period Matrix

We fix a canonical basis $\alpha_1, \ldots, \alpha_{\tilde{g}}, \beta_1, \ldots, \beta_g \in H^1(X_0, \mathbb{Z})$ in the sense of Lemma 13.1. Because of the isomorphism $H^1(X, \mathbb{Z}) \to H^1(X_0, \mathbb{Z})$ these classes can be thought of as existing on all of X. So they restrict to classes $\alpha_i(s)$, $\beta_j(s) \in H^1(X_s, \mathbb{Z})$, $s \neq 0$. The map $\pi^*\colon H^2(X_0, \mathbb{Z}) \to H^2(X_s, \mathbb{Z}) = \mathbb{Z}$ is just the map from Sect. 13. So the intersection-product formulae in Proposition 13.1, (ii) hold also for $\alpha_1(s), \ldots, \alpha_{\tilde{g}}(s)$, $\beta_1(s), \ldots, \beta_g(s)$, if $s \neq 0$. The exact sequence (18) shows that $H^1(X, \mathbb{Z})$ is a direct summand in $H^1(X_s, \mathbb{Z})$. Therefore we can find, for fixed $s \neq 0$, classes $\alpha_{\tilde{g}+1}(s), \ldots, \alpha_g(s)$ such that $\alpha_1(s), \ldots, \alpha_g(s)$, $\beta_1(s), \ldots, \beta_g(s)$ form a basis in $H^1(X_s, \mathbb{Z})$. The intersection product being unimodular on the space spanned by $\alpha_1(s), \ldots, \alpha_{\tilde{g}}(s)$, $\beta_1(s), \ldots, \beta_{\tilde{g}}(s)$, by Lemma I, 2.3 we may choose $\alpha_{\tilde{g}+1}(s), \ldots, \alpha_g(s)$ such that $(\alpha_1(s), \ldots, \alpha_g(s), \beta_1(s), \ldots, \beta_g(s))$ is a canonical basis for $H^1(X_s, \mathbb{Z})$.

On Δ we consider the following direct image sheaves:

$\mathscr{E} = f_{*1}\mathcal{O}_X$ which is locally free of rank g by Lemma 11.2, and
$L = f_{*1}\mathbb{Z}_X$.

The exponential sequence on X induces an exact sequence

$$f_*\mathcal{O}_X \xrightarrow{e} f_*\mathcal{O}_X^* \to L \xrightarrow{p} \mathscr{E}$$

making L into a subsheaf of $\mathscr{E}$. If E is the vector bundle corresponding to $\mathscr{E}$, then L defines a lattice in each $E(s)$, $s \neq 0$. The natural morphism $H^1(X, \mathbb{Z}) \to H^0(\Delta, f_{*1}\mathbb{Z}_X) = H^0(\Delta, L)$ is bijective (consider the beginning of Leray's spectral sequence), so the classes $p\alpha_1, \ldots, p\alpha_{\tilde{g}}$, $p\beta_1, \ldots, p\beta_g$ are holomorphic sections in $L \subset \mathscr{E}$. The additional classes $\alpha_{\tilde{g}+1}(s), \ldots, \alpha_g(s)$ extend locally on Δ^* to holomorphic sections in $L \subset \mathscr{E}$, but globally on Δ^* we obtain only multivalued sections, because of non-trivial monodromy.

The *period matrix* $Z(s)$, $s \in \Delta^*$, is the holomorphic multivalued matrix defined by

$$(p\alpha_1(s), \ldots, p\alpha_g(s)) = -(p\beta_1(s), \ldots, p\beta_g(s))\, Z(s). \tag{19}$$

It satisfies the Riemann period relations I, (16) in each $s \in \Delta^*$. If we write it in blocks

$$Z(s) = \begin{pmatrix} Z_{11}(s) & Z_{12}(s) \\ Z_{12}^t(s) & Z_{22}(s) \end{pmatrix} \begin{matrix} \}\, \tilde{g} \\ \}\, g-\tilde{g} \end{matrix}$$

then Z_{11} is a symmetric $\tilde{g} \times \tilde{g}$ matrix, holomorphic and univalued on all of Δ. Also Z_{12} is holomorphic and univalued on Δ, but Z_{22} changes under monodromy, say $T\colon Z_{22}(s) \to Z_{22}(s) + \Lambda$.

(15.1) **Lemma.** i) *The matrix Λ is constant with integral entries.*

ii) $\Lambda = \Lambda^t$

iii) $\det \Lambda \neq 0$.

Proof. i) Consider for $j = \tilde{g}+1, \ldots, g$ the "variation" $T\alpha_j - \alpha_j$. Since $T\beta_i = \beta_i$, we have $(T\alpha_j, \beta_i) = (\alpha_j, \beta_i)$ for all i, so $T\alpha_j - \alpha_j$ must be an integral linear combination of $\beta_1, \ldots, \beta_g$ in $H^1(X_s, \mathbb{Z})$.

ii) Z and TZ both are symmetric.

iii) If Λ were not of maximal rank $g - \tilde{g}$, there would be an integral linear combination $\alpha \neq 0$ of $\alpha_{\tilde{g}+1}(s), \ldots, \alpha_g(s)$ invariant under monodromy. By Theorem 14.1, this α would come from $H^1(X, \mathbb{Z})$ and it would thus be an integral linear combination of $\alpha_1, \ldots, \alpha_{\tilde{g}}$, $\beta_1, \ldots, \beta_g$, a contradiction! □

Next we consider on Δ^* the single-valued holomorphic matrix

$$Z'_{22}(s) = Z_{22}(s) - (2\pi\sqrt{-1})^{-1} \log s \cdot \Lambda.$$

(15.2) **Theorem.** *Z'_{22} extends holomorphically over* 0.

Proof. Im $Z_{22}(s)$ is positive definite. So for each diagonal entry z_{jj} of this matrix, we have $|\mathfrak{e}(z_{jj})|<1$, and by Riemann's extension theorem $\mathfrak{e}(z_{jj})$ extends to a holomorphic function h on all of Δ. If we write $h(s)=s^m g(s)$ with $m\in\mathbb{N}$ and $g(0)\neq 0$, then $z_{jj}-m(2\pi\sqrt{-1})^{-1}\log s$ is holomorphic and single-valued on Δ. The integer m must be the diagonal entry λ_{jj} of Λ. So the diagonal entries of Z'_{22} extend over 0.

If $i\neq j$, then applying the form $\operatorname{Im} Z_{22}$ to the difference e_i-e_j of basis vectors yields

$$\operatorname{Im}(z_{ii}+z_{jj})>2\operatorname{Im} z_{ij}.$$

Hence the preceding argument also applies to the function $z_{ii}+z_{jj}-2z_{ij}$ showing that $z_{ij}-\lambda_{ij}(2\pi\sqrt{-1})^{-1}\log s$ extends too. □

(15.3) **Corollary.** $\Lambda>0$.

Proof. $\operatorname{Im} Z_{22}(s)=\operatorname{Im} Z'_{22}(s)-\dfrac{1}{2\pi}\log|s|\cdot\Lambda$ is positive definite and $Z'_{22}(s)$ is bounded near $s=0$. So necessarily $\Lambda\geq 0$, but $\det\Lambda\neq 0$, as was already observed (Lemma 15.1). □

Finally we consider a fibration $f\colon X\to S$ *without singular fibres*. We choose a base point $0\in S$ and a canonical basis $\alpha_1,\dots,\alpha_g,\beta_1,\dots,\beta_g\in H^1(X_0,\mathbb{Z})$. If S is simply-connected, the direct image sheaf $L=f_{*1}\cdot\mathbb{Z}_X$ is trivial, and the canonical basis extends uniquely to a basis of sections in L all over S. But then the period matrix Z is globally a well-defined holomorphic map $S\to\mathfrak{H}_g$.

(15.4) **Theorem.** *Let $f\colon X\to S$ be a fibration without singular fibres for which S is compact and either rational or elliptic. Then f is locally trivial.*

Proof. If S is simply connected and compact, the period map $S\to\mathfrak{H}_g$ is constant. If S is elliptic, we still have a well-defined period map $\mathbb{C}\to\mathfrak{H}_g$ on its universal covering. Since $\mathfrak{H}_g$ is a bounded domain, this map is constant. It follows in both cases, that all curves X_s have the same period point in $\mathfrak{H}_g$, so they are biholomorphically equivalent by Torelli's theorem (compare [Mar]). The assertion then follows from Theorem I.10.1. □

Remark. If S is rational and the fibration of genus ≥ 2, then it is is even *globally* trivial.

16. Extending the Period Map

By assigning to $s\in\Delta^*$ the class of $Z(s)$ in the period domain $\mathfrak{H}_g/\Gamma_g$, we obtain a holomorphic map $\Delta^*\to\mathfrak{H}_g/\Gamma_g$, the *period map*.

Satake's compactification $\overline{\mathfrak{H}_g/\Gamma_g}$ of the period domain ([B-B]) is a normal projective variety, which is stratified by subvarieties

$$\overline{\mathfrak{H}_g/\Gamma_g} = \mathfrak{H}_g/\Gamma_g \cup \mathfrak{H}_{g-1}/\Gamma_{g-1} \cdots \mathfrak{H}_1/\Gamma_1 \cup pt.$$

To describe the topology of Satake's compactification we need the notion of *Siegel sets.* Every real symmetric matrix $Y>0$ has a unique decompositon $Y = W^t D W$ with D diagonal and

$$W = \begin{pmatrix} 1 & w_{12} & \cdots & w_{1g} \\ 0 & \ddots & \ddots & \vdots \\ \vdots & \ddots & \ddots & w_{g-1,g} \\ 0 & \cdots & 0 & 1 \end{pmatrix}.$$

It is easy to see that the entries of D and W depend continuously on Y. For real u, let us denote by $\mathfrak{H}_g(u) \subset \mathfrak{H}_g$ the set of all matrices

$$Z = X + \sqrt{-1}\, Y, \qquad Y = W^t D W$$

satisfying

$$\begin{array}{lll} |x_{ij}| < u, & |w_{ij}| < u & 1 \le i,j \le g, \\ 1 < u d_{11}, & d_{ii} < u d_{i+1,i+1} & 1 \le i \le g-1. \end{array}$$

For fixed $\tilde{g} < g$ we write

$$Z = \begin{matrix} \tilde{g}\{ \end{matrix}\!\begin{pmatrix} \overbrace{Z_{11}}^{\tilde{g}} & Z_{12} \\ Z_{12}^t & Z_{22} \end{pmatrix} \quad D = \begin{pmatrix} D_1 & 0 \\ 0 & D_2 \end{pmatrix} \quad W = \begin{pmatrix} W_1 & W_{12} \\ 0 & W_2 \end{pmatrix}.$$

A sequence converges in Satake's compactification to the point represented by $Z_0 \in \mathfrak{H}_{\tilde{g}}$, if there is a representing sequence of matrices $Z^{(\nu)} \in \mathfrak{H}_g$ such that

a) there is some u with $Z^{(\nu)} \in \mathfrak{H}_g(u)$ for all ν,
b) $Z_{11}^{(\nu)} \to Z_0$ in the usual sense,
c) $D_2^{(\nu)} \to \infty$, i.e., all entries go to infinity.

(16.1) **Theorem.** *The period map $\Delta^* \to \mathfrak{H}_g/\Gamma_g$ extends holomorphically to a map $\Delta \to \overline{\mathfrak{H}_g/\Gamma_g}$. If*

$$Z(s) = \begin{pmatrix} Z_{11}(s) & Z_{12}(s) \\ Z_{12}^t(s) & Z_{22}'(s) \end{pmatrix} + (2\pi\sqrt{-1})^{-1} \log s \begin{pmatrix} 0 & 0 \\ 0 & A \end{pmatrix},$$

then the class of $Z_{11}(0)$ in $\mathfrak{H}_{\tilde{g}}/\Gamma_{\tilde{g}} \subset \overline{\mathfrak{H}_g/\Gamma_g}$ is the image of $0 \in \Delta$ under the extended map.

Proof. Let $(2\pi\sqrt{-1})^{-1} \log s = \xi + \sqrt{-1}\,\eta$. We may restrict our attention to one branch of $\log s$, say the one given by $|\xi| \le 1$, because the other branches of $Z(s)$ represent in $\mathfrak{H}_g/\Gamma_g$ the same class. In the strip $|\xi| \le 1$, the matrix $X = \operatorname{Re} Z(s)$ is bounded and for $Y = \operatorname{Im} Z$ we have

$$Y = \begin{pmatrix} Y_{11} & Y_{12} \\ Y_{12}^t & Y_{22} \end{pmatrix} + \eta \begin{pmatrix} 0 & 0 \\ 0 & A \end{pmatrix}$$

with Y_{11}, Y_{12}, Y_{22} extending continuously for $\eta\to\infty$. In fact $Y_{11}(0)>0$, because $Z_{11}(0)$ is the direct sum of the period matrices of all components of the normalisation $\tilde{X}_0$ (see Sect. 13).

Comparing Y with

$$W^t DW=\begin{pmatrix} W_1^t D_1 W_1 & W_1^t D_1 W_{12} \\ W_{12}^t D_1 W_1 & W_{12}^t D_1 W_{12}+W_2^t D_2 W_2 \end{pmatrix}$$

we see that W_1, D_1 extend continuously for $s\to 0$ as is the case for $W_{12}=D_1^{-1}(W_1^t)^{-1}Y_{12}$. Hence $W_2^t\cdot\frac{1}{\eta}D_2\cdot W_2$ converges to Λ if $\eta\to\infty$. So W_2 is bounded and $D_2\to\infty$. Conditions b) and c) now are satisfied. To meet condition a) you just have to choose u sufficiently large. This shows that the period map extends continuously, hence holomorphically for $s\to 0$. □

17. The Degree of $f_*\omega_{X/S}$

Now let $f\colon X\to S$ be a stable fibration with S compact and connected. We cover S by open sets $U_i\cong\Delta$ on which there is a canonical basis $\alpha_1^i(s),\ldots,\beta_g^i(s)$, $s\in U_i$, as in Sect. 15. Either U_i contains no critical value, and this basis is univalued, or U_i contains one critical value and $\alpha_{\tilde{g}_i}^i(s),\ldots,\alpha_g^i(s)$ are multi-valued. In any case, the sections $p\beta_1^i,\ldots,p\beta_g^i\in\Gamma(U_i,f_{*1}\mathcal{O}_X)$ form a basis, and the (multi-valued, holomorphic) period matrix $Z^i(s)$ on U_i is defined by (19). If we denote by Λ^i the $g\times g$ monodromy-matrix (vanishing if U_i contains no critical value), then monodromy acts as (cf. Sect. 15):

$$Z^i\mapsto Z^i+\Lambda^i$$

$$(\alpha_1^i,\ldots,\beta_g^i)\mapsto(\alpha_1^i,\ldots,\beta_g^i)\begin{pmatrix}1 & 0\\ -\Lambda^i & 1\end{pmatrix}.$$

Next let $\sigma^{ji}=\begin{pmatrix}A^{ji} & B^{ji}\\ C^{ji} & D^{ji}\end{pmatrix}\in Sp(g,\mathbb{Z})$ be a matrix transforming the canonical bases over $U_i\cap U_j$ by

$$(\alpha_1^i,\ldots,\beta_g^i)=(\alpha_1^j,\ldots,\beta_g^j)\,\sigma^{ji}.$$

By I, (18) we have on $U_i\cap U_j$:

$$Z^j=(D^{ji}Z^i+C^{ji})(B^{ji}Z^i+A^{ji})^{-1}.$$

Notice that monodromy changes σ^{ji} into

$$\begin{pmatrix}1 & 0\\ \Lambda^j & 1\end{pmatrix}\sigma^{ji}\begin{pmatrix}1 & 0\\ -\Lambda^i & 1\end{pmatrix}=\begin{pmatrix}A^{ji}-B^{ji}\Lambda^i & B^{ji}\\ * & *\end{pmatrix},$$

but $B^{ji}Z^i+A^{ji}$ is unchanged. So

$$c^{ji}=\det(B^{ji}Z^i+A^{ji})\in\Gamma(U_i\cap U_j,\mathcal{O}_S)$$

is a well-defined function.

(17.1) **Proposition.** *The system* $\{c^{ij}\}$ *is a cocycle defining the line bundle* $\det(f_* \omega_{X/S})$. *I.e., a section h in this bundle is given by functions* $h^i \in \Gamma(U_i, \mathcal{O}_S)$ *satisfying* $h^i = c^{ij} h^j$.

Proof. Consider the basis $p\beta_1^i, \ldots, p\beta_g^i$ in $\Gamma(U_i, f_{*1}\mathcal{O}_X)$. Its dual basis (under relative duality) in $\Gamma(U_i, f_* \omega_{X/S})$ consists of sections $f_*\omega_1^i, \ldots, f_*\omega_g^i$ with $\omega_1^i, \ldots, \omega_g^i \in \Gamma(f^{-1}(U_i), \omega_{X/S})$. Then $c^i = f_*\omega_1^i \wedge \ldots \wedge f_*\omega_g^i$ generates $\det(f_*\omega_{X/S})$ over U_i. We have

$$\sigma^{ij} = (\sigma^{ji})^{-1} = \begin{pmatrix} D^{ji} & -C^{ji} \\ -B^{ji} & A^{ji} \end{pmatrix}^t,$$

$$(p\beta_1^j, \ldots, p\beta_g^j) = (p\alpha_1^i, \ldots, p\beta_g^i)\begin{pmatrix} B^{ij} \\ D^{ij} \end{pmatrix} = (p\beta_1^i, \ldots, p\beta_g^i)(B^{ji}Z^i + A^{ji})^t,$$

$$(f_*\omega_1^j, \ldots, f_*\omega_g^j) = (f_*\omega_1^i, \ldots, f_*\omega_g^i)\cdot(B^{ji}Z^i + A^{ji})^{-1},$$
$$c^{ji} = c^i / c^j. \quad \square$$

(17.2) **Proposition.** *Unless the period map maps S to a point in* $\mathfrak{H}_g/\Gamma_g$, *the line bundle* $\det(f_*\omega_{X/S})$ *on S has degree* >0.

Proof. We use modular forms of some weight $m > 0$ to construct a nontrivial section $h \in \Gamma(\det(f_*\omega_{X/S})^{\otimes m})$ with zeros. By definition a modular form of weight m is a holomorphic function φ on $\mathfrak{H}_g$ satisfying for all $Z \in \mathfrak{H}_g$, $\begin{pmatrix} A & B \\ C & D \end{pmatrix} \in Sp(g, \mathbb{Z})$ the relation

$$\varphi((AZ+B)(CZ+D)^{-1}) = \det(CZ+D)^m \cdot \varphi(Z).$$

(If $g=1$, it is additionally required that φ stays bounded for $\operatorname{Im} z \to \infty$.) The modular forms form a finitely generated graded ring. Its associated projective variety is just the Satake compactification $\overline{\mathfrak{H}_g/\Gamma_g}$, see [B-B]. So there is a very ample line bundle $\mathcal{O}(1)$ on $\overline{\mathfrak{H}_g/\Gamma_g}$, whose sections come from modular forms of some weight m. If $\varphi_0, \ldots, \varphi_N$ form a basis for the vector space of these forms, then the projective embedding of $\overline{\mathfrak{H}_g/\Gamma_g}$ is on $\mathfrak{H}_g/\Gamma_g$ defined by

$$Z \bmod \Gamma_g \to (\varphi_0(Z): \ldots : \varphi_N(Z)) \in \mathbb{P}_N(\mathbb{C}).$$

Assume now that the image of S in $\overline{\mathfrak{H}_g/\Gamma_g}$ has dimension one. Then φ_0 may be chosen in such a way that the hyperplane $\mathbb{P}_{N-1} \subset \mathbb{P}_N$ corresponding to $\varphi_0 = 0$ intersects the image of S in a discrete set, not only consisting of critical values for f. A nontrivial section h in $(\det f_*\omega_{X/S})^{\otimes m}$ is now constructed as follows:

If we put $h^i(s) = \varphi_0(Z^i(s))$ on U_i, then $h^i(s)$ is holomorphic for s not a critical value. Since $\begin{pmatrix} D & C \\ B & A \end{pmatrix}$ belongs to $Sp(g, \mathbb{Z})$ if $\begin{pmatrix} A & B \\ C & D \end{pmatrix}$ does, the relation between Z^j and Z^i, and the transformation property of φ_0 show that

$$h^j = \varphi_0((D^{ji}Z^i + C^{ji})(B^{ji}Z^i + A^{ji})^{-1})$$
$$= \det(B^{ji}Z^i + A^{ji})^m \varphi_0(Z^i) = (c^{ji})^m h^i.$$

So the h^i patch together and yield a section in $(\det f_*\omega_{X/S})^{\otimes m}$ over $S \setminus \{$critical values$\}$. If $s \in U_i$ converges to some critical value, we have seen in the proof of

Theorem 16.1, that we may choose $Z^i(s)$ with all its values in a Siegel set $\mathfrak{H}_g(u)$. It is a rather elementary fact, that on such a set any modular form is bounded (see [Maa], Theorem 1 on p. 185). So h^i extends holomorphically to all of U_i. □

We recall that the degree of a vector bundle on a curve is defined as the degree of its determinant bundle (II, Sect. 2).

(17.3) **Theorem.** *Let $f\colon X \to S$ be a fibration whose singular fibres are reduced with at worst nodes and do not contain (-1)-curves. Then* $\deg(f_* \omega_{X/S}) > 0$, *unless f is locally trivial (and hence all fibres of f are nonsingular and isomorphic).*

Proof. If the period map is not constant, $\deg(f_* \omega_{X/S})$ is positive by Proposition 17.2 above. So let us assume that the image of the period map $S \to \overline{\mathfrak{H}_g/\Gamma_g}$ is a point. This point, contained in $\mathfrak{H}_g/\Gamma_g$ will be the period point of all the nonsingular fibres. By Torelli's theorem (see [Mar]) all nonsingular fibres will be isomorphic. The assertion then follows from Theorem I.10.1, provided that we prove the absence of singular fibres.

So let $X_s = \sum C_i$, $s \in S$ be a singular fibre. By Theorem 16.1 it has the same period matrix as all the nonsingular fibres. If g_i is the genus of the normalisation $\tilde{C}_i$, then $\tilde{g} = \sum g_i$ equals g, the genus of a nonsingular fibre. So Theorem 11.1 shows $\tilde{g} = h^1(\mathcal{O}_{X_s})$.

The cohomology sequence of II, (5) implies that the components C_i are nonsingular and also that they form a tree. This tree cannot have a rational curve at one of its ends, because such a curve would be a (-1)-curve. So there must be at least two different irrational components C_i, and the period matrix Z of X_s decomposes as direct sum of at least two blocks. The non-singular fibres have the same period matrix, so their jacobian will be a product of at least two factors and the theta-divisor on this jacobian will be reducible. But for a nonsingular curve, the theta-divisor, an image of the $(g-1)$-fold symmetric product of the curve, is always irreducible. This contradiction proves the theorem. □

18. Iitaka's Conjecture $C_{2,1}$

The aim of this section is to prove Iitaka's conjecture $C_{2,1}$, which will appear as Theorem 18.4 below. But first some preliminaries.

(18.1) **Lemma.** *Let F be a nonsingular, compact, connected Riemann surface of genus $g > 0$. If there is a nontrivial finite subgroup $G \subset \mathrm{Aut}(F)$ operating trivially on $H^1(\mathcal{O}_F)$, then F must be elliptic and G must consist of translations.*

Proof. The quotient $Q = F/G$ is nonsingular with $H^1(\mathcal{O}_Q) \subset H^1(\mathcal{O}_F)$ the subgroup of G-invariants (compare [Gk 2]). If G acts trivially on $H^1(\mathcal{O}_F)$, then $g(Q) = g(F)$ and from Hurwitz' formula

$$2g(F) - 2 \geq |G|(2g(Q) - 2)$$

we deduce that not only both F and Q are elliptic, but also that the map $F \to Q$ is unramified. So G consists of translations only, because all other automorphisms of F have fixed points. □

This lemma and the Stable Reduction Theorem 10.3 now enable us to generalise Theorem 17.3 to the case of non-stable fibrations.

(18.2) **Theorem.** *Let X and S be compact, and $f\colon X \to S$ a relatively minimal fibration with strictly positive fibre genus. Then*

$$\deg(f_* \omega_{X/S}) \geq 0$$

and this degree vanishes if and only if

- *either f is locally trivial (hence smooth)*
- *or $g=1$ and the only singular fibres are multiples of nonsingular elliptic curves.*

Proof. Instead of $f_* \omega_{X/S}$ we consider its dual $f_{*1} \mathcal{O}_X$. We take a stable reduction

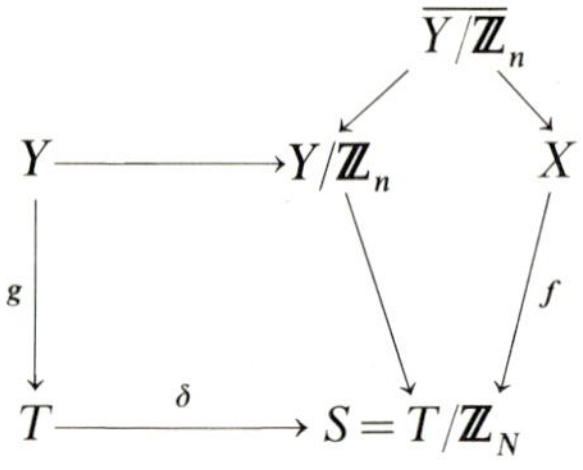

as constructed in Sect. 10. In Proposition 11.3 we observed that there is an injection of $\mathcal{O}_T$-sheaves $\iota\colon \delta^*(f_{*1}\mathcal{O}_X) \to g_{*1}\mathcal{O}_Y$ which is an isomorphism outside of the ramification points. Both sheaves are locally free of the same rank, so Theorem 17.1 shows that

$$\deg(f_* \omega_{X/S}) = -\deg(f_{*1}\mathcal{O}_X) = -\frac{1}{n}\deg \delta^*(f_{*1}\mathcal{O}_X) > 0$$

unless $\deg(g_{*1}\mathcal{O}_Y)=0$ and ι is an isomorphism. By Theorem 17.3, g is locally trivial.

Now we consider some ramification point $t \in T$ lying over a critical value $s \in S$ with local ramification group $\mathbb{Z}_\nu$, $\nu | n$. Then $\mathbb{Z}_\nu$ acts nontrivially on Y_t, because otherwise $Y/\mathbb{Z}_\nu$ were nonsingular over s, hence $X = \overline{Y/\mathbb{Z}_\nu} = Y/\mathbb{Z}_\nu$ locally trivial near X_s. Since ι is an isomorphism in t, all germs in $(g_{*1}\mathcal{O}_Y)_t$ are invariant under $\mathbb{Z}_\nu$. By base-change it follows that $\mathbb{Z}_\nu$ acts trivially on $H^1(\mathcal{O}_{Y_t})$. So by Lemma 18.1 above, Y_t is elliptic and $\mathbb{Z}_\nu$ acts by translations. This implies that X_s is a multiple of a non-singular elliptic curve. □

(18.3) **Proposition.** *If* $\deg(f_*\omega_{X/S})=0$ *in the situation of Theorem* 18.2, *then there is a finite unramified covering $\sigma\colon T \to S$ such that the pull-back $\sigma^*(f_*\omega_{X/S})$ is trivial (hence* $\det(f_*\omega_{X/S})$ *is a torsion bundle).*

Proof. If $g \geq 2$, then the automorphism group $\mathrm{Aut}(X_s)$ of the fibre is finite. So f is locally trivial with $\mathrm{Aut}(X_s)$ as structure group and the structure group of the bundle $f_* \omega_{X/S}$ can be reduced to this same group.

If $g=1$, we take a cyclic covering $\delta_1\colon T_1 \to S$ such that the pull-back $f_1\colon X_1 = X \times_S T_1 \to T_1$ is locally trivial (stable reduction). The automorphism group of the fibre is an extension

$$0 \to \mathbb{C}/\Gamma \to \mathrm{Aut}(X_s) \to \mathbb{Z}_n \to 0$$

with $\mathbb{C}/\Gamma$ operating trivially on $H^0(\omega_{X_s})$. So the structure group of $(f_1)_* \omega_{X_1/T_1} = \delta_1^*(\omega_{X/S})$ can be reduced to $\mathbb{Z}_n$, and there is an unramified n-fold covering $\delta_2\colon T_2 \to T_1$ such that the pull-back of $f_* \omega_{X/S}$ to T_2 is trivial. The line bundle $f_* \omega_{X/S}$ represents an element in the kernel of $\mathrm{Pic}^0(S) \to \mathrm{Pic}^0(T_2)$. This kernel must be a proper subgroup of $\mathrm{Pic}^0(S)$ and by Lemma I.16.2 it consists of torsion elements only. Applying the unbranched covering trick (I.18.1), the result follows. □

(18.4) **Theorem** (Iitaka's conjecture $C_{2,1}$). *Let $f\colon X \to S$ be a fibration with X minimal and compact. Then the inequality*

$$\mathrm{kod}(X_s) + \mathrm{kod}(S) \leq \mathrm{kod}(X)$$

holds for a general fibre X_s.

Proof. Without loss of generality we can assume $g = g(X_s) > 0$ and $g(S) > 0$. Then Theorem 18.2 and Proposition 18.3 leave us with the following cases:

a) $\deg(f_* \omega_{X/S}) > 0$. After passing to an unramified double covering we may assume $\deg(f_* \omega_{X/S}) \geq 2$. So by Riemann-Roch on S

$$h^0(f_* \mathcal{K}_X) - h^1(f_* \mathcal{K}_X) = \deg(f_* \omega_{X/S}) + g(g(S)-1),$$

and we conclude $h^0(\mathcal{K}_X) \geq 2$, and in particular $\mathrm{kod}(X) \geq 1$. There is a pencil $C_\lambda + F \in |K_X|$ of effective canonical divisors, with F the fixed part and C_λ, $\lambda \in \mathbb{P}_1$, variable. By Proposition 2.3 and minimality we have

$$K_X^2 = C_\lambda^2 + FC_\lambda + K_X F \geq 0.$$

Since $K_X^2 > 0$ together with $p_g(X) \geq 2$ would imply $\mathrm{kod}(X) = 2$, we may assume $K_X^2 \leq 0$, i.e., $C_\lambda^2 = FC_\lambda = K_X F = 0$. So $K_X C_\lambda = 0$, and using Proposition 2.3 again we conclude $K_X D = 0$ for each irreducible component $D \subset C_\lambda$. This, the vanishing of C_λ^2, and Stein factorisation imply that there exists a connected holomorphic map from X onto a curve T such that the general fibre is elliptic. Then either the general fibre X_s of f is elliptic, or (by Lüroth's theorem) S is elliptic. In both cases the asserted inequality holds.

b) $g > 1$ and f is locally trivial. Since $\mathrm{Aut}(X_s)$ is finite, there is some finite unramified covering $T \to S$ such that the pull-back of X becomes the product $X_s \times T$. Then

$$\begin{aligned} \mathrm{kod}(X) &= \mathrm{kod}(X_s \times T) && \text{(by Theorem I.7.4)} \\ &= \mathrm{kod}(X_s) + \mathrm{kod}(T) && \text{(by Theorem I.7.3)} \\ &= \mathrm{kod}(X_s) + \mathrm{kod}(S) && \text{(by Theorem I.7.4).} \end{aligned}$$

c) $g = 1$ and $f_* \omega_{X/S}$ is a torsion bundle. Consider an unramified covering $\gamma\colon T \to S$, where the pullback of $f_* \omega_{X/S}$ becomes trivial. If $g\colon Y = X \times_S T \to T$, then $g_* \omega_{Y/T} = \gamma^*(f_* \omega_{X/S})$ is trivial, and there will be a non-zero map $g^* \omega_T \to \mathcal{K}_Y$. This implies $\mathrm{kod}(Y) \geq \mathrm{kod}(T)$. But $\mathrm{kod}(Y) = \mathrm{kod}(X)$ and $\mathrm{kod}(T) = \mathrm{kod}(S)$ by Theorem I.7.4. □

The methods to deal with rational singularities, described in Sect. 3, are essentially due to M. Artin, see [An2], [An3]. In the treatment of Hirzebruch-Jung singularities (Sect. 5) we follow Hirzebruch in [Hir1]. For the history of the famous question how to desingularise a surface we refer to Lipman's survey [Li].

For stable fibrations we referred already to [D-M]. Relative duality (Sect. 12) is due to Grothendieck. A simple treatment of the analytic case in relative dimension one does not seem to be available (see however [R-R-V]).

The results in Sects. 15 and 16 are due to A. Mayer, Mumford, Griffiths and others. We follow the treatment given in the secret seminar [Cl]. Theorems 15.2 and 16.1 are due to A. Mayer, the proof of Theorem 15.2 follows Jambois in [Cl].

Theorem 17.3 is due to Paršin ([Pa]) and Arakelov ([Ar]). For the algebraic case compare also Astérisque 86, Exp. 3. Theorem 18.2 is contained in [Ue3]. Proofs of Iitaka's conjecture $C_{2,1}$ which are independent of the classification of surfaces were first given by Ueno ([Ue3]) and Viehweg ([Vie]). Our proof is a mixture of these two proofs.

IV. Some General Properties of Surfaces

In this chapter a surface is always a connected 2-dimensional complex manifold.

1. Meromorphic Maps, Associated to Line Bundles

The considerations in this section apply mutatis mutandis to complex manifolds of any dimension, but since we shall use resolution of singularities we shall restrict ourselves to the 2-dimensional case which is the only one we shall need.

Let X be a compact surface, and $\mathscr{L}=\mathcal{O}_X(D)$ a holomorphic line bundle on X with $h^0(\mathscr{L})\geq 2$. There exists a maximal divisor $V\geq 0$ with $D-V\geq 0$ for all $D\in|D|$. This divisor V is called the *fixed part* of $|D|$. There is an obvious isomorphism between $|D-V|$ and $|D|$. The line bundle $\mathscr{M}=\mathcal{O}_X(D-V)$ has the property that there exists a finite number of *base points* $b_1,\dots,b_k$ on X such that $B=\bigcup_{i=1}^{k} b_i$ is exactly the subset of X where all sections of $\Gamma(\mathscr{M})$ vanish.

Let $\gamma_1,\dots,\gamma_{N+1}$ be a base for $\Gamma(\mathscr{M})$. Each point $x_0\in X$ has a neighbourhood U with local coordinates $u=(u_1,u_2)$ on U such that, with respect to a trivialisation of $\mathscr{M}$ over U, the section γ_i can be given by a holomorphic function $f_i(u)$. There exists an analytic subset $A\subset X\times\mathbb{P}_N$ which on each $U\times\mathbb{P}_N$ is given by the equations

$$f_i(u)\,z_j-f_j(u)\,z_i=0,$$

where $(z_1:\dots:z_{N+1})$ are homogeneous coordinates on $\mathbb{P}_N$ and $1\leq i,j\leq N+1$.

Over $X\setminus B$ the set A is nothing but a section S in $X\times\mathbb{P}_N$, i.e. a holomorphic map $s\colon X\setminus B\to\mathbb{P}_N$ which has the property that $s^*(\mathcal{O}_{\mathbb{P}_N}(1))=\mathscr{M}|X\setminus B$, as is easily checked. Let Y be the irreducible component of A which contains S. Then Y is a 2-dimensional reduced complex space which can be seen as a "meromorphic map" from X into $\mathbb{P}_N$. This is by definition the *meromorphic map associated to* $\mathscr{M}$ and the base $\gamma_1,\dots,\gamma_{N+1}$. Replacing this base by another one can be interpreted as a change of homogeneous coordinates in $\mathbb{P}_N$. Since for many purposes such a change is irrelevant, we shall often speak of *the* meromorphic map $f_{\mathscr{M}}$ associated to $\mathscr{M}$. The meromorphic map $f_{\mathscr{L}}$ associated to the original line bundle $\mathscr{L}$ is by definition $f_{\mathscr{M}}$ again. The image $f_{\mathscr{L}}(X)$ is defined as the projection of Y in $\mathbb{P}_N$ which is the same as $\overline{s(X\setminus B)}$.

Remarks. 1) If X is a projective algebraic surface and $\mathscr{L}$ an algebraic line bundle on X (by GAGA every vector bundle on an algebraic variety is analytically equivalent to an algebraic one), then Y is algebraic and $f_{\mathscr{L}}$ is the *rational map*, associated to $\mathscr{L}$.

2) The preceding considerations apply equally well to the case that you don't take all of $\Gamma(X, \mathscr{L})$, but only a subspace of dimension at least 2.

Let $\bar{Y}$ be the minimal desingularisation of Y (compare Theorem III.6.2) and let $\pi: \bar{Y} \to X$, $\rho: \bar{Y} \to \mathbb{P}_N$ be the natural projections. There are two possibilities:

(i) $\dim \rho(\bar{Y}) = 2$. In this case Bertini's Theorem I.20.2 yields that for a general section $\beta \in \Gamma(\mathbb{P}_N, \mathcal{O}_{\mathbb{P}_N}(1))$ the induced section $\rho^*(\beta)$ vanishes on a smooth irreducible curve. We claim that the images by π of the zero divisors of such sections $\rho^*(\beta)$ are exactly the divisors in $|D|$. Indeed, if E is such an image, then certainly $\mathcal{O}_{X \setminus B}(E) = \mathcal{O}_{X \setminus B}(D)$, but if a line bundle is trivial outside B, it is trivial on X (the section 1 can be extended by Riemann's extension theorem to a section on X that nowhere vanishes). Conversely, every element of $|D|$ appears in this way for dimension reasons. We conclude that in this case the general member of $|D|$ is irreducible and smooth outside of the base points.

(ii) $\dim \rho(\bar{Y}) = 1$, i.e. $\rho(\bar{Y})$ is an irreducible curve Z. We can apply Stein factorisation to ρ, i.e. $\rho = \sigma\tau$ where $\tau: \bar{Y} \to W$ is a connected map onto the smooth irreducible curve W and where $\sigma: W \to Z$ is finite. There is an integer $l \in \mathbb{N}$ such that $\rho^*(\mathcal{O}_{\mathbb{P}_N}(1)) = \mathcal{O}_{\bar{Y}}(F_1 + \ldots + F_l)$, where $F_1, \ldots, F_l$ are general fibres of τ. On the other hand, by construction $\rho^*(\mathcal{O}_{\mathbb{P}_N}(1)) = \mathcal{O}_{\bar{Y}}(\pi^*(D))$ (mod exceptional divisors over B). The images on X of the fibres F form a 1-dimensional, irreducible system of effective divisors P, all connected and in general irreducible. If W is rational, then this system is easily seen to be a pencil in the usual sense (1-dimensional linear system), otherwise it is called an irrational pencil (of genus equal to genus (W)). Multiplicity taken into account, the fibres D are sums of l curves P. The situation is described by saying that D is *composed with a (rational or irrational) pencil*. In the case of an irrational pencil all exceptional curves over B must be contained in fibres of ρ by Lüroth's theorem, i.e. $B = \emptyset$ in this case.

In particular, if we take $\mathscr{L} = \mathscr{K}_X^{\otimes m}$, then the meromorphic map $f_{\mathscr{L}}$ is called the *m-th pluricanonical map of* X and denoted by f_m (it needn't exist of course). If $f_m(X)$ is a surface, it is called the *m-th canonical model of* X. *If* $P_m(X) \geq 2$ for at least one m, it is known that

$$\operatorname{kod}(X) = \max_{m \geq 1} \dim f_{\mathscr{K}^{\otimes m}}(X)$$

(see [Ue1], p. 86). This remains true if X is replaced by a connected compact complex manifold of any dimension.

2. Hodge Theory on Surfaces

Let X be a compact Kähler manifold. If we consider $H^k(X, \mathbb{C})$ as a de Rham space, then the subspace of those elements which can be represented by a d-closed form of type (p, q) is naturally isomorphic to the Dolbeault group

$H^{p,q}(X)$, and thus there is a decomposition $H^k(X,\mathbb{C})\cong \bigoplus_{p+q=k} H^{p,q}(X)$ (see Corollary I.13.3). If X is not kählerian, in general such a decomposition does not exist. However, if X is a non-kählerian *surface*, then there always exists such a decomposition for $k=2$, whereas for $k=1$ the existence of such a decomposition is assured as soon as $b_1(X)$ is even. All this will be explained in the present section. The main point is the fact that, like for Kähler manifolds, the Fröhlicher spectral sequence degenerates at E_1-level for any compact surface.

(2.1) **Lemma.** *Every holomorphic differential form on a compact surface X is closed.*

Proof. The assertion for 0- and 2-forms being trivial, we consider a holomorphic 1-form ω. By Stokes' theorem we have

$$\int_X d\omega\wedge d\bar{\omega}=\int_X d(\omega\wedge d\bar{\omega})=0.$$

Upon writing locally $d\omega=f dz_1\wedge dz_2$ and

$$d\omega\wedge d\bar{\omega}=-|f|^2 dz_1\wedge d\bar{z}_1\wedge dz_2\wedge d\bar{z}_2=|f|^2 dx_1\wedge dy_1\wedge dx_2\wedge dy_2$$

(with $z_j=x_j+\sqrt{-1}\,y_j$, $j=1,2$) we see that the vanishing of the integral implies $d\omega=0$. □

Remark. The same proof works for holomorphic $(n-1)$-forms on a n-dimensional compact complex manifold. But if the manifold is non-kählerian, then holomorphic forms of lower dimension need not be closed (in fact there exist 3-manifolds with non-closed holomorphic 1-forms, compare [Ue1], § 17).

(2.2) **Lemma.** (i) *If for $\omega\in H^{1,0}$ there is some $f\in C^\infty(X)$ with $\omega=\partial f$, then $\omega=0$;* (ii) *if for $\sigma\in H^{2,0}$ there is some $\varphi\in\Gamma(\mathscr{D}^{1,0})$ with $\sigma=\partial\varphi$, then $\sigma=0$.*

Proof. (i) Since $\bar{\partial}\partial f=\bar{\partial}\omega=0$, we have that f is a harmonic function, hence constant by the maximum principle.
(ii) Since $\bar{\partial}\varphi\wedge\bar{\sigma}=0$, we have $\sigma\wedge\bar{\sigma}=d\varphi\wedge\bar{\sigma}=d(\varphi\wedge\bar{\sigma})$, so by Stokes $\int_X \sigma\wedge\bar{\sigma}=0$, and arguing as in the proof of Lemma 2.1 we find $\sigma=0$. □

Using Lemma 2.1 we have first of all homomorphisms

$$H^{1,0}(X)\to H^{0,1}(X):\ \omega\to\bar{\omega}$$
$$H^{2,0}(X)\to H^{0,2}(X):\ \sigma\to\bar{\sigma}$$

which are injective by Lemma 2.2 (so the second map is an isomorphism by Serre duality). Next we define $H^{1,0}(X)\to H^1(X,\mathbb{C})$ and $H^{2,0}(X)\to H^2(X,\mathbb{C})$ by sending a holomorphic form to its de Rham class. Again by Lemma 2.2 these homomorphisms are injective, and we shall identify $H^{1,0}(X)$ and $H^{2,0}(X)$ with their images.

On $H^k(X,\mathbb{C})=H^k(X,\mathbb{R})\otimes\mathbb{C}$ $(k=1,2)$ one has complex conjugation. As usual, we denote by $\overline{V}$ the conjugate of any subspace $V\subset H^k(X,\mathbb{C})$ with respect to the complex conjugation on $H^k(X,\mathbb{C})=H^k(X,\mathbb{R})\otimes\mathbb{C}$.

(2.3) **Proposition.** $H^{k,0}(X)\cap\overline{H^{k,0}(X)}=0$ $(k=1,2)$.

Proof. A class in $H^{1,0}(X)\cap\overline{H^{1,0}(X)}$ can be written as $[\omega_1]=[\bar\omega_2]$, with ω_1, $\omega_2\in\Gamma(\Omega^1_X)$. So $\omega_1-\bar\omega_2=df$ for some $f\in C^\infty(X)$, hence $\omega_1=\partial f=0$ by Lemma 2.2. As to the case $k=2$, any class in $H^{2,0}(X)\cap\overline{H^{2,0}(X)}$ can be written as $[\sigma_1]=[\bar\sigma_2]$, with $\sigma_1,\sigma_2\in\Gamma(\Omega^2_X)$. So $\int_X\sigma_1\wedge\bar\sigma_1=\int_X\bar\sigma_2\wedge\bar\sigma_1=0$, and exactly as in the proof of Lemma 2.1 we find $\sigma_1=0$. □

(2.4) *Remark. Let V be a real vector space and W a complex subspace of $V\otimes_{\mathbb{R}}\mathbb{C}$, which as a subspace is invariant under conjugation. Then W is the complexification of $W\cap V$.*

Proof. The space W is the direct sum of the subspace of elements, invariant under conjugation and the subspace of elements, which are anti-invariant under conjugation. Since the first of these subspaces is just $W\cap V$ and since the second one can be obtained from the first one by multiplication with $\sqrt{-1}$, the statement follows. □

(2.5) **Lemma.** *For every compact surface X the following inequalities hold:*
(i) $2h^{1,0}\le b_1\le h^{0,1}+h^{1,0}\le 2h^{0,1}$
(ii) $2p_g\le b^+$.

Proof. The inequality $2h^{1,0}\le b_1$ is an immediate consequence of Proposition 2.3.

If for a moment we let $\mathscr{S}$ be the sheaf of closed holomorphic 1-forms on X, then there is an exact sequence

$$0\to\mathbb{C}_X\to\mathcal{O}_X\to\mathscr{S}\to 0.$$

Using Lemma 2.1 we obtain from this an exact sequence

$$0\to H^0(\Omega^1_X)\to H^1(X,\mathbb{C})\to H^1(\mathcal{O}_X), \tag{1}$$

which proves $h^{1,0}+h^{0,1}\ge b_1$, hence $b_1\le 2h^{0,1}$ because of Lemma 2.1.

To prove (ii) we consider the linear subspace of $H^2(X,\mathbb{R})$, consisting of the classes $[\sigma+\bar\sigma]$ with $\sigma\in H^{2,0}$, i.e. by Lemma 2.3 $(H^{2,0}\oplus\overline{H^{2,0}})\cap H^2(X,\mathbb{R})$. On this space the intersection form is

$$(\sigma_1+\bar\sigma_1,\sigma_2+\bar\sigma_2)=\int_X(\sigma_1+\bar\sigma_1)\wedge(\sigma_2+\bar\sigma_2)=\int_X\sigma_1\wedge\bar\sigma_2+\bar\sigma_1\wedge\sigma_2,$$

so this restriction is positive definite and (ii) follows by Remark 2.4 ($H^{2,0}\oplus\overline{H^{2,0}}$ is the complexification of its intersection with $H^2(X,\mathbb{R})$, so this intersection is of dimension $2h^{2,0}=2p_g$ by Lemma 2.3). □

(2.6) **Theorem.** *Let X be a compact surface. Then*

(i) $b_1(X)=h^{1,0}+h^{0,1}$;
(ii) *if* $b_1(X)$ *is even, then* $h^{1,0}=h^{0,1}$ *and* $b^+(X)=2p_g(X)+1$;
(iii) *if* $b_1(X)$ *is odd, then* $h^{1,0}=h^{0,1}-1$ *and* $b^+(X)=2p_g(X)$;
(iv) $q(X)$ *and* $p_g(X)$ *are topological invariants,* $q(X)$ *of the non-oriented, and* p_g *of the oriented underlying manifold.*

Proof. Eliminating $c_1^2(X)$ from the index theorem of Thom-Hirzebruch (Theorem I.3.1)

$$\tau(X)=\tfrac{1}{3}(c_1^2(X)-2c_2(X))$$

and the Riemann-Roch formula

$$1-q+p_g=\tfrac{1}{12}(c_1^2(X)+c_2(X))$$

we obtain

$$4-4q+4p_g-b^++b^-=c_2(X)=e(X)=2-2b_1+b^++b^-,$$

which means

$$(b^+-2p_g)+(2q-b_1)=1.$$

Since by Lemma 2.5 both terms within brackets are non-negative, we are left with only two possibilities, namely (ii) and (iii) of the theorem. Both imply (i). □

Remark. Noether's formula, together with Theorem 2.6, (iv) shows that $c_1^2(X)$ is a topological invariant of the underlying oriented manifold, but this is of course already clear from the index theorem.

(2.7) **Theorem.** *For any compact surface the Fröhlicher spectral sequence degenerates at* E_1*-level.*

Proof. If we consider $H^p(\Omega_X^q)$ as a Dolbeault group, then at E_1-level the derivatives are given by the ∂-operator:

$$H^0(\mathcal{O}_X)\xrightarrow{\partial}H^0(\Omega_X^1)\xrightarrow{\partial}H^0(\Omega_X^2),$$

$$H^1(\mathcal{O}_X)\xrightarrow{\partial}H^1(\Omega_X^1)\xrightarrow{\partial}H^1(\Omega_X^2),$$

$$H^2(\mathcal{O}_X)\xrightarrow{\partial}H^2(\Omega_X^1)\xrightarrow{\partial}H^2(\Omega_X^2).$$

The first map in the first row clearly vanishes, and the second map in the first row vanishes because of Lemma 2.1. Furthermore, Theorem 2.6 and the sequence (1) imply that $H^1(\mathbb{C})\to H^1(\mathcal{O}_X)$ is surjective, so $H^1(\mathcal{O}_X)\to H^1(\Omega_X^1)$ vanishes. The compatibility of Serre duality with exterior products (Proposition I.12.1) then implies the vanishing of all other derivatives at E_1-level. Similarly it can be shown that also at E_2-level all derivatives vanish. □

Once we have this result we can try to mimic the procedure of I, Sect. 13. So we start from the Hodge filtrations defined on $H^1(X,\mathbb{C})$ and $H^2(X,\mathbb{C})$ by the Fröhlicher spectral sequence:

$$F^1(H^1)=\{[\omega];\omega\in\Gamma(\mathscr{D}_X^{1,0}),d\omega=0\}$$
$$F^1(H^2)=\{[\sigma];\sigma\in\Gamma(\mathscr{D}_X^{2,0}\oplus\mathscr{D}_X^{1,1}),d\sigma=0\}$$
$$F^2(H^2)=\{[\sigma];\sigma\in\Gamma(\mathscr{D}_X^{2,0}),d\sigma=0\},$$

and show

(2.8) **Proposition.** *There are formal Hodge decompositions*

(i) $H^1=F^1(H^1)\oplus\overline{F^1(H^1)}$ *(if b_1 is even)*
(ii) $H^2=F^2(H^2)\oplus(F^1(H^2)\cap\overline{F^1(H^2)})\oplus\overline{F^2(H^2)}$ *(always).*

Proof. The case (i) being very similar, we restrict ourselves to the case (ii).

The degeneracy of the Fröhlicher spectral sequence implies that $\dim F^1(H^2)=\dim\overline{F^1(H^2)}=h^{2,0}+h^{1,1}$ and also that $\dim H^2=h^{2,0}+h^{1,1}+h^{0,2}$. Hence $\dim(F^1(H^2)\cap\overline{F^1(H^2)})\geq h^{1,1}$. So (ii) follows as soon as we have proved that the three summands have intersection 0, in other words that $F^2(H^2)\cap\overline{F^1(H^2)}=0$. But a class in $F^2(H^2)\cap\overline{F^1(H^2)}$ can be represented by some $\alpha^{2,0}\in\Gamma(\Omega_X^2)$ and at the same time by a form $\sigma^{1,1}+\sigma^{0,2}\in\Gamma(\mathscr{D}_X^{1,1})\oplus\Gamma(\mathscr{D}_X^{0,2})$. Then $\sigma^{1,1}+\sigma^{0,2}=\alpha^{2,0}+d\omega$ for some 1-form ω. Writing $\omega=\omega^{1,0}+\omega^{0,1}$ one finds $\alpha^{2,0}=\partial\omega^{1,0}=0$ by Lemma 2.2. □

These formal decompositions are in fact the usual Hodge decompositions. This is clear for the first one. As to (ii), by definition $F^2(H^2)$ consists of those de Rham classes which can be represented by a form of type (2,0). Consequently, $\overline{F^2(H^2)}$ consists of those classes, representable by a (0,2)-form. We claim that $F^1(H^2)\cap\overline{F^1(H^2)}$ consists of the de Rham classes which are representable by a (closed) form of type (1,1). Indeed, by definition $[\alpha]\in F^1(H^2)\cap\overline{F^1(H^2)}$ if and only if there are closed forms α_1 and α_2, with $[\alpha_1]=[\alpha_2]=[\alpha]$, such that $\alpha_1=\alpha_1^{1,1}+\alpha_1^{2,0}$ and $\alpha_2=\alpha_2^{1,1}+\alpha_2^{0,2}$ with $\alpha_k^{i,j}\in\Gamma(\mathscr{D}_X^{i,j})$. Now $\alpha_1-\alpha_2=d\beta$, for some 1-form β. If $\beta=\beta^{1,0}+\beta^{0,1}$ is the type decomposition of β, we have that $\alpha_1-d\beta^{1,0}=\alpha_2+d\beta^{0,1}$ is a d-closed (1,1)-form of class $[\alpha]$. Since on the other hand every element of $H^2(X,\mathbb{C})$ which can be represented by a closed (1,1)-form lies in $F^1(H^2)\cap\overline{F^1(H^2)}$, we have proved our claim.

The Fröhlicher spectral sequence being degenerated, we obtain natural isomorphisms

$$H^{2,0}=E_\infty^{2,0}\cong F^2(H^2)$$
$$H^{1,1}=E_\infty^{1,1}\cong F^1(H^2)/F^2(H^2)\cong F^1(H^2)\cap\overline{F^1(H^2)}$$
$$H^{0,2}=E_\infty^{0,2}\cong H^2(X,\mathbb{C})/F^1(H^2)\cong\overline{F^2(H^2)}.$$

So we finally obtain

(2.9) **Theorem.** *Let X be a compact surface. Then, always for $p+q=2$ and if $b_1(X)$ is even also for $p+q=1$, the Dolbeault group $H^{p,q}(X)$ is naturally isomor-*

phic to the subspace of $H^{p+q}(X,\mathbb{C})$, *whose elements can be represented by a d-closed form of type* (p,q). *In this way one obtains natural decompositions* $H^k(X,\mathbb{C})=\bigoplus_{p+q=k} H^{p,q}(X)$.

In particular, every class in $H^2(X,\mathbb{C})$ has a unique decomposition into types.

We give three applications of the preceding considerations; the first one about the Albanese torus, the second one about Lefschetz's theorem on (1,1)-classes and the third one about the index theorem.

We have seen that $H^1(X,\mathbb{C})$ is generated by holomorphic 1-forms and their conjugates as soon as $b_1(X)$ is even. This gives, exactly like in the kählerian case (I, Sect. 14), a Hodge structure of weight 1 and

(2.10) **Corollary.** *If X is a compact surface with* $b_1(X)$ *even, then* Alb(X) *is a torus of dimension* $q(X)=h^{1,0}(X)=\frac{1}{2}b_1(X)$.

Next we shall prove Lefschetz's theorem on (1,1)-classes for any compact surface. We identify the sheaf cohomology group $H^2(X,\mathbb{C})$ with the de Rham group of closed 2-forms in the canonical way, and we also identify $H^2(X,\mathcal{O}_X)$ with the subspace of (0,2)-classes in $H^2(X,\mathbb{C})$ in the same way as described above. Then we have

(2.11) **Proposition.** *In 2-cohomology the injection* $\mathbb{C}_X\to\mathcal{O}_X$ *induces the projection onto the* (0,2)-*component.*

Proof. Sending for $k=0,1,2$ a complex k-form to its $(0,k)$-component defines a morphism of complexes

$$j^{\bullet}\colon \Gamma(\mathcal{D}_X^{\bullet})\to\Gamma(\mathcal{D}_X^{0,\bullet})$$

which extends the injection of sheaves $j\colon \mathbb{C}_X\to\mathcal{O}_X$. In view of Theorem 2.9 the induced map $j^*\colon H^2(X,\mathbb{C})\to H^2(X,\mathcal{O}_X)$ is of course nothing but the projection. □

Now the Lefschetz theorem on (1,1)-classes answers the following question: if $\delta\colon H^1(X,\mathcal{O}_X^*)\to H^2(X,\mathbb{Z})$ is the boundary homomorphism in the exponential cohomology sequence, and if $i^*\colon H^2(X,\mathbb{Z})\to H^2(X,\mathbb{C})$ is induced by the embedding $i\colon \mathbb{Z}_X\to\mathbb{C}_X$, then what is $i^*\circ\delta(H^1(X,\mathcal{O}_X^*))$, i.e. what is the image of Pic(X) in $H^2(X,\mathbb{C})$?

(2.12) **Theorem** (Lefschetz theorem on (1,1)-classes). *Let X be a compact surface. Then the image of* Pic(X) *in* $H^2(X,\mathbb{C})$ *is* $H^{1,1}(X)\cap i^*(H^2(X,\mathbb{Z}))$. *In other words: an element of* $H^2(X,\mathbb{C})$ *is in the image of* Pic(X) *if and only if it is "integral" and can be represented by a real closed* (1,1)-*form.*

Proof. If $k\colon \mathbb{Z}_X\to\mathcal{O}_X$ is the natural injection, then we have a commutative diagram

$$\begin{array}{ccccc} \mathrm{Pic}(X)=H^1(X,\mathcal{O}_X^*) & \xrightarrow{\delta} & H^2(X,\mathbb{Z}) & \xrightarrow{k^*} & H^2(X,\mathcal{O}_X) \\ & & {\scriptstyle i^*}\searrow & & \nearrow{\scriptstyle j^*} \\ & & & H^2(X,\mathbb{C}) & \end{array}$$

It follows that $\mathrm{im}(i^*\circ\delta)=(\ker j^*)\cap(\mathrm{im}\, i^*)$, hence the image of $\mathrm{Pic}(X)$ in $H^2(X,\mathbb{C})$ coincides with

$$(\ker j^*)\cap(\mathrm{im}\, i^*)=(\ker j^*)\cap\overline{(\ker j^*)}\cap\mathrm{im}(i^*)=H^{1,1}(X)\cap\mathrm{im}\, i^*.$$

As to the second formulation, we have to prove that every element of $H^{1,1}(X)\cap H^2(X,\mathbb{R})$ can be represented by a *real* d-closed $(1,1)$-form. But such an element can be represented by a d-closed $(1,1)$-form $\rho=\sigma+d\tau$, with $[\sigma]\in H^2(X,\mathbb{R})$. So it is also represented by the real d-closed $(1,1)$-form $\frac{1}{2}(\rho+\bar{\rho})=\sigma+d(\tau+\bar{\tau})$. □

Remark. The group $H^{1,1}(X)\cap i^*(H^2(X,\mathbb{Z}))$ is the *Néron-Severi group* of X (I, Sect. 6); its rank is the *Picard number* $\rho(X)$.

Actually, $H^{1,1}(X)$ is the complexification of $H^{1,1}_{\mathbb{R}}(X)=H^{1,1}(X)\cap H^1(X,\mathbb{R})$. This follows from Remark 2.4.

We come to our last application, the important

(2.13) **Theorem** (Signature theorem). *Let X be a compact surface. Then the cupproduct form on $H^2(X,\mathbb{R})$, restricted to $H^{1,1}_{\mathbb{R}}(X)$, is non-degenerate of type $(1,h^{1,1}-1)$ if $b_1(X)$ is even and of type $(0,h^{1,1})$ if $b_1(X)$ is odd.*

Proof. The complex subspace $H^{2,0}(X)\oplus H^{0,2}(X)\subset H^2(X,\mathbb{C})$ is the complexification of $(H^{2,0}(X)\oplus H^{0,2}(X))\cap H^2(X,\mathbb{R})$ by Remark 2.4. On this $2p_g$-dimensional subspace of $H^2(X,\mathbb{R})$ the cupproduct-form is positive definite. Using Theorem 2.9 we see that its orthogonal complement in $H^2(X,\mathbb{R})$ is nothing but $H^{1,1}_{\mathbb{R}}(X)$. So the result follows from Theorem 2.6 (ii) and (iii). □

(2.14) **Corollary.** *Let X be a compact surface with $b_1(X)$ even, and let h be an element of $H^{1,1}_{\mathbb{R}}(X)$ with $h^2>0$. Then the cupproduct-form is negative definite on the orthogonal complement in $H^{1,1}_{\mathbb{R}}(X)$ of the line, determined by h.*

The signature theorem is mostly used in the following form:

(2.15) **Corollary** (Algebraic index theorem or Hodge index theorem). *Let D, E be divisors with rational coefficients on the algebraic surface X. If $D^2>0$ and $DE=0$, then $E^2\leq 0$ and $E^2=0$ if and only if E is homologous to 0 (in rational homology).*

This follows from Corollary 2.14, Theorem 2.12 and the fact that an algebraic surface, being kählerian, has always an even first Betti number.

Warning: The reader should not confuse the algebraic index theorem above with the topological index theorem of Thom-Hirzebruch (Theorem I.3.1). Unfortunately, both names are well-established.

3. Deformations of Surfaces

First we formulate a result which is particularly useful for studying deformations of non-minimal surfaces.

(3.1) **Proposition.** *Let $f\colon X \to S$ be a complex analytic family of compact surfaces.*

(i) *If for some point $0 \in S$ the fibre X_0 contains a (-1)-curve E_0, then there exists an open neighbourhood U of 0 in S and a closed and connected submanifold E of $f^{-1}(U)$ such that $E \cap X_0 = E_0$ and such that $E \cap X_t$ is a (-1)-curve for every $t \in U$.*

(ii) *If there exists a closed submanifold E of X such that $E_s = E \cap X_s$ is a (-1)-curve for all $S \in U$, then there exists a family $g\colon X' \to S$ and a commutative triangle*

$$\begin{array}{ccc} X & \xrightarrow{h} & X' \\ & {\scriptstyle f}\searrow \quad \swarrow{\scriptstyle g} & \\ & S & \end{array}$$

such that $h|X_s \to X'_s$ is the blowing down of E_s.

A proof of the first property – called *stability of (-1)-curves* – can be found in [Ko3], Theorem 5, whereas the second property (simultaneous blowing down of (-1)-curves in a family) is proved in [Ii], Appendix I.

Let $f\colon X \to S$ be a complex-analytic family of compact surfaces, and let $b = b_2(X_s)$, $p_g = p_g(X_s)$ (constant by Proposition 2.6, (iv)). Furthermore, let $\mathscr{H} = f_{*2}\,\mathbb{C}_X \otimes_{\mathbb{C}_S} \mathcal{O}_S$.

By Theorem I.9.5 the $\mathcal{O}_S$-module $f_{*2}\mathcal{O}_X$ is a locally free sheaf on S of rank p_g with fibres $H^{0,2}(X_s)$.

The canonical injection $\mathbb{C}_X \to \mathcal{O}_X$ gives a mapping of $\mathcal{O}_S$-modules $p\colon \mathscr{H} \to f_{*2}\mathcal{O}_X$ which is surjective, since it is surjective on each fibre (Proposition 2.11). Each space $H^{0,2}(X_s)$ is Serre-dual to $H^{2,0}(X_s)$. The vector spaces $H^{2,0}(X_s)$ form the fibres of $f_*\Omega^2_{X/S}$, and the compatibility of Serre duality and Poincaré duality implies that p is dual to an injective map of $\mathcal{O}_S$-modules $f_*\Omega^2_{X/S} \to \mathscr{H}$, which on the fibres is the inclusion $H^{2,0}(X_s) \to H^2(X_s, \mathbb{C})$. This proves

(3.2) **Theorem.** *With the above notations,*

$$\mathscr{H}^{2,0} = \bigcup_{s \in S} H^{2,0}(X_s)$$

is a holomorphic subbundle of $\mathscr{H}$ of rank $p_g = p_g(X_s) = h^{2,0}(X_s)$, $s \in S$.

Now assume that a euclidean lattice L (see I, Sect. 2) of rank b is fixed with the property that L is isometric to any $H^2(X_s,\mathbb{Z})/\mathrm{Tors}\,H^2(X_s,\mathbb{Z})$.

We associate to L a *period-domain* in the following way. We extend the bilinear form (,) on L to $L_{\mathbb{C}}=L\otimes_{\mathbb{Z}}\mathbb{C}$ in a $\mathbb{C}$-bilinear fashion and set

$D=D(L)=\{P\in Gr(p_g,L_{\mathbb{C}});\ P$ is totally isotropic and $(p,\bar{p})>0$ for all $p\in P,\ p\neq 0\}$;
$G=G_{\mathbb{Z}}$, the group of isometries of L;
$G_{\mathbb{C}}$= group of isometries of $L_{\mathbb{C}}$.

It is easy to show that $G_{\mathbb{C}}$ acts transitively on the set of isotropic p_g-dimensional complex subspaces of $L_{\mathbb{C}}$, so in particular D is a manifold.

A trivialisation

$$\varphi\colon\ f_{*2}\mathbb{Z}_X \xrightarrow{\sim} L_S \quad (\text{mod torsion})$$

with the property that it is an isometry on each fibre is called a *marking*. It induces a map

$$\tau\colon\ S\to \mathrm{Gr}(p_g,L_{\mathbb{C}})$$

by sending $s\in S$ to the image of $H^{2,0}(X_s)$ under the $\mathbb{C}$-linear extension of φ.

(3.3) **Theorem.** *The map τ is holomorphic and has its image in the period domain D.*

Proof. The first statement is a consequence of Theorem 3.2. As to the second assertion, for any two classes $[\omega_1]$, $[\omega_2]$ in $H^{2,0}(X_s)$ we have

$$([\omega_1],[\omega_2])=\int_{X_s}\omega_1\wedge\omega_2=0$$

and similarly for $[\omega]\in H^{2,0}(X_s)$

$$([\omega],[\bar{\omega}])=\int_{X_s}\omega\wedge\bar{\omega}>0 \quad \text{if } \omega\neq 0. \quad \square$$

We close this section with a proposition which will be used in Chap. VI (Sect. 7).

(3.4) **Proposition.** *If V is a surface with $\mathscr{K}_V^{\otimes m}=\mathscr{O}_V$ for some $m\geq 1$, $\mathscr{K}_V^{\otimes n}\neq\mathscr{O}_V$ $(0<n<m)$, then the same holds for every deformation of V.*

Proof. We first prove the proposition for $m=1$. We have to show two things:
(i) $\mathscr{K}_{X_t}=\mathscr{O}_{X_t}$ for every small deformation $f\colon X\to S$ of $X_0=V$, $0\in S$.
(ii) If $\{X_t\}$ is a complex analytic family over $|t|<\varepsilon$ and if there is a sequence $t_i\to 0$ such that $\mathscr{K}_{X_{t_i}}=\mathscr{O}_{X_{t_i}}$ for $t\neq 0$, then $\mathscr{K}_{X_0}=\mathscr{O}_{X_0}$.

As to (i), we observe that $p_g(X_t)$ being constant (Proposition 2.6, (iv)), the natural map $(f_*\mathscr{K}_X)_t\to H^{2,0}(X_t)$ is surjective by Theorem I.8.5, (iv). Since $\mathscr{K}_{X_0}$ contains a nowhere vanishing section s, we therefore can find a section $\tilde{s}$ of $f_*\mathscr{K}_X$ in a neighbourhood of 0 and $\tilde{s}(t)$ gives a nowhere zero section of $\mathscr{K}_{X_t}$ for t close to 0.

Part (ii) is an immediate consequence of the upper-semi-continuity of both $p_g(X_t)$ and $\dim H^0(\mathscr{K}_{X_t}^{-1})$.

Next, we prove the proposition for arbitrary $m \geq 1$. Here we have to prove three things

(i) Every small deformation X_t of $X_0 = V$ occurring in a family $f\colon X \to S$ has $\mathscr{K}_{X_t}^{\otimes m} = \mathcal{O}_{X_t}$.

(ii) If $\{X_t\}$ is a complex-analytic family over $|t| < \varepsilon$, and if there is a sequence $t_i \to 0$, such that $\mathscr{K}_{X_{t_i}}^{\otimes m} = \mathcal{O}_{X_0}$, then $\mathscr{K}_{X_0}^{\otimes m} = \mathcal{O}_{X_0}$.

(iii) If $f\colon X \to S$ is a family of surfaces with $\mathscr{K}_{X_t}^{\otimes m} = \mathcal{O}_{X_t}$, $t \in S$, $m > 0$, and $\mathscr{K}_{X_0}^{\otimes n} = \mathcal{O}_{X_0}$ for $0 < n < m$, then $\mathscr{K}_{X_t}^{\otimes n} = \mathcal{O}_{X_t}$ for all $t \in S$.

Now (ii) is proved in the same way as the corresponding statement for $m = 1$, whereas (iii) directly follows from (i) and (ii), so the only thing left to prove is (i). We reduce it to the corresponding statement for $m = 1$ as follows. First, we may assume that f is differentiably trivial with simply-connected base, so that the inclusion induces an isomorphism $\pi_1(X_0) \xrightarrow{\sim} \pi_1(X)$. Then the m-fold unramified covering $X_0' \to X_0$ corresponding to $\mathscr{K}_{X_0}$ extends to an m-fold unramified covering $\rho\colon X' \to X$ fitting into a commutative diagram

$$\begin{array}{ccc} X' & \xrightarrow{\ \rho\ } & X \\ & {\scriptstyle f'}\searrow \quad \swarrow{\scriptstyle f} & \\ & S & \end{array}$$

where $f'\colon X' \to S$ is a complex-analytic family. By construction $\rho^*(\mathscr{K}_X | X_0')$ is trivial. But this bundle is precisely $\mathscr{K}_{X_0'}$ and so by the corresponding statement for $m = 1$ we see that $\mathscr{K}_{X_t'} = \mathcal{O}_{X_t'}$ for small t. Since $\rho_t = \rho | X_t'$ is an m-fold unramified covering and since for every line bundle $\mathscr{L}$ on X_t with $\rho_t^*(\mathscr{L}) = \mathcal{O}_{X_t'}$ one always has $\mathscr{L}^{\otimes m} = \mathcal{O}_{X_t}$ (by Lemma I.16.2), it follows that $\mathscr{K}_{X_t}^{\otimes m} = \mathcal{O}_X$ for those t. This completes the proof of the proposition. □

4. Some Inequalities for Hodge Numbers

Apart from their own interest, the following simple observations are quite useful, in particular with respect to the question which Chern numbers a surface can have.

(4.1) **Proposition.** *If on the compact surface X there are two linearly independent holomorphic 1-forms ω_1 and ω_2 with $\omega_1 \wedge \omega_2 \equiv 0$, then there exist a smooth curve R of genus ≥ 2, a connected holomorphic map $k\colon X \to R$ from X onto R and 1-forms α_1, α_2 on R, such that $\omega_1 = k^*(\alpha_1)$ and $\omega_2 = k^*(\alpha_2)$.*

Proof. Let us agree that we shall consider connected coordinate neighbourhoods only. If in local coordinates (z_1, z_2) the 1-form ω_i is given by

$$\omega_i = f_i(z_1, z_2)\, dz_1 + g_i(z_1, z_2)\, dz \qquad (i = 1, 2),$$

with $f_2(z_1, z_2) \not\equiv 0$, then the quotient f_1/f_2 is a meromorphic function, which is independent of the coordinate system and thus the restriction of a global meromorphic function h on X. After blowing up the points of indeterminacy of h, we get a surface $\bar{X}$ and a holomorphic map $h: \bar{X} \to \mathbb{P}_1$, such that h is constant along the fibres of h. Using Stein factorisation we then obtain a connected holomorphic map $k: \bar{X} \to R$, where R is a Riemann surface, and where h is still constant along the fibres of k.

We denote by B the union of all curves on $\bar{X}$ which are mapped onto points by the canonical projection from $\bar{X}$ onto X. Let $p \in \bar{X} \setminus B$, such that there exist local coordinates (z_1, z_2), $|z_1| < A$, $|z_2| < A$, covering a neighbourhood U of p, with

a) $U \cap B = \emptyset$;
b) the restriction $k|U$ is given by $t = z_1$;
c) if for $i = 1, 2$ the restriction $\omega_i|U$ is given by $\omega_i = f_i dz_1 + g_i dz_2$, then f_1 and f_2 don't vanish on U.

On all but a finite number of fibres of k (namely, the fibres for which each component has multiplicity ≥ 2) there are points p, satisfying this condition.

Now let the point p and the coordinates (z_1, z_2) be as just described. Then, using the identities $f_1 = h f_2$, $g_1 = h g_2$ and the fact that ω_1 and ω_2 are closed (Lemma 2.1), we obtain

$$0 = f_2 \frac{\partial h}{\partial z_2} = \frac{\partial f_1}{\partial z_2} - h \frac{\partial f_2}{\partial z_2}$$

$$= \frac{\partial g_1}{\partial z_1} - h \frac{\partial g_2}{\partial z_1} = g_2 \frac{\partial h}{\partial z_1}$$

i.e. g_2 vanishes on U. Since $\dfrac{\partial f_2}{\partial z_2} = \dfrac{\partial g_2}{\partial z_1}$, we find that $\omega_2|U = f_2(z_1) dz_1$, that is, there exists a holomorphic 1-form α_2 on $k(U)$, such that $k^*(\alpha_2) = \omega_2$ on $k^{-1}(k(U))$. Hence there are finitely many points on R: $a_1, \ldots, a_n$, such that there is a holomorphic 1-form α_2 on $R \setminus \bigcup_{i=1}^{n} a_i$, with $k^*(\alpha_2) = \omega_2$ on $k^{-1}\left(R \setminus \bigcup_{i=1}^{n} a_i\right)$. What we have to do is to show that α_2 can be extended to a holomorphic 1-form on all of R.

Given any i, $i = 1, \ldots, n$, there is a point $p \in k^{-1}(a_i)$ with the property that in a neighbourhood of p there is a connected non-singular curve C, such that with respect to suitable local coordinates on C and R the restriction $k|C$ is given by

$$t = u^n,$$

with p being $(0, v)$, say, and $t = 0$ being the point a_i. Outside of a_i we have that

$$\alpha_2 = f(t)\, dt$$

with f holomorphic outside a_i. We know that $(k|C \setminus p)^*(\alpha_2)$ can be extended to a holomorphic form on all of C, i.e.

$$n f(u^n) u^{n-1} du$$

is holomorphic around $u = 0$. But this implies that also for $t = 0$ the function f is holomorphic.

In the same way we find a holomorphic 1-form α_1 on R with $k^*(\alpha_1)=\omega_1$. Finally, k factorises through X because of Lüroth's theorem. □

(4.2) **Proposition.** *If the compact surface X does not admit a holomorphic map onto a curve of genus ≥ 2, then $h^{2,0}(X) \geq 2h^{1,0}(X)-3$.*

Proof. The preceding proposition implies that the kernel of the natural homomorphism from $\wedge^2 H^{1,0}(X)$ into $H^{2,0}(X)$ meets the cone of decomposable elements only in 0. This cone has dimension $2h^{1,0}-3$, so $2h^{1,0}-3 \leq h^{2,0}$. □

Remark. If X is kählerian, then we can rewrite this inequality as $p_g(X) \geq 2q(X) - 3$.

(4.3) **Proposition.** *If the compact surface X with $h^{1,0}(X) \geq 2$ does not admit any holomorphic map onto a curve of genus ≥ 2, then $h^{1,1}(X) \geq 2h^{1,0}(X)-1$.*

Proof. By Theorem 2.9 we can identify $H^{1,1}(X)$ with the space of d-closed (1,1)-forms modulo d-boundaries.

Let h: $H^{1,0}(X) \times \overline{H^{1,0}(X)} \to H^{1,1}(X)$ be defined by $h(\omega, \rho) = \omega \wedge \rho$.

We claim that for fixed $\omega \neq 0$ and also for fixed $\rho \neq 0$ the restrictions $h|\omega \times \overline{H^{1,0}(X)}$ and $h|H^{1,0}(X) \times \rho$ are injective. To see this, let $h(\omega, \bar{\theta})=0$, with θ a holomorphic 1-form. Then we have $\omega \wedge \bar{\theta} = d\alpha$, with α a complex 1-form on X. It follows that

$$\omega \wedge \theta \wedge \bar{\omega} \wedge \bar{\theta} = d\beta,$$

with β a 3-form on X. Therefore, by Stokes' theorem we find

$$\int_X \omega \wedge \theta \wedge \bar{\omega} \wedge \bar{\theta} = 0.$$

But this implies that the holomorphic 2-form $\omega \wedge \theta$ vanishes identically on X (compare the proof of Lemma 2.1). Because of Proposition 4.1 we thus find that θ is linearly dependent on ω. However, this being the case, we conclude from $\omega \wedge \bar{\omega} = d\alpha'$ that $\omega \wedge \eta \wedge \bar{\omega} \wedge \bar{\eta} = d\beta'$, with η independent of ω (such an η exists by assumption). Using Stokes' theorem again we are led to a contradiction.

The injectivity of $h|H^{1,0}(X) \times \rho$ is proved in the same way. Thus h induces a regular map

$$\mathbb{P}(H^{1,0}(X)) \times \mathbb{P}(\overline{H^{1,0}(X)}) \to \mathbb{P}(H^{1,1}(X))$$

which is injective on the fibres in both directions. Since each holomorphic map from a product of projective spaces onto any complex space of strictly lower dimension factorises through one of the projections ([R-V2], p. 155) we see that the image of $\mathbb{P}(H^{1,0}(X)) \times \mathbb{P}(\overline{H^{1,0}(X)})$ in $\mathbb{P}(H^{1,1}(X))$ has the same dimension as the first of these manifolds. Consequently $h^{1,1}(X)-1 \geq 2(h^{1,0}(X)-1)$, that is $h^{1,1}(X) \geq 2h^{1,0}(X)-1$. □

(4.4) **Corollary.** *If the compact Kähler surface X does not admit any connected fibration with base genus ≥ 2, then $h^{1,1}(X) \geq 2h^{1,0}(X)-1$.*

5. Projectivity of Surfaces

Theorem 5.2 below and its corollaries play a very crucial role in this book, for they are essential in handling non-projective surfaces. We obtain the results as simple consequences of Grauert's criterion I.19.3, together with the Riemann-Roch theorem for surfaces. Firstly, we observe that as a special case of Theorem I.19.3 we have

(5.1) **Theorem** (Grauert's ampleness criterion for surfaces). *A line bundle $\mathcal{L}$ on the compact surface X is ample if and only if it has the following two properties:*

(i) *for some $n \geq 1$ there is an effective divisor D with $\mathcal{L}^{\otimes n} \cong \mathcal{O}_X(D)$,*

(ii) *given any irreducible curve C on X, then for some $n \geq 1$ (depending on C) there is a section in $\mathcal{L}^{\otimes n} | C$, which vanishes somewhere on C, but not everywhere on C.*

Using the Riemann-Roch formula of II, Sect. 3 for an irreducible curve we see that to prove the ampleness of a line bundle $\mathcal{L}$ on a compact surface X it is sufficient to exhibit on X an effective divisor B with $\mathcal{L} = \mathcal{O}_X(B)$ such that $BD > 0$ for every irreducible curve D. This is exactly what we shall do in the proof of Theorem 5.2.

(5.2) **Theorem.** *A compact surface X is projective if and only if there exists on X a line bundle $\mathcal{L}$ with $c_1^2(\mathcal{L}) > 0$.*

Proof. In one direction the theorem is trivial: if $\mathcal{L}$ is very ample then $c_1^2(\mathcal{L}) > 0$. So let X be a compact surface and $\mathcal{L}$ a line bundle on X with $c_1^2(\mathcal{L}) > 0$. We start by applying Riemann-Roch to the line bundle $\mathcal{L}^{\otimes n}$:

$$h^0(\mathcal{L}^{\otimes n}) - h^1(\mathcal{L}^{\otimes n}) + h^2(\mathcal{L}^{\otimes n}) = \frac{n}{2}(n\, c_1(\mathcal{L}) + c_1(X))\, c_1(\mathcal{L}) + \chi(X).$$

By Serre duality $h^2(\mathcal{L}^{\otimes n}) = h^0(\mathcal{K}_X \otimes \mathcal{L}^{\otimes(-n)})$, so for n large we have that either $h^0(\mathcal{L}^{\otimes n}) \geq 2$ or $h^0(\mathcal{K}_X \otimes \mathcal{L}^{\otimes(-n)}) \geq 2$. Since not only $c_1^2(\mathcal{L}^{\otimes n}) > 0$, but also $c_1^2(\mathcal{K}_X \otimes \mathcal{L}^{\otimes(-n)}) > 0$ for large n, there always exists on X an effective divisor D with $D^2 > 0$ and $\dim |D| \geq 1$.

By Sect. 1 there always is a rational or irrational pencil $\mathcal{P}$ on X with the following properties:

(i) a general member C of $\mathcal{P}$ is an irreducible curve;

(ii) there exists a positive integer k_0 and a non-negative divisor F (the fixed part of $|D|$) with $k_0 C + F$ homologous to D for all $C \in \mathcal{P}$.

From here on we distinguish between two cases:

a) $\mathcal{P}$ has at least one base point, i.e. $C^2 > 0$. Let $E_1, \ldots, E_m$ be the (possibly empty) set of all those components of members of $\mathcal{P}$ which don't pass through any base point. By Zariski's Lemma III.8.2 and Corollary I.2.11 there exist non-negative integers $a_1, \ldots, a_m$ such that $\left(\sum_{i=1}^{m} a_i E_i\right) E_j < 0$ for all $j = 1, \ldots, m$. We observe: if $n \geq n_0$, then $(nC - \sum a_i E_i)G > 0$ for every irreducible component G of any member of $\mathscr{P}$. On the other hand if $n \geq n_1$, then $|nC - \sum a_i E_i|$ can be represented by an effective divisor, consisting only of fibre components of elements of $\mathscr{P}$, and containing with some positive multiplicity an irreducible C, hence $(nC - \sum a_i E_i)G > 0$ for any irreducible curve G not contained in a member of $\mathscr{P}$. Application of Grauert's criterion 5.1 to $\mathcal{O}_X(nC - \sum a_i E_i)$ with $n \geq \max(n_0, n_1)$ now completes the proof.

b) $\mathscr{P}$ has no base points. If $F = \sum b_k F_k$, then there must be at least one curve F_j, which is not contained in a fibre of $\mathscr{P}$; otherwise we would have $D^2 = (k_0 C + F)^2 \leq 0$ by Zariski's lemma. We may assume that F_1 is such a curve. Let $E_1, \ldots, E_m$ be all those components of members of $\mathscr{P}$ which don't meet F_1. As above there are integers $a_1, \ldots, a_m$ such that $\left(\sum_{i=1}^{m} a_i E_i\right) E_j < 0$ for $j = 1, \ldots, m$. And as before we find that we can apply Grauert's criterion to $\mathcal{O}_X(nC + mF_1 - \sum a_i E_i)$ with n, m large and $n > -\frac{F_1^2}{CF_1} m$, to obtain the desired result. □

(5.3) **Corollary.** *Every compact surface X with $c_1^2(X) = \mathscr{K}_X^2 > 0$ is projective.*

(5.4) **Corollary** (Nakai's criterion). *A line bundle $\mathscr{L}$ on a compact surface X is ample if and only if $c_1^2(\mathscr{L}) > 0$ and $(\mathscr{L}, D) > 0$ for every effective divisor D on X.*

Proof. By Theorem 5.2 we know already that X is projective. We take a very ample divisor H on X and fix an n_0, such that $(H, \mathscr{K}_X \otimes \mathscr{L}^{\otimes(-n)}) < 0$ for $n \geq n_0$. This is possible by assumption. On the other hand, by the Riemann-Roch theorem for surfaces we know that either $\mathscr{L}^{\otimes n}$ or $\mathscr{K}_X \otimes \mathscr{L}^{\otimes(-n)}$ has sections for n large enough. But the second possibility is excluded, since $(H, \mathscr{K}_X \otimes \mathscr{L}^{\otimes(-n)}) < 0$ for $n \geq n_0$, whereas $HD > 0$ for every effective D. So $\mathscr{L}^{\otimes n}$ has a section and we can apply the remark following Theorem 5.1. □

(5.5) **Corollary.** *A compact surface X is projective if and only if it has algebraic dimension* 2.

Proof. Let X be a surface with $a(X) = 2$. Take a meromorphic function on X and consider the associated pencil $\mathscr{P}$ with general member C. If $C^2 > 0$ we are ready by Theorem 5.2. If $C^2 = 0$, then $\mathscr{P}$ defines a regular map $f\colon X \to \mathbb{P}_1$. Let $f = hg$ be its Stein factorisation. If every meromorphic function on X would be constant on the fibres of g, then $a(X)$ would be 1. Hence there is an irreducible curve D on X, which is not contained in any fibre of g. If G denotes a general fibre of g, then $(nG + D)^2 > 0$ for n sufficiently large. Hence X is algebraic by Theorem 5.2. □

Remark. The preceding result is no longer true if X is allowed to have singularities. The following example of Hironaka (see [Gr2]) shows much more: it is possible to obtain from a smooth projective surface a normal non-projective complex space by blowing down a smooth (non-rational!) curve.

Let $C \subset \mathbb{P}_2$ be a smooth cubic, and $p_1, \ldots, p_{10}$ ten points on C, such that there is no curve $D \subset \mathbb{P}_2$ with the property that set-theoretically $C \cap D \subset p_1 \cup \ldots \cup p_{10}$. (The possibility of such a choice of points on C follows from the classical geometrical interpretation of the addition on C, once this curve is made into an abelian variety by choosing a base point, compare [Wal], p. 192.) Then let X be obtained from $\mathbb{P}_2$ by blowing up $p_1, \ldots, p_{10}$. If $\bar{C}$ denotes the proper transform of C on X, we have $\bar{C}^2 = -1$, and by Theorem III.2.1, $\bar{C}$ can be blown down in X, such that the result is a normal complex space Y. If $q \in Y$ is the image of $\bar{C}$, then there are two independent meromorphic functions on $Y \setminus q = X \setminus \bar{C}$, hence on Y by Levi's theorem (Theorem I.8.7). But Y is not projective. For if Y were projective, then there would be a curve on Y, not containing q; in other words there would be a curve D on $\mathbb{P}_2$, such that $C \cap D \subset p_1 \cup \ldots \cup p_{10}$, which is impossible by construction.

An *abstract algebraic* variety is defined by the usual patching procedure, using affine open sets and rational morphisms. An abstract algebraic variety need not be projective. However, Corollary 5.5 immediately yields

(5.6) **Corollary.** *Every (smooth) compact abstract algebraic surface is projective.*

Because of this result it is not dangerous to call a smooth projective-algebraic surface simply an algebraic surface, as we shall often do (and did!).

(5.7) **Corollary.** *Let X be a compact surface and Y be obtained from X by blowing up a point. Then X is projective if and only if Y is projective.*

(5.8) **Theorem.** *Let X, Y be compact surfaces and $f\colon X \to Y$ a finite map. Then X is projective if and only if Y is projective.*

Proof. If Y is projective, we can take a line bundle $\mathscr{L}$ on Y with $c_1^2(\mathscr{L}) > 0$. Then $c_1^2(f^*(\mathscr{L})) = (\deg f)\, c_1^2(\mathscr{L}) > 0$, and X is projective by Theorem 5.2. Conversely, if X is projective, we can take a divisor D on X with $DC > 0$ for every effective divisor C on X. Then, for every effective divisor E on Y we have $f_*(D)\, E = D f^*(E) > 0$, so Y is algebraic by Nakai's criterion 5.4. □

6. Surfaces of Algebraic Dimension Zero

Let X be a compact surface. The fact that $\mathrm{a}(X) = 0$ readily implies several noteworthy properties, some of which we shall use later on in this book.

(6.1) **Proposition.** *Let X be a compact surface with* $\mathrm{a}(X) = 0$. *Then*

(i) $h^0(\mathscr{L}) \leq 1$ *for every line bundle $\mathscr{L}$ on X, and in particular $p_g(X) = h^0(\Omega_X^2) \leq 1$;*
(ii) $h^{1,0}(X) \leq 2$.

Proof. (i) Trivial.

(ii) Let ω_1, ω_2 and ω_3 be three linearly independent holomorphic 1-forms. Then $\omega_1 \wedge \omega_2$ and $\omega_1 \wedge \omega_3$ don't vanish identically, otherwise $a(X)$ would be at least 1 (Proposition 4.1). But (i) implies that for some $\lambda, \mu \neq 0$ we have $\lambda \omega_1 \wedge \omega_2 + \mu \omega_1 \wedge \omega_3 \equiv 0$, i.e. $\omega_1 \wedge (\lambda \omega_2 + \mu \omega_3) \equiv 0$ and we would find again that $a(X) \geq 1$, contrary to our assumption. So $h^{1,0}(X) \leq 2$. □

Remark. All of these bounds are sharp, as is clear from the fact that there exist 2-tori of algebraic dimension 0.

(6.2) **Theorem.** *If X is a compact surface with* $a(X)=0$, *then the number of irreducible curves on X is finite and at most equal to* $h^{1,1}(X)+2$.

Proof. If $\mathcal{M}_X^1$ denotes the sheaf of germs of meromorphic 1-forms on X, then there is a natural injection $\Omega_X^1 \hookrightarrow \mathcal{M}_X^1$. We denote the quotient sheaf by $\mathcal{Q}_X$. To each curve C on X we can attach an element of $\Gamma(\mathcal{Q}_X)$ by writing C locally as $f=0$ and then consider $\frac{df}{f}$. It is easily verified that the images in $\Gamma(\mathcal{Q}_X)$ of a finite number of different irreducible curves are linearly independent. So the exact cohomology sequence

$$\ldots \to \Gamma(\mathcal{M}_X^1) \to \Gamma(\mathcal{Q}_X) \to H^1(\Omega_X^1) \to \ldots$$

shows that it is sufficient to prove that $\dim \Gamma(\mathcal{M}_X^1) < 3$, i.e. that every three sections of $\mathcal{M}_X^1$ are linearly dependent. But if ω_1 and ω_2 are meromorphic 1-forms on X, independent over $\mathbb{C}$, then ω_1 and ω_2 must be independent over the field of meromorphic functions $\mathcal{M}(X)$, since $a(X)=0$. It follows that every third meromorphic form must be dependent on ω_1 and ω_2 over $\mathcal{M}(X)$, and since $a(X)=0$, it must be dependent on ω_1 and ω_2 over $\mathbb{C}$. □

Remark. Again this inequality is sharp, as follows for example from the existence of Hopf surfaces, without meromorphic functions, with two curves (see Proposition V.18.2).

7. Almost-Complex Surfaces without any Complex Structure

In this section we shall show how the preceding results easily yield compact, oriented 4-dimensional differentiable manifolds, which admit almost-complex structures, but no complex structure.

We shall use the concept of a connected sum of two differentiable manifolds. We refer for a formal treatment to [B-J], § 10, and just say the following.

Let X and Y be two oriented, connected n-dimensional differentiable manifolds. Let $D \subset \mathbb{R}^n$ be the unit disc, and let $f_1: \mathbb{R}^n \to X$, $f_2: \mathbb{R}^n \to Y$ be orientation-preserving embeddings. The connected sum $X \# Y$ (with respect to f_1 and f_2) is the naturally oriented n-dimensional differentiable manifold, obtained from the disjoint union

$$(X - f_1(\tfrac{1}{3}D)) \cup (Y - f_2(\tfrac{1}{3}D))$$

by identifying $f_1(tx)$ with $f_2((1-t)x)$ for all $\frac{1}{3}<t<\frac{2}{3}$, $x \in S^{n-1}$.

If X and Y are compact, then so is $X \# Y$. Since the special choice of f_1 and f_2 plays no role here, we simply denote by $X \# Y$ some connected sum of X and Y (they are all diffeomorphic anyway).

What we need about the topology of connected sums is contained in the following simple result.

(7.1) **Proposition.** *Let G be either $\mathbb{Z}$ or a field. Then for all i, $1 \leq i \leq n-1$, there are natural isomorphisms*

$$\lambda_i\colon\ H^i(X,G) \oplus H^i(Y,G) \xrightarrow{\sim} H^i(X \# Y, G)$$

such that if $a \in H^k(X,G)$, $b \in H^{n-k}(X,G)$, then $\lambda_k(a)\,\lambda_{n-k}(b) = ab$.

Now we are ready for the examples.

(7.2) **Theorem.** *A connected sum $W = S^1 \times S^3 \# S^1 \times S^3 \# \mathbb{P}_2$ admits almost-complex structures, but no complex structure.*

The existence of almost-complex structures on W is a consequence of the following criterion ([Wu], p. 74, together with Theorem I.3.1).

(7.3) **Proposition.** *Let X be an oriented compact connected 4-dimensional differentiable manifold, and let $h \in H^2(X,\mathbb{Z})$. Then there exist almost-complex structures $\mathscr{A}$ on X with $c_1(\mathscr{A}) = h$ (and of course $c_2(\mathscr{A}) = e(X)$) if and only if the following two conditions are satisfied:*

(i) $h \equiv w_2(X) \pmod 2$,

(ii) $h^2 = 3\tau(X) + 2e(X)$.

Proof of Theorem 7.2. From Proposition 7.1 we find $H^2(W,\mathbb{Z}) \cong \mathbb{Z}$, such that if g is a generator of $H^2(W,\mathbb{Z})$, then $g^2 = 1$. Proposition 7.1 also yields that $H^2(W,\mathbb{Z}_2) \cong \mathbb{Z}_2$. Since $w_2 g \equiv g^2 \pmod 2$ by [M-S], p. 732, we see that $w_2 \neq 0$, so $w_2 \equiv g \pmod 2$. Applying Proposition 7.1 once more we find $e(W) = -1$, and since $\tau(W) = 1$, the conditions of Proposition 7.3 are satisfied (with $h = g$). Consequently there exist almost-complex structures $\mathscr{A}$ on W with $c_1(\mathscr{A}) = g$. Conversely, every almost-complex structure on X has g as its first Chern class.

It remains to be seen that there is no complex structure on W. If there were one, it would be a projective one, for, as already observed, it would have Chern class g, hence it would be projective by Theorem 5.2. Furthermore, $b_1(W) = 2$ by Proposition 7.1, so Alb(W) would be an elliptic curve, and there would exist a surjective map $f\colon W \to \mathrm{Alb}(W)$. If F is any fibre of f, then $F^2 = 0$, but F is not homologous to 0 (Lemma I.13.1). This, however, is impossible since $H^2(W,\mathbb{Z}) \cong \mathbb{Z}$. Hence W has no complex structure. □

There are many examples of this type, like $S^1 \times S^3 \# S^1 \times S^3 \# S^2 \times S^2$, or the connected sum of $2k+1$, $k \geq 1$, copies of a smooth surface of degree 4 in $\mathbb{P}_3$.

This last example is simply-connected, since $\pi_1(X \# Y)$ is the free product of $\pi_1(X)$ and $\pi_1(Y)$, whereas a smooth surface in $\mathbb{P}_3$ is simply-connected by Lefschetz' Theorem I.20.4.

If X is a compact complex surface, then it follows from Noether's formula that $c_1^2(X)+c_2(X)\equiv 0$ (12). This remains true if X is any almost-complex surface ([Hir3], p. 125). Conversely, given any ordered pair (p,q) of integers, with $p+q\equiv 0$ (12), there always exists a compact differentiable 4-manifold with an almost-complex structure $\mathscr{A}$, such that $c_1^2(\mathscr{A})=p$, $c_2(\mathscr{A})=q$. To see this, it is sufficient to apply Propositions 7.1 and 7.3 to a suitable connected sum

$$\mathbb{P}_2 \# \ldots \# \mathbb{P}_2 \# \overline{\mathbb{P}}_2 \# \ldots \# \overline{\mathbb{P}}_2 \# \mathbb{P}_1 \times R \# \ldots \# \mathbb{P}_1 \times R,$$

where $\overline{\mathbb{P}}_2$ is $\mathbb{P}_2$ with the orientation reversed, and R a curve of genus 2 ([Ve1], p. 1625).

On the other hand, as we shall see in Chap. VII, there are many pairs (p,q), with $p+q\equiv 0$ (12), such that there is no compact complex surface with these Chern numbers. In this way many more examples of the type above can be produced. A deeper study yields still more examples, but up to now these examples, however abundant, reveal no pattern as to which almost-complex structures can be deformed into an integrable one. And there is not a single example of a (compact) higher-dimensional differentiable manifold with almost-complex structures, but no complex structure. Few doubt that these exist, but there is no method to handle the higher-dimensional case. In particular, a simple projectivity criterion like Theorem 5.2 does not exist in higher dimensions.

8. The Vanishing Theorems of Ramanujam and Mumford

Ramanujam's theorem is a criterion for the vanishing of $H^1(\mathcal{O}_X(-D))$, where D is an effective divisor on a compact Kähler surface X. We shall use it frequently in the rest of this book.

(8.1) **Theorem** (Ramanujam's vanishing theorem). *Let X be a compact Kähler surface and D an effective divisor on X, such that*

(i) *D is 1-connected,*
(ii) *$h^0(\mathcal{O}_X(nD))\geq 2$ for some $n\geq 1$,*
(iii) *the linear system $|nD|$ is not composed with an irrational pencil.*

Then $h^1(\mathcal{O}_X(-D))=0$.

For the concept of "being composed with an irrational pencil" we refer to Sect. 1.

(8.2) **Corollary.** *Let X be a compact Kähler surface and D an effective, 1-connected divisor on X with $D^2>0$. Then $h^1(\mathcal{O}_X(-D))=0$.*

Proof of the Corollary. By Riemann-Roch, the effectiveness of D and $D^2>0$ imply that $h^0(\mathcal{O}_X(nD))$ grows like cn^2, $c>0$ for n large enough. So the linear system $|nD|$ can't be composed with a pencil, for n large. □

Before we come to the proof of Theorem 8.1, we have to deal with a few auxiliary results.

(8.3) **Lemma.** *Let $C\subset X$ be an effective divisor and* $\mathrm{restr}_C\colon H^1(\mathcal{O}_X)\to H^1(\mathcal{O}_C)$ *the restriction map. If η is an element in $H^1(\mathcal{O}_X)$ with* $\mathrm{restr}_C(\eta)=0$, *then already $\eta|U=0$ for an open neighbourhood U of C in X. In particular, the kernel of* $\mathrm{restr}_{nC}\colon H^1(\mathcal{O}_X)\to H^1(\mathcal{O}_{nC})$ *is independent of n.*

Proof. By Proposition II.2.1 the subgroup $H^1(C,\mathbb{Z})\subset H^1(\mathcal{O}_C)$ is discrete. Since X is Kähler, $H^1(X,\mathbb{Z})$ spans $H^1(\mathcal{O}_X)$ over $\mathbb{R}$ (I, Sect. 13). It follows that the kernel of restr_C is spanned over $\mathbb{R}$ by the kernel of $H^1(X,\mathbb{Z})\to H^1(C,\mathbb{Z})$. This is the same as the kernel of $H^1(X,\mathbb{Z})\to H^1(U,\mathbb{Z})$, where U is a suitable open neighbourhood of C in X (cf. Theorem I.8.8). □

(8.4) **Lemma.** *Let D be an effective divisor on X with linear system $|D|$ free of base points. Then*

$$\mathrm{restr}_D\colon H^1(\mathcal{O}_X)\to H^1(\mathcal{O}_D)$$

is injective if and only if the system $|D|$ is not composed with an irrational pencil.

Proof. Let $\varphi_D=\varphi\colon X\to\mathbb{P}_N$, $N=\dim|D|$, be the map determined by $|D|$. Like in Sect. 1, we distinguish between two cases:

The case $\dim\varphi(X)=1$. Let $\varphi=\rho\psi$, with $\psi\colon X\to S$ and S a non-singular Riemann surface, be the Stein factorisation of φ. Then $\psi_*\mathcal{O}_X=\mathcal{O}_S$ ([Rem2], p. 363), and Leray's spectral sequence gives an exact sequence

$$0\to H^1(\mathcal{O}_S)\xrightarrow{\psi^*}H^1(\mathcal{O}_X)\to H^0(\psi_{*1}\mathcal{O}_X)\to 0.$$

If S has genus ≥ 1, then $\psi^*(H^1(\mathcal{O}_S))\subset H^1(\mathcal{O}_X)$ is a non-trivial subspace in the kernel of restr_D. If S is rational, we consider a class $\eta\in H^1(\mathcal{O}_X)$ with $\mathrm{restr}_D(\eta)=0$. Let X_s, $s\in S$, be a connected component of D. Then η vanishes on X_s too, and by Lemma 8.3, η even vanishes in a neighbourhood of X_s. By Corollary III.11.2 the direct image sheaf $\psi_{*1}\mathcal{O}_X$ is locally free. The image of η in $H^0(\psi_{*1}\mathcal{O}_X)$ vanishes on a neighbourhood of s, i.e. it is the zero section. But now $H^1(\mathcal{O}_X)\to H^0(\psi_{*1}\mathcal{O}_X)$ is injective and we find $\eta=0$.

The case $\dim\varphi(X)=2$. We use Bertini's theorem I.20.2 saying in particular that each $D'\in|D|$ is connected. We pick D' such that $\mathrm{supp}(D)\cap\mathrm{supp}(D')$ is finite, and blow up this intersection by way of $\tau\colon\bar{X}\to X$ such that the meromorphic map $X\to\mathbb{P}_1$ defined by the 1-dimensional system $|\lambda D+\mu D'|$, $\lambda,\mu\in\mathbb{C}$, induces a holomorphic map $\psi\colon\bar{X}\to\mathbb{P}_1$. This ψ is connected, and one fibre $\bar{X}_s$, $s\in\mathbb{P}_1$, is contained in the total transform $\tau^*(D)$ of D. By the commutativity of

$$\begin{array}{ccccc} H^1(\mathcal{O}_X) & \xrightarrow{\text{restr}} & H^1(\mathcal{O}_D) & & \\ \| & & \downarrow{\scriptstyle \tau^*} & & \\ H^1(\mathcal{O}_{\bar{X}}) & \xrightarrow{\text{restr}} & H^1(\mathcal{O}_{\tau^* D}) & \xrightarrow{\text{restr}} & H^1(\mathcal{O}_{\bar{X}_s}) \end{array}$$

it suffices to show that restr: $H^1(\mathcal{O}_{\bar{X}}) \to H^1(\mathcal{O}_{\bar{X}_s})$ is injective, which can be done as in the case $\dim \varphi(X)=1$ above. □

Proof of Theorem 8.1. We consider the exact sequence

$$0 \to \Gamma(\mathcal{O}_X) \to \Gamma(\mathcal{O}_D) \to H^1(\mathcal{O}_X(-D)) \to H^1(\mathcal{O}_X) \to H^1(\mathcal{O}_D).$$

Since D is 1-connected, $\Gamma(\mathcal{O}_D) \cong \mathbb{C}$ by Lemma II.12.2, and it suffices to prove injectivity of restr_D: $H^1(\mathcal{O}_X) \to H^1(\mathcal{O}_D)$. By Lemma 8.3 this is equivalent to the injectivity of restr_{nD} for sufficiently large $n \in \mathbb{N}$. We choose n as in the theorem and put $nD = F + C$, with F the fixed part of $|nD|$. Since restr_C factors through restr_{nD}, it suffices to prove injectivity for restr_C. Next we blow up the fixed points of $|C|$ via σ: $\bar{X} \to X$ such that $\sigma^*(C) = \bar{F} + B$ with $\bar{F}$ the fixed part of $|\sigma^*(C)|$ and $|B|$ without base points. Just as in the proof of Lemma 8.4 above it suffices to show the injectivity of restr_B: $H^1(\mathcal{O}_X) \to H^1(\mathcal{O}_B)$, which is again obtained by the argument used to prove Lemma 8.4. □

Mumford's vanishing theorem is a criterion for the vanishing of $H^1(\mathcal{L}^\vee)$, where the line bundle $\mathcal{L}$ on X is not necessarily defined by an effective divisor.

(8.5) **Theorem** (Mumford's vanishing theorem). *Let X be a compact Kähler surface and $\mathcal{L}$ a line bundle on X. If for $n \gg 0$ the bundle $\mathcal{L}^{\otimes n}$ is globally generated and has three algebraically independent sections, then*

$$H^1(\mathcal{L}^\vee) = 0.$$

Proof. If $\mathcal{L}$ is of the form $\mathcal{O}_X(C)$, with C a non-singular connected curve, then the assertion follows from Ramanujam's theorem. The general situation can be reduced to this special case by the following trick. For $n \gg 0$, the bundle $\mathcal{L}^{\otimes n}$ will admit sections vanishing simply on a non-singular curve $C \subset X$, which by Bertini's theorem I.20.2 is connected. Since $\mathcal{L}^{\otimes n} = \mathcal{O}_X(C)$, there exists an n-fold cyclic covering π: $Y \to X$ branched exactly over C. If $D \subset Y$ is the ramification curve, then π: $D \to C$ is an isomorphism and $\pi^*(\mathcal{L}) = \mathcal{O}_Y(D)$. So by Corollary 8.2 $H^1((\pi^*(\mathcal{L}))^\vee) = 0$.

Now it is sufficient to observe that $\mathcal{O}_X \subset \pi_* \mathcal{O}_Y$ is a direct summand and so $\mathcal{L}^\vee \subset \pi_*(\pi^*(\mathcal{L}^\vee)) = \mathcal{L}^\vee \otimes \pi_* \mathcal{O}_Y$ is a direct summand too. Finally, by Leray's spectral sequence

$$H^1(\mathcal{L}^\vee) \subset H^1(\pi_*(\pi^*(\mathcal{L}^\vee))) = H^1(\pi^*(\mathcal{L}^\vee)) = 0. \quad \square$$

Mumford's vanishing theorem contains as a special case a theorem, which is the algebraic formulation of Kodaira's vanishing theorem (for surfaces).

(8.6) **Theorem.** *Let X be a compact surface and $\mathcal{L}$ an ample line bundle on X. Then $h^1(\mathcal{L}^\vee)=0$.*

Practically all results in Sects. 2, 5 and 6 are due to Kodaira, though he never mentions explicitly the Fröhlicher spectral sequence. Most of his results in this direction appear in [Ko4] part I.

Propositions 4.1 and 4.2 go back to Castelnuovo and de Franchis ([Cas2]), whereas Proposition 4.3 can be found in [Ve1] or [Ve2].

Corollary 5.5 is due to Chow and Kodaira (see [C-K]).

A first bound for the number of curves on a surface X with $a(X)=0$ was given in [Ko2] (Theorem 5.0); for the bound given in Sect. 6 see [Kr] and [F-F].

The first examples of almost-complex surfaces without complex structure appeared in [Ve1], and other examples can e.g. be found in [Y1], [Bro], [Bra].

For the results of Sect. 8 we refer to [Ram], [Bom2] and [Mu2].

V. Examples

In this chapter a surface will again mean a connected 2-dimensional complex manifold.

Some Classical Examples

1. The Projective Plane $\mathbb{P}_2$

A universally known and in many ways most basic compact complex surface is the projective plane. Some obvious questions concerning $\mathbb{P}_2$ have fascinated many a geometer, in particular the question (raised by Severi in [Sev]) whether a surface which is homeomorphic to $\mathbb{P}_2$, is also isomorphic to $\mathbb{P}_2$. The last and very difficult step towards the affirmative answer was done only recently by S.-T. Yau. The striking point is that the only known proof uses analysis (hidden in Riemann-Roch for non-algebraic surfaces) and differential geometry as well as the methods of analytic and algebraic geometry.

(1.1) **Theorem.** *Let X be a compact surface with $b_1(X)=0$, $b_2(X)=1$. Then X is algebraic. If furthermore either* (i) $P_2(X)=0$ *or* (ii) $\pi_1(X)$ *is finite, then X is (algebraically) isomorphic to* $\mathbb{P}_2$.

Proof. We either have $b^+(X)=0$ or $b^+(X)=1$. The first possibility is excluded since the index formula would give $c_1^2(X)=3$, leading to a violation of $c_1^2(X)+e(X)\equiv 0$ (12). If $b^+(X)=1$, the index formula yields $c_1^2(X)=9$, and hence X is algebraic by Corollary IV.5.3.

Now let $P_2(X)=0$, and let g be a generator for the free part of $H_2(X,\mathbb{Z})$, such that for some $n>0$ the class ng is the class of a hyperplane section H. Then there is no effective divisor homologous (mod torsion) to ag with $a\leq 0$. We have (again by the index formula) that $c_1(X)=\pm 3g$ (mod torsion). The case $c_1(X)=-3g$ (mod torsion) is excluded since here Riemann-Roch would imply $P_2(X)>0$. In the case that $c_1(X)=3g$ (mod torsion), we take a line bundle $\mathcal{L}$ with $c_1(\mathcal{L})=g$ (this is possible by the exponential cohomology sequence, since $h^{2,0}(X)=0$), and we put $\mathcal{L}=\mathcal{O}_X(L)$. Then Riemann-Roch yields that $\dim|L|\geq 2$. No element of $|L|$ can be reducible, otherwise there would be an effective divisor on X, homologous (mod torsion) to ag, with $a\leq 0$. So $|L|$

has no fixed components. Since $L^2=1$ and $\dim|L|\geq 2$ there can't be any fixed points either. Hence $f_{\mathscr{L}}$ is everywhere defined and maps X birationally onto a surface of degree 1, i.e. onto the projective plane (it follows that $\dim|L|=2$). If $f_{\mathscr{L}}$ would map any curve onto a point, we would have on X two curves which don't intersect, which is impossible since $b_2(X)=1$. So $f_{\mathscr{L}}$ is an isomorphism by Lemma III.4.3. This proves (i). To prove (ii) it is sufficient to show that if $P_2(X)>0$, then $\pi_1(X)$ is infinite. In this case $c_1(X)=-3g$ (mod torsion), and there exists a positive integer m, such that $-mc_1(X)$ is a hyperplane class H' (with torsion taken into account). Therefore by Theorem I.15.2 and Corollary I.15.5 the universal covering of X is the unit ball in $\mathbb{C}^2$, so $\pi_1(X)$ is infinite. □

(1.2) *Remark.* As was shown recently by Mumford in [Mu6], there exists at least one other algebraic surface X with $b_1(X)=0$, $b_2(X)=1$. It follows as before that X is a quotient of the unit ball $E\subset\mathbb{C}^2$. But Mumford has obtained his example in a totally different way and it is not yet known how to obtain it as a quotient of E. As we shall prove later on (Theorem VII.5.2) $f_{\mathscr{K}^{\otimes 5}}$ provides an embedding of X as a surface of degree $25c_1^2(X)=225$ in $\mathbb{P}_{90}$. Since (by a theorem of Calabi and Vesentini, see [C-V]) any small deformation of a quotient of E is trivial, and since the subset on a Chow scheme representing smooth varieties consists of finitely many connected components, we see that in any case there exists only a finite number of surfaces X with $b_1(X)=0$, $b_2(X)=1$. But there might be many "fake projective planes"!

(1.3) **Example.** *Every deformation of* $\mathbb{P}_2$ *is isomorphic to* $\mathbb{P}_2$.

This fact is of course an immediate consequence of Theorem 1.1, (ii) but it can also be proved without making use of Yau's results. In fact, by upper-semi-continuity all plurigenera of a small deformation of $\mathbb{P}_2$ vanish, hence every small deformation of $\mathbb{P}_2$ is isomorphic to $\mathbb{P}_2$ by the elementary part (i) of Theorem 1.1. To complete the proof it is sufficient to show that if we have a 1-dimensional deformation of a surface X, say X_t, with $X_0=X$, such that X_{t_i} is isomorphic to $\mathbb{P}_2$ for a sequence $t_i\to 0$, then also X_0 is isomorphic to $\mathbb{P}_2$. This follows again from semi-continuity, but this time applied to $h^0(\mathscr{K}_{X_t}^{\vee})$, which does not vanish for $t=t_i$, hence does not vanish for $t=0$, so $P_2(X_0)=0$.

(1.4) **Example.** *Let* X *be a compact surface, and* $f\colon\mathbb{P}_2\to X$ *a non-constant holomorphic map. Then* f *is finite and* X *is again* $\mathbb{P}_2$.

Proof. The image $f(\mathbb{P}_2)\subset X$, which is a connected analytic subset of X by Remmert's theorem (Theorem I.8.4) can't be a curve, otherwise we would obtain two curves in $\mathbb{P}_2$ (namely two fibres), which don't intersect. So $f(\mathbb{P}_2)=X$. By Corollary I.1.2 we know that $b_1(X)=0$ and $b_2(X)=0$ or 1. The case $b_2(X)=0$ is not possible: we would have $c_1^2(X)=0$, $c_2(X)=2$ and $c_1^2(X)+c_2(X)\not\equiv 0$ (12). In the second case we find that X is $\mathbb{P}_2$ by Theorem 1.1, (i), since $P_2(X)\leq P_2(\mathbb{P}_2)=0$. Finally, f must be finite, otherwise we could again find a couple of non-intersecting curves on $\mathbb{P}_2$. □

The fact that every surface X with $b_1(X)=0$, $b_2(X)=1$ is algebraic has been known since Kodaira proved our Corollary IV.5.3 in [Ko4]. Theorem 1.1,(i) was proved for Kähler surfaces already in [H-K], whereas Theorem 1.1,(ii) was proved in [Y2]. Example 1.3 is due to Kodaira and Spencer and Example 1.4 was given in [R-V1].

2. Complete Intersections

Let $d_i \in \mathbb{Z}$, $d_i \geq 2$ for $i=1,\ldots,n-2$. A smooth complete intersection of type $(d_1,\ldots,d_{n-2})$ in $\mathbb{P}_n$ is a surface X which is the transversal intersection of $n-2$ hypersurfaces of degree $d_1,\ldots,d_{n-2}$ respectively. (So the hypersurfaces may have singularities, but not along X.) Repeated application of Bertini's theorem I.20.2 yields that $n-2$ "general hypersurfaces" of degree $d_1,\ldots,d_{n-2}$ meet in such a smooth complete intersection.

(2.1) **Proposition.** *If X is a smooth complete intersection of type $(d_1,\ldots,d_{n-2})$, then*

(i) $\pi_1(X)=0$

(ii) $\mathscr{K}_X=\mathcal{O}_X\left(\sum_{i=1}^{n-2} d_i-(n+1)\right)$

(iii) $c_1^2(X)=(\sum d_i-(n+1))^2 \prod_{i=1}^{n-2} d_i$

(iv) $e(X)=\left[\binom{n+1}{2}-(n+1)\sum d_i+\sum d_i^2+\sum_{i\neq j} d_i d_j\right]\prod_{i=1}^{n-2} d_i.$

Proof. Property (i) is an immediate consequence of Lefschetz's theorem I.20.4. The properties (ii), (iii) and (iv) following from the adjunction formula since $\mathcal{N}_{X/\mathbb{P}_n} \cong \mathcal{O}_X(d_1)\oplus\ldots\oplus\mathcal{O}_X(d_{n-2})$. $\square$

Since $h^{0,1}=h^{1,0}=\frac{1}{2}b_1(X)=0$ by (i), and since we know $\chi(X)=\frac{1}{12}(c_1^2(X)+e(X))$ by (iii) and (iv), we can use Noether's formula to obtain $h^{2,0}(X)=h^{0,2}(X)$. From this, together with (i) and (iv) we then can find $h^{1,1}(X)$.

From (ii) we see that

$$\operatorname{kod}(X)=\begin{cases} -\infty & \text{if } X \text{ is of type (2), (3) or (2, 2)} \\ 0 & \text{if } X \text{ is of type (4), (2, 3) or (2, 2, 2)} \\ 2 & \text{otherwise.} \end{cases}$$

A smooth complete intersection of type (2) is a quadric in $\mathbb{P}_3$, hence isomorphic to $\mathbb{P}_1 \times \mathbb{P}_1$ and rational. As is well-known, all smooth intersections of type (3) in $\mathbb{P}_3$ and (2, 2) in $\mathbb{P}_4$ are also rational, and they can be obtained from $\mathbb{P}_2$ by blowing up six and five points respectively.

Reference: [Hir4], Sect. 22.1, [Ha2], p. 395 and [G-H], p. 480 and 550.

3. Tori of Dimension 2

Tori in general were already discussed in Chap. I. Here we shall only prove a special result for the 2-dimensional case, which will later be needed.

Let L be a free $\mathbb{Z}$-module of rank 4. An orientation for L is an isomorphism $\det: \bigwedge^4 L \xrightarrow{\sim} \mathbb{Z}$. It gives rise to a (symmetric) integral bilinear form on $\bigwedge^2 L$ by $(u,v)=\det(u \wedge v)$ $(u, v \in \bigwedge^2 L)$. This form is unimodular.

(3.1) **Proposition.** *Let L and L' be two oriented free $\mathbb{Z}$-modules of rank 4, and let $\varphi: \bigwedge^2 L \to \bigwedge^2 L'$ be an isometry. If the* mod-2 *reduction of φ is of the form $\psi_2 \wedge \psi_2$ for some isomorphism $\psi_2: L \otimes \mathbb{F}_2 \to L' \otimes \mathbb{F}_2$, then $\varphi = \pm \psi \wedge \psi$, where $\psi: L \to L'$ is an isomorphism.*

Proof. For any field k we denote $L \otimes_{\mathbb{Z}} k$ by L_k, $\varphi \otimes id_k = \varphi_k$, etc.

The isotropic lines in $\bigwedge^2 L_{\mathbb{Q}}$ are exactly the lines in $\bigwedge^2 L_{\mathbb{Q}}$ which as points of $\mathbb{P}(\bigwedge^2 L_{\mathbb{Q}})$ are the points of the Plücker quadric $Gr(1, \mathbb{P}(L_{\mathbb{Q}}))$. So φ induces a projective linear map $\mathbb{P}(\varphi_{\mathbb{Q}})$ from $\mathbb{P}(\bigwedge^2 L_{\mathbb{Q}})$ onto $\mathbb{P}(\bigwedge^2 L'_{\mathbb{Q}})$ which maps $Gr(1, \mathbb{P}(L_{\mathbb{Q}}))$ onto $Gr(1, \mathbb{P}(L'_{\mathbb{Q}}))$. Now on these Grassmann varieties there are two systems of 2-dimensional linear spaces: those given by the lines through a point of $\mathbb{P}(L_{\mathbb{Q}})$ or $\mathbb{P}(L'_{\mathbb{Q}})$, and those given by the lines in a plane of $\mathbb{P}(L_{\mathbb{Q}})$ or $\mathbb{P}(L'_{\mathbb{Q}})$. The map $\mathbb{P}(\varphi_{\mathbb{Q}})$ either maps the planes of the first system onto the planes of the first system or it maps the planes of the first system onto the planes of the second system. The "main theorem of projective geometry" tells us that in the first case $\mathbb{P}(\varphi_{\mathbb{Q}})$ is induced by an isomorphism from $L_{\mathbb{Q}}$ onto $L'_{\mathbb{Q}}$, whereas in the second case $\mathbb{P}(\varphi_{\mathbb{Q}})$ is induced by an isomorphism from $L_{\mathbb{Q}}$ onto $(L'_{\mathbb{Q}})^{\vee}$. Since by assumption we are in the first case when we reduce modulo 2, we must also be in the first case after tensoring with $\mathbb{Q}$. In other words: there exists an isomorphism $\tilde{\psi}: L_{\mathbb{Q}} \to L'_{\mathbb{Q}}$, such that $\tilde{\psi} \wedge \tilde{\psi} = \mu \varphi_{\mathbb{Q}}$, with $\mu \in \mathbb{Q}$. Replacing $\tilde{\psi}$ by $n\tilde{\psi}$ $(n \in \mathbb{N})$ we can achieve that $\tilde{\psi}(L) \subset L'$. Now given a sublattice M' of a lattice M, there is always a sublattice M'' of M, with $M' = mM''$ $(m \in \mathbb{N})$, such that M'' contains a primitive vector. So multiplying $\tilde{\psi}$ again with a suitable rational number, we obtain an isomorphism from $L_{\mathbb{Q}}$ onto $L'_{\mathbb{Q}}$ which we denote again by $\tilde{\psi}$, such that

(i) $\tilde{\psi} \wedge \tilde{\psi} = \lambda \varphi$, with $\lambda \in \mathbb{Q}$,
(ii) $\tilde{\psi}(L) \subset L'$,
(iii) L' contains a primitive vector $\tilde{\psi}(e_1)$.

Then e_1 is primitive itself, and therefore can be complemented to a base $e_1, \ldots, e_4$ of L. Since $\tilde{\psi}(e_1) \wedge \tilde{\psi}(e_i) = \lambda \varphi(e_1 \wedge e_i)$, and since $\varphi(e_i \wedge e_j)$ is primitive, we have that $\lambda \in \mathbb{Z}$. Now $\tilde{\psi}(e_1) \wedge \tilde{\psi}(e_i) \in \lambda(\bigwedge^2 L')$, so after replacing e_i by a suitable sum $e_i + p_i e_1$ $(p_i \in \mathbb{Z})$ we have $\tilde{\psi}(e_i) \in \lambda L'$. Hence $\tilde{\psi}(e_2) \wedge \tilde{\psi}(e_3) = \lambda \varphi(e_2 \wedge e_3) \in \lambda^2 \bigwedge^2 L'$. But $\varphi(e_2 \wedge e_3)$ is primitive, therefore $\lambda = \pm 1$. To conclude that $\tilde{\psi} = \psi_{\mathbb{Q}}$, with ψ as described in the proposition, it remains to be shown that $\tilde{\psi}(L) = L'$. In other words we still have to show that all elementary divisors of $\tilde{\psi}$, considered as a homomorphism from L into L', are equal to 1. This, however, is an immediate consequence of the fact that all elementary divisors of φ are equal to 1. □

For a 2-torus $T=V/\Gamma$, the lattice $H^1(T,\mathbb{Z})$ carries a natural orientation, induced by the complex structure, for $\bigwedge^4 H^1(T,\mathbb{Z}) \cong H^4(T,\mathbb{Z})$. In fact, the abstract form on $\bigwedge^2 H^1(T,\mathbb{Z})$ given by the orientation is nothing but the usual quadratic form on $H^2(T,\mathbb{Z})$.

(3.2) **Theorem.** *Let T, T' be 2-tori, and let $\varphi: H^2(T',\mathbb{Z}) \to H^2(T,\mathbb{Z})$ be an isometry which preserves the Hodge decomposition. If there exists an isomorphism $\psi_2: H^1(T',\mathbb{Z}_2) \to H^1(T,\mathbb{Z}_2)$, such that $\psi_2 \wedge \psi_2$ equals the* mod-2 *reduction of φ, then $\pm\varphi$ is induced by an isomorphism from T onto T'.*

Proof. From the remarks above and from Proposition 3.1 it follows that $\varphi = \pm\psi \wedge \psi$, where $\psi: H^1(T',\mathbb{Z}) \to H^1(T,\mathbb{Z})$ is an isomorphism. We claim that ψ induces an isomorphism $\psi_{\mathbb{C}}: H^1(T',\mathbb{C}) \to H^1(T,\mathbb{C})$, which preserves the Hodge decomposition. This immediately follows from the fact that $\psi_{\mathbb{C}} \wedge \psi_{\mathbb{C}}$ maps the point on $Gr(1,\mathbb{P}(H^1(T',\mathbb{C})))$ corresponding to $H^{1,0}(T')$, to the point on $Gr(1,\mathbb{P}(H^1(T,\mathbb{C})))$ corresponding to $H^{1,0}(T)$. Indeed, these points correspond to $H^{2,0}(T')$, resp. $H^{2,0}(T)$, and $\psi_{\mathbb{C}} \wedge \psi_{\mathbb{C}} = \varphi_{\mathbb{C}}$ preserves the Hodge decomposition. So we can apply the Torelli theorem I.14.2 to obtain the desired result. □

The results of this section have been known for some time and they appear in [L-P], Sect. 3.

Fibre Bundles

4. Ruled Surfaces

A *ruled surface* is a compact surface which admits a ruling, more precisely, a compact surface which (in at least one way) is the total space of an analytic fibre bundle with fibre $\mathbb{P}_1$ and structural group $PGL(2,\mathbb{C})$ over a smooth, connected curve B. Actually, as we shall presently see, such an analytic fibre bundle is always equivalent to an algebraic one.

Remark. Some authors use "ruled surface" for ruled surfaces embedded in a projective space, such that all fibres are lines.

It will be proved later that, with the single exception of $\mathbb{P}_1 \times \mathbb{P}_1$, a surface admits at most one ruling.

Obvious examples are provided by the projective bundles $\mathbb{P}(\mathscr{V})$ of algebraic 2-vector bundles over a smooth, connected, compact curve. In fact there are no other examples, as follows from

(4.1) **Proposition.** *Any analytic fibre bundle with fibre $\mathbb{P}_n$ and structural group $PGL(n+1,\mathbb{C})$ over a smooth, compact curve B is isomorphic to $\mathbb{P}(\mathscr{V})$, where $\mathscr{V}$ is an algebraic $(n+1)$-vector bundle over B.*

Proof. The general theory of fibre bundles yields an "exact sequence" of cohomology sets

$$\to H^1(B, \mathcal{GL}(n+1, \mathbb{C})) \to H^1(B, \mathcal{PGL}(n+1, \mathbb{C})) \to H^2(B, \mathcal{O}_B^*),$$

where $\mathcal{GL}(n+1, \mathbb{C})$, $\mathcal{PGL}(n+1, \mathbb{C})$ and $\mathcal{O}_B^*$ are the sheaves of germs of analytic maps from B into $GL(n+1, \mathbb{C})$, $PGL(n+1, \mathbb{C})$ and $\mathbb{C}^*$ respectively, and where $H^1(B, \mathcal{GL}(n+1, \mathbb{C}))$ (resp. $H^1(B, \mathcal{PGL}(n+1, \mathbb{C}))$) classify analytic $(n+1)$-vector bundles (resp. $\mathbb{P}_n$-bundles) over B. Since B is a curve, we immediately find from the exponential cohomology sequence that $H^2(B, \mathcal{O}_B^*)=0$. The proposition follows from this and from GAGA (every analytic vector bundle over B is algebraic). □

As a consequence, every ruled surface with base B is birationally equivalent to $B \times \mathbb{P}_1$.

Remark. The preceding proof also works if dim $B \geq 2$, provided that $H^2(B, \mathcal{O}_B) = H^3(B, \mathbb{Z}) = 0$ (the proposition is however not true for any smooth base).

Actually, a ruling can be characterised as a fibering over a smooth connected, compact curve B, such that all fibres are isomorphic to $\mathbb{P}_1$. In that case the projection is everywhere of maximal rank, and the statement is an immediate consequence of the Grauert-Fischer theorem (Theorem I.10.1). A direct proof can be given in the following way.

Let $f: X \to B$ be a fibering such that all fibres are isomorphic to $\mathbb{P}_1$. The adjunction formula implies that all fibres are smooth, and from the differentiable point of view, f is a locally trivial fibration by Ehresmann's theorem (compare [M-K], p. 19). So, by elementary obstruction theory there is a section. This implies that topologically there is a complex line bundle $\mathcal{L}$, such that $\mathcal{L}|F \cong \mathcal{O}_F(1)$ for every fibre F. Since $p_g(X) = h^2(\mathcal{O}_X) = 0$ (a canonical divisor has negative intersection number with all fibres) the dual cohomology class of this section is the Chern class of a holomorphic line bundle by the exponential cohomology sequence (I, Sect. 6). Let $\mathcal{M}$ be some very ample line bundle on B and let $\mathcal{L}_n = f^*(\mathcal{M}^{\otimes n}) \otimes \mathcal{L}$. We claim that for n large enough $f_{\mathcal{L}_n}: X \to \mathbb{P}_N$ is everywhere defined, mapping every F isomorphically onto a line in $\mathbb{P}_N$. This will follow as soon as we know that $H^1(X, \mathcal{L}_n \otimes \mathcal{O}_X(-F)) = 0$. By Leray's spectral sequence this last group is equal to $H^1(B, f_* \mathcal{L} \otimes \mathcal{O}_X(-F) \otimes \mathcal{M}^{\otimes n})$ and vanishes for n large enough by Theorem B (see [Se2]). This means that we obtain a regular map from B into the Grassmann variety $Gr(1, N)$, such that X as a fibre space is the pull-back of the universal subbundle on $Gr(1, N)$. So f is locally trivial as claimed.

If $B = \mathbb{P}_1$, then by a theorem of Grothendieck ([Gk1]) every algebraic vector bundle over B is isomorphic to a direct sum of line bundles. So in this case every ruled surface over $\mathbb{P}_1$ is of the form $\mathbb{P}(\mathcal{O}_{\mathbb{P}_1} \oplus \mathcal{O}_{\mathbb{P}_1}(n))$ for some $n \geq 0$ (since Pic $(\mathbb{P}_1) \cong \mathbb{Z}$ and $\mathbb{P}(\mathcal{V} \otimes \mathcal{L}) \cong \mathbb{P}(\mathcal{V})$ for any algebraic line bundle $\mathcal{L}$ on B). As is customary, we denote the surface $\mathbb{P}(\mathcal{O}_{\mathbb{P}_1} \oplus \mathcal{O}_{\mathbb{P}_1}(n))$ by Σ_n and call it the

n-th *Hirzebruch surface*. The surfaces Σ_n are birationally equivalent to $\mathbb{P}_1 \times \mathbb{P}_1$, hence to $\mathbb{P}_2$, so they are all rational. The surfaces Σ_n can be characterised in many ways. For example, Σ_0 is $\mathbb{P}_1 \times \mathbb{P}_1$, Σ_1 is $\mathbb{P}_2$ blown up in one point, and Σ_n, $n \geq 2$, is obtained by desingularising the cone in $\mathbb{P}_{n+1}$ over a rational normal curve spanning $\mathbb{P}_n$ (see [G-H], p. 523).

Let $n \geq 1$ and $C_n = \mathbb{P}(\mathcal{O}_{\mathbb{P}_1}(n)) \subset \Sigma_n$. We claim that $C_n^2 = -n$. To see this it is sufficient to prove that $\mathcal{N}_{C_n/\Sigma_n} \cong \mathcal{O}_{C_n}(-n)$ (Proposition I.6.2). This follows by restricting the standard sequence I, (13) to C_n, since the tautological bundle, restricted to C_n, is $\mathcal{O}_{C_n}(n)$, whereas the bundle along the fibres restricted to C_n is $\mathcal{N}_{C_n/\Sigma_n}$. We furthermore claim that any irreducible curve D on Σ_n with $D^2 \leq 0$ is either C_n or a fibre F. Indeed, such a curve D is homologous to $rC_n + sF$, with $r, s \in \mathbb{Z}$. Now if D is neither C_n nor a fibre, then $(rC_n + sF)\, F > 0$ and $(rC_n + sF)\, C_n \geq 0$, but these inequalities are incompatible with $(rC_n + sF)^2 \leq 0$.

In particular we have

(4.2) **Proposition.** *The Hirzebruch surfaces* Σ_n, $n \geq 0$, *are all biregularly distinct. Except for* $\Sigma_0 = \mathbb{P}_1 \times \mathbb{P}_1$, *each such surface has only one ruling.*

Remark. Since any surface with two different rulings is a $\mathbb{P}_1$-bundle over $\mathbb{P}_1$ by Lüroth's theorem, we have shown at the same time that $\mathbb{P}_1 \times \mathbb{P}_1$ is the only surface with more than one ruling.

If B is elliptic, then there are 2-bundles on B which don't split (that is, which are not direct sum of two 1-subbundles). To see this we start from an *extension*

$$0 \to \mathcal{O}_B \xrightarrow{i} \mathcal{V} \to \mathcal{O}_B \to 0$$

which does not split (this is possible since these extensions are classified by $H^1(B, \mathcal{O}_B \otimes \mathcal{O}_B^\vee) = H^1(B, \mathcal{O}_B) \cong \mathbb{C}$, compare [Gk1]) and show that the *bundle* $\mathcal{V}$ does not split either. Indeed, let us assume that $\mathcal{V} \cong \mathcal{L}_1 \oplus \mathcal{L}_2$, with $\mathcal{L}_1$ and $\mathcal{L}_2$ 1-subbundles of $\mathcal{V}$. If, say $\mathcal{L}_1$, were $i(\mathcal{O}_B)$, then $\mathcal{L}_2$ would also be isomorphic to $\mathcal{O}_B$ (for $\mathcal{L}_1 \otimes \mathcal{L}_2 \cong \mathcal{O}_B$) and the extension would split. If neither $\mathcal{L}_1$ nor $\mathcal{L}_2$ were $i(\mathcal{O}_B)$, then, since $\mathcal{L}_1 \otimes \mathcal{L}_2 \cong \mathcal{O}_B$, and both line bundles would admit non-trivial homomorphisms onto $\mathcal{O}_B$, we would have $\mathcal{L}_1 \cong \mathcal{L}_2 \cong \mathcal{O}_B$. It would follow that, say, $\mathcal{L}_1$ and $i(\mathcal{O}_B)$ span $\mathcal{V}$ everywhere, which is impossible.

Similarly there exists an extension

$$0 \to \mathcal{O}_B \to \mathcal{W} \to \mathcal{L} \to 0,$$

such that the bundle $\mathcal{W}$ does not split, where now $\mathcal{L}$ is some line bundle of degree 1. Atiyah has shown in [At1] that $\mathbb{P}(\mathcal{V})$ and $\mathbb{P}(\mathcal{W})$ are the only ruled surfaces over an elliptic curve B which are not the projective bundle of a splitting vector bundle of rank 2.

For the classification of ruled surfaces over a base of genus ≥ 2 we refer to [Tj1], [Tj2].

(4.3) **Proposition.** *Let X be a compact surface and C a smooth rational curve on X.*

(i) *if $C^2=0$, then there exists a modification $\varphi: X \to Y$, where Y is ruled, such that C meets no exceptional curve of φ, and $\varphi(C)$ is a fibre of Y;*
(ii) *if $C^2>0$, then X is either $\mathbb{P}_2$, a Hirzebruch surface, or a blown-up of one of these surfaces.*

Proof. (i) We start by observing that X is algebraic by Theorem IV.5.2, for $c_1^2(\mathcal{O}_X(nC)\otimes\mathcal{K}_X^\vee)>0$ for n large enough.

This said, we distinguish between two cases: $q(X)=0$ and $q(X)>0$.

In the first of these cases we see immediately, considering the exact cohomology sequence of

$$0\to\mathcal{O}_X\to\mathcal{O}_X(C)\to\mathcal{O}_C(C)\to 0$$

that $h^0(\mathcal{O}_X(C))=2$. Since the irreducible curve C is a member of $|C|$, also the general member of $|C|$ is irreducible, and $f=f_{\mathcal{O}_X}(C)$ provides a regular map onto $\mathbb{P}_1$. The general fibre of f is again rational, since its intersection number with K_X *is* -2. Now suppose there is a non-smooth fibre, say $\sum_{i=1}^{m} c_i C_i$. Because of the adjunction formula we must have $m\geq 2$. Since $K_X(\sum c_i C_i)=-2$, there must be a C_i, say C_1, such that $KC_1<0$. By Zariski's Lemma III.8.2 we have $C_1^2<0$, and we find from the adjunction formula that $K_X C_1=C_1^2=-1$, i.e. C_1 is a (-1)-curve (Proposition III.2.2). So after blowing down X a finite number of times, we obtain a surface Y admitting a holomorphic map onto $\mathbb{P}_1$, such that all fibres are smooth rational curves, i.e. a Hirzebruch surface.

On the other hand, if $q(X)>0$ we use the Albanese map $f': X\to \mathrm{Alb}(X)$ (since X is algebraic, $\dim \mathrm{Alb}(X)=q(X)>0$ and f' is not a constant map). Let $f'=h'g'$ be the Stein factorisation of f'. If the general fibre of g' would be a point, then C^2 would be strictly negative by Theorem III.2.1. So f' maps X onto a curve $D'\subset \mathrm{Alb}(X)$. If D is the desingularisation of D', then f' can be lifted to $f: X\to D$. By Zariski's Lemma the fibre of f containing C must be of the form aC, $a\geq 1$, but the adjunction formula immediately yields $a=1$. Consequently, also the general fibre of f is rational. The proof can now be completed in the same way as in the case $q(X)=0$.

(ii) If X contains a smooth rational curve A with $A^2=0$, then X is a blown-up ruled surface by (i). Since C is mapped onto the base of any ruling, by Lüroth's theorem this base is rational, so in this case we are ready.

Now suppose that X does not contain any smooth rational A with $A^2=0$. Let $k=\min(B^2)$, B smooth rational, $B^2\geq 1$, and let D be smooth rational with $D^2=k$. Without limiting the generality it may be assumed that all (-1)-curves on X meet D. We blow up k points $x_1,\ldots,x_k\in D$, obtaining (-1)-curves $E_1,\ldots,E_k$ on the resulting surface $\bar{X}$. Since for the proper transform $\bar{D}$ of D we have $\bar{D}^2=0$, the surface $\bar{X}$ is blown-up ruled with $\bar{D}$ a fibre of a ruling $p: \bar{X}\to R$. Each E_i is a section of this ruling, so R is rational. If all fibres of p are irreducible, then $\bar{X}$ is a surface Σ_m. Since Σ_m only contains a (-1)-curve if $m=1$, we must have $m=k=1$ and X is $\mathbb{P}_2$. On the other hand, if p would have a reducible fibre, then this fibre would contain an exceptional curve E which would meet at least one E_i, otherwise we would have on X a (-1)-curve not meeting D. But if E would meet some E_i, then it would meet this curve transversally in one point, so the image F of E on X would be smooth rational with $0 \leq F^2 \leq k-1$, which is impossible by assumption. □

5. Elliptic Fibre Bundles

If E is an elliptic curve, we denote by $A(E)$ the group of its biholomorphic automorphisms. After fixing an origin $0 \in E$ this group can be described in the following way: E, acting on itself by translations, forms a normal subgroup of $A(E)$, and the quotient $A(E)/E$ can be identified with the group of automorphisms leaving 0 fixed. So this quotient is the cyclic group $\mathbb{Z}_n$ of order

$$\begin{aligned} n&=4 && \text{if } E=\mathbb{C}/\mathbb{Z}\oplus\mathbb{Z}\sqrt{-1}, \\ n&=6 && \text{if } E=\mathbb{C}/\mathbb{Z}\oplus\mathbb{Z}\omega, \quad \omega=\mathrm{e}(1/6), \\ n&=2 && \text{in all other cases.} \end{aligned}$$

Then $A(E)$ is the semi-direct product $E \times \mathbb{Z}_n$. We write its elements as (e, z), $e \in E$, $z \in \mathbb{Z}_n$,

$$(e, z)\colon x \to e + zx, \qquad x \in E.$$

The group operation is

$$(e, z)(e', z') = (e + ze', zz').$$

The translation group E is described by the universal covering sequence

(1) $$0 \to \Gamma \xrightarrow[j]{} \mathbb{C} \to E \to 0, \qquad \Gamma = \mathbb{Z}\oplus\mathbb{Z}.$$

If B is any smooth, compact, connected curve, then the holomorphic fibre bundles with typical fibre E and base B are classified by the cohomology set $H^1(\mathscr{A}_B)$ and there is an exact sequence of coholomogy *sets*

(2) $$H^1(\mathscr{E}_B) \to H^1(\mathscr{A}_B) \to H^1(B, \mathbb{Z}_n).$$

(Here $\mathscr{A}_B$, resp. $\mathscr{E}_B$, is the sheaf of germs of local holomorphic maps from B to $A(E)$, resp. E.) We shall call a bundle $X \to B$ a *principal* bundle if its structure group can be reduced to E. To describe $H^1(\mathscr{E}_B)$ we use the exact cohomology sequence:

(3) $$H^1(B, \Gamma) \to H^1(\mathcal{O}_B) \to H^1(\mathscr{E}_B) \xrightarrow{c} H^2(B, \Gamma) \to 0$$

which is induced by (1).

(5.1) **Lemma.** a) *The bundle $X \to B$ with typical fibre E is principal if and only if X admits an action of the group E which on all fibres X_b, $b \in B$, induces the translation group.*

b) *Two principal E-bundles defined by cocycles $\xi = \{\xi_{ij}\}$ and $\xi' = \{\xi'_{ij}\}$ are isomorphic as $A(E)$-bundles if and only if there is some $z \in \mathbb{Z}_n$ such that $\xi' = z\xi$ in $H^1(\mathscr{E}_B)$.*

c) *A principal E-bundle with class $\xi \in H^1(\mathscr{E}_B)$ can be defined by a locally constant cocycle if and only if $c(\xi) = 0$.*

Proof. a) E being abelian, the "only if" part is trivial. So let $\xi_{ij}=(e_{ij},z_{ij})$ be a cocycle with values in $\mathscr{A}_B$ defining X, and assume that the structure group cannot be reduced to E. Because of (2) this means that the cocycle $\{z_{ij}\}$ defines a non-zero class $z\in H^1(B,\mathbb{Z}_n)$. Any automorphism t of X, leaving invariant all fibres and acting as a translation on each of them, is given by a collection $t_i\colon U_i\to E$ such that

$$t_i\,\xi_{ij}=\xi_{ij}\,t_j, \quad \text{i.e.,}$$
$$(t_i+e_{ij},z_{ij})=(e_{ij}+z_{ij}\,t_j,z_{ij}), \quad \text{i.e.,}$$
$$t_i=z_{ij}\,t_j.$$

So the collection $\{t_i\}$ is nothing but a section in the bundle $E^{(z)}$ obtained from the trivial bundle $E\times B$ by twisting with the cocycle $\{z_{ij}\}$. Now $\mathbb{Z}_n$ operates equivariantly on the sequence (1). Therefore, this sequence may be treated by $\{z_{ij}\}$, which yields an exact sequence

$$0\to\Gamma^{(z)}\to\mathcal{O}_B^{(z)}\to E^{(z)}\to 0.$$

Here $\mathcal{O}_B^{(z)}\in\operatorname{Pic}(B)$ is a non-trivial torsion bundle, because $H^1(B,\mathbb{Z}_n)\to H^1(\mathcal{O}_B^*)$ is injective. So $H^0(\mathcal{O}_B^{(z)})=0$ and the group of sections in $E^{(z)}$ must be discrete. This means that X does not admit enough automorphisms to induce all translations on the fibres.

b) ξ and ξ' define the same class in $H^1(\mathscr{A}_B)$ if and only if there are maps $h_i\colon U_i\to A(E)$ such that $h_i\,\xi_{ij}=\xi'_{ij}\,h_j$. If we write $h_i=(e_i,z_i)$ as above, then these equations become

$$(e_i+z_i\,\xi_{ij},z_i)=(\xi'_{ij}+e_j,z_j).$$

This means $z_i=z_j=z\in\mathbb{Z}_n$ and $\xi'_{ij}-z\,\xi_{ij}=e_i-e_j$.

c) Let as usual $\mathbb{C}_B$, resp. E_B, be the sheaf of germs of locally constant maps $B\to\mathbb{C}$, resp. $B\to E$. The diagram of sheaves

$$\begin{array}{ccccccccc}
0 & \longrightarrow & \Gamma & \longrightarrow & \mathbb{C}_B & \longrightarrow & E_B & \longrightarrow & 0\\
 & & \| & & \downarrow & & \downarrow & & \\
0 & \longrightarrow & \Gamma & \longrightarrow & \mathcal{O}_B & \longrightarrow & \mathscr{E}_B & \longrightarrow & 0
\end{array}$$

leads to this diagram in cohomology:

$$\begin{array}{ccccccc}
H^1(B,\mathbb{C}) & \longrightarrow & H^1(B,E) & \xrightarrow{\ \gamma\ } & H^2(B,\Gamma) & \longrightarrow & H^2(B,\mathbb{C})\\
\downarrow & & \downarrow & & \| & & \\
H^1(\mathcal{O}_B) & \longrightarrow & H^1(\mathscr{E}_B) & \longrightarrow & H^2(B,\Gamma) & &
\end{array}$$

Since $H^2(B,\mathbb{Z})$ is free of torsion, the map $H^2(B,\Gamma)\to H^2(B,\mathbb{C})$ is injective and $\gamma=0$. Because $H^1(B,\mathbb{C})\to H^1(\mathcal{O}_B)$ is surjective, one finds that $c(\xi)=0$ if and only if ξ comes from $H^1(B,E)$. □

Next we show that every principal bundle with typical fibre E admits as unramified covering a holomorphic $\mathbb{C}^*$-bundle. There is a canonical isomorphism $\Gamma \to H^2(B,\Gamma) = H^2(B,\mathbb{Z}) \otimes_{\mathbb{Z}} \Gamma$. For $\xi \in H^1(\mathscr{E}_B)$ given, we have a primitive embedding $\mathbb{Z} \xrightarrow{i} \Gamma$ such that $c(\xi)$ is in the image of the induced map $H^2(B,\mathbb{Z}) \to H^2(B,\Gamma)$. Chasing the cohomology diagram of

$$\begin{array}{ccccccccc} 0 & \longrightarrow & \mathbb{Z} & \longrightarrow & \mathbb{C} & \xrightarrow{\mathfrak{e}} & \mathbb{C}^* & \longrightarrow & 0 \\ & & \big\downarrow{\scriptstyle i} & & \big\| & & \big\downarrow{\scriptstyle q} & & \\ 0 & \longrightarrow & \Gamma & \longrightarrow & \mathbb{C} & \longrightarrow & E & \longrightarrow & 0 \end{array}$$

i.e.

$$\begin{array}{ccccccc} H^1(\mathcal{O}_B) & \longrightarrow & H^1(\mathcal{O}_B^*) & \longrightarrow & H^2(B,\mathbb{Z}) & \longrightarrow & 0 \\ \big\| & & {\scriptstyle H^1(q)}\big\downarrow & & \big\downarrow & & \\ H^1(\mathcal{O}_B) & \longrightarrow & H^1(\mathscr{E}_B) & \longrightarrow & H^2(B,\Gamma) & \longrightarrow & 0, \end{array}$$

where q is the quotient with respect to some subgroup $\mathbb{Z}.\tau$, $|\tau|>1$, of $\mathbb{C}^*$, we obtain

(5.2) **Proposition.** *For every $\xi \in H^1(\mathscr{E}_B)$ there is some $\eta \in H^1(\mathcal{O}_B^*)$ with $\xi = H^1(q)\eta$, i.e. the bundle space X of ξ is a $\mathbb{Z}$-quotient of the total space of η where the $\mathbb{Z}$-action is generated by multiplication with τ.*

Since the embedding $i: \mathbb{Z} \to \Gamma$ is primitive, we may complement it to obtain an isomorphism $i \times j: \mathbb{Z} \times \mathbb{Z} \to \Gamma$. Topologically, this induces a splitting $E = S^1 \times S^1$. The topological version of (3) shows that, topologically, principal E-bundles are classified by their class in $H^2(B,\Gamma)$. In our case this implies that topologically $X = C \times S^1$, where C is the S^1-bundle over B determined by η.

The bundle $C \to B$ has a Gysin sequence ([M-S], p. 143)

$$\begin{array}{ccccccccc} 0 & \longrightarrow & H^1(B,\mathbb{Z}) & \longrightarrow & H^1(C,\mathbb{Z}) & \longrightarrow & H^0(B,\mathbb{Z}) & \xrightarrow{\delta} & \\ & \longrightarrow & H^2(B,\mathbb{Z}) & \longrightarrow & H^2(C,\mathbb{Z}) & \longrightarrow & H^1(B,\mathbb{Z}) & \longrightarrow & 0 \end{array}$$

where δ is multiplication with the Chern class $c(\eta) = c(\xi)$.

(5.3) **Proposition.** (i) *If $c(\xi)=0$, i.e. the bundle $X \to B$ is topologically trivial, then $b_1(X) = b_1(B)+2$ and $b_2(X) = 2b_1(B)+2$,*

(ii) *If $c(\xi) \neq 0$ (so $X \to B$ is not topologically a product), then $b_1(X) = b_1(B)+1$ and $b_2(X) = 2b_1(B)$.*

Proof. (i) is an immediate consequence of the Künneth formula. To prove (ii) we deduce from the Gysin sequence $b_1(C) = b_1(B)$ and $b_2(C) = b_2(B) + b_1(B) - 1 = b_1(B)$. Applying the Künneth formula to $C \times S^1$ we obtain the result. □

These preliminaries being out of the way, we describe explicitly the situation for $B = \mathbb{P}_1$ or B an elliptic curve.

A) The Case $B = \mathbb{P}_1$

Since B is simply connected, by (2) the structural group reduces to the translation group, i.e. every bundle is a principal bundle. Furthermore $H^1(\mathcal{O}_{\mathbb{P}_1}) = 0$, so (3) shows that the topological and analytical classification for these bundles is the same. In particular, in Proposition 5.3 we may replace the term "topologically" by "analytically".

(5.4) **Theorem.** *Any elliptic fibre bundle over* $\mathbb{P}_1$ *is either a product or a Hopf surface* (see Sect. 18).

Proof. Let $X \to \mathbb{P}_1$ be an elliptic fibre bundle with $c(\xi) \neq 0$. We have to show that X is a Hopf surface. By Proposition 5.2 we may assume that $X \to \mathbb{P}_1$ comes from a $\mathbb{C}^*$-bundle $Y \to \mathbb{P}_1$. The surface Y is an unramified covering of X. Since a Hopf surface is defined as a surface for which the universal covering is isomorphic to $W = \mathbb{C}^2 \setminus \{0\}$, it thus suffices to exhibit an unramified covering $W \to Y$. However, W is nothing but the bundle space of the $\mathbb{C}^*$-bundle of degree -1 over $\mathbb{P}_1$, and this bundle space can be mapped onto Y fibre-wise by $z \to z^{-d}$, where d is the degree of $Y \to \mathbb{P}_1$, which does not vanish since $X \to \mathbb{P}_1$ is not the product bundle. □

B) The Case of an Elliptic Base B

We distinguish between the following cases.

BI) Principal bundles:
BIa) bundles defined by cocycles $\xi \in H^1(\mathscr{E}_B)$ with $c(\xi) = 0$,
BIb) bundles defined by cocycles $\xi \in H^1(\mathscr{E}_B)$ with $c(\xi) \neq 0$.
BII) Non-principal bundles.

BIa) Because of Lemma 5.1.c), the bundle $X \to B$ is defined over some covering $\{U_i\}$ of B by patching the pieces $E \times U_i$ according to a locally constant cocycle $\{\xi_{ij}\}$ with $\xi_{ij} \in E$. On each product $E \times U_i$ we consider two kinds of holomorphic tangent vector fields: those coming from E and the ones coming from $B \supset U_i$. Since the cocycle is locally constant, both kinds of vector fields are respected by the patching procedure. This means they extend to all of X and span T_X everywhere. The two one-parameter subgroups of $\mathrm{Aut}(X)$, generated by the two types of vector fields, commute, so they generate an abelian complex Lie group operating transitively on X. This shows that all surfaces of the form considered are quotients of $\mathbb{C}^2$ by a lattice, i.e. are complex tori.

BIb) *Bundles Defined by Cocycles* ξ *with* $c(\xi) \neq 0$.
Such surfaces are called primary Kodaira surfaces. Their invariants are

$$H^1(X, \mathbb{Z}) = \mathbb{Z}^3, \quad H^2(X, \mathbb{Z}) = \mathbb{Z}^4 \quad \text{or} \quad \mathbb{Z}^4 \oplus \mathbb{Z}_m,$$
$$e(X) = 0, \quad h^1(\mathcal{O}_X) = 2, \quad h^2(\mathcal{O}_X) = 1, \quad \mathscr{K}_X = \mathcal{O}_X.$$

The topological invariants follow from Proposition 5.3. As to the other invariants, consider the direct image sheaf $f_{*1}\mathcal{O}_X$ on B, where $f: X \to B$ is the projection. Since translations operate trivially on $H^1(\mathcal{O}_E)$, the line bundle $f_{*1}\mathcal{O}_X$ on B is trivial. From the exact sequence

$$0 \longrightarrow H^1(\mathcal{O}_B) \xrightarrow{f^*} H^1(\mathcal{O}_X) \longrightarrow H^0(f_{*1}\mathcal{O}_X) \longrightarrow 0 \tag{4}$$

we find $h^1(\mathcal{O}_X)=2$ and from

(5) $$0\to f^*(\omega_B)\to\Omega_X\to\omega_{X/B}\to 0$$

we deduce that $\mathcal{K}_X$ is trivial, because there are the isomorphisms

$$\mathcal{K}_X=\omega_{X/B}=f^*(f_*\,\omega_{X/B})=f^*(f_{*1}\,\mathcal{O}_X),$$

the last one being relative duality (Theorem III.12.3).

Notice that because of $b_1(X)=3$, Kodaira surfaces are *not kählerian.*

Sometimes a primary Kodaira surface admits a finite, freely-operating group of automorphisms. The smooth quotients thus obtained are called *secondary Kodaira surfaces.* An example can be constructed in the following way.

Let B be any elliptic curve, and $p\in B$. We consider the line bundle $\mathcal{O}_B(p)$, and denote by L the total space of the associated principal $\mathbb{C}^*$-bundle. Let $a\in\mathbb{C}^*$, $|a|\neq 1$, and let $g_a\colon L\to L$ be the automorphism obtained by multiplication with a in each fibre. The quotient $X=L/\langle g_a\rangle$ is an elliptic fibre bundle over B. Since $c_1(\mathcal{O}_B(p))\neq 0$, the diagram preceding Proposition 5.2 shows that X is a primary Kodaira surface. We want to construct a fixpoint-free involution $\rho\colon X\to X$. To do this, we start from an involution $\iota\colon B\to B$ which has p as a fixpoint. Since $\iota^*(\mathcal{O}_B(p))\cong\mathcal{O}_B(p)$, there exists a biholomorphic map $\alpha\colon L\to L$, covering ι. Upon multiplication with a suitable automorphism of L we obtain a biholomorphic map $\beta\colon L\to L$, covering ι, with $\beta^2=\mathrm{id}_L$. Now we can simply take $\rho=g_{\sqrt{a}}\circ\beta$.

A secondary Kodaira surface Y satisfies $b_1(Y)=1$, $b_2(Y)=0$, $q(Y)=1$, $p_g(Y)=0$. This will follow from our considerations in VI, Sect. 4.

BII) *Non-Principle Bundles*

Now we consider a bundle $X\to B$ given by a class $\xi\in H^1(\mathcal{A}_B)$ which has a non-trivial image $\zeta\in H^1(B,\mathbb{Z}_n)$ under (1). To study X, the first thing to do is to form the cyclic covering $B'\to B$ of order $m, m|n$, killing ζ. The pull-back X' of X to B' is a principal bundle over B'. Since X' is an unramified cyclic m-fold covering of X, it admits a cyclic group $\mathbb{Z}_m$ operating compatibly with the fibration $X'\to B'$. Let $\sigma\in\mathrm{Aut}(X')$ be a generator of $\mathbb{Z}_m$ and let τ be a translation on B' induced by σ. Then σ induces an isomorphism h between the two principal bundles X' and $\tau^*(X')$ over B'. This h cannot commute with the E-actions on X' and $\tau^*(X')$, because otherwise this E-action would descend to $X\to B$, which is impossible by Lemma 5.1, a) and the assumption that X is not a principal bundle. This means that h does not respect the structure of principal bundles. By Lemma 5.1, b) there is some $z\in\mathbb{Z}_n$, $z\neq 1$, defining h. For $\xi'\in H^1(\mathcal{E}_{B'})$, the class defining X', this implies $c(\xi')=0$: indeed, τ^* operates trivially on $H^2(B',\Gamma)$, whereas z has no fixed point $\neq 0$ in Γ. So necessarily X' is a torus containing E as closed subgroup with $B'=X'/E$. The translation τ on the base B' is induced by some group element $t\in X'$. Then $\sigma=t\circ h$. Consider the closed subgroup

$$C = \{x \in X' : \sigma(x) = t + x\}$$

of fixed points of h. Since h acts on all the fibres of X' as an automorphism which is not a translation, C intersects each fibre in a finite non-empty set. So C is an elliptic curve projected onto B'. The torus X' is isogeneous with $E \times C$ and if we pull back X' under $C \to B'$, it becomes the product $E \times C$. Altogether this proves:

There is an elliptic curve C such that $B = C/G$, where G is a finite subgroup of the translation group of C. The surface X is the quotient $E \times C/G$, with G acting on C by translations and on E by some representation $G \to A(E)$, which has its image not in the group of translations only.

Such a surface X is called *hyper-elliptic* (because the points of each projective image of X are parametrised by abelian functions in two variables, which were called hyperelliptic functions when these surfaces were discovered). The classification of these surfaces is easy, see [B-M1], p. 36–37. One finds the following classical list for $E = \mathbb{C}/\Gamma$ and $G \subset C$:

Type	Γ	G	Action on E of generators of G
a1)	arbitrary	$\mathbb{Z}_2$	$e \to -e$
a2)	arbitrary	$\mathbb{Z}_2 \oplus \mathbb{Z}_2$	$e \to -e$, $e \to e + e_1$, where $2e_1 = 0$
b1)	$\mathbb{Z} \oplus \mathbb{Z}\omega$	$\mathbb{Z}_3$	$e \to \omega e$
b2)	$\mathbb{Z} \oplus \mathbb{Z}\omega$	$\mathbb{Z}_3 \oplus \mathbb{Z}_3$	$e \to \omega e$ $e \to e + e_1$, where $\omega e_1 = e_1$
c1)	$\mathbb{Z} \oplus \mathbb{Z}\sqrt{-1}$	$\mathbb{Z}_4$	$e \to \sqrt{-1}e$
c2)	$\mathbb{Z} \oplus \mathbb{Z}\sqrt{-1}$	$\mathbb{Z}_4 \oplus \mathbb{Z}_2$	$e \to \sqrt{-1}e$ $e \to e + e_1$, $\sqrt{-1}e_1 = e_1$
d)	$\mathbb{Z} \oplus \mathbb{Z}\omega$	$\mathbb{Z}_6$	$e \to -\omega e$

List of hyperelliptic surfaces from [B-M1] ($\omega = \mathbb{e}(1/6)$)

Since the covering $E \times C$ is projective, by Theorem IV.5.8 every hyperelliptic surface X is projective. Its invariants are

$$h^{1,0} = 1, \qquad h^{2,0} = 0, \qquad h^{1,1} = 2,$$

and $\mathcal{K}_X$ is a torsion bundle of order 2, 3, 4 and 6 for the type a, b, c and d respectively.

To see this, we observe that G acts non-trivially (with its first factor) on $H^1(\mathcal{O}_E)$. So $f_{*1}\mathcal{O}_X$ is a torsion line bundle on B of the orders as given above in the different cases. The sequence (4) then shows $h^1(\mathcal{O}_X) = 0$. Similarly, from

$$H^2(\mathcal{O}_B) \to H^2(\mathcal{O}_X) \to H^1(f_{*1}\mathcal{O}_X)$$

one deduces $h^2(\mathcal{O}_X) = 0$. Finally, $\omega_{X/B} = f^*(f_{*1}\mathcal{O}_X)$, again by relative duality. So (5) shows that $\mathcal{K}_X$ is a torsion bundle of the order as stated above.

Furthermore $e(X)=c_2(\Omega_X)=0$ and

$$h^{1,1}=e(X)+2b_1(X)-2=2. \quad \square$$

6. Higher Genus Fibre Bundles

If D is a smooth, compact, connected curve of genus ≥ 2, then $\mathrm{Aut}(D)$ is a finite group. Consequently, every D-fibre bundle over a curve C is given by a representation $\pi_1(C)\to\mathrm{Aut}(D)$, and the classification of D-bundles becomes a purely algebraic problem. We shall only need the fact that, given a D-bundle over C, there exists a finite unramified covering $\tilde{C}$ of C such that the lifted bundle on $\tilde{C}$ is isomorphic to $\tilde{C}\times D$.

As to the product case $X=C\times D$, we have (as a special case of a much more general result)

(6.1) **Proposition.** *If C and D are smooth curves and $X=C\times D$, then $q(X)=g(C)+g(D)$ and $P_n(X)=P_n(C)\cdot P_n(D)$. In particular we have*

$$\mathrm{kod}(X)=\begin{cases} -\infty & \text{if } C \text{ or } D \text{ is rational}\\ 0 & \text{if } C \text{ and } D \text{ are elliptic}\\ 1 & \text{if } g(C)=1,\ g(D)\geq 2 \\ & \text{or conversely}\\ 2 & \text{if } g(C)\geq 2,\ g(D)\geq 2.\end{cases}$$

Proof. Let $f_1\colon X\to C$ and $f_2\colon X\to D$ be the projections. Then

$$\mathscr{K}_X=f_1^*(\mathscr{K}_C)\otimes f_2^*(\mathscr{K}_D),$$

hence (by the Künneth formula)

$$\Gamma(\mathscr{K}_X^{\otimes n})\cong\Gamma(\mathscr{K}_C^{\otimes n})\otimes\Gamma(\mathscr{K}_D^{\otimes n}).$$

This gives the formula for $P_n(X)$. And since $H^1(\mathcal{O}_X)\cong H^1(\mathcal{O}_C)\oplus H^1(\mathcal{O}_D)$ again by Künneth, we have $q(X)=g(C)+g(D)$. $\square$

Elliptic Fibrations

We recall that by an elliptic fibration of a surface X we mean a proper, connected holomorphic map $f\colon X\to S$, such that the general fibre X_s ($s\in S$) is non-singular elliptic (the holomorphic structure may depend on s). Unless otherwise stated we shall always assume that f is (relatively) minimal, i.e. all fibres are free of (-1)-curves.

7. Kodaira's Table of Singular Fibres

Let $f\colon X \to \Delta$ be an elliptic fibration over the unit disk Δ, such that all fibres X_s, $s \neq 0$, are smooth. We shall list all possibilities for the nature of X_0. The result is embodied in Table 3. We use Kodaira's original notation, which is now generally accepted.

We distinguish between three cases:

a) X_0 is irreducible. In this case the adjunction formula immediately yields that X_0 is either smooth elliptic, or rational with a node, or rational with a cusp (types I_0, I_1 and II);

b) X_0 is reducible, but not multiple. We claim that every component C_i of $X_0 = \sum n_i C_i$ is a (-2)-curve. For we have $0 = (\mathscr{K}, X_0) = \sum n_i (\mathscr{K}, C_i) = \sum n_i(-C_i^2 + 2g(C_i) - 2)$, whereas $C_i^2 \leq -1$ by Zariski's lemma, and the case $C_i^2 = -1$, $g(C_i) = 0$ is excluded by our assumption about relative minimality. Applying again Zariski's lemma, we see that for every two different components C_i, C_j we must have that $C_i C_j$ is either 0, 1 or 2. In the last case we must have $X_0 = C_i + C_j$, thus obtaining the possibilities I_1 and III. Otherwise the intersection graph Γ has the property that any two vertices are joined by at most one edge, and we can consider the associated quadratic form $Q(\Gamma)$. Using once more Zariski's lemma we find that this form is positive semi-definite, so we can apply Lemma I.2.12, obtaining as only possibilities the graphs $\tilde{A}_n$, $\tilde{D}_{n+4}$, $\tilde{E}_6$, $\tilde{E}_7$ and $\tilde{E}_8$. The graph $\tilde{A}_n$, $n \geq 3$ leads to I_n, whereas for $\tilde{A}_2$ there are two possibilities, namely I_2 and IV. The graphs $\tilde{D}_{n+4}$, $\tilde{E}_6$, $\tilde{E}_7$ and $\tilde{E}_8$ give the types I_n^*, II^*, III^* and IV^* respectively.

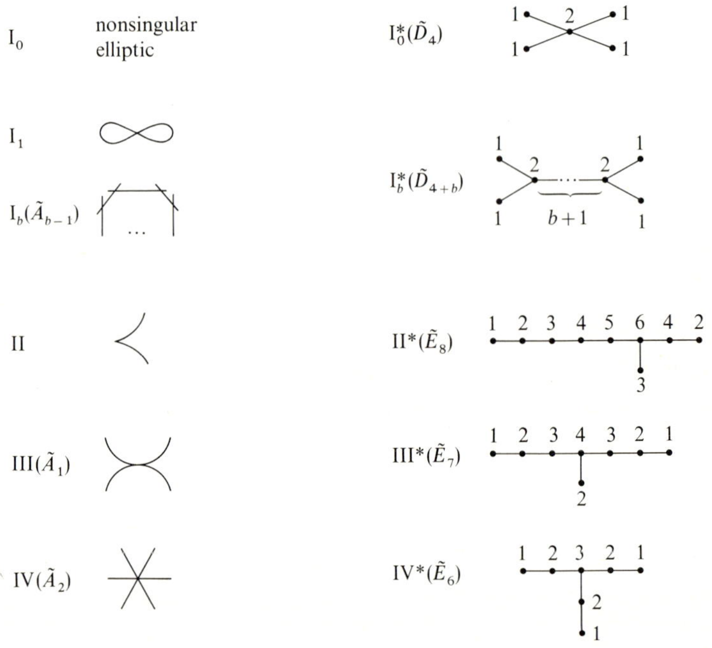

Table 3. Kodaira's table of singular elliptic fibres

c) X_0 is a multiple fibre. Exploiting Zariski's lemma in the same way as we did earlier, we find that $X_0 = m X'_0$, where X'_0 is one of the types described before. As a consequence of Lemma III.8.3, however, X'_0 can't be simply-connected, hence we are left with the possibilities $X'_0 = \mathrm{I}_0$, I_1 or I_b, leading to the types ${}_m\mathrm{I}_0$, ${}_m\mathrm{I}_1$ and ${}_m\mathrm{I}_b$.

Remark 1. The existence of all types will be established later (Sects. 8 and 10).

Remark 2. The table shows that only the fibres of type I_b, $b \geq 0$, are stable, a fact that was already mentioned before (III, Sect. 10).

Kodaira's table is given in [Ko2], Theorem 6.2.

8. Stable Fibrations

In this section we give the construction of some stable elliptic fibrations $f\colon X \to \Delta$ over the unit disk and compute their monodromy and period map.

Type I_0. Let $z(s)$ be an arbitrary holomorphic function on Δ with $\operatorname{Im} z(s) > 0$. Let $\mathbb{Z} \times \mathbb{Z}$ act on $\mathbb{C}_c \times \Delta_s$ by

$$(m, n)(c, s) = (c + m + n z(s), s).$$

The quotient $X = (\mathbb{C} \times \Delta)/(\mathbb{Z} \times \mathbb{Z})$ is a non-singular surface fibred over Δ, such that X_s is an elliptic curve with periods 1, $z(s)$. Conversely, every elliptic fibration which has only smooth fibres, can locally be obtained in this way. Obviously, the monodromy is trivial. The period domain is $\mathfrak{H}_1/\Gamma_1 = \mathbb{C}$, where $\mathfrak{H}_1 = \{z \in \mathbb{C};\ \operatorname{Im} z > 0\}$ is the upper half plane and $\Gamma_1 = SL(2, \mathbb{Z})/\{\pm 1\}$ is the modular group acting by $z \to \dfrac{az+b}{cz+d}$. The isomorphism $\mathfrak{H}_1/\Gamma_1 \to \mathbb{C}$ is given explicitly by the j-function. Following Kodaira, we shall use the convention in which

$$\begin{aligned} j(z) &= 0 \quad \text{if } z \text{ equivalent with } \mathbb{e}(\tfrac{1}{6}) \text{ under } \Gamma_1, \\ j(z) &= 1 \quad \text{if } z \text{ equivalent with } \sqrt{-1} \text{ under } \Gamma_1. \end{aligned}$$

So, wherever j vanishes in the upper half-plane, it vanishes there to the third order, and when $j=1$, then $j-1$ vanishes to the second order.

The period map $J(s) = j(z(s))$ was called the functional invariant by Kodaira. We see that for a smooth fibration $f\colon X \to \Delta$, the functional invariant always satisfies

$$\text{(6)} \qquad \begin{cases} \text{if } J(0) = 0, \text{ then } \ J \ \text{ vanishes at 0 to an order } h \equiv 0\ (3), \\ \text{if } J(0) = 1, \text{ then } J-1 \text{ vanishes at 0 to an order } h \equiv 0\ (2). \end{cases}$$

If one has any fibration in Weierstrass normal form

$$X = \{((z_0 : z_1 : z_2), s) \in \mathbb{P}_2 \times \Delta;\ z_0 z_2^2 = 4 z_1^3 - g_2(s) z_0^2 z_1 - g_3(s) z_0^3\},$$

then

$$J(s) = \frac{g_2^3}{g_2^3 - 27 g_3^2}$$

and X_s is non-singular if and only if $g_2^3(s) \neq 27 g_3^2(s)$.

Type I_1. An example in Weierstrass normal form can be constructed using $g_2(s)=3-s$, $g_3(s)=1-s$, hence

$$X=\{((z_0:z_1:z_2),s)\in \mathbb{P}_2\times\Delta;\ z_0 z_2^2=4z_1^3+(s-3)z_0^2 z_1+(s-1)z_0^3\}.$$

It is easily checked that the surface X is non-singular, that X_0 is irreducible rational with node at $(1:-\frac{1}{4}:0)$, and that X_s is non-singular elliptic for $s\neq 0$.

The functional invariant is

$$J(s)=\frac{(3-s)^3}{s(27-18s-s^2)}$$

and has at $s=0$ an ordinary pole. (In fact, in this case Satake's compactification is $\mathbb{P}_1$.)

We claim that the monodromy in integral homology can be given by

$$T=\begin{pmatrix}1 & 1\\ 0 & 1\end{pmatrix}.$$

Proof. For real s, $0<s<1$, the polynomial $4x^3+(s-3)x+s-1$ has three real roots $s_1<s_2<s_3$ (as its discriminant is $\neq 0$ for $0<s<1$, it suffices to check this for $s=1$). If $s\to 0$, then both s_1 and s_2 tend to $-\frac{1}{4}$, whereas s_3 tends to 1. The projection $((z_0:z_1:z_2),s)\to(z_0:z_1)$ exhibits the curve X_s as a double covering of $\mathbb{P}_1$, ramified in s_1,s_2,s_3 and ∞. The part of X_s over the real axis consists of the four "circles" a, b, a', b' as in Fig. 4b.

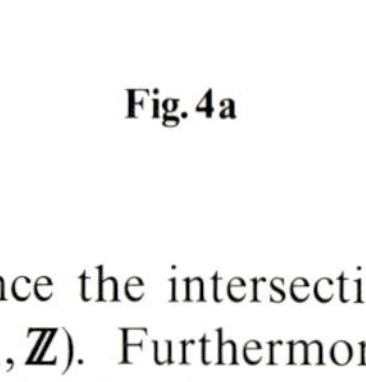

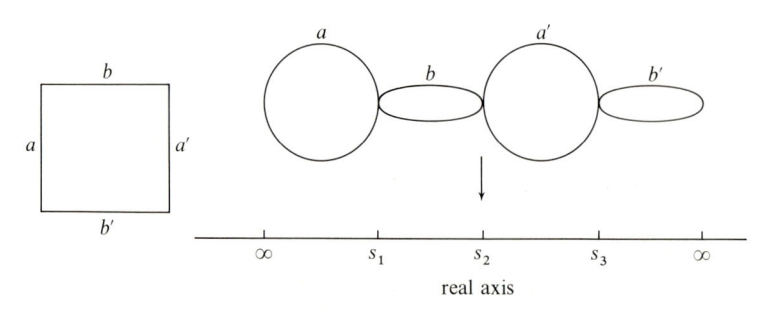

Fig. 4a

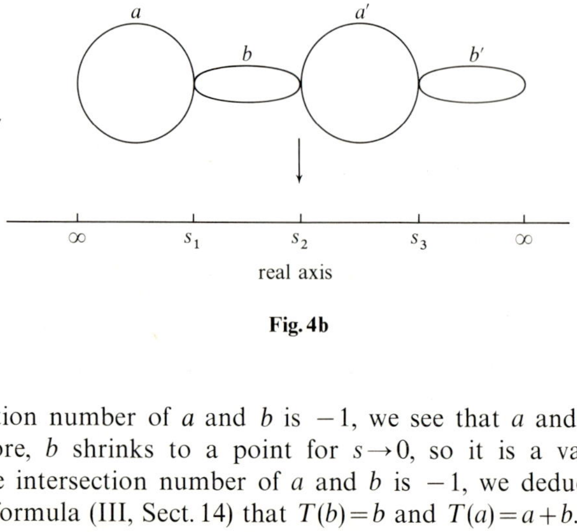

Fig. 4b

Since the intersection number of a and b is -1, we see that a and b span $H_1(X_s,\mathbb{Z})$. Furthermore, b shrinks to a point for $s\to 0$, so it is a vanishing cycle. Again since the intersection number of a and b is -1, we deduce from the Picard-Lefschetz formula (III, Sect. 14) that $T(b)=b$ and $T(a)=a+b$.

Type I_b, $b>1$. Let f: $X\to\Delta$ be the fibration constructed above with X_0 of type I_1, and let $f^{(b)}$: $X^{(b)}\to\Delta$ be its b-th root fibration (III, Sect. 9).

There is the diagram

$$\begin{array}{ccccc} X^{(b)} & \xrightarrow{\tau'} & X' & \xrightarrow{\tau} & X \\ {\scriptstyle f^{(b)}}\downarrow & & \downarrow & & \downarrow{\scriptstyle f} \\ \Delta & =\!= & \Delta & \xrightarrow[\delta_b:\, s\to s^b]{} & \Delta \end{array}$$

where X' is the normalisation of $X \times_\Delta \Delta$ (the fibre product with respect to δ_b). This X' has one singularity of type A_b over the double point of X_0. The map τ is a b-fold covering map ramified along $C'_1 = \tau^{-1}(X_0)$. The singularity in X' is resolved under τ' by a string $C_2, \ldots, C_b$ of (-2)-curves such that $C_1, C_2, \ldots, C_b$ form a cycle (C_1 is the proper transform of C'_1). So the singular fibre in $X^{(b)}$ is of type I_b.

Outside of $s=0$, the fibration $f^{(b)}$ is the pull-back of f under δ_b. This shows that its functional invariant at $s=0$ has a *pole of order* b and that its monodromy is given by

$$\begin{pmatrix} 1 & 1 \\ 0 & 1 \end{pmatrix}^b = \begin{pmatrix} 1 & b \\ 0 & 1 \end{pmatrix}.$$

The examples in this section are from [Ko2], p. 598–600.

9. The Jacobian Fibration

In this section let $f\colon X \to S$ be an elliptic fibration without multiple fibres. We describe the construction of $\mathrm{Jac}(f)$, a fibre space over S with fibres $\mathrm{Jac}(f)_s$, the one-dimensional complex Lie groups $\mathrm{Pic}^0(X_s)$. The sheaf of holomorphic sections of $\mathrm{Jac}(f)$ will be the sheaf $f_{*1}\mathcal{O}_X / f_{*1}\mathbb{Z}_X$.

As in III, Sect. 15, let E be the line bundle over S corresponding to the invertible sheaf $f_{*1}\mathcal{O}_X$ and $L \subset E$ the subset corresponding to the subsheaf $f_{*1}\mathbb{Z}_X$. If X_s is non-singular then L_s is a lattice in $E_s \cong \mathbb{C}$, and L is a closed submanifold of E near s. The whole point is to show that $L \subset E$ is closed everywhere. Then it is straight-forward to form $\mathrm{Jac}(f)$ as the quotient E/L. We refer to Kodaira ([Ko2], Sections 8 and 9) for details.

To prove that L is indeed a closed submanifold of E we argue in the following way. The assertion being local with respect to S, we fix some $s_0 \in S$ with X_{s_0} singular, and distinguish between several cases.

Case 1: X_{s_0} is stable. By Proposition III.13.1 and Theorem III.15.2 near s_0 the set L is spanned over $\mathbb{Z}$ by two sections β and $\alpha = z \cdot \beta$, where

$$z = \tilde{z} + \frac{\lambda}{2\pi\sqrt{-1}} \log s, \qquad 0 < \lambda \in \mathbb{N},$$

with $\tilde{z}$ holomorphic in s_0 and $\tilde{z}(s_0) = 0$.

If we take any sequence $g_1, g_2, \ldots$, with

$$g_n = \mu_n \alpha(s_n) + \nu_n \beta(s_n) = (\mu_n z(s_n) + \nu_n)\, \beta(s_n) \in L,$$

then convergence of g_n implies $\mu_n = 0$, hence $\nu_n = \nu$ is constant for $n \gg 0$ (this follows since β is holomorphic near s_0, but $\operatorname{Im} z(s_n) \to \infty$). So $g_n \to \nu\, \beta(s_0) \in L$.

Case 2: X_{s_0} is unstable. We choose a stable reduction

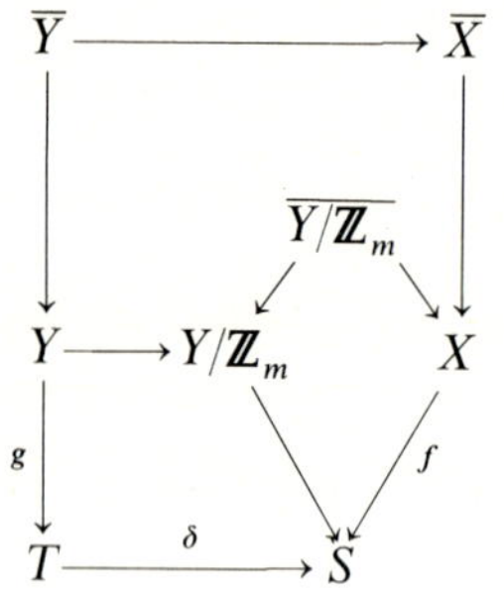

as in III, Sect. 10. It suffices to show that $\mathbb{Z}_m$ acts non-trivially on $H^1(\mathcal{O}_{Y_{t_0}})$, where $\delta(t_0) = s_0$, because then there is the diagram

$$\begin{array}{ccc} (\delta^*(f_{*1}\mathbb{Z}_X) & \longrightarrow & g_{*1}\mathbb{Z}_Y \\ \downarrow & & \downarrow \\ \delta^*(f_{*1}\mathcal{O}_X) & \longrightarrow & g_{*1}\mathcal{O}_Y \end{array}$$

in which the lower horizontal arrow vanishes at $t_0 \in T$, the point over $s_0 \in S$ (Proposition III.11.3). So if a sequence $g_n \in L = f_{*1}\mathbb{Z}_X$ converges in $E = f_{*1}\mathcal{O}_X$, by case 1, $\delta^*(g_n) = 0$ for $n \gg 0$, hence $g_n = 0$ for these n. Let us now consider the action of $\mathbb{Z}_m$ on Y. It must have fixed points on Y_{t_0} (otherwise $X_0 = Y_{t_0}/\mathbb{Z}_m$ would be a multiple fibre), but cannot act trivially on Y_{t_0} (because then $X_0 = Y_{t_0}$ would be stable). This implies already that $\mathbb{Z}_m$ acts non-trivially on $H^1(\mathcal{O}_{Y_{t_0}})$ if Y_{t_0} is non-singular. And if Y_{t_0} is of type I_b, $b > 0$, it is sufficient to notice that the generator of $\mathbb{Z}_m$ cannot preserve the orientation of the cycle of rational curves, because then X_0 would not be simply connected. □

(9.1) **Proposition.** *Assume that* $f\colon X \to S$ *admits a section* $\sigma\colon S \to X$, *and let* $X^\sigma = X \setminus \{$*irreducible components of fibres* X_s *not meeting* $\sigma(S)\}$. *Then there is a canonical fibre-preserving isomorphism from* $\operatorname{Jac}(f)$ *onto* X^σ, *mapping the zero-section in* $\operatorname{Jac}(f)$ *onto* σ.

Proof. For $x \in X^\sigma$, $f(x) = s$, let $\mathcal{O}_{X_s}(x - \sigma(s)) \in \operatorname{Pic}(X_s)$. From Serre duality and the Riemann-Roch theorem II.3.1 it follows that

(i) $\mathcal{O}_{X_s}(x - \sigma(s)) \in \operatorname{Pic}^0(X_s) = \operatorname{Jac}(f)_s$,
(ii) $\mathcal{O}_{X_s}(x - \sigma(s)) = \mathcal{O}_{X_s}(x' - \sigma(s))$ if and only if $x = x'$,
(iii) for $d \in \operatorname{Jac}(f)_s$, there is always some $x \in X_s \cap X^\sigma$ with $d = \mathcal{O}_{X_s}(x - \sigma(s))$.

This shows that there is a bijective map $\mathrm{Jac}(f) \to X^\sigma$ defined by $d \to x$ if $d = \mathcal{O}_{X_s}(x - \sigma(s))$. It is a formality to verify that this map is holomorphic. □

Remark. Proposition 9.1 implies that X is a "properification" of its Jacobian fibration. This can be used to prove uniqueness theorems. For example, consider a fibration $f\colon X \to \Delta$ with X_0 of type I_b, $b>0$. By construction, $\mathrm{Jac}(f) = \mathbb{C} \times \Delta/L$ where L is spanned over $\mathbb{Z}$ by 1 and $z = \tilde{z} + \frac{\lambda}{2\pi\sqrt{-1}} \log s$, $0 < \lambda \in \mathbb{N}$, with $\tilde{z}$ holomorphic near 0. Replacing s by $s\,\mathrm{e}\left(\frac{\tilde{z}}{\lambda}\right)$, we can achieve $z = \frac{\lambda}{2\pi\sqrt{-1}} \log s$. This means that all those fibrations which have the same λ are biholomorphically equivalent (cf. Proposition III.8.5). But λ is determined by the monodromy, and for the fibration $f^{(b)}$ constructed above, $\lambda = b$. So every fibration with X_0 of type I_b is (at the point 0) locally equivalent with the fibration $f^{(b)}$, and with the properification of $\mathbb{C} \times \Delta/L$, where $L = \mathbb{Z} + \mathbb{Z}\frac{b}{2\pi\sqrt{-1}} \log s$.

Finally, let $X^\#$ be the subset of X where $df \neq 0$, and let $X_s^\#$ be the intersection $X_s \cap X^\#$. Denote by $\mathcal{O}(X^\#)$ the sheaf of local holomorphic sections of $f\colon X \to S$. For any two sections $\sigma, \tau\colon S \to X$ there is (by Proposition 9.1 and Proposition III.8.5) a unique biholomorphic automorphism of X mapping σ to τ and acting as a translation on every non-singular fibre X_s. So, upon specifying one section $\sigma\colon S \to X$ as zero section, the sheaf $\mathcal{O}(X^\#)$ becomes a sheaf of commutative groups. There is an exact sequence

$$0 \to f_{*1}\mathcal{O}_X / f_{*1}\mathbb{Z}_X \to \mathcal{O}(X^\#) \to F \to 0,$$

where F is concentrated at the critical values $s \in S$ and F_s is finite.

Since each fibre X_s which is not a multiple fibre admits local sections (cf. Table 3), the curve $X_s^\#$ carries the structure of a 1-dimensional complex Lie group. If in particular X_s is of type I_b, $b>0$, then $F_s = \mathbb{Z}_b$ and $X_s^\# = \mathbb{C}^* \times \mathbb{Z}_b$.

We have closely followed [Ko2], Sect. 9.

10. Stable Reduction

In this section we keep an earlier promise (Sect. 7, Remark 1) and give examples of elliptic fibrations $f\colon X \to \Delta$ with unstable singular fibres X_0. It turns out that all types in Table 3 actually occur. For the types which are not multiple fibres we list in Table 6 below the behaviour of the functional invariant $J(s)$ and the monodromy in the examples constructed.

Type $_mI_0$. We start with a non-singular fibration $Y = \mathbb{C} \times \Delta/L$, where $L = \mathbb{Z} + \mathbb{Z} \cdot z(s)$ with z holomorphic on Δ and $z(s) = z(0) + \mathrm{const} \cdot s^{mh}$, $h \in \mathbb{N}$. The automorphism of $\mathbb{C}\times\Delta$

$$(c,s)\to\left(c+\frac{1}{m},\ \mathrm{e}\left(\frac{1}{m}\right)s\right)$$

generates a group $\mathbb{Z}_m$ acting on Y without fixed points. The quotient has over 0 a singular fibre of type ${}_m\mathrm{I}_0$.

Type ${}_mI_b$, $b>0$. Here we start with a fibration $f\colon Y\to\Delta$ where

$$\mathrm{Jac}(f)=\mathbb{C}\times\Delta/L \quad\text{with}\quad L=\mathbb{Z}+\mathbb{Z}\frac{mb}{2\pi\sqrt{-1}}\log s.$$

So Y_0 is of type I_{mb}, say $Y_0=C_1+\ldots+C_{mb}$. The automorphism

$$(c,s)\to\left(c,\ \mathrm{e}\left(\frac{1}{m}\right)s\right)$$

of $\mathbb{C}\times\Delta$ induces on Y a fibre-preserving automorphism μ of order m, which commutes with all translations from $\mathcal{O}(Y^{\#})_0$. Since $Y_0^{\#}=\mathbb{C}^*\times\mathbb{Z}_{mb}$, there is some section $t\in\mathcal{O}(Y^{\#})_0$ acting as $C_i\to C_{i+b}$ on Y_0 and as translation of order m on the nearby regular fibres. So μt generates a group $\mathbb{Z}_m$ acting on Y without fixed points, such that the quotient has a singular fibre of type ${}_m\mathrm{I}_b$.

Type I_0^*. This time we start with a non-singular fibration $Y=\mathbb{C}\times\Delta/L$, where $L=\mathbb{Z}+\mathbb{Z}z(s)$ with $z(s)=z(0)+s^{2h}$, $h\in\mathbb{N}$. The map

$$(c,s)\to(-c,-s)$$

defines an involution ι on Y having four isolated fixed points on Y_0. In $Y/\{\mathrm{id},\iota\}$ these four points give rise to four A_1-singularities, located on the image of Y_0 which is a non-singular rational curve of multiplicity two. Resolution of the four singularities leads to a fibration with singular fibre of type I_0^*.

Type I_b^*. Here we begin with a fibration $f\colon Y\to S$ of type I_{2b}, where $\mathrm{Jac}(f)=\mathbb{C}\times\Delta/L$ and $L=\mathbb{Z}+\mathbb{Z}\frac{2b}{2\pi\sqrt{-1}}\log s$. The map

$$(c,s)\to(-c,-s)$$

induces on $\mathrm{Jac}(f)$ an involution ι. By Proposition III.8.5, this ι extends to Y, where it interchanges C_i and C_{2b+2-i} and has on both C_1 and C_{b+1} two fixed points. The image of Y_0 in $Y/\{\mathrm{id},\iota\}$ is a string of b non-singular rational curves of multiplicity two. Resolution of the A_1-singularities on the first and the last of these curves leads to a fibration with singular fibre of type I_b^*.

Table 5

type	II	II*	III	III*	IV	IV*
h	2 mod 6	4 mod 6	2 mod 4		2 mod 3	1 mod 3
$z(s)$	$(\mathbb{e}(1/3)-\mathbb{e}(2/3)s^h)/(1-s^h)$		$(\sqrt{-1}+\sqrt{-1}s^h)/(1-s^h)$		$(\mathbb{e}(1/3)-\mathbb{e}(2/3)s^h)/(1-s^h)$	
m	6		4		3	
$z(\mathbb{e}(1/m)s)$	$-z(s)^{-1}-1$	$-(z(s)+1)^{-1}$	$-z(s)^{-1}$		$-(z(s)+1)^{-1}$	$-z(s)^{-1}-1$
$\tilde{\mu}(c)$	$-z(s)^{-1}c$	$(z(s)+1)^{-1}c$	$-z(s)^{-1}c$	$z(s)^{-1}c$	$-(z(s)+1)^{-1}c$	$z(s)^{-1}c$
$\mu \mid Y_0$	$\mathbb{e}(1/6)$	$\mathbb{e}(5/6)$	$\sqrt{-1}$	$-\sqrt{-1}$	$\mathbb{e}(1/3)$	$\mathbb{e}(2/3)$
fixed points with weights	μ: $(1,1)$ μ^2: $2\times(1,1)$ μ^3: $3\times(1,1)$	μ: $(5,1)$ μ^2: $2\times(2,1)$ μ^3: $3\times(1,1)$	μ: $2\times(1,1)$ μ^2: $2\times(1,1)$	μ: $2\times(3,1)$ μ^2: $2\times(1,1)$	$3\times(1,1)$	$3\times(2,1)$
quotient singularities	$A_{6,1}A_{3,1}A_1$	$A_5A_2A_1$	$2A_{4,1}A_1$	$2A_3A_1$	$3A_{3,1}$	$3A_2$

For *all the other types* II–IV* we can start with a non-singular fibration $Y=\mathbb{C}\times\Delta/L$, where $L=\mathbb{Z}+\mathbb{Z}\,z(s)$, and $z(s)-z(0)$ vanishes at $s=0$ to a certain order $h\in\mathbb{N}$. In Table 5 we give for all types this order h and the form, to which $z(s)$ can be normalised (perhaps after shrinking Δ) by changing the coordinate s. Then a group $\mathbb{Z}_m$ of order $m=6$, 4 or 3 acting on Y is constructed as follows. On s, the generator μ acts as multiplication with $\mathbb{e}\left(\frac{1}{m}\right)$. So it transforms $z(s)$ into $z\left(\mathbb{e}\left(\frac{1}{m}\right)s\right)$, which is computed in row 5 of Table 5. Then on Y μ is induced by the automorphism of $\mathbb{C}\times\Delta$

$$(c,s)\to\left(\tilde{\mu}(c),\ \mathbb{e}\left(\frac{1}{m}\right)s\right)$$

with $\tilde{\mu}(c)$, $c\in\mathbb{C}$, defined in row 6. So μ acts on Y_0 by multiplication with an easily computed root of unity (row 7).

In $Y/\mathbb{Z}_m$ the quotient $Y_0/\mathbb{Z}_m$ becomes a rational curve of multiplicity m. On this curve, there are some singularities of $Y/\mathbb{Z}_m$ coming from non-trivial isotropy groups. These fixed points, and the weights by which the group acts, as well as the quotient singularities, are easily determined (III, Sect. 5). The minimal resolution of these singularities leads to the fibration X. In the cases II*, III*, IV* no further blowing down is necessary, whereas in the other cases the final step is described below (numbers without brackets are multiplicities, and numbers within brackets denote self-intersections):

type II:

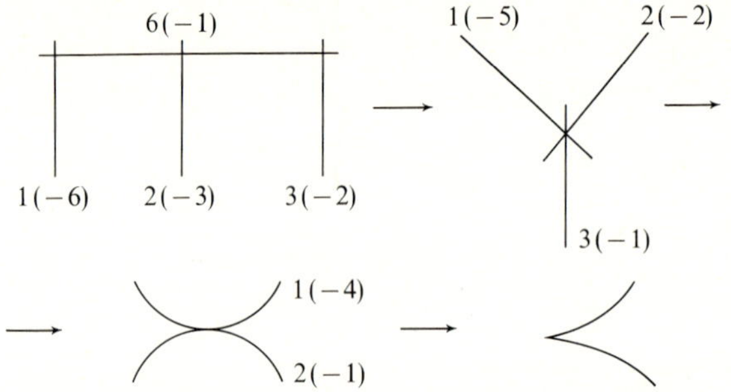

type III:

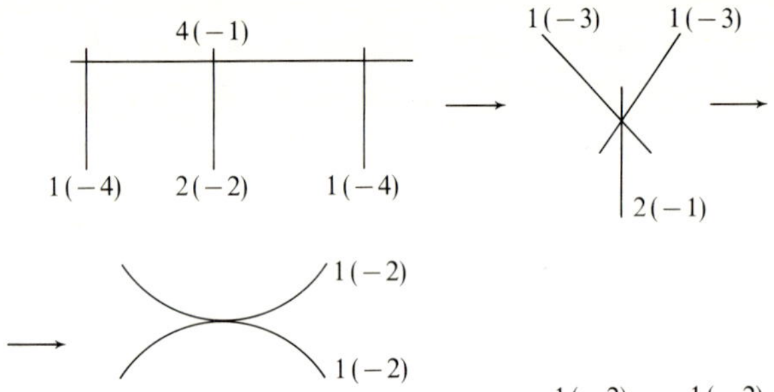

type IV:

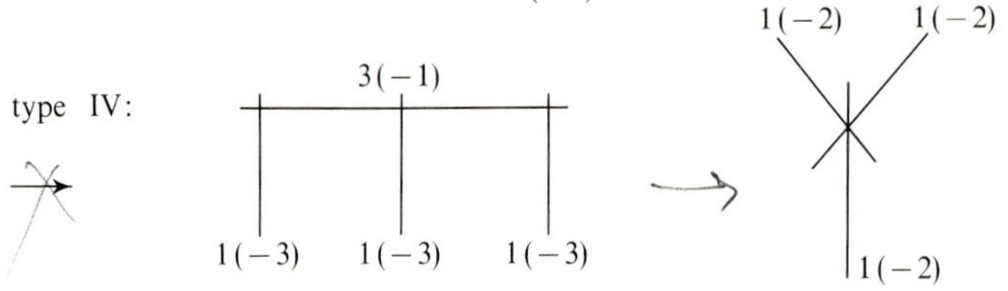

In the examples constructed above, the functional invariant J and the monodromy T are as in Table 6, p. 159.

Proof. The behaviour of J is easily deduced from the behaviour of J for the stable reduction. The monodromy of I_0 is trivial and that of I_b was computed in Sect. 8. The monodromy for I_b^* must be a matrix T with

$$T^2 = \begin{pmatrix} 1 & 2b \\ 0 & 1 \end{pmatrix}, \quad T\begin{pmatrix} 1 \\ 0 \end{pmatrix} = \begin{pmatrix} -1 \\ 0 \end{pmatrix}.$$

This leaves us with the possibility $T = -\begin{pmatrix} 1 & b \\ 0 & 1 \end{pmatrix}$. For all other types, the stable reduction Y_0 is non-singular. So T is just the action on $H_1(Y_0, \mathbb{Z}) = H_1(Y, \mathbb{Z})$ by the generator μ^{-1} of the ramification group $\mathbb{Z}_m$, $m = 2, 3, 4$ or 6.

The results in this section are due to Kodaira ([Ko2], Sects 8, 9, [Ko4], Sect. 4).

Table 6

type	$J(s)$	T
I_0	subject to conditions (6)	$\begin{pmatrix}1 & 0\\0 & 1\end{pmatrix}$
I_0^*		$-\begin{pmatrix}1 & 0\\0 & 1\end{pmatrix}$
I_b	pole of order b	$\begin{pmatrix}1 & b\\0 & 1\end{pmatrix}$
I_b^*		$-\begin{pmatrix}1 & b\\0 & 1\end{pmatrix}$
II	s^h, $h\equiv 1$ (3)	$\begin{pmatrix}1 & 1\\-1 & 0\end{pmatrix}$
II*	s^h, $h\equiv 2$ (3)	$\begin{pmatrix}0 & -1\\1 & 1\end{pmatrix}$
III	$1+s^h$, h odd	$\begin{pmatrix}0 & 1\\-1 & 0\end{pmatrix}$
III*		$\begin{pmatrix}0 & -1\\1 & 0\end{pmatrix}$
IV	s^h, $h\equiv 2$ (3)	$\begin{pmatrix}0 & 1\\-1 & -1\end{pmatrix}$
IV*	s^h, $h\equiv 1$ (3)	$\begin{pmatrix}-1 & -1\\1 & 0\end{pmatrix}$

11. Classification

Let $f\colon X\to S$ be an elliptic fibration and let $s_1,\dots,s_k\in S$ be the critical values and $S^*=S\setminus\{s_1,\dots,s_k\}$. Then $L=f_{*1}\mathbb{Z}_X$ is a locally constant sheaf on S^* with fibre $\mathbb{Z}\oplus\mathbb{Z}$. This sheaf, or equivalently its global monodromy which is an equivalence class of representations $R\colon \pi_1(S^*)\to SL(2,\mathbb{Z})$, was called by Kodaira the *homological invariant* of the fibration.

In this section it is essential to distinguish carefully between $SL(2,\mathbb{Z})$ and the modular group $\Gamma_1=SL(2,\mathbb{Z})/\{\pm 1\}$. Any representation R as above induces by composition a representation $r\colon \pi_1(S^*)\to\Gamma_1$. The equivalence class of the representation r is already uniquely determined by the functional invariant $J(s)$. For let $\mathfrak{H}^*$ be the upper half plane $\{z\in\mathbb{C};\operatorname{Im} z>0\}$ with all points in the orbits $\Gamma_1\cdot\mathfrak{e}(\frac{1}{3})$ and $\Gamma_1\cdot\sqrt{-1}$ removed. Then the covering $j\colon \mathfrak{H}^*\to\mathbb{C}\setminus\{0,1\}$ makes $\mathfrak{H}^*$ into a Γ_1-principal bundle over $\mathbb{C}\setminus\{0,1\}$, thus fixing a class of representations

$\pi_1(\mathbb{C}\setminus\{0,1\})\to\Gamma_1$. Any meromorphic function J on S defines a map of $S^* = S\setminus J^{-1}(\{0,1,\infty\})\to\mathbb{C}\setminus\{0,1\}$, hence uniquely determines a class of representations r: $\pi_1(S^*)\to\Gamma_1$ by pull-back. In terms of the holomorphic multivalued period function $z(s)$, this r is the monodromy imposed on z by the condition $j(z(s))=J(s)$. Kodaira has said it this way: a homological invariant R: $\pi_1(S^*)\to SL(2,\mathbb{Z})$ *belongs to* J, if it induces the representation class of r: $\pi_1(S^*)\to\Gamma_1$ defined by J.

(11.1) **Theorem** (Classification theorem for elliptic fibrations without multiple fibres). *Let S be a smooth compact connected curve, $S^*=S\setminus\{s_1,\dots,s_k\}$, and J a meromorphic function on S with $J(s)\neq 0,1,\infty$ for $s\in S^*$.*

(a) *If $k\geq 1$, then there are exactly $2^{g(S)+k-1}$ inequivalent homological invariants L belonging to J.*

(b) *Given J and a homological invariant L belonging to J, there is exactly one elliptic fibration f: $X\to S$ with these invariants, admitting a section.*

(c) *All elliptic fibrations, without multiple fibres, with given invariants J and L form a set $\mathcal{F}(J,L)$ parametrised by the abelian group $H^1(\mathcal{J}ac(f))$. Here* $\mathrm{Jac}(f)$ *is the jacobian fibration for one, hence by* (b), *for all members of $\mathcal{F}(J,L)$ and $\mathcal{J}ac(f)$ is the sheaf (of abelian groups) of local holomorphic sections in* $\mathrm{Jac}(f)$.

Proof. (a) Let $l_1,\dots,l_{2g}$ be a canonical set of generators for $\pi_1(S)$ and $l_{2g+1},\dots,l_{2g+k}$ be loops about the points $s_1,\dots,s_k$. Then $\pi_1(S^*)$ is the group with generators $l_1,\dots,l_{2g+k}$ and one relation

$$l_1 l_2 l_1^{-1} l_2^{-1} \dots l_{2g-1} l_{2g} l_{2g-1}^{-1} l_{2g}^{-1} l_{2g+1} \dots l_{2g+k}=1.$$

So for any given $r(l_i)\in\Gamma_1$, $i=1,\dots,2g+k-1$, a representative $R(l_i)\in SL(2,\mathbb{Z})$ can be chosen arbitrarily, but for $R(l_{2g+k})$ there is no choice.

(b) *Local existence.* By inspection of Table 6, it is possible to choose for each point $s\in S$ a neighbourhood $N\subset S$ and a holomorphic injection $(N,s)\to(\Delta,0)$ such that the function $J|N$ is transformed into one which is given in column 2 of this table. Even the two different possibilities for T, differing by sign, are realised by a row of this table.

Local uniqueness. It was observed already (Remark after Proposition 9.1), that a stable fibration with section is locally determined uniquely by its functional invariant $J(s)$. Now let f: $X\to\Delta$ be a fibration with section σ and functional invariant $J\neq 0,1,\infty$ for $s\in\Delta^*$, and let g: $Y\to\Delta$ be its stable reduction. Then X is bimeromorphically equivalent with $Y/\mathbb{Z}_m$, where the generator μ of $\mathbb{Z}_m$ acts on Y by fibre-preserving automorphisms, and on Δ by $s\to \mathrm{e}\left(\frac{1}{m}\right)\cdot s$. Now g is determined uniquely by the pull-back of J. The map μ is determined by the monodromy up to a translation on the fibres. But, since μ respects the pull-back of the section σ, it is totally determined by the monodromy of f.

Global existence. Choose a covering $S=\bigcup\Delta_i$ by unit disks such that *either* Δ_i does not contain any one of the points $s_1,\dots,s_k$ *or* Δ_i contains exactly one of the points $s_1,\dots,s_k$, this point being its centre.

The $\Delta_i \subset S$ may be chosen so small that by Table 6, there is an elliptic fibration f_i: $X_i \to \Delta_i$ admitting a section σ_i, having the functional invariant $J|\Delta_i$, and $f_{*i}\,\mathbb{Z}_{X_i} = L|\Delta_i$. Over an intersection $\Delta_i \cap \Delta_j$, both fibrations f_i and f_j are smooth with the same functional invariant and trivial monodromy. So there is an isomorphism c_{ij}: $X_j|\Delta_i \cap \Delta_j \to X_i|\Delta_i \cap \Delta_j$ preserving the fibration, transforming $\sigma_j|\Delta_i \cap \Delta_j$ into $\sigma_i|\Delta_i \cap \Delta_j$, and inducing the identity on $f_{*1}\,\mathbb{Z}_{X_i} = L|\Delta_i \cap \Delta_j = f_{*1}\,\mathbb{Z}_{X_j}$. For any three indices i, j, k the composition $c_{ij}\,c_{jk}\,c_{ki}$ is an automorphism of $X_i|\Delta_i \cap \Delta_j \cap \Delta_k$ preserving the section σ and acting trivially on the homology of the fibres. So $c_{ij}\,c_{jk}\,c_{ki} = \mathrm{id}$, and the c_{ij} can be used to define a fibration f: $X \to S$ by patching the X_i together.
Global uniqueness. This follows from the construction as given before, since the only freedom in the construction of X is the choice of the Δ_i.

(c) If X is any element of $\mathscr{F}(J, L)$, there is a covering $S = \bigcup \Delta_i$ as above with isomorphisms c_i: $X_i \to X|\Delta_i$ inducing the identity on $L|\Delta_i$. It follows that $c_j^{-1} c_i$ is a cocycle with values in $\mathrm{Jac}(f)$ which glues the collection of X_i together to build X. Visibly, cohomologous $c_j^{-1} c_i$ give isomorphic X. □

For this section, we refer to [Ko2], Sect. 10.

12. Invariants

We start with Kodaira's formula for the canonical bundle of an elliptic surface (with or without multiple fibres).

(12.1) **Theorem** (Canonical bundle formula for elliptic fibrations). *Let f: $X \to S$ be a relatively minimal elliptic fibration, such that its multiple fibres are $X_{s_1} = m_1 F_1, \ldots, X_{s_k} = m_k F_k$. Then*

$$\mathscr{K}_X = f^*(\mathscr{K}_S \otimes (f_{*1}\,\mathscr{O}_X)^\vee) \otimes \mathscr{O}_X(\textstyle\sum (m_i - 1) F_i).$$

Proof. We have seen (Proposition III.12.1) that $f_*\,\mathscr{K}_X$ is locally free of rank one. By relative duality

$$f_*\,\mathscr{K}_X = f_*\,\omega_{X/S} \otimes \mathscr{K}_S = \mathscr{K}_S \otimes (f_{*1}\,\mathscr{O}_X)^\vee.$$

This proves that

$$\mathscr{K}_X = f^*(\mathscr{K}_S \otimes (f_{*1}\,\mathscr{O}_X)^\vee) \otimes \mathscr{O}_X(D)$$

where D is the zero divisor of the canonical morphism

$$\lambda\colon\ f^*(f_*\,\mathscr{K}_X) \to \mathscr{K}_X.$$

This λ is an isomorphism on each non-singular fibre X_s. On a singular fibre the map λ cannot vanish identically, i.e., to the same order as the multiplicity of the fibre, for the following reason. Locally with respect to S, any 2-form on X is of the type $f^*(\omega)$, with ω a section in $f_*\,\mathscr{K}_X$. If now $\lambda|X_s$ were zero, there would be on X a 2-form $(f^*(\omega))/g$, with g an equation for X_s, which is not of the type $f^*(\omega)$.

Let D_s be the part of the divisor D over s. It is a canonical divisor on a neighbourhood of X_s. By Kodaira's table we have $D_s \cdot C = 0$ for all irreducible components $C \subset X_s$. In particular $D_s^2 = 0$, and by Zariski's lemma $D_s = r X_s$ with $r \in \mathbb{Q}$, $r < 1$. So if $D_s \neq 0$, then X_s must be a multiple fibre.

We see therefore that $D = \sum n_i F_i$, where the summation is taken over all multiple fibres and $n_i < m_i$. To determine the precise value of n_i, we use the adjunction formula

$$\omega_{F_i} = \mathcal{K}_X \otimes \mathcal{O}_{F_i}(F_i) = \mathcal{O}_{F_i}(F_i)^{\otimes (n_i+1)}.$$

Since F_i is a curve of type I_b, $b \geq 0$, there is an isomorphism $\omega_{F_i} = \mathcal{O}_{F_i}$. But $\mathcal{O}_{F_i}(F_i)$ is a torsion bundle of order exactly m_i (Lemma III.8.3). So m_i divides $n_i + 1$, and $n_i = m_i - 1$. □

Remark. By Theorem III.18.2 we have

$$\deg(f_{*1}\mathcal{O}_X)^\vee = \deg(f_* \omega_{X/S}) > 0$$

unless all the smooth fibres of f are isomorphic and the singular fibres are of type ${}_m\mathrm{I}_0$ only

(12.2) **Proposition.** $\deg(f_{*1}\mathcal{O}_X)^\vee = \chi(\mathcal{O}_X)$.

Proof. By Leray's spectral sequence for f and $\mathcal{O}_X$ there are isomorphisms

$$H^0(\mathcal{O}_X) \cong H^0(f_* \mathcal{O}_X), \qquad H^2(\mathcal{O}_X) \cong H^1(f_{*1}\mathcal{O}_X)$$

and an exact sequence

$$0 \to H^1(f_* \mathcal{O}_X) \to H^1(\mathcal{O}_X) \to H^0(f_{*1}\mathcal{O}_X) \to 0.$$

So by Riemann-Roch on S

$$\chi(\mathcal{O}_X) = \chi(f_* \mathcal{O}_X) - \chi(f_{*1}\mathcal{O}_X) = -\deg(f_{*1}\mathcal{O}_X). \quad \square$$

In many cases a weak consequence of 12.1 and 12.2 is sufficient:

(12.3) **Corollary.** *Let f: $X \to S$ be an elliptic fibration, such that its multiple fibres are $X_{s_1} = m_1 F_1, \ldots, X_{s_k} = m_k F_k$. Then*

$$\mathcal{K}_X = f^*(\mathcal{L}) \otimes \mathcal{O}_X\left(\sum (m_i - 1) F_i\right)$$

where $\mathcal{L}$ is a line bundle of degree $\chi(\mathcal{O}_X) - 2\chi(\mathcal{O}_S)$ on S.

(12.4) **Corollary.** *If a compact surface X admits a relatively minimal elliptic fibration over $\mathbb{P}_1$, then*

(i) $\mathrm{kod}(X) = 0$ *if and only if* $\mathcal{K}_X^{\otimes 12} \cong \mathcal{O}_X$,
(ii) $\mathrm{kod}(X) = -\infty$ *if and only if* $P_{12}(X) = 0$.

Proof. (i) If $\mathcal{K}_X^{\otimes 12} \cong \mathcal{O}_X$, then $\mathrm{kod}(X) = 0$. To prove the converse, we consider a minimal elliptic fibration $X \to \mathbb{P}_1$ with k multiple fibres of multiplicity

$m_1, \ldots, m_k$ respectively. From Corollary 12.3 we infer that $\operatorname{kod}(X)=0$ implies

$$\chi(\mathcal{O}_X)-2+k-\sum_{i=1}^{k} 1/m_i=0.$$

(If this number is strictly positive, then $P_n(X)\geq 2$ for n large enough and hence $\operatorname{kod}(X)\geq 1$, whereas if this number is strictly negative, K_X would be rationally homologous to a strictly negative multiple of a fibre and then $\operatorname{kod}(X)=-\infty$.) Since $\chi(\mathcal{O}_X)\geq 0$ by Proposition 12.2 and the remark preceding it, we see that there remain only the following possibilities:

$$\begin{aligned}
&\chi(\mathcal{O}_X)=1,\ k=2,\ m_1=m_2=2\\
&\chi(\mathcal{O}_X)=2,\ k=4,\ m_1=m_2=m_3=m_4=4\\
&\chi(\mathcal{O}_X)=2,\ k=3,\ m_1=m_2=m_3=3\\
&\qquad\qquad\qquad\qquad\ \ m_1=2,\ m_2=m_3=4\\
&\qquad\qquad\qquad\qquad\ \ m_1=2,\ m_2=3,\ m_3=6.
\end{aligned}$$

In all cases we have $\mathcal{K}_X^{\otimes 12}\cong\mathcal{O}_X$.

(ii) If $\operatorname{kod}(X)=-\infty$, then $P_{12}(X)=0$. For the other direction, let $X\to\mathbb{P}_1$ be a relatively minimal elliptic fibration, with general fibre F and multiple fibres $F_1, \ldots, F_k$, of multiplicity $m_1, \ldots, m_k$ respectively, such that $P_{12}(X)=0$. There exist non-negative integers l_i, r_i with $0\leq r_i\leq m_i$, such that

$$m_i l_i=12+r_i, \qquad i=1, \ldots, k.$$

(Hence $l_i\leq 6$ and $l_i\neq 5$.) By Corollary 12.3 we have

$$\mathcal{K}_X^{\otimes 12}=\mathcal{O}_X\left\{\left(12\chi(\mathcal{O}_X)-24+\sum_{i=1}^{k}(12-l_i)\right)F+\sum_{i=1}^{k} r_i F_i\right\}.$$

So $\sum_{i=1}^{k}(12-l_i)<24-12\chi(\mathcal{O}_X)$. Since $\chi(\mathcal{O}_X)\geq 0$, this implies already that $k\leq 3$. The case $k=2$ being trivial, it is sufficient to consider the case $k=3$. Assuming $l_1\geq l_2\geq l_3$, we find two possibilities: $l_1=l_2=6$ and $l_1=6$, $l_2=4$, $l_3\leq 3$. In the first case we obtain $m_1=m_2=6$ and in the second case $m_1=2$, $m_2=3$, $m_3\leq 5$. In both cases $\operatorname{kod}(X)=-\infty$. □

(12.5) **Proposition.** *Let $f\colon X\to S$ be a relatively minimal elliptic fibration with X compact. Then always* $\operatorname{kod}(X)\leq 1$ *and* $\operatorname{kod}(X)=1$ *if one of the following holds:*

(i) *for some $n\geq 0$ there is an effective n-canonical divisor,*

(ii) $g(S)\geq 2$,

(iii) $g(S)=1$ *and f is not locally trivial.*

Proof. Let m be the l.c.m. of all the multiplicities m_i appearing in Theorem 12.1. Then for $\mu\in\mathbb{N}$

$$\begin{aligned}
\mathcal{K}_X^{\otimes\mu m}&=f^*(\mathcal{K}_S\otimes(f_{*1}\mathcal{O}_X)^\vee)^{\otimes\mu m}\otimes\mathcal{O}_X(\textstyle\sum(m_i-1)\,m F_i)^{\otimes\mu}\\
&=f^*(\mathcal{K}_S\otimes(f_{*1}\mathcal{O}_X)^\vee)^{\otimes m}\otimes\mathcal{O}_X(\textstyle\sum(m_i-1)(m/m_i)\,X_{s_i})^{\otimes\mu}\\
&=f^*(D^{\otimes\mu})
\end{aligned}$$

where $D=\mathcal{K}_S^{\otimes m}\otimes(f_{*1}\mathcal{O}_X)^{\otimes -m}\otimes\mathcal{O}_S(\sum(m_i-1)(m/m_i)s_i)$ is a line bundle on S. So $h^0(\mathcal{K}_X^{\otimes\mu m})$ cannot grow faster than linearly in μ. And if $\deg(D)>0$, then by Riemann-Roch on S this growth is linear. If there is an effective n-canonical divisor, then there is also a μmn-canonical one and $\deg(D)>0$.

If $g(S)\geq 2$, then $\deg(\mathcal{K}_S)>0$ and $\deg((f_{*1}\mathcal{O}_X)^\vee)\geq 0$ enforce $\deg(D)>0$.

If $g(S)=1$, then $\deg(D)=0$ is only possible if both $\deg(f_{*1}\mathcal{O}_X)$ and $\sum(m_i-1)(m/m_i)$ vanish. The vanishing of $\deg(f_{*1}\mathcal{O}_X)$ ensures that f is locally trivial outside of the multiple fibres (Theorem III.18.2), whereas $\sum(m_i-1)(m/m_i)$ $=0$ implies that no multiple fibres are present. □

For the results in Sect. 12 we refer to [Ko2], Theorem 12.1 and [Ko4], Theorem 12. Corollary 12.4 is more or less equivalent to [Ko4], Theorems 28, 35. Our treatment of the canonical bundle formula is inspired by [B-Hu]. It has the advantage that it can be generalised immediately to algebraic elliptic fibrations in positive characteristic.

13. Logarithmic Transformations

A logarithmic transformation is a means of replacing any given fibre of type I_b in an elliptic fibration by a multiple fibre of type ${}_m\mathrm{I}_b$, $m\geq 2$. To be more precise, given an elliptic fibration $X\to S$ and a point $0\in S$, over which the fibre is of type I_b, a new elliptic fibration $X'\to S$ is constructed, such that X'_0 is of type ${}_m\mathrm{I}_b$, whereas $X\setminus X_0$ and $X'\setminus X'_0$ are isomorphic as fibre spaces over $S\setminus\{0\}$. This process, still depending on the choice of a suitable $k\in\mathbb{N}$, is called a *logarithmic transformation of order m at* 0 *or with centre* 0. The process can be reversed in the sense that given $X'\to S$ with X'_0 of type ${}_m\mathrm{I}_b$, then a fibration $X\to S$ can be constructed with X_0 of type I_b, such that $X'\to S$ can be derived from $X\to S$ by applying a logarithmic transformation at 0.

i) The case $b=0$.

We start with $X\to S$ and $0\in S$ as above. Since X_0 is smooth, over a coordinate disk Δ_t, centred at 0, this fibration, is isomorphic to a fibration

$$Y=\mathbb{C}\times\Delta/L,$$

where $L=\mathbb{Z}+\mathbb{Z}\omega(t)$, with ω holomorphic on Δ and $\operatorname{Im}\omega(t)\neq 0$. Now we imitate the procedure of Sect. 10 to construct an elliptic fibre space over Δ whose only singular fibre lies over 0 and is of type ${}_m\mathrm{I}_0$. Let $t=s^m$. We take a $k\in\mathbb{N}$, with $(k,m)=1$ and form the smooth quotient Y' of $\mathbb{C}\times\Delta_s/\mathbb{Z}+\mathbb{Z}\omega(s^m)$ by the cyclic group of order m, generated by

$$(c,s)\to\left(c+k/m,\ \mathrm{e}\left(\frac{1}{m}\right)s\right).$$

If we define $Y'\to\Delta_t$ by $(c,s)\to s^m=t$, then $Y'\to\Delta_t$ has indeed only one singular fibre, namely Y'_0, which is of type ${}_m\mathrm{I}_0$. The map given by

$$(c,s)\to(c-(k/2\pi\sqrt{-1})\log s,\ s^m)$$

induces a biholomorphic fibre-preserving map $f\colon Y'\setminus Y'_0\xrightarrow{\sim}Y\setminus Y_0$. If we identify $Y'\setminus Y'_0$ with the part of X, lying over $\Delta\setminus\{0\}$, by way of f, then the union $(X\setminus X_0)\cup Y'$ becomes a Hausdorff space, which in fact is an elliptic fibration $X'\to S$ with the two required properties: $X'\setminus X_0'\cong X\setminus X_0$ and $X_0'\cong Y_0'$ is of type ${}_m\mathrm{I}_0$.

Conversely, to show that it is always possible to replace a fibre of type ${}_m\mathrm{I}_0$ by a smooth fibre, it is clearly sufficient to prove that a neighbourhood of it can be obtained as the quotient of an elliptic fibre space

$$\mathbb{C}\times\Delta/\mathbb{Z}+\mathbb{Z}\,\omega(s^m)$$

by a cyclic group

$$(c,s)\to\left(c+k/m,\ \mathrm{e}\left(\frac{1}{m}\right)s\right)$$

for suitable k with $(k,m)=1$. So let $Y'\to\Delta$ be an elliptic fibration with a single singular fibre Y_0 of type ${}_m\mathrm{I}_0$. Let $Y''\to\Delta$ be its m-th root fibration (III, Sect. 9). Then Y'' is smooth and $Y''\to Y'$ is an unramified cyclic covering, whose covering group G is generated by an automorphism g of Y of order m, such that $g\,Y_t''=Y_{\rho t}''$, $\rho=\mathrm{e}\left(\frac{1}{m}\right)$. Now $Y''=\mathbb{C}\times\Delta/\mathbb{Z}+\mathbb{Z}\,\omega'(s^m)$, with ω' holomorphic, and g can be given as

$$(c,s)\to\left(c+\beta(s),\ \mathrm{e}\left(\frac{1}{m}\right)s\right)$$

for some holomorphic function β on Δ. Since the order of g is m we have

$$\sum_{\nu=0}^{m-1}\beta(g^\nu s)=p+q\,\omega'(s^m),\qquad p,q\in\mathbb{Z}.$$

If $\beta=\sum_{n=0}^{\infty}\beta_n s^n$, we define $\gamma(s)=\sum_{n=0}^{\infty}\sum_{\nu=1}^{m-1}\frac{\beta_{nm+\nu}}{1-\rho^\nu}s^{mn+\nu}$. Then γ is holomorphic, and the preceding relation implies that

$$\gamma(\rho\,s)-\gamma(s)=-\beta(s)+\frac{1}{m}(p+q\,\omega'(s^m)).$$

Let $(p,q)=k$, and let $a,b\in\mathbb{Z}$ satisfy $a\,p-b\,q=k$. If we put $p=k\,p'$, $q=k\,q'$, then

$$\begin{pmatrix}a & b\\ p' & q'\end{pmatrix}\in SL_2(\mathbb{Z}),$$

so introducing the new fibre coordinate $c'=(c-\gamma(s))/q'\,\omega'(s^n)+p'$ the lattice $\mathbb{Z}+\mathbb{Z}\,\omega'(s^m)$ becomes $\mathbb{Z}+\mathbb{Z}\,\omega(s^m)$, where

$$\omega(s^m)=(a\,\omega'(s^m)+b)/(q'\,\omega'(s^m)+p').$$

Since

$$(\beta(s)-\gamma(\rho\,s)-\gamma(s))/(p'+q'\,\omega'(s^m))=k/m,$$

we find that g can be given by

$$(c',s)\to(c'+k/m,\ \rho\,s),$$

as required.

ii) The case $b\neq 0$.

We start by describing a (small) neighbourhood of a fibre of type I_n, $n\geq 2$. Since such neighbourhoods are unique up to fibre preserving automorphisms (cf. the proof of Theorem 11.1), it is sufficient to exhibit one such neighbourhood. We take

$$U_k = \{(u_k, v_k) \in \mathbb{C}^2; |u_k v_k| < 1\}, \qquad k \in \mathbb{Z},$$

and then glue U_k and U_{k+1} together by identifying (u_{k+1}, v_{k+1}) with $(v_k^{-1}, u_k v_k^2)$ in as far as this makes sense. The infinite cyclic group operates on the resulting manifold by $(u_k, v_k) \to (u_{k+n}, v_{k+n})$ and the quotient Z admits a surjective holomorphic map $Z \to \Delta$, given by $(u_k, v_k) \to s = u_k v_k$. Clearly, for $s \neq 0$ the fibre $Z_s \cong \mathbb{C}^*/\{s^{nk}\}_{k \in \mathbb{Z}}$, so it is a smooth elliptic curve, whereas Z_0 is of type I_n. If $n = mb$, then $\mathbb{Z}_m$ acts on Z by

$$(u_k, v_k) \to (\varepsilon \rho^{-k+1} u_{k+b}, \varepsilon \rho^k v_{k+b}), \qquad \varepsilon = \mathbb{e}\left(\frac{b(m-1)}{2m}\right),$$

and the quotient is an elliptic fibre space $Y' \to \Delta$ with Y'_0 of type ${}_m\mathrm{I}_b$. If we start with an elliptic fibration $X \to B$ with X_0 of type I_b, we can glue $X \setminus X_0$ and Y' over $\Delta \setminus \{0\}$, where we identify Δ with a coordinate disc on C centred at 0. In fact the gluing is given by the biholomorphic map

$$Y'|\Delta \setminus \{0\} \to X|\Delta \setminus \{0\}, \qquad \text{induced by } (u_k, v_k) \to ((\varepsilon u_k)^m, (\varepsilon v_k)^m).$$

As in the case $b=0$ the preceding process can be reversed, but our coordinatisation is not very suitable for showing this, so we refer to Kodaira ([Ko4], part I, p. 768–771).

A logarithmic transformation should by no means be seen as something close to a birational transformation. It can completely change the topological as well as the analytic nature of an elliptic surface, as may be illustrated by the following examples.

(13.1) **Example.** Let E be a smooth elliptic curve and $X = \mathbb{P}_1 \times E$, seen as an elliptic fibre space $X \to \mathbb{P}_1$. It follows from the canonical bundle formula that if we apply logarithmic transformations to sufficiently many fibres $p \times E$, then we obtain a surface Y with $\mathrm{kod}(Y) = 1$. Since $\mathrm{kod}(X) = -\infty$, we see that *application of logarithmic transformations can change the Kodaira dimension.*

(13.2) **Example.** Let X be as in the preceding example and let Y be obtained from X by applying a single logarithmic transformation. We claim: *Y is a non-algebraic surface.* To see this, let p: $Y \to \mathbb{P}_1$ be the projection and F the multiple fibre of p. Since $c_1^2(Y) = p_g(Y) = 0$ by Corollary 12.3 and $c_2(Y) = c_2(X) = 0$, we find by the Todd-Hirzebruch formula that $q(Y) = 1$. Hence Theorem IV.2.6 yields that either $b_1(Y) = 2$, $h^{1,0}(Y) = 1$ or $b_1(Y) = 1$, $h^{1,0}(Y) = 0$. So it is sufficient to show that $h^{1,0}(Y) = 0$. We prove this by deriving a contradiction from the assumption that $h^{1,0}(Y) = 1$. If $h^{1,0}(Y) = 1$, then there is a surjective Albanese map f: $Y \to \mathrm{Alb}(Y)$, where $\mathrm{Alb}(Y)$ is a 1-dimensional torus. Let C be a smooth component of a general fibre of f. Then C is not a fibre of p, otherwise f would factorise through p, which is impossible by Lüroth's theorem. So $p|C \to \mathbb{P}_1$ is a ramified covering, which has degree ≥ 2, since p: $Y \to \mathbb{P}_1$ has a multiple fibre by construction. We claim that $p|C$ can only be ramified in the points of $F \cap C$. In other words, we assert that in all other points $y \in C$ we have that C intersects $p^{-1}(p(y))$ transversally. Since outside of F the map p is everywhere of maximal rank, this is however an immediate consequence of the fact that the restriction of f to a fibre $p^{-1}(p(y))$ is an *unramified* covering map onto E. So, $\mathbb{P}_1 \setminus p(F)$ being simply-connected, we have arrived at our contradiction. Thus we conclude: *application of a logarithmic transformation may change an algebraic surface into a non-algebraic one.*

Logarithmic transformations have been introduced by Kodaira in [Ko4], part I. We have closely followed him for the case $b=0$, whereas our approach for $b \neq 0$ has been motivated by toroidal embeddings.

Kodaira Fibrations

14. Kodaira Fibrations

A connected fibration f: $X \to C$ of the compact surface X over the smooth curve C is called a *Kodaira fibration* if f is everywhere of maximal rank but *not* a complex analytic fibre bundle map. In view of the Grauert-Fischer theorem I.10.1 this means that, though all fibres are smooth curves, their complex structure varies.

It follows immediately from the uniqueness of $\mathbb{P}_1$ as a curve of genus 0 and the existence of the J-fibration (Sect. 9) that the fibre genus of a Kodaira fibration is at least 2. Theorem III.15.4 implies that this also holds for the base genus.

Every surface X admitting a Kodaira fibration is algebraic: if F is a fibre, then $\mathcal{K}_X F \geq 2$, hence $c_1^2(\mathcal{K}_X \otimes \mathcal{O}_X(nF)) > 0$ for n large enough, and X is algebraic by Theorem IV.5.2.

A surface admitting at least one Kodaira fibration is sometimes called a Kodaira surface. However, as we have seen in Sect. 5, this name is also used for a totally different type of surface. In this book we shall reserve the name Kodaira surface for the surfaces described in Sect. 5.

It follows readily from the projectivity of the Satake compactification for the variety of moduli that there exist Kodaira fibrations with any given genus $g, g \geq 3$. We want to give here however a more direct construction.

In fact, for certain values of g we shall construct connected fibrations f: $X \to C$ of fibre genus g which are everywhere of maximal rank such that the index $\tau(X)$ is strictly positive. Then f is automatically a Kodaira fibration. For if f were a locally trivial bundle map, then by Sect. 7 there would exist an unramified covering Y of X, which is the product of two curves, and $\tau(X)$ would vanish.

Let C and D be smooth, compact, connected curves of genus ≥ 2, $h\colon C \to D$ a regular map and G a finite group operating analytically and without fixed points on D. If B_g denotes the graph of $g \circ h$, $g \in G$, then $B = \bigcup_{g \in G} B_g$ is a smooth curve on $C \times D$, consisting of $|G|$ disjoint components. Suppose that for some $r \geq 2$ there exists an r-fold cyclic covering X of $C \times D$ which is ramified exactly over B. The surface X admits a natural projection $f\colon X \to C$ and we claim that it is everywhere of maximal rank and connected, whereas $\tau(X) > 0$. The first claim is clear since f admits local sections everywhere. As to the connectedness of f, if there would be a non-connected fibre, then over $c \times D$ ($c \in C$) there would lie on X a non-connected smooth curve, which is impossible since there are points on $C \times D$, over which there is only one point of X. So we only have to calculate $\tau(X)$.

Let $p = g(C)$, $q = g(D)$ and d the degree of h. Then we have first of all

$$\begin{aligned} e(X) &= e(C)(r\,e(D) - (r-1)\,|G|) \\ &= 4r(p-1)(q-1) + 2(p-1)(r-1)\,|G|. \end{aligned}$$

Furthermore, by Lemma I.17.1 there is a line bundle $\mathscr{L}$ on $C \times D$ with $\mathscr{O}(B) = \mathscr{L}^{\otimes r}$. So we have by Lemma I.17.1, (iii) that

$$\mathscr{K}_X = k^*(\mathscr{K}_{C \times D} \otimes \mathscr{L}^{\otimes (r-1)})$$

where $k\colon X \to C \times D$ denotes the projection. Hence if $C_0 = c_0 \times D$ and $D_0 = C \times d_0$, we find

$$K_X^2 = r\left[(2q-2)\,C_0 + (2p-2)\,D_0 + \frac{r-1}{r}\,B\right]^2.$$

Since $B^2 = \sum_{g \in G} B_g^2 = -2d\,|G|(q-1)$ by the adjunction formula, it follows that

$$K_X^2 = 8r(p-1)(q-1) + 4(r-1)(p-1)\,|G| + 2(r-1) \cdot \frac{r+1}{r} \cdot d\,|G|(q-1).$$

So $\tau(X) = \frac{1}{3}(K_X^2 - 2e(X)) > 0$.

Since $K_X^2 > 0$ and $\Gamma((\mathscr{K}_X^\vee)^{\otimes n}) = 0$ for all $n \geq 1$ we find, using Riemann-Roch, that $\mathrm{kod}(X) = 2$ for our example (it can be proved that this holds for any surface X admitting a Kodaira fibration).

If we put $g(D/G) = s$, we find

$$\frac{K_X^2}{e(X)} = \frac{c_1^2(X)}{c_2(X)} = 2 + \frac{1 - \dfrac{1}{r^2}}{(2s-1) - \dfrac{1}{r}}$$

and when we fix s, which has to be at least 2, we find that $\lim\limits_{r \to \infty} \dfrac{c_1^2(X)}{c_2(X)} = 2 + \dfrac{1}{2s-1}$. In particular we see that we always have $2 < \dfrac{c_1^2(X)}{c_2(X)} < 7/3$.

We still have to show that there really exist sets (C, D, G, h, r) as described above. By I, Sect. 18 it is sufficient to produce sets (C, D, G, h, r) with C, D curves, with G a finite analytic fixed-point-free transformation group of D, and with $h\colon C \to D$ a morphism, such that $\mathcal{O}(B)$ is divisible by r in $\mathrm{Pic}(C \times D)$. This can be accomplished, even if r and s are given in advance (but not C and D), by the following method.

Let $r, s \in \mathbb{N}$, $r, s \geq 2$. We start with a smooth curve D_0 with $g(D_0) = s$ and take a normal unramified covering $D \to D_0$ of degree kr $(k \geq 1)$. Then our G will of course be the group of this covering. Let us now assume for a moment that we have another curve C and a regular map $h\colon C \to D$ with the property that

(*) all elements of $h^*(H^1(D, \mathbb{Z}))$ are r-divisible in $H^1(C, \mathbb{Z})$.

We claim that then $\mathcal{O}(B)$ is r-divisible in $\mathrm{Pic}(C \times D)$. Indeed, since $\mathrm{Pic}^0(C \times D)$ is r-divisible anyhow, it is sufficient to show that $c_1(\mathcal{O}(B))$ is r-divisible in $H^2(C \times D, \mathbb{Z})$. The cupproduct induces on this torsion-free group a perfect pairing with itself and it is therefore sufficient to prove that

$$(c_1(\mathcal{O}(B)), \alpha) \equiv 0 \pmod r \qquad \text{for all } \alpha \in H^2(C \times D, \mathbb{Z}).$$

By Künneth's formula this will be true if we prove it in three cases: if $\alpha \in p_1^*(H^2(C, \mathbb{Z}))$, if $\alpha \in p_2^*(H^2(D, \mathbb{Z}))$ and if $\alpha = p_1^*(\beta) \cdot p_2^*(\gamma)$, where $p_1\colon C \times D \to C$ and $p_2\colon C \times D \to D$ are the projections. The first two cases follow from the fact that the intersection number of B with both "horizontal" and "vertical" fibres is divisible by $|G|$, hence by r. And in the last case we have by the projection formula

$$\begin{aligned}(c_1(\mathcal{O}_{C \times D}(B)), \alpha) &= \sum_{g \in G} (c_1(\mathcal{O}_{C \times D}(B_g)), p_1^*(\beta) \cdot p_2^*(\gamma)) \\ &= \sum_{g \in G} (\beta, p_1!(c_1(\mathcal{O}_X(B_g)) \cdot p_2^*(\gamma))) = \sum_{g \in G} (\beta, g^* h^*(\gamma)) \equiv 0 \pmod r\end{aligned}$$

by the assumption (*).

It remains to convince the reader that the assumption (*) can be satisfied. In other words, given a curve D and $r \in \mathbb{N}$, we have to find a curve C and a morphism $h\colon C \to D$, such that (*) holds. But that is rather easy. For if we consider $H_1(D, \mathbb{Z}_r)$ as a quotient of $\pi_1(D)$ by way of the canonical surjections

$$\pi_1(D) \to H_1(D, \mathbb{Z}) \quad \text{and} \quad H_1(D, \mathbb{Z}) \to H_1(D, \mathbb{Z}_r),$$

then we can take the unbranched covering $h\colon C \to D$ with covering group $H_1(D, \mathbb{Z}_r)$, obtaining an exact sequence

$$0 \to \pi_1(C) \to \pi_1(D) \to \mathbb{Z}_r \to 0.$$

It follows that $h_*(H_1(C, \mathbb{Z}))$ is r-divisible in $H_1(D, \mathbb{Z})$, so by transposition $h^*(H^1(D, \mathbb{Z}))$ is r-divisible in $H^1(C, \mathbb{Z})$.

Remark. If, as in the case just described, the map $h\colon C \to D$ is an unramified covering, then X admits a natural fibration in two ways: over C *and* over D.

References: [Ko5], [At3], [Ks]. In this last paper it is proved that the fibre genus of any Kodaira fibration is at least 3. Kas also proved that every small deformation of a Kodaira surface is again a Kodaira surface. His result was extended to the case of any deformation by Jost and Yau (see [J-Y]).

Finite Quotients

15. The Godeaux Surface

Let $(\xi_0:\ldots:\xi_3)$ be a system of homogeneous coordinates in $\mathbb{P}_3$. We define an action of the cyclic group $G=\mathbb{Z}_5$ by putting

$$(1_G)(\xi_0:\ldots:\xi_3)=(\xi_0:\rho\,\xi_1:\ldots:\rho^3\,\xi_3),$$

where $\rho=\mathbb{e}(\frac{1}{5})$. The Fermat surface Y, given by $\sum_{i=0}^{3}\xi_i^5=0$, is G-invariant and does not pass through any of the four fixpoints. Consequently, the quotient Y/G is a smooth surface X, which is algebraic by Theorem IV.5.8. By Proposition 2.1 we have $\pi_1(Y)=0$, so $\pi_1(X)\cong\mathbb{Z}_5$, and $q(X)=\frac{1}{2}b_1(X)=0$. Since $\chi(X)=\frac{1}{5}\chi(Y)=1$ (Proposition 2.1) we have $p_g(X)=0$. Furthermore, if $p\colon Y\to X$ denotes the projection, then

$$H^i(\mathcal{K}_Y^{\otimes n})=H^i(p_*\mathcal{K}_Y^{\otimes n})\supset H^i(\mathcal{K}_X^{\otimes n})\qquad\text{(Lemma I.17.2)},$$

so $H^i(\mathcal{K}_X^{\otimes n})=0$ for $i=1,2$ as soon as $H^i(\mathcal{K}_Y^{\otimes n})=0$ for $i=1,2$. Since $\mathcal{K}_Y=\mathcal{O}_Y(1)$, we find by Serre duality

$$H^i(\mathcal{K}_Y^{\otimes n})=H^{2-i}(\mathcal{O}_Y(-n)),$$

so using Theorem IV.8.6 we obtain

$$H^i(\mathcal{K}_X^{\otimes n})=0\qquad\text{for } i=1,2 \text{ and } n\geq 2.$$

Riemann-Roch then yields $P_n(X)=1+\frac{1}{2}n(n-1)$ for $n\geq 2$. In particular we have that $\operatorname{kod}(X)=2$.

The Godeaux surface provides a historically important example of a surface X with $q(X)=p_g(X)=0$ which is not rational. (In VII, Sect. 11 we shall give some more information about these surfaces.)

The Godeaux surface appears in [Gx1]. In [Gx2] Godeaux constructed a similar example, with Y an intersection of four quadrics in $\mathbb{P}_6$, and $G\cong\mathbb{Z}_8$.

16. Kummer Surfaces

Let T be a 2-torus on which a base point has been chosen. The involution $\iota\colon T\to T$, defined by $\iota(x)=-x$ has exactly sixteen fixed points, namely the points of order 2 on T. At each of these the exponents are $(-1,-1)$, so $T/\langle 1,\iota\rangle$ has sixteen ordinary double points (of type A_1). Resolving the double points we obtain a smooth surface X, the Kummer surface $Km(T)$ of T, or associated to T. We let $g\colon X\to T/\langle 1,\iota\rangle$ be the projection.

Let $\tilde{T}$ be obtained from T by blowing up the sixteen fixpoints of ι. Then ι can be lifted to an involution $\tilde{\iota}$ of $\tilde{T}$, which leaves the sixteen exceptional curves point-wise invariant but has otherwise no fixpoints. So the quotient is a smooth surface Y, admitting a map onto $T/\{1,\iota\}$ which is $1-1$, except that over each singularity there lies a (-2)-curve. It follows that Y is isomorphic to X in such a way, that the obvious diagram

$$\begin{array}{ccc} \tilde{T} & \longrightarrow & X \\ {\scriptstyle f}\Big\downarrow & & \Big\downarrow{\scriptstyle g} \\ T & \longrightarrow & T/\{1,\iota\} \end{array}$$

is commutative. Since $H^1(\tilde{T},\mathbb{Q})=f^*(H^1(T,\mathbb{Q}))$ and since the only element in $H^1(T,\mathbb{Q})$, invariant under ι, is the zero element, we have $H^1(X,\mathbb{Q})=0$. Similarly we see that $p_g(X)=1$. Now a 2-form on X could only vanish on components of the ramification divisor, but then we would have on $\tilde{T}$ a 2-form vanishing to a higher order on (-1)-curves, which is impossible. Hence $\mathscr{K}_X\cong\mathscr{O}_X$.

Kummer surfaces are special K3-surfaces and as such will play an important role in Chap. VIII. They are all kählerien by [Fu].

17. Quotients of Products of Curves

We shall give here an example in which one of the curves is elliptic.

Let E be an elliptic curve, C some other smooth compact, connected curve, G a finite subgroup of $\mathrm{Aut}(C)$, which also acts as a translation group of E. So G acts without fixed points on $C\times E$, and if $X=C\times E/G$, then the natural projection $f\colon X\to D=C/G$ is an elliptic fibration. If $g\colon C\to D$ is the projection, then all fibres, except those lying over branchpoints of g, are smooth elliptic curves, isomorphic to E/G, and if $g^{-1}(d)$ contains $m<|G|$ points, then $g^{-1}(d)$ is a fibre of type ${}_n\mathrm{I}_0$, with $n=\dfrac{|G|}{m}$.

As to the invariants of X, we have for example

$$H_1(X,\mathbb{Q})\cong H_1(E\times C,\mathbb{Q})^G\cong H_1(E,\mathbb{Q})\oplus H_1(C,\mathbb{Q})^G\cong H_1(E,\mathbb{Q})\oplus H_1(D,\mathbb{Q}),$$

and

$$\Gamma(\mathscr{K}_X^{\otimes n})\cong\Gamma(\mathscr{K}_E^{\otimes n}\otimes\mathscr{K}_C^{\otimes n})^G\cong\Gamma(\mathscr{K}_C^{\otimes n})^G\cong\Gamma(\mathscr{K}_D^{\otimes n}).$$

Also $c_1^2(X)=\dfrac{1}{|G|}c_1^2(C\times E)=0$, $c_2(X)=\dfrac{1}{|G|}c_2(C\times E)=0$.

There exist many examples of this type. Given a smooth compact, connected curve D and natural integers a, b, it is always possible to construct a surface as above with this D and $G\cong\mathbb{Z}_a\oplus\mathbb{Z}_b$. For if x, y and z are any points on D, there exists a covering C of D of degree ab, with group G (which is ramified exactly over x, y and z), whereas G can always be taken as a translation group on an elliptic curve.

The existence of the covering C can be proved in the following way. If we set $g(D)=p$, then $\pi_1(D\setminus\{x,y,z\})$ is generated by elements $\alpha_1,\dots,\alpha_{2p}$, β_1,β_2 and

β_3, with one relation, namely

$$\alpha_1 \alpha_2 \alpha_1^{-1} \alpha_2^{-1} \ldots \alpha_{2p-1}^{-1} \alpha_{2p}^{-1} \beta_1^{-1} \beta_2^{-1} \beta_3^{-1} = 1,$$

where β_1, β_2 and β_3 are represented by suitable loops around x, y and z. There is a homomorphism h from $\pi_1(D \setminus \{x, y, z\})$ onto $\mathbb{Z}_a \oplus \mathbb{Z}_b$ with $h(\alpha_j) = 0$, $j = 1, \ldots, 2p$, $h(\beta_1) = (1, 0)$, $h(\beta_2) = (0, 1)$ and $h(\beta_3) = (-1, -1)$. This homomorphism h determines a covering over $D \setminus \{x, y, z\}$ as described before. Restricting the covering to punctured disks around x, y and z it becomes readily clear that it can be extended to a ramified covering over all of D.

Infinite Quotients

18. Hopf Surfaces

Let W be the complex surface obtained from $\mathbb{C}^2$ by deleting the origin. A compact complex surface is called a *Hopf surface* if its universal covering is (analytically) isomorphic to W.

The original Hopf surface, as defined in [Hop1] by Hopf in 1948, is the quotient of W by the infinite cyclic group, generated by the homothety

$$(z_1, z_2) \to (\tfrac{1}{2} z_1, \tfrac{1}{2} z_2).$$

Historically, this surface H has played an important role for several reasons. For example, it was the first example of a compact surface that does not admit any Kähler metric. This follows from Corollary I.13.4, for H is diffeomorphic to $S^1 \times S^3$, hence $b_1(H) = 1$. The surface H is an elliptic fibre bundle over $\mathbb{P}_1$ (compare Sect. 5), and it is homogeneous.

Hopf's construction can be generalised immediately to the case where W is divided out by the infinite cyclic group G which is generated by the automorphism

$$(z_1, z_2) \to (\alpha_1 z_1, \alpha_2 z_2),$$

with $0 < |\alpha_1| \leq |\alpha_2| < 1$. Putting $\alpha = (\alpha_1, \alpha_2)$, we shall denote W/G by H_α.

(18.1) **Proposition.** *The surfaces H_α have the following properties:*

(i) *H_α is diffeomorphic to $S^1 \times S^3$, and therefore does not admit any Kähler metric;*

(ii) $h^{1,0}(H_\alpha) = h^{2,0}(H_\alpha) = h^{0,2}(H_\alpha) = h^{1,1}(H_\alpha) = 0$, *and* $h^{0,1}(H_\alpha) = 1$;

(iii) $\mathrm{Pic}(H_\alpha) \cong \mathbb{C}^*$.

Proof. (i) This is a consequence of the fact that H_α can be deformed differentiably into H.

(ii) Since $b_2(H_\alpha)=0$, we find from Theorem IV.2.9, that

$$h^{2,0}(H_\alpha)=h^{0,2}(H_\alpha)=h^{1,1}(H_\alpha)=0.$$

Combining this with $c_1^2(H_\alpha)=e(H_\alpha)=0$, we find from the Todd-Hirzebruch formula that $h^{0,1}(H_\alpha)=1$. Finally, Theorem IV.2.9 yields $h^{1,0}(H_\alpha)=0$.

(iii) It is obvious from the exponential cohomology sequence that the map $H^1(H_\alpha,\mathbb{Z})\to H^1(H_\alpha,\mathcal{O}_{H_\alpha})$ is injective, so

$$\mathrm{Pic}(H_\alpha)=H^{0,1}(H_\alpha)/H^1(H_\alpha,\mathbb{Z})\cong\mathbb{C}^*. \quad \square$$

Though a surface H_α always contains two elliptic curves (namely the images C and D of the punctured z_1- and z_2-axes), it is not always elliptic.
In fact we have

(18.2) **Proposition.** *H_α is an elliptic fibre space over $\mathbb{P}_1$ if and only if $\alpha_1^k=\alpha_2^l$ for some $k,l\in\mathbb{Z}$. Otherwise, H_α contains exactly two irreducible curves.*

Proof. If the condition is satisfied, then $z_1^k z_2^{-l}$ is a non-constant meromorphic function on H_α, defined in every point. This function gives a surjective holomorphic map $f'\colon H_\alpha\to\mathbb{P}_1$. Using Stein factorisation, we obtain from f' a connected map $f\colon H_\alpha\to B$, where B is some smooth curve. This curve is again rational, since otherwise there would be a holomorphic 1-form on H_α, which is impossible by Proposition 18.1, (ii). Finally, the adjunction formula shows immediately that the fibration is an elliptic one.

Conversely, if H_α is in some way a fibre space, then mC is a fibre for some $m\in\mathbb{Z}$, $m\geq 1$; otherwise C would intersect some other curve in a strictly positive number of points, which is impossible, since every curve is homologous to 0. So there exists a meromorphic function g' on W, which has the z_1-axis, taken with multiplicity m as polar divisor, whereas g' is left invariant by the automorphism $(z_1,z_2)\to(\alpha_1 z_1,\alpha_2 z_2)$. Then the function $g=z_1^m g'$ is a holomorphic function on W, hence on $\mathbb{C}^2$ (Theorem I.8.7), satisfying $g(\alpha_1 z_1,\alpha_2 z_2)=\alpha_1^m g(z_1,z_2)$. But a straightforward calculation shows that this is only possible if $\alpha_1^k=\alpha_2^l$ for suitable $k,l\in\mathbb{N}$.

So if $\alpha_1^k\neq\alpha_2^l$ for all $k,l\in\mathbb{N}$, there is no meromorphic function at all on H_α, i.e. H_α has algebraic dimension 0. Then, by Theorem IV.6.2 the number of irreducible curves on H_α is finite. Now given any two points in $H_\alpha\setminus(C\cup D)$, there exists an automorphism of H_α (induced by a linear automorphism of $\mathbb{C}^2$), carrying the first of these points into the second. So if there were any (closed) irreducible curves on H_α, different from C and D, there would be an infinity of such curves. Consequently, in the case that H_α has no elliptic fibration, there are no irreducible curves on H_α but C and D. $\square$

(18.3) *Remark.* This example shows that the bound given in Theorem IV.6.2 is sharp, for $h^{1,1}(H_\alpha)=0$ by Proposition 18.1, (ii).

There are many Hopf surfaces which are not of the type H_α, or not even diffeomorphic to $S^1\times S^3$. For example, you can take the quotient of H by a suitable fibre preserving cyclic group of order n to obtain a Hopf surface X, which is still an elliptic fibre bundle over $\mathbb{P}_1$, with $\pi_1(X) \cong \mathbb{Z} \oplus \mathbb{Z}_n$. — or by a copy of [illegible] — obtaining $S^1 \times$ fake S^3.

Kodaira has treated Hopf surfaces extensively in [Ko4], part II and III. We mention some of his results.

Every Hopf surface has Kodaira dimension $-\infty$.

A Hopf surface X is called *primary* if $\pi_1(X) \cong \mathbb{Z}$. Any Hopf surface admits a finite, unramified covering which is a primary Hopf surface.

(18.4) **Theorem.** *A compact surface X is a primary Hopf surface if and only if either one of the following (equivalent) conditions is satisfied:*

(i) *X is homeomorphic to $S^1 \times S^3$,*

(ii) *$b_2(X)=0$ and $\pi_1(X) \cong \mathbb{Z}$.*

Kodaira has furthermore shown

(18.5) **Theorem.** *A minimal compact surface X with $\mathrm{a}(X)=1$ is a Hopf surface if and only if $P_{12}(X)=0$.*

We shall see in the next chapter that every surface X with $\mathrm{a}(X)=1$ is elliptic. Combining this fact with Corollary 12.4, (ii), Proposition 12.5 and the results of Sect. 5 we obtain:

(18.6) **Theorem.** *The minimal compact surfaces X with $\mathrm{a}(X)=1$, $\mathrm{kod}(X)=-\infty$ are exactly the Hopf surfaces.*

As to the case $\mathrm{a}(X)=0$, we have

(18.7) **Theorem.** *A compact surface X with $\mathrm{a}(X)=0$ is a Hopf surface if and only if $b_1(X)=1$, $b_2(X)=0$ and there is a curve on X.*

Additional references: [Ka1], [Weh] and [Da]. The last paper deals with the existence of global moduli spaces.

19. Inoue Surfaces

Let $M=(m_{i,j}) \in SL(3,\mathbb{Z})$, and suppose that the eigenvalues of M satisfy the following conditions: one of them, say α, is real with $\alpha>1$, and the other two, β and $\bar{\beta}$, are not real, i.e. $\beta \neq \bar{\beta}$. For example, we can take for M a matrix

$$\begin{pmatrix} 0 & 1 & 0 \\ n & 0 & 1 \\ 1 & 1-n & 0 \end{pmatrix}$$

with $n \in \mathbb{Z}$. Starting from such a matrix M, a compact surface S_M can be constructed in the following way.

Let (a_1, a_2, a_3) be a real eigenvector of M corresponding to α, and (b_1, b_2, b_3) an eigenvector of M corresponding to β. Since (a_1, a_2, a_3), (b_1, b_2, b_3) and $(\bar{b}_1, \bar{b}_2, \bar{b}_3)$ are independent over $\mathbb{C}$, the vectors $\begin{pmatrix} a_1 \\ b_1 \end{pmatrix}$, $\begin{pmatrix} a_2 \\ b_2 \end{pmatrix}$ and $\begin{pmatrix} a_3 \\ b_3 \end{pmatrix}$ are independent over $\mathbb{R}$.

Now let as usual $\mathfrak{H}$ be the upper half plane, and G_M the group of analytic automorphisms of $\mathfrak{H} \times \mathbb{C}$, generated by g_0, g_1, g_2, g_3, where

$$g_0(w, z) = (\alpha w, \beta z)$$
$$g_i(w, z) = (w + a_i, z + b_i), \qquad i = 1, 2, 3.$$

Furthermore, let $G \subset G_M$ be the subgroup generated by g_1, g_2 and g_3. The independence of $\begin{pmatrix} a_1 \\ b_1 \end{pmatrix}$, $\begin{pmatrix} a_2 \\ b_2 \end{pmatrix}$ and $\begin{pmatrix} a_3 \\ b_3 \end{pmatrix}$ over $\mathbb{R}$ implies that the group G, which is isomorphic to the free abelian group $\mathbb{Z}^3$, acts properly and discontinuously on $\mathfrak{H} \times \mathbb{C}$, without fixpoints. In fact, G transforms any real 3-dimensional affine linear variety

$$(w_0, z_0) + (\mathbb{R}(a_1, b_1) \oplus \mathbb{R}(a_2, b_2) \oplus \mathbb{R}(a_3, b_3))$$

into itself. Since these spaces are parametrised by $\operatorname{Im}(w_0)$, we see that $\mathfrak{H} \times \mathbb{C}/G$ is a real 3-torusbundle over $\mathbb{R}^+$ (in fact the product bundle). Because of

$$\begin{pmatrix} \alpha a_j \\ \beta b_j \end{pmatrix} = \sum_{k=1}^{3} m_{j,k} \begin{pmatrix} a_k \\ b_k \end{pmatrix}, \qquad j = 1, 2, 3$$

the transformation g_0 acts in a fibre-preserving way on $\mathfrak{H} \times \mathbb{C}/G$, and since $\alpha > 1$ by assumption, the quotient is a 3-torusbundle over S^1, at least from the real point of view. Thus G_M acts properly and discontinuously without fixpoints on $\mathfrak{H} \times \mathbb{C}$, so by [Car] the quotient is a compact complex surface. This is the announced surface S_M.

From the relations

$$g_i g_j = g_j g_i \qquad (i = 1, 2, 3)$$
$$g_0 g_i g_0^{-1} = g_1^{m_{i,1}} g_2^{m_{i,2}} g_3^{m_{i,3}}$$

we see that

$$H_1(S_M, \mathbb{Z}) \cong \pi_1(S_M)/[\pi_1(S_M), \pi_1(S_M)] \cong G_M/[G_M, G_M]$$
$$\cong \mathbb{Z} \oplus \mathbb{Z}_{e_1} \oplus \mathbb{Z}_{e_2} \oplus \mathbb{Z}_{e_3},$$

where e_1, e_2 and e_3 are the elementary divisors of $M - \mathbb{1}_3$.

Since $e(S_M) = 0$, we find that also $b_2(S_M) = 0$.

(19.1) **Proposition.** *The surface S_M does not contain any (closed) curve.*

Proof. Since $b_2(S_M) = 0$, the adjunction formula immediately yields that any irreducible curve on S_M is either smooth elliptic, or rational with one ordinary double point or one cusp.

Let D be any irreducible compact curve on S_M, and let p: $\mathfrak{H} \times \mathbb{C} \to S_M$ be the projection. Then $\mathfrak{H} \times \mathbb{C}$ contains at least one irreducible curve $\tilde{D}$, such that p: $\tilde{D} \to D$ exhibits $\tilde{D}$ as an unramified covering of D. This fact already excludes the rational cases, for otherwise $\mathfrak{H} \times \mathbb{C}$ would contain a compact curve. For the same reason, if D is elliptic, then $\tilde{D}$ has to be non-compact, so isomorphic to $\mathbb{C}$ or $\mathbb{C}^*$. But every holomorphic map from $\mathbb{C}$ or $\mathbb{C}^*$ into $\mathfrak{H}$ is constant, so $\tilde{D}$ would have to be contained in a fibre $w_0 \times C$, $w_0 \in \mathfrak{H}$. Then $\tilde{D}$ would be dense in $w_0 \times \mathbb{C}$, so $p(w_0 \times \mathbb{C})$ would be D again.

Now we observe that $w_0 \times \mathbb{C}$ is contained in the real affine space

$$L = (w_0, 0) + \sum_{i=1}^{3} \mathbb{R}(a_i, b_i).$$

Indeed, given any point (w_0, z), the system of equations

$$\sum_{i=1}^{3} \lambda_i a_i = 0, \qquad \sum_{i=1}^{3} \lambda_i b_i = z$$

has a real solution, as follows from the independence of $\binom{a_1}{b_1}$, $\binom{a_2}{b_2}$ and $\binom{a_3}{b_3}$ over $\mathbb{R}$.

The description of S_M as a real 3-torus bundle over S^1 tells us that $p|w_0 \times \mathbb{C}$ is the restriction to $w_0 \times \mathbb{C}$ of the projection from L onto the torus L/G. So as soon as we can prove that the rank over $\mathbb{Q}$ of

$$(w_0 \times \mathbb{C}) \cap \left((w_0, 0) + \sum_{i=1}^{3} \mathbb{Z}(a_i, b_i)\right)$$

is at most one, we are ready, for then it would follow that D is not compact. This rank statement follows once we prove that the solution space $V \subset \mathbb{R}^3(x) = \mathbb{R}^3(x_1, x_2, x_3)$ of the equation $\sum_{i=1}^{3} a_i x_i = 0$ has rank at most 1 over $\mathbb{Q}$. Suppose that this rank were 2. Since $({}^t M x, a) = (x, M a) = \alpha(x, a)$, we see that ${}^t M$ leaves this V invariant, i.e. if $v = 0$ is an equation for V, then $M v = \lambda v$, with $\lambda \in \mathbb{Q}$. This however would imply that M has a rational eigenvalue, which is actually not the case (by assumption, M has only one real eigenvalue $\alpha > 1$, which is clearly irrational). □

(19.2) **Proposition.** *For any surface* S_M *we have* $p_g(S_M) = 0$, $q(S_M) = 1$, $h^{1,0}(S_M) = 0$.

Proof. Since $b_2(S_M) = 0$, Theorem IV.2.9 gives that

$$h^{1,1}(S_M) = h^{2,0}(S_M) = h^{0,2}(S_M) = p_g(S_M) = 0.$$

Furthermore $c_1^2(S_M) = 0$ (since $b_2(S_M) = 0$) and $e(S_M) = 0$ (since S_M is a torus-bundle), so Noether's formula yields $q(S_M) = 1$. Applying IV.2.9 again, we find that $h^{1,0}(S_M) = 0$. □

For a long time the only compact surfaces X known with $\mathrm{kod}(X) = -\infty$, $b_1(X) = 1$ were the Hopf surfaces (Sect. 18). In 1972 Inoue and Bombieri independently found the example described above. For more details and more examples we refer to [In1] and [In2]. All these surfaces have $b_2 = 0$. In 1974 (see [In2]) Inoue and later Hirzebruch constructed examples with $b_2 > 0$ (Inoue-Hirzebruch surfaces). For more recent work in this direction see [Ka2] and [En].

20. Quotients of Bounded Domains in $\mathbb{C}^2$

Whereas in most cases the examples in this chapter are either elementary or based on general theorems, we shall use in this and the next section some specific deep results which can only be quoted.

Let D be a bounded symmetric domain in $\mathbb{C}^2$. Then D is isomorphic to either the unit ball E (with group $G_D = SU(2,1)$) or the product P of two disks (with $G_D \cong PSL(2,\mathbb{R}) \times PSL(2,\mathbb{R})$).

Suppose that $G \subset G_D$ operates freely and properly discontinuously on D, such that the quotient is a compact surface X.

According to Hirzebruch's proportionality theorem ([Hir2]) the Chern numbers of X are proportional, with a strictly positive proportionality factor, to the Chern numbers of the dual homogeneous complex manifold of D. This dual manifold is $\mathbb{P}_2$ if $D = E$, and it is $\mathbb{P}_1 \times \mathbb{P}_1$ if $D = P$.

A theorem of Borel (valid for all bounded symmetric domains) ([Bor]) yields for $D = E$ the existence of many arithmetical subgroups of G_D satisfying the condition above. (Recently Mostow ([Mo1], [Mo2]) has found non-arithmetical groups with compact quotient.)

As a consequence we have

(20.1) **Theorem.** *For infinitely many $a \in \mathbb{N}$ there exists a surface X which has the unit ball in $\mathbb{C}^2$ as its universal covering, such that $c_1^2(X) = 3a$, $c_2(X) = a$.*

Borel's theorem does not give any specific value of a which actually occurs.

These surfaces with $c_1^2(X) = 3c_2(X)$ (all of them algebraic by Cor. IV.5.3) play an important role as extreme cases in the geography of Chern numbers. We shall return to this point in Chap. VII.

As to the case of P, every product of two smooth compact curves, both of genus ≥ 2, is a quotient of P, but there are infinitely many others. In particular, there is a "fake quadric" X due to Kuga ([Kug]), with $c_1^2(X) = 8$, $c_2(X) = 4$ and $b_1(X) = 0$.

21. Hilbert Modular Surfaces

For the many facts, some of them highly non-trivial, which are only indicated in this section we mention [Hir5], [H-V1], [H-Z] as general references.

Let p be a square-free natural number and let $\mathfrak{o}$ be the ring of algebraic integers in the field $\mathbb{Q}(\sqrt{p})$. To avoid some technical points – we are only giving examples anyway – we shall restrict ourselves to the case p prime and $p \equiv 1(4)$.

The group $SL_2(\mathfrak{o})$ operates on the product $\mathfrak{H} \times \mathfrak{H}$ of the upper half plane by itself:

$$\begin{pmatrix} a & b \\ c & d \end{pmatrix}(z_1, z_2) = \left(\frac{a z_1 + b z_2}{c z_1 + d z_2}, \frac{a' z_1 + b' z_2}{c' z_1 + d' z_2}\right),$$

where $a', \ldots$ are the conjugates of $a, \ldots$ in $\mathbb{Q}(\sqrt{p})$. This action becomes effective if we divide out by $\pm$ (identity). It is properly discontinuous, so by [Car] the quotient is a normal complex space $X(p)$. This non-compact space has only a finite number of singular points, called the *quotient singularities*. They arise from the points on $\mathfrak{H} \times \mathfrak{H}$ with non-trivial isotropy group in $PSL_2(\mathfrak{o})$. But for the case $p=5$, where the order 5 occurs, all these points have order 2 or 3. The corresponding quotient singularities are of type A_1 if the order is 2 and either of type $A_{3,1}$ or $A_{3,2} = A_2$ if the order is 3. So their minimal resolution consist of a (-2)-curve, a (-3)-curve, and two (-2)-curves intersecting transversally in one point, respectively. The number $a_2(p)$ of quotient singularities of type A_1 equals $h(-p)$, whereas of each of the types $A_{3,1}$ and $A_{3,2}$ there are $\frac{1}{2} a_3(p) = \frac{1}{2} h(-3p)$ quotient singularities. Here $h(q)$ denotes the class number of $\mathbb{Q}(\sqrt{q})$.

The space $X(p)$ can be compactified to an analytic space $\bar{X}(p)$ by adding a finite number of points, the *cusps*, in a specific way. In fact, there is a natural 1-1-correspondence between these cusps and the ideal classes in the ring $\mathfrak{o}$. If c is any of the cusps, then its minimal resolution can be described in the following way. The cusp c determines a $\mathbb{Z}$-module $M = \mathbb{Z} \cdot 1 + \mathbb{Z} \cdot w$ in $\mathbb{Q}(\sqrt{p})$ with $0 < w' < 1 < w$. The number w admits an expansion as a continued fraction:

$$w = [[b_0 b_1 b_2 \ldots]] = b_0 - 1 \rfloor b_1 - 1 \rfloor b_2 \ldots, \qquad b_i \in \mathbb{Z},\ b_i \geq 2$$

for $i \geq 1$ with $b_{i+r} = b_i$ for all $i \geq 0$, i.e. w is purely periodic with period r. We denote this by $w = [[\overline{b_0 b_1 b_2 \ldots b_r}]]$. If $p > 5$ the minimal resolution of c consists of an r-gon of smooth rational curves, with self-intersection $-b_0, -b_1, \ldots, -b_{r-1}$ in this order (For $p=5$ there is one cusp, which is resolved by a rational curve with a node.) In particular, if $M = \mathbb{Z} \cdot 1 + \mathbb{Z} \cdot w_0$, with $0 < w_0' < 1 < w_0$, is a principal ideal, then w_0 has a continued fractional expansion of the form

$$w_0 = [[\overline{b_0 b_1 \ldots b_t b_t b_{t-1} \ldots b_1}]],$$

where $b_0 = \{\sqrt{p}\}$ (the smallest odd integer $\geq \sqrt{p}$). The corresponding curves in the cusp resolution are denoted by $S_0, S_1, \ldots, S_t, S_{-t}, \ldots, S_{-1}$, so $S_{\pm i}^2 = -b_i$ (compare [H-V1], p. 14).

Let $Y(p)$ be obtained from $\bar{X}(p)$ by resolving its singularities (both the quotient singularities and the cusps). The surface $Y(p)$ is called the *Hilbert modular surface* associated to p.

By Levi's extension Theorem I.8.7 the field of meromorphic functions on $Y(p)$ is isomorphic to that of $\bar{X}(p)$ which on its turn is isomorphic to that of $X(p)$. This last field is isomorphic to the field of meromorphic functions on $\mathfrak{H} \times \mathfrak{H}$ which are automorphic with respect to the action of $SL_2(\mathfrak{o})$. It is known that this field is an algebraic function field of transcendency degree 2. In fact, in

[B-B] it is proved that suitable automorphic forms yield an embedding of $\bar{X}(p)$ in some projective space.

On $Y(p)$ we have several special types of curves:

1) the curves resolving the quotient singularities;

2) the curves resolving the cusps;

3) the curves F_N. They are obtained as follows. Let N be a strictly positive integer. Consider the set of points (z_1, z_2) satisfying any equation of the form

$$a\sqrt{p}\, z_1 z_2 + \lambda z_2 - \lambda' z_1 + b\sqrt{p} = 0,$$

with $a, b \in \mathbb{Z}$, $\lambda \in \mathfrak{o}$, a, b, λ primitive (i.e. not divisible by a common integral factor) and $\lambda\lambda' + abp = N$. This point set is left invariant by $SL_2(\mathfrak{o})$ and its projection on $X(p)$ is a curve which is either compact or becomes a closed curve on $\bar{X}(p)$ by adding some of the cusps. Its proper transform on $Y(p)$ is the curve F_N; it may be reducible and it may be empty. The curves F_1, F_4 and, if present, F_2, F_3 are irreducible, smooth and rational. In many cases it is possible to determine the nature of F_N (number of components, singularities, genus). It is always possible to calculate F_N^2 and the intersection behaviour of F_N with respect to the curves of type 1) and 2) and with respect to the other curves F_N (see [H-Z]). For example, F_1 is always a (-1)-curve, and F_p is smooth and irreducible, with $F_p^2 = -(p+1)/6 + g(F_p)$, where the genus $g(F_p)$ is given by a classical formula from the theory of modular curves. It turns out that F_p is rational if $p = 5$, 13, 17, 29 and 41.

The configuration formed by the curves 1), the curves 2), some of the curves F_N (namely F_1, F_4, F_p and, if present, F_2 and F_3), is called the basic configuration. In Fig. 7 the basic configuration is shown for the case that $p > 5$ and there is only one cusp. The curves B_1, B_2, E, L, C_1, $C_1', \dots, C_k$, C_k', $D_1, \dots, D_l, U_1, U_1', \dots, U_m$, U_m', V_1, $V_1', \dots, V_n, V_n'$ are the resolutions of the quotient singularities.

The numbers δ and ε are defined by

$$\delta = \begin{cases} 0 & \text{if } p \equiv 5 \pmod 8 \\ 1 & \text{if } p \equiv 1 \pmod 8 \end{cases}$$

$$\varepsilon = \begin{cases} 0 & \text{if } p \equiv 2 \pmod 3 \\ 1 & \text{if } p \equiv 1 \pmod 3 \end{cases}$$

and the numbers $k, \dots, n$ are given by

$$\begin{aligned} k &= \tfrac{1}{2} a_3(p) \\ l &= \tfrac{1}{2} a_2(p) \\ m &= \tfrac{1}{4} a_2(p) - \tfrac{1}{2}(1+\delta) \\ n &= \tfrac{1}{4} a_3(p) - \tfrac{1}{2}(1+\varepsilon). \end{aligned}$$

All curves are smooth, all except possibly F_p are rational, all intersections are transversal, self-intersection numbers as indicated, and the only intersections not shown are some intersections of the curves F_N with the curves

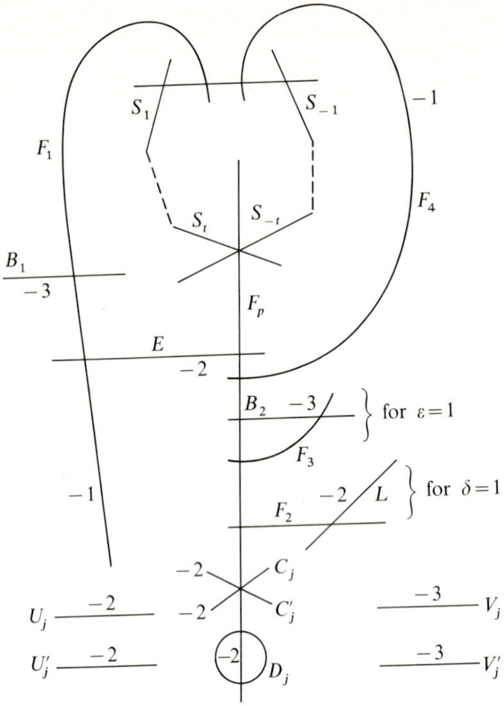

Fig. 7

$C_1, C'_1, \ldots, V_n, V'_n$. Intersection points which are different in the figure are different on $Y(p)$.

In several cases the information contained in the basic configuration is already more than sufficient to determine the nature of the surface $Y(p)$. For example, if $p=17$, we obtain the configuration of Fig. 8. So we can blow down F_1, then (the image of) E, then B_1, then F_{17} and finally C_1. After all this blowing down, the curve F has become a smooth rational curve with self-intersection $+1$. Hence $Y(17)$ is a rational surface by Proposition 4.3.

As a corollary we obtain the fact that $Y(17)$ is simply connected. Actually, this fact holds for any surface $Y(p)$ ([Sv]).

Given p, we can always blow down the smooth rational curves F_1, E, B_1, F_4 and then F_2, L, F_3 if they are present. The smooth surface thus obtained is denoted by $Y^0(p)$. It is conjectured that if $Y^0(p)$ is not rational, i.e. if $p \neq 5$, 13, 17, then this surface is minimal. The conjecture has been verified for many p's ([G-V], [Hir6]), but a general proof seems difficult.

Independent of the conjecture it is possible to give all surfaces $Y(p)$ their place in the Enriques-Kodaira classification of surfaces ([H-V1],

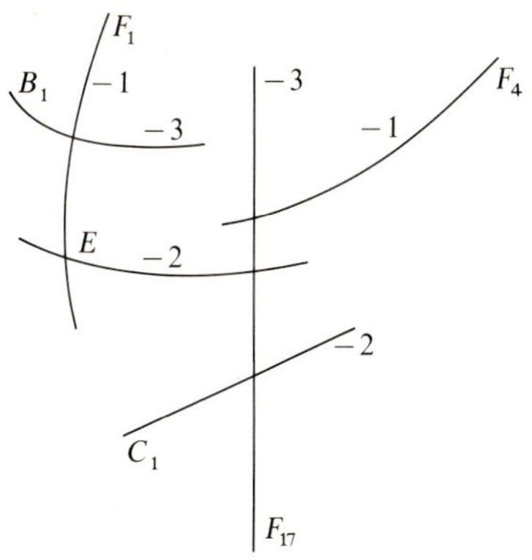

Fig. 8

Theorem III.1). Apart from the basic configuration and the fact that $b_1(Y(p)) = 0$ one needs also the values of $e(Y(p))$ and $\chi(Y(p))$. They are given by

$$e(Y(p)) = 2\zeta_{\mathbb{Q}(\sqrt{p})}(-1) + \tfrac{3}{2}h(-p) + \tfrac{13}{6}h(-3p) + l(p)$$
$$\chi(Y(p)) = \tfrac{1}{4}(2\zeta_{\mathbb{Q}(\sqrt{p})}(-1) + \tfrac{3}{2}h(-p) + \tfrac{13}{6}h(-3p)),$$

where $l(p)$ denotes the total number of all curves occurring in the cusp resolutions. (Compare [H-V1], p. 12 and 20.)

We would like to illustrate this by considering the case $p=29$. In this case one finds $\chi(Y^0(29))=2$ and $c_1^2(Y^0(29))=0$. It is easily verified that on $Y^0(29)$ the image of S_0 satisfies $KS_0=S_0^2=0$, whereas S_1 has remained a (-2)-curve intersecting S_0 transversally in one point (thus we find a confirmation of the fact that $Y^0(29)$ is projective, for $(2S_0+S_1)^2>0$, and this is enough by Theorem IV.5.2). We want to show that $Y^0(29)$ is an *elliptic* K3-*surface*. Applying Riemann-Roch we find

$$\dim|S_0| + \dim|K - S_0| \geq 1.$$

If $\dim|K-S_0|$ would be strictly positive, then also $\dim|K|$ would be strictly positive, whereas

$$\chi(Y^0(29))=2 \quad \text{and} \quad q(Y^0(29)) = \tfrac{1}{2}b_1(Y^0(29)) = \tfrac{1}{2}b_1(Y(29)) = 0$$

yield $\dim|K| = p_g(Y^0(29)) - 1 = 0$. So $\dim|S_0| \geq 1$. Since S_0 is irreducible this linear system can't have any fixed components or base points and since $S_0^2=0$ it provides a map onto $\mathbb{P}_1$, the general fibre of which is elliptic by the adjunction formula. This elliptic fibration is relatively minimal since otherwise $c_1^2(Y^0(29))$ would be strictly negative by Theorem 12.3, and it can't have any multiple fibres because of the existence of the section S_1. So Theorem 12.3 tells us that $\mathcal{K}_{Y^0(29)}$ is trivial. Combined with $b_1(Y^0(29))=0$ this yields the desired result.

In the situations above (for $p=17$ and 29) we have used only a small part of the available information. This will make it clear, on the one hand, that there are numerous other ways to arrive at the same conclusion, and, on the other hand, that much and much more can be said about the surfaces in question.

Double Coverings

22. Invariants

We recall the following facts from I, Sect. 17 and III, Sect. 7. A reduced divisor B on the compact surface Y, such that $\mathcal{O}_Y(B) = \mathcal{L}^{\otimes 2}$ for some $\mathcal{L} \in \mathrm{Pic}(Y)$, determines a double covering $\pi: X \to Y$ which is ramified exactly over B. The surface X is normal, and if B has a simple singularity at $y_1 \in Y$, then X has a rational singularity of the same type at $\pi^{-1}(y_1)$.

If $\sigma: \bar{X} \to X$ is the canonical resolution of singularities, and $p = \pi \circ \sigma$, we always have

(7) $$p_* \mathcal{O}_{\bar{X}} = \mathcal{O}_Y \oplus \mathcal{L}^{-1}, \quad p_{*i}(\mathcal{O}_{\bar{X}}) = 0 \quad \text{for } i \geq 1.$$

And if moreover B has at most simple singularities, then

(8) $$\mathcal{K}_{\bar{X}} = p^*(\mathcal{K}_Y \otimes \mathcal{L}).$$

Formula (8) has been established in III, Sect. 7. As to (7), the case of a smooth B has already been treated in I, Sect. 17. The general case can be dealt with in the following way. If

$$\begin{array}{ccc} \bar{X} & \xrightarrow{\sigma} & X \\ {\scriptstyle\bar{\pi}}\downarrow & {\scriptstyle p}\searrow & \downarrow{\scriptstyle\pi} \\ \bar{Y} & \longrightarrow & Y \end{array}$$

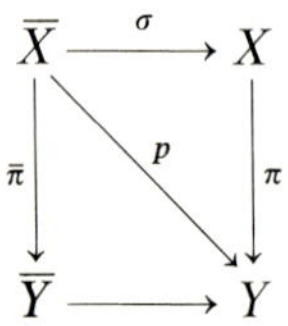

is the canonical resolution diagram (Theorem III.7.2), then the involution of $\bar{X}$, corresponding to $\bar{\pi}$ induces a canonical splitting

$$p_* \mathcal{O}_{\bar{X}} = \mathcal{O}_Y \oplus \mathcal{M},$$

where $\mathcal{M}$ is a line bundle on Y. Since outside of the finitely many singular points of B we are in the smooth case and hence there $\mathcal{M} \cong \mathcal{L}^{-1}$, we must have $\mathcal{M} \cong \mathcal{L}^{-1}$ everywhere on Y. This proves the first part of (7). As to the second part, this follows from the corresponding statement for σ and the very definition of (higher) direct image sheaves.

(22.1) *Remark.* It is an immediate consequence of Corollary IV.5.5 that $\bar{X}$ is algebraic as soon as Y is.

Now let B have at most simple singularities. From (7), the Leray spectral sequence, Riemann-Roch and Serre duality, we obtain

$$(9)\quad \begin{aligned} &\chi(\bar{X})=2\chi(Y)+\tfrac{1}{2}(\mathscr{L},\mathscr{K}_Y)+\tfrac{1}{2}(\mathscr{L},\mathscr{L})\\ &p_g(\bar{X})=p_g(Y)+h^0(Y,\mathscr{K}_Y\otimes\mathscr{L})\\ &c_1^2(\bar{X})=2c_1^2(Y)+4(\mathscr{L},\mathscr{K}_Y)+2(\mathscr{L},\mathscr{L})\\ &c_2(\bar{X})=2c_2(Y)+2(\mathscr{L},\mathscr{K}_Y)+4(\mathscr{L},\mathscr{L}).\end{aligned}$$

For example, if we take $Y=\mathbb{P}_2$, and for B a curve of degree $2m$ with at most simple singularities, we obtain:

$$\begin{aligned} &p_g(\bar{X})=1+\tfrac{1}{2}m(m-3)\\ &q(\bar{X})=0\\ &c_1^2(\bar{X})=2(m-3)^2\\ &c_2(\bar{X})=4m^2-6m+6.\end{aligned}$$

If $m\geq 3$, then the surface $\bar{X}$ is always minimal, since $(\mathscr{K}_{\bar{X}},\mathscr{O}_{\bar{X}}(C))\geq 0$ for every curve C on $\bar{X}$, whereas $(\mathscr{K}_{\bar{X}},\mathscr{O}_{\bar{X}}(E))$ would be -1 for a (-1)-curve on $\bar{X}$ (If $m=1$, then $\bar{X}$ is either $\mathbb{P}_1\times\mathbb{P}_1$ or Σ_2, and for $m=2$ one obtains a non-minimal rational surface). Formula (8) also immediately yields that

$$\operatorname{kod}(\bar{X})=\begin{cases}-\infty & \text{if } m=1,2\\ 0 & \text{if } m=3\\ 2 & \text{if } m\geq 4.\end{cases}$$

Next we investigate the effect on the invariants of the minimal resolution $\bar{X}$ when B acquires certain non-simple singularities. Suppose that the point $y_1\in Y$ is an ordinary d-fold point of B. Let $\sigma\colon \bar{Y}\to Y$ be the blowing-up of Y at y_1. The proper transform $\bar{B}$ of B on $\bar{Y}$ is smooth, and there exists a 2-fold covering $p\colon \bar{X}\to\bar{Y}$, branched over B_1, where $B_1=\bar{B}$ if $d=2m$ and $B_1=\bar{B}+E$, if $d=2m+1$ ($E=\sigma^{-1}(y_1)$). In both cases the branch curve has at most simple singularities, and the formula above apply.

More generally, we may assume that y_1 is a singular point of B, such that $\bar{B}$ has only simple singularities. If the curve B has r of these points, say $y_1,\dots,y_r$, and if $\sigma\colon \bar{Y}\to Y$ is the blowing-up of Y at $y_1\cup\dots\cup y_r$, then there is again a 2-fold covering of $\bar{Y}$ with branch curve the union of $\bar{B}$ and some of the exceptional curves. If y_j is a point of multiplicity d_j on B, with $d_j=2m_j$ or $d_j=2m_j+1$, then application of (9) yields:

$$(10)\quad \begin{cases}\chi(\bar{X})=2\chi(Y)+\tfrac{1}{2}(\mathscr{L},\mathscr{K}_Y\otimes\mathscr{L})-\sum_{j=1}^{r}\tfrac{1}{2}m_j(m_j-1)\\ c_1^2(\bar{X})=2c_1^2(Y)+4(\mathscr{L},\mathscr{K}_Y)+2(\mathscr{L},\mathscr{L})-2\sum_{j=1}^{r}(m_j-1)^2.\\ \text{And if } p_g(Y)=0\text{:}\\ p_g(\bar{X})=\text{dimension of the subspace of } \Gamma(\mathscr{K}_Y\otimes\mathscr{L}), \text{ consisting of those}\\ \text{sections, vanishing of order at least } m_j-1 \text{ in } y_j,\ j=1,\dots,r.\end{cases}$$

It will be clear that taking double coverings, as described in this section, provides a rich source for examples of all possible sorts of surfaces. This method has been applied very frequently in recent years to obtain surfaces with given invariants (compare VII, Sects. 10 and 11).

23. An Enriques Surface

Preserving the notation introduced in Sect. 22, we now take for Y the quadric $\mathbb{P}_1 \times \mathbb{P}_1$ and for B any curve of bidegree (4, 4) with at most simple singularities. Using (8) and (9) we find that the resulting surface $\bar{X}$ satisfies:

(i) $\mathcal{K}_{\bar{X}} = \mathcal{O}_{\bar{X}}$,
(ii) $q(\bar{X}) = 0$.

So $\bar{X}$ is a K3-surface.

Let $(x_0 : x_1,\ y_0 : y_1)$ be bihomogeneous coordinates on $\mathbb{P}_1 \times \mathbb{P}_1$, and let i be the involution on $\mathbb{P}_1 \times \mathbb{P}_1$ given by $i(x_0 : x_1, y_0 : y_1) = (x_0 : -x_1, y_0 : -y_1)$. It has four isolated fixpoints, namely $p_1 = (0:1, 0:1)$, $p_2 = (0:1, 1:0)$, $p_3 = (1:0, 0:1)$ and $p_4 = (1:0, 1:0)$. The i-invariant polynomials of bidegree (4, 4) form a 13-dimensional vector space with a basis consisting of the nine elements $x_0^{2k} x_1^{4-2k} y_0^{2j} y_1^{4-2j}$ $(0 \le k, j \le 2)$ and the four elements $x_0^3 x_1 y_0^3 y_1$, $x_0 x_1^3 y_0^3 y_1$, $x_0^3 x_1 y_0 y_1^3$, $x_0 x_1^3 y_0 y_1^3$. The corresponding linear system of curves with bidegree (4, 4) has no base points, hence by Bertini's theorem I.20.2 its general member is smooth and irreducible. So there exist plenty of i-invariant curves B of bidegree (4, 4) not passing through any of the points p_i and having at most simple singularities. We fix such a curve B and form $\bar{X}$ as before. We claim that i lifts to a fixed-point-free involution j of $\bar{X}$. To see this, we note that i operates on the total space F of the line bundle $\mathcal{F} = \mathcal{K}^{\vee}_{\mathbb{P}_1 \times \mathbb{P}_1}$ in which X is embedded as the pull-back under the squaring map $\mathcal{F} \to \mathcal{F}^{\otimes 2}$ of an i-invariant section. On F the fibres over p_j are pointwise fixed under i, so $i|X$ has eight fixed points and the involution which is the composition of $i|X$ and the covering involution of $X \to \mathbb{P}_1 \times \mathbb{P}_1$ has no fixed points on X. It lifts to a fixed-point-free involution j of the minimal resolution $\bar{X}$ of X. We let $\pi\colon \bar{X} \to Y$ be the quotient map. Then Y is a compact surface with the following properties

(i) $\mathcal{K}_Y \neq \mathcal{O}_Y$
(ii) $\mathcal{K}_Y^{\otimes 2} = \mathcal{O}_Y$
(iii) $q(Y) = 0$.

To prove this, we first observe that $\chi(\mathcal{O}_Y) = \frac{1}{2}\chi(\mathcal{O}_{\bar{X}}) = 1$, so – since $q(Y) \le q(\bar{X}) = 0$ – it follows that $p_g(Y) = q(Y) = 0$. In particular $\mathcal{K}_Y$ is not trivial. Since however $\pi^*(\mathcal{K}_Y) = \mathcal{K}_{\bar{X}} = \mathcal{O}_{\bar{X}}$, we see that $\mathcal{K}_Y$ is the 2-torsion bundle defining π, so $\mathcal{K}_Y^{\otimes 2} = \mathcal{O}_Y$.

A compact surface with properties (i)–(iii) is called an *Enriques surface*. In Chap. VIII we shall deal intensively with these surfaces. Here we only mention two properties:

(i) An Enriques surface is projective. In fact, since $p_g(Y) = 0$ it follows from the exponential cohomology sequence that for every $c \in H^2(Y, \mathbb{Z})$ there is a line

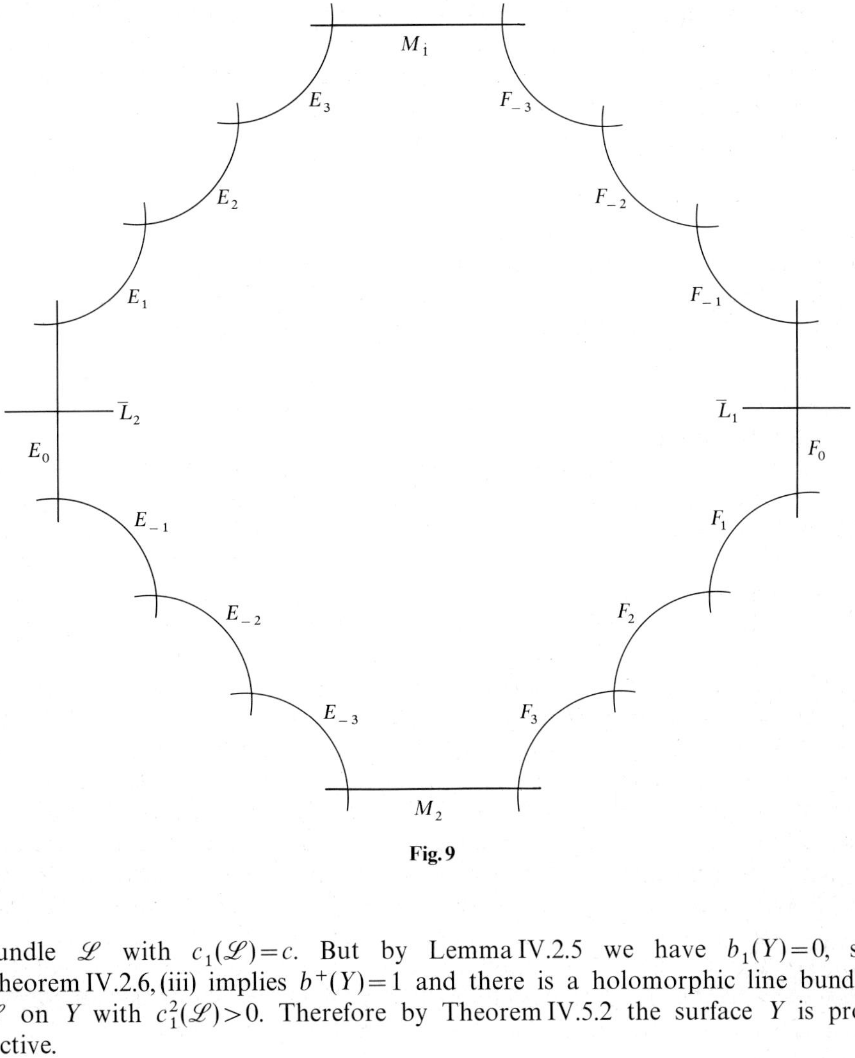

Fig. 9

bundle $\mathscr{L}$ with $c_1(\mathscr{L})=c$. But by Lemma IV.2.5 we have $b_1(Y)=0$, so Theorem IV.2.6, (iii) implies $b^+(Y)=1$ and there is a holomorphic line bundle $\mathscr{L}$ on Y with $c_1^2(\mathscr{L})>0$. Therefore by Theorem IV.5.2 the surface Y is projective.

(ii) An Enriques surface is a minimal surface, since $\mathscr{K}_Y$ restricted to a curve always has degree zero.

In Chap. VIII we need an Enriques surface having an E_8-configuration of curves. We close this section by constructing an example having this property.

In the preceding construction we take for B the reducible curve consisting of the lines $L_1=\{y_0=y_1\}$, $L_2=\{y_0=-y_1\}$ and the curve C with the equation

$$a x_1^4(y_0^2-y_1^2)+b x_0^2 x_1^2(y_0^2-y_1^2)+x_0^4(c y_0^2+d y_1^2)=0 \tag{11}$$

where a, b, c and d have to be chosen suitably, namely such that C is smooth and does not pass through the points p_j. This last condition is satisfied if $a \neq 0$, $c \neq 0$ and $d \neq 0$. Then also C is smooth at the points $q_1=(0:1, 1:-1)$ and $q_2=(0:1, 1:1)$. Since (11) defines a linear system having q_1 and q_2 as base points, the general member C will be smooth except possibly at q_1 and q_2, but if $a \neq 0$, $c \neq 0$ and $d \neq 0$ this is automatically the case, as we have seen. The curve $B = C \cup L_1 \cup L_2$ has two A_7-singularities, namely at q_1 and q_2. If we apply the results of Chap. III, Table 1 we obtain the configuration of (-2)-curves on X shown in Fig. 9. The curve $\bar{L}_i$ is the pre-image on X of the proper transform of the curve L_i $(i=1,2)$, and $M_1 \cup M_2$ is the pre-image of $\{x_0=0\}$. The E_i give the A_7-singularity at q_1 and the F_j the A_7-singularity at q_2. The involution interchanges $\bar{L}_1$ and $\bar{L}_2$, M_1 and M_2, E_i and F_i. If we omit E_{-3} and F_{-3} we obtain two E_8-configurations interchanged by j, so their image on Y is an E_8-configuration.

VI. The Enriques-Kodaira Classification

In this chapter a surface will be a compact, connected 2-dimensional complex manifold. As defined in II, Sect. 1, a curve on a surface is always a closed 1-dimensional subvariety, locally given by one equation (essentially, an effective divisor).

1. Statement of the Main Result

In I, Sect. 7 it has been explained that for given n the n-dimensional compact, connected complex manifolds X can be classified according to their Kodaira dimension $\mathrm{kod}(X)$, which can assume the values $-\infty, 0, 1, \ldots, n$. In the case $n = 2$ the surfaces in the classes $\mathrm{kod}(X) = -\infty$ or $\mathrm{kod}(X) = 0$, and to a lesser extent those with $\mathrm{kod}(X) = 1$, can be classified much more in detail. Thus, starting from the rough classification by Kodaira dimension, surfaces are divided into ten classes. This classification is called the Enriques-Kodaira classification and is embodied by the following central result.

(1.1) **Theorem.** *Every surface has a minimal model in exactly one of the classes* 1) *to* 10) *of Table* 10. *This model is unique (up to isomorphisms) except for the surfaces with minimal models in the classes* 1) *and* 3).

Contrary to the Kodaira dimension, the algebraic dimension $a(X)$ is not the same for all surfaces in one class; the values that occur are shown in the table. The table also gives some more details about the plurigenera $P_n(X)$ for the case $\mathrm{kod}(X) = 0$, and about the first Betti number $b_1(X)$ wherever this is possible. This information is sufficient to characterize the different classes by the plurigenera and the first Betti number. (Blowing up changes neither of these, compare Theorem I.9.1, (iv) and (viii).)

For convenience we give below the definitions of all these classes, though practically all of them have appeared earlier. These definitions are the standard ones, except perhaps for the classes 5) and 6) and in particular class 2). They vary widely in explicitness: sometimes (e.g. for tori) they are as explicit as anybody can ask for; in other cases (e.g. for K3-surfaces) they are very formal.

The surfaces in several classes are minimal by definition. The minimality of the surfaces in class 3) is an easy consequence of Lüroth's theorem for curves (the image of a rational curve is again rational), whereas the minimality in the classes 4)–8) is due to the fact that $(\mathcal{K}, E) = -1$ for a (-1)-curve E.

Table 10

Class of X	kod(X)	smallest $n>0$ with $\mathcal{K}_X^{\otimes n}=\mathcal{O}_X$	$b_1(X)$	possible value of a(X)	c_1^2	c_2
1) minimal rational surfaces	$-\infty$		0	2	8 or 9	4 or 3
2) minimal surfaces of class VII	$-\infty$		1	0, 1	≤ 0	≥ 0
3) ruled surfaces of genus $g \geq 1$	$-\infty$		$2g$	2	$8(1-g)$	$4(1-g)$
4) Enriques surfaces	0	2	0	2	0	12
5) hyperelliptic surfaces	0	2, 3, 4, 6	2	2	0	0
6) Kodaira surfaces	0					
a) primary	0	1	3	1	0	0
b) secondary	0	2, 3, 4, 6	1	1	0	0
7) K3-surfaces	0	1	0	0, 1, 2	0	24
8) tori	0	1	4	0, 1, 2	0	0
9) minimal properly elliptic surfaces	1			1, 2	0	≥ 0
10) minimal surfaces of general type	2		$\equiv 0(2)$	2	>0	>0

A rational surface is a surface that is birationally equivalent to $\mathbb{P}_2$. Apart from $\mathbb{P}_2$ we have described in V, Sect. 4 an infinite sequence of other minimal rational surfaces, namely the Hirzebruch surfaces Σ_n, $n=0,2,3,\ldots$. It will be shown later (Remark 2.4) that there are no others.

A surface of class VII is a surface X with $\operatorname{kod}(X)=-\infty$ and $b_1(X)=1$. (Minimal surfaces in this class are often called of class VII_0.) We have met two types of examples, namely Hopf surfaces (V, Sect. 18) and Inoue surfaces (V, Sect. 19). We have mentioned already that the minimal surfaces X with $\mathrm{a}(X)=1$ which are contained in class VII are exactly the Hopf surfaces of algebraic dimension 1. For surfaces X with $\mathrm{a}(X)=0$ there are, apart from the Hopf surfaces and the Inoue surfaces, other examples known (compare the references at the end of V, Sect. 19), but a complete classification is still lacking.

It follows from Theorem IV.2.6 that $q(X)=1$, $h^{1,0}(X)=0$ for every surface of class VII.

The name "class VII" comes from Kodaira's presentation of the Enriques classification ([Ko4], part I and part IV). However, our "class VII" is not Kodaira's class VII, (which contains surfaces of different Kodaira dimension) but a subclass of it, namely Kodaira's class 7). We have chosen this name since on the one hand we like to keep the traditional name "class VII" for Hopf surfaces, Inoue surfaces, ..., whereas on the other hand it is not possible to have a class containing surfaces of different Kodaira dimension.

A *ruled surface* of genus g is a surface X, admitting a ruling, i.e. an analytically locally trivial fibration with fibre $\mathbb{P}_1$ and structural group

$PGL(2,\mathbb{C})$ over a smooth curve of genus g. Ruled surfaces, which are all algebraic, have been discussed in V, Sect. 4. At that place we have also seen that the fibration is always equivalent to an *algebraically* locally trivial one.

An *Enriques surface* is a surface X with $q(X)$ (or equivalently $b_1(X)$)$=0$, for which $\mathcal{K}_X^{\otimes 2} \cong \mathcal{O}_X$, but $\mathcal{K}_X \neq \mathcal{O}_X$. Such a surface appeared in V, Sect. 23, where it was also proved that any Enriques surface is projective. The second part of Chap. VIII will deal with the classification of Enriques surfaces.

A *hyperelliptic surface* is a surface X with $b_1(X)=2$, admitting a holomorphic, locally trivial fibration over an elliptic curve with an elliptic curve as typical fibre. Their classification has been described in V, Sect. 5. In particular, every hyperelliptic surface is algebraic.

A *primary Kodaira surface* (V, Sect. 5) is a surface with $b_1(X)=3$, admitting a holomorphic locally trivial fibration over an elliptic curve with an elliptic curve as typical fibre.

A *secondary Kodaira surface* (V, Sect. 5) is a surface, other than a primary Kodaira surface, admitting a primary Kodaira surface as unramified covering. They are elliptic fibre spaces over rational curves, with first Betti number equal to 1.

A K3-*surface* is a surface X with $q(X)=0$ and $\mathcal{K}_X = \mathcal{O}_X$. Examples of such surfaces appeared several times in Chap. V, namely as complete intersections of type (4), (2, 3) and (2, 2, 2) in V, Sect. 2, as Kummer surfaces in V, Sect. 16 and as double coverings of $\mathbb{P}_2$, ramified over a curve of degree 6 with simple singularities in V, Sect. 22. A large part of Chap. VIII will be devoted to the classification theory of these surfaces.

A *torus* is a surface, isomorphic to the quotient of $\mathbb{C}^2$ by a lattice of real rank 4.

A *properly elliptic surface* is an elliptic surface X (V, Sect. 7) with $\mathrm{kod}(X)=1$. A very simple example is provided by the product of two curves, one elliptic and the other of genus ≥ 2 (V, Sect. 6).

A *surface of general type* is a surface X with $\mathrm{kod}(X)=2$. Examples of such surfaces are manifold: complete intersections of sufficiently high degree (V, Sect. 2), products of curves of genus ≥ 2 (V, Sect. 6), Kodaira fibrations (V, Sect. 14), quotients of symmetric domains (V, Sect. 20) and "practically all" ramified double coverings of $\mathbb{P}_2$ (V, Sect. 22). These surfaces are general in the same sense as are curves of genus ≥ 2. Since always $a(X) \geq \mathrm{kod}(X)$, every surface of general type is algebraic by Corollary IV.5.5. Chapter VII deals with the classification of surfaces of general type.

Remarks. 1) The "size" of the various classes is quite different. The surfaces in the classes 4), 7) and 8) "form one irreducible family", so they would form one main class in any classification. But other classes consist of an infinite number of families.

2) It might seem more natural to take the algebraic dimension $a(X)$ as the primary invariant instead of $\operatorname{kod}(X)$. But in doing so you tear apart the "families" of tori and K3-surfaces as well as certain "families" of properly elliptic surfaces, which is a rather cruel thing to do.

The proof of Theorem 1.1 will be based on

(i) the results of Chap. IV, in particular Sects. 2, 5 and 6;

(ii) the basic facts about fibrations, in particular Iitaka's conjecture $C_{2,1}$ (Theorem III.18.4) and the canonical bundle formula for elliptic fibrations (Theorem V.12.1);

(iii) the Castelnuovo criterion.

The Enriques classification of algebraic surfaces seems to appear for the first time in [Enr1], but the classical reference is [Enr2]. Also [Ge] should be mentioned. The extension to non-algebraic surfaces is due to Kodaira who at the same time gave a modern (and much extended) treatment of the complex-algebraic case ([Ko4], part IV, p. 1062). The case of characteristic 0 has also been the subject of seminars by Zariski and by Šafarevič ([Saf]).

The Enriques classification for characteristic $p>0$ is due to Mumford and Bombieri ([Mu5], [B-M1], [B-M2]). Other treatments for the case $p=0$ (avoiding an a priori proof of $C_{2,1}$) can be found for example in [Be1], [G-H], [Bad] and [Kur].

2. The Castelnuovo Criterion

The main theorem of this section is

(2.1) **Theorem** (Castelnuovo's criterion). *An algebraic surface X is rational if and only if $q(X)=P_2(X)=0$.*

In one direction this statement is an immediate consequence of the invariance of q and P_2 under bimeromorphic transformations (Corollary III.6.4). In the other direction the proof will follow from Proposition V.4.3 together with some auxiliary results, which we now shall prove.

(2.2) **Proposition.** *Let X be a minimal algebraic surface with $q(X)=P_2(X)=0$. If D is any divisor on X, then $|D+nK_X|$ is empty for n large enough.*

Proof. We distinguish between the cases $K_X^2\geq 0$ and $K_X^2<0$.

In the first of these cases, if we feed the information $q(X)=P_2(X)=0$ into the Riemann-Roch theorem, applied to $-K_X$, we find that $\dim|-K_X|\geq 0$. Now $-K_X$ cannot be 0, for then we would also have $2K_X=0$ and $P_2(X)=1$. So $-K_X$ contains an effective divisor and $|D+nK_X|$ is empty for n sufficiently large.

In the second case we start by observing that $(D+nK_X)K_X<0$ for $n\geq n_0>0$. So if $|D+nK_X|$ is not empty for n large enough, then there exists an $n_1>0$ such that at the same time $K_X(D+n_1K_X)<0$ and $|D+n_1K_X|$ contains an effective divisor $\sum a_i C_i$. From $K_X(D+n_1 K_X)<0$ we find that there is at least one C_i, say C_1, such that $K_X C_1<0$. If $C_1^2<0$, then by Proposition III.2.2 this curve C_1 would be a (-1)-curve. Since we have assumed X to be minimal, we find $C_1^2\geq 0$. Then $C_1(D+nK_X)\geq 0$ as soon as $|D+nK_X|$ is not empty. But $C_1(D+nK_X)<0$ for n large enough since $C_1 K_X<0$. Hence also here $|D+nK_X|$ is empty for large n. □

(2.3) **Proposition.** *Let X be a minimal algebraic surface with $q(X)=P_2(X)=0$. Then X contains a smooth rational curve C with $C^2\geq 0$.*

Proof. If D is any very ample divisor on X, then by Proposition 2.2 there is an $n=n(D)\geq 0$, such that $|D+(n+1)K_X|$ is empty, but $|D+nK_X|$ is not. So either we have for all very ample D that there is an $n=n(D)>0$ with $D+nK_X=0$, or there is a very ample D, such that $|D+nK_X|$ contains an effective divisor, while $|D+(n+1)K_X|$ is empty.

The first of these possibilities is excluded for topological reasons. Indeed, since any divisor is linearly equivalent to the difference of two very ample ones ([Se2]) we have in this case that for every divisor D (very ample or not) there is an m, such that $D=mK_X$. This means that every line bundle on X is isomorphic to $\mathcal{K}_X^{\otimes m}$ for some $m\in\mathbb{Z}$. So from $q(X)=p_g(X)=0$ and the exponential cohomology sequence we find that $H^2(X,\mathbb{Z})\cong\mathbb{Z}$, with the first Chern class $c_1(X)$ as a generator. By Poincaré duality we have $c_1^2(X)=\pm 1$. Since $b_1(X)=2q(X)=0$ we see that the Euler characteristic $e(X)=3$. But then $c_1^2(X)+e(X)\not\equiv 0$ (12), which is impossible by Noether's formula I, (4).

Now let D be very ample, such that $|D+nK_X|$ contains the effective divisor $C=\sum a_i C_i$, whereas $|D+(n+1)K_X|$ is empty. Substituting $\dim|-C_i|=\dim|K_X+C_i|=-1$ in the Riemann-Roch formula, applied to $-C_i$, we find that $g(C_i)=0$ for all i. Hence all C_i are smooth rational curves. The only thing we still have to do is to exclude the possibility that $C_i^2<0$ for all i. If all $C_i^2<0$, then the adjunction formula and the minimality of X again imply that $K_X C_i\geq 0$ for all i, hence $K_X C\geq 0$, and $K_X C+C^2=(n+1)K_X C+CD>0$. On the other hand, if we apply Riemann-Roch to $-C$, we obtain

$$\dim|-C|+\dim|K_X+C|\geq \tfrac{1}{2}(C^2+K_X C)-1,$$

i.e. $C^2+K_X C\leq -2$, a contradiction. So there is at least one C_i with $C_i^2\geq 0$. □

Combining this last result with Proposition V.4.3 we obtain at once that every minimal surface X with $q(X)=P_2(X)=0$ is either $\mathbb{P}_2$ or a surface Σ_n, $n\neq 1$, and in any case a rational surface. So Theorem 2.1 has been proved.

(2.4) *Remark. It follows in particular that $\mathbb{P}_2$ and the Σ_n, $n\neq 1$, are the only minimal rational surfaces.*

As an application of Castelnuovo's criterion we shall prove that every unirational surface is rational. A surface X is called *unirational* if there exists a meromorphic map from $\mathbb{P}_2$ onto X, i.e. if on the product $X\times\mathbb{P}_2$ there exists a compact, irreducible 2-dimensional subspace Y (not necessarily smooth), projecting onto X and $\mathbb{P}_2$, such that the projection from Y onto $\mathbb{P}_2$ is generically one-to-one.

(2.5) **Theorem.** *Every unirational surface is rational.*

Proof. Let Z be a smooth model for Y (III, Sect. 4). Then Z is bimeromorphically equivalent to $\mathbb{P}_2$, so by Theorems I.6.3 and I.9.1, (iv) and (viii), we have $b_1(Z)=b_1(\mathbb{P}_2)=0$ and $P_2(Z)=P_2(\mathbb{P}_2)=0$. From Remark I.1.3 and Theorem I.7.4 we then derive that $b_1(X)=P_2(X)=0$. Since $P_2(X)=0$ implies $p_g(X)=0$, we derive from Theorem IV.2.6 that $b^+(X)=1$. By the exponential cohomology sequence there exists on X a holomorphic line bundle $\mathscr{L}$ with $c_1^2(\mathscr{L})>0$. Consequently, X is projective by Theorem IV.5.2, and its rationality follows from Castelnuovo's criterion 2.1. □

Castelnuovo's criterion remains one of the corner stones of surface theory. It was proved by Castelnuovo in 1896 (Sulle superficie di genero 0, Mem. Soc. Ital. Sci, ser. III, X).

Remark 2.4 is due to Andreotti (see [Ad]).

We shall show that every minimal surface belongs to one of the classes 1) to 10), leaving it to the reader to verify that two different classes never contain birationally equivalent surfaces.

3. The Case a$(X)=2$

The Case $\mathrm{a}(X)=2$, $\mathrm{kod}(X)=-\infty$.

Since $\mathrm{a}(X)$ and $\mathrm{kod}(X)$ are birational invariants, we have for every rational surface X that $\mathrm{a}(X)=2$, $\mathrm{kod}(X)=-\infty$. The same holds for any ruled surface by Proposition V.4.1. So what we have to show is this: a minimal algebraic surface X with $\mathrm{kod}(X)=-\infty$ is either rational, or ruled of genus ≥ 1.

Let X be a minimal algebraic surface with $\mathrm{kod}(X)=-\infty$. If $q(X)=0$, then X is rational by the Castelnuovo criterion. If $q(X)\geq 1$, then X admits a non-constant map f into its Albanese torus $\mathrm{Alb}(X)$ (compare I, Sect. 13). If $\dim f(X)\geq 2$, then we could find a holomorphic 2-form ω on $\mathrm{Alb}(X)$, such that $f^*(\omega)$ would be a holomorphic 2-form on X, different from the 0-form. But that would mean $p_g(X)\geq 1$, contrary to the assumption that $\mathrm{kod}(X)=-\infty$. Hence f maps X onto a curve C, which is contained in $\mathrm{Alb}(X)$ and therefore has genus ≥ 1. By Stein factorisation we obtain a connected map g from X onto a smooth curve of genus ≥ 1. The Iitaka conjecture $C_{2,1}$ implies that the general fibre of g has genus 0, so Proposition V.4.3 yields that X is a ruled surface of genus ≥ 1.

The Case $\mathrm{a}(X)=2$, $\mathrm{kod}(X)=0$

Since the converse is clear from the definitions we have only to show that every minimal algebraic surface X with $\mathrm{kod}(X)=0$ is either an Enriques surface, a hyperelliptic surface, a K3-surface or a torus.

From Proposition III.2.3 we derive

(3.1) **Proposition.** *If X is a minimal surface with* $\mathrm{kod}(X)\geq 0$, *then* $K_X^2\geq 0$.

(3.2) **Proposition.** *If X is a minimal surface with* $\mathrm{kod}(X)=0$ *or* 1, *then* $K_X^2=0$.

Proof. By the preceding result we have that $K_X^2\geq 0$. If $K_X^2>0$, then the Riemann-Roch theorem shows that $P_n(X)$ grows like n^2, i.e. that $\mathrm{kod}(X)=2$ (see I, Sect. 7). □

Now let X be a minimal algebraic surface with $\mathrm{kod}(X)=0$. If $q(X)=0$, then $P_2(X)>0$, otherwise X would be rational by Castelnuovo's criterion (Theorem 2.1) and $\mathrm{kod}(X)$ would be $-\infty$. Hence if $q(X)=0$, there are three possibilities: (i) $\mathcal{K}_X=0$, (ii) $\mathcal{K}_X\neq 0$, $\mathcal{K}_X^{\otimes 2}=0$, (iii) there is an effective divisor in $|2K_X|$. In the first of these cases X is a K3-surface by definition, and in the second case X is an Enriques surface by definition. We shall now show that case (iii) does not occur. In fact, since $q(X)=0$, we have $\chi(X)=1-q(X)+p_g(X)\geq 1$. Furthermore, since there is an effective divisor in $|2K_X|$, the linear system $|-nK_X|$ is empty for $n\geq 1$. So if we apply Riemann-Roch to $|nK_X|$, $n\geq 2$, and take into account that $K_X^2=0$ by Proposition 3.2, then we find that $P_n(X)=1$ for all $n\geq 2$. In other words, there exists exactly one effective n-canonical divisor $D^{(n)}$ for $n\geq 2$. If $D^{(2)}=\sum d_i D_i$ and $D^{(3)}=\sum e_j E_j$, then $3D^{(2)}=2D^{(3)}$, that is, $D^{(2)}=2D$ and $D^{(3)}=3D$, where D is an effective divisor on X. By subtraction we find that D is a canonical divisor, hence $p_g(X)=1$. But then $\chi(X)=2$, and applying again Riemann-Roch to $|nK_X|$, $n\geq 2$, we find $P_n(X)\geq 2$ for $n\geq 2$, which is impossible since $\mathrm{kod}(X)=0$.

Next we turn to the algebraic surfaces X, with $\mathrm{kod}(X)=0$, for which $q(X)>0$. Here there are two possibilities: if $f\colon X\to \mathrm{Alb}(X)$ denotes a standard map, then either

(i) $\dim f(X)=2$ or
(ii) $\dim f(X)=1$.

In case (i) we have $p_g(X)=1$ since there is a 2-form on X, induced from $\mathrm{Alb}(X)$. Using $K_X^2=0$ (Proposition 3.2), we thus obtain from $12(1-q(X)+p_g(X))=K_X^2+e(X)=e(X)$ and $e(X)=2-4q(X)+b_2(X)$ that $q(X)\leq 2$. But on the other hand we have two independent holomorphic 1-forms on X, induced from $\mathrm{Alb}(X)$. Hence $q(X)=h^{0,1}(X)=h^{1,0}(X)\geq 2$. So $q(X)=2$ and f is surjective. Since $b_2(X)=2b_1(X)-2=4q(X)-2=6=b_2(\mathrm{Alb}(X))$, we have that $f_*\colon H_2(X,\mathbb{Q})\to H_2(\mathrm{Alb}(X),\mathbb{Q})$ is an isomorphism by Corollary I.1.2 and hence f can't blow down any curves, so it is a finite map. If f is an unramified covering, then X is a torus (and, a fortiori, f is an isomorphism). So suppose from now on that f is ramified with ramification divisor $R=\sum r_i R_i$ on X. The divisor R is a canonical divisor by the Hurwitz formula, hence $K_X R=0$ by Proposition 3.2. Since $KR_i\geq 0$ by Proposition III.2.3, we must have $KR_i=0$ for every curve R_i. If R_i^2 were strictly positive, then Riemann-Roch would yield that $\dim|nR_i|\geq 1$ for large n, hence $P_n(X)\geq 2$ for large n, contradicting $\mathrm{kod}(X)$

$=0$. Therefore $R_i^2 \leq 0$, and the adjunction formula leaves only the possibility $R_i^2 = 0$, $g(R_i)=1$. Since every holomorphic map from a rational curve into a torus is constant, it follows that there are smooth elliptic curves on $\mathrm{Alb}(X)$. Hence $\mathrm{Alb}(X)$, and therefore also X, admit a morphism onto an elliptic curve. By Stein factorisation X admits a connected map onto some curve C of genus ≥ 1. The genus of the general fibre can't be 0, otherwise $\mathrm{kod}(X)$ would be $-\infty$. So the Iitaka conjecture $C_{2,1}$ (Theorem III.18.4) leaves us with only one possibility: X is an elliptic fibre space over an elliptic curve. Using Proposition V.12.5, (i) we see that the existence of an effective canonical divisor would imply $\mathrm{kod}(X) \geq 1$, which contradicts the assumption $\mathrm{kod}(X)=0$. Consequently, f can't be ramified and so we have only tori in case (i).

In the case (ii) we find in the same way as a few lines earlier that X is an elliptic fibre space over an elliptic curve. By the universal property of the Albanese torus this elliptic base must be $\mathrm{Alb}(X)$ itself, hence $q(X)=1$ and $\chi(X) \geq 0$. If $\chi(X)$ were strictly positive, then by Corollary V.12.3 we would have $P_n(X) \geq 2$ for n large. So $\chi(X) = p_g(X) = 0$. For the same reason there cannot be any multiple fibre. Furthermore $e(X) = \frac{\chi(X)}{12} - K_X^2 = 0$ (Proposition 3.2), so there cannot be any singular fibres at all by Remark III.11.5. It follows from Theorem III.15.4 that X is a locally trivial fibre bundle with fibre an elliptic curve over an elliptic curve. Since $b_1(X) = 2q(X) = 2$, we find that X is a hyperelliptic surface.

The Case $\mathrm{a}(X)=2$, $\mathrm{kod}(X)=1$

In this case there exists an index $n_0 \geq 1$, such that $P_{n_0}(X) \geq 2$. This means that there is a (not necessarily complete) 1-dimensional linear system $F + D_\lambda$, $\lambda \in \mathbb{P}_1$, of n-canonical divisors. Here F denotes the fixed part of the pencil (compare IV, Sect. 1). We have $K_X(F+D_\lambda) = n_0 K_X^2 = 0$ by Proposition 3.2, and also $K_X F \geq 0$, $K_X D_\lambda \geq 0$ by Proposition III.2.3. So $K_X D_\lambda = K_X F = 0$. On the other hand we have

$$(F+D_\lambda)^2 = F(F+D_\lambda) + FD_\lambda + D_\lambda^2 = n_0 K_X F + FD_\lambda + D_\lambda^2 = FD_\lambda + D_\lambda^2 = 0.$$

Now $FD_\lambda \geq 0$, $D_\lambda^2 \geq 0$, hence $D_\lambda^2 = 0$. The linear system $|D_\lambda|$ has neither base points nor fixed components and therefore defines a holomorphic map onto some curve. Using Stein factorisation we obtain a regular connected map from X onto a smooth curve Y, the general fibre of which is elliptic by the adjunction formula. So X is a properly elliptic surface and we are ready again.

The Case $\mathrm{a}(X)=2$, $\mathrm{kod}(X)=2$

Because of the definition of a surface of general type there is nothing to prove here.

4. The Case $\mathrm{a}(X)=1$

(4.1) **Proposition.** *Every surface of algebraic dimension* 1 *is elliptic.*

Proof. The fact that $\mathrm{a}(X)=1$ implies that there is a non-constant meromorphic function m on X. This function gives rise to a surjective holomorphic map

$f\colon X\setminus\bigcup_{i=1}^{k} p_i \to \mathbb{P}_1$, where $p_1,\ldots,p_k$ are the points of indeterminacy of m. So, according to IV, Sect. 1, after (repeatedly) blowing up X in $p_1,\ldots,p_k$, we obtain a surface $\tilde{X}$ and a surjective holomorphic map $\tilde{f}\colon \tilde{X}\to\mathbb{P}_1$, such that $\tilde{f}$ agrees with f outside of the exceptional trees. Using Stein factorisation we obtain a connected map $h\colon \tilde{X}\to Y$, where Y is a smooth curve. We claim that a general fibre F of h is an elliptic curve. For if $g(\tilde{F})\neq 1$, then $(\mathscr{K}_X, F)\neq 0$ (by the adjunction formula), and $c_1^2(\mathscr{K}_{\tilde{X}}\otimes\mathcal{O}_{\tilde{X}}(nF))>0$ for suitable $n\in\mathbb{Z}$, which is impossible by Theorem IV.5.2 and Corollary IV.5.7. In the same way we see that every curve C on $\tilde{X}$ is mapped to a point by h, for otherwise $C\tilde{F}$ would be strictly positive and $(C+nF)^2>0$ for large n. This means that h maps the exceptional trees above $p_1,\ldots,p_k$ onto points of Y. Hence also X admits a holomorphic map onto Y, such that the general fibre is elliptic (from the fact that all curves on X are contained in a fibre of this map it follows that this elliptic fibration is unique). □

Now that we know that every surface X with $\mathrm{a}(X)=1$ admits an elliptic fibration, we also know for the case that X is minimal:

(i) $K_X^2=0$ by Corollary V.12.3;
(ii) $\chi(X)\geq 0$ by (i), Proposition III.11.4, and the Todd-Hirzebruch formula.

Since always $\mathrm{kod}(X)\leq \mathrm{a}(X)$ we have $\mathrm{kod}(X)\leq 1$. We want to single out those surfaces X for which $\mathrm{kod}(X)$ is either 0 or $-\infty$. In fact, it follows from Table 10 that we have to show: $\mathrm{kod}(X)=0$ only occurs for Kodaira surfaces, K3-surfaces and tori, whereas $\mathrm{kod}(X)=-\infty$ implies that X is of class VII.

If g denotes the genus of Y, then the canonical bundle formula (Corollary V.12.3) yields immediately that $\mathrm{kod}(X)=1$ as soon as either $g\geq 2$, or $g=1$, $\chi(X)\geq 1$, or $\chi(X)\geq 3$. So we are left with the following cases:

(i) $g=1$, $\chi=0$
(ii) $g=0$, $\chi=2$
(iii) $g=0$, $\chi=1$
(iv) $g=0$, $\chi=0$,

which we shall treat separately.

In case (i) we find again by Corollary V.12.3 that there can't be any multiple fibres. Furthermore, $e(X)=\frac{1}{12}\chi(X)-c_1^2(X)=0$, so f has no singular fibres at all. By Theorem III.15.4, f is a locally trivial fibre bundle and hence by V, Sect. 5 either a torus or a primary Kodaira surface, since every hyperelliptic surface is projective.

As to case (ii), we see immediately that there can't be any multiple fibres, hence $\mathscr{K}_X$ is trivial and $p_g(X)=1$. It follows that $q(X)=0$, so X is a K3-surface by definition.

Case (iii) is easily excluded. Since $p_g\leq 1$, we have either $q(X)=1$ or $q(X)=0$. In the first case, by the unbranched covering trick there would be an unramified covering X' of X with $a(X')=a(X)=1$, $\operatorname{kod}(X')=\operatorname{kod}(X)$, with $\chi(X)\geq 3$, which is impossible by the remarks above. If $q(X)=p_g(X)=0$, we would have $b^+(X)=0$, otherwise X would be algebraic, but this leads via Theorem IV.2.6 to the absurdity $h^{1,0}(X)=-1$.

Finally the case (iv). The restriction $p_g(X)\leq 1$ leads to $p_g(X)=1$, $q(X)=2$ or $p_g(X)=0$, $q(X)=1$. If $p_g(X)=1$, then it follows from the canonical bundle formula that there can't be an effective canonical divisor, for a multiple of it would be of the form $f^*(D)$, with D effective on Y, and $\operatorname{kod}(X)$ would be 1. But $\mathcal{K}_X$ trivial is also impossible here, since in that case there would be multiple fibres by Corollary V.12.3, whereas by Lemma III.8.3 and again Corollary V.12.3 the restriction of $\mathcal{K}_X$ to the reduction of a multiple fibre can't be trivial.

The case $p_g(X)=0$, $q(X)=1$ can actually occur. In this case we have $b^+(X)=0$ (otherwise the exponential cohomology sequence, together with Theorem V.5.2 and $p_g(X)=0$ would imply that X is algebraic), hence $b_1(X)=1$ by Theorem IV.2.6. If there is a positive n_0 such that $\mathcal{K}_X^{\otimes n_0}\cong\mathcal{O}_X$, then there is a finite unramified covering X' of X with $\mathcal{K}_{X'}$ trivial. This obviously minimal surface X' must belong to one of the types of surfaces with $a(X')=0$, $\mathcal{K}_{X'}$ trivial we have encountered before: tori, K3-surfaces and primary Kodaira surfaces. But X' can't be a torus, otherwise X would also be a torus, and X' can't be a K3-surface for $b_1(X')\geq b_1(X)=1$. So X' must be a primary Kodaira surface and X is a secondary Kodaira surface by definition. On the other hand, if there is no $n_0\neq 0$ with $\mathcal{K}_X^{\otimes n_0}\cong\mathcal{O}_X$ then Corollary V.12.4 implies that $P_n(X)=0$ for all $n\geq 1$. So X is a surface of class VII.

5. The Case $a(X)=0$

We know already that $\operatorname{kod}(X)\leq a(X)=0$. Furthermore we have

(j) $p_g(X)\leq 1$ and $h^{1,0}(X)\leq 2$ (Proposition IV.6.1)

Combining the second inequality with Theorem IV.2.6 we obtain:

(jj) $q(X)\leq 3$.

Combining (j) and (jj) we find

(jjj) $|\chi(X)|\leq 2$.

Using the unbranched covering trick and we obtain from this:

(jv) If $q(X)>0$, then $\chi(X)=0$.

Taking into account Theorem IV.2.6 we see from (j)–(jv) that we are left with the following possibilities:

a) $q(X)=p_g(X)=0$

b) $q(X)=0$, $p_g(X)=1$

c) $q(X)=1$, $p_g(X)=0$

d) $q(X)=2$, $p_g(X)=1$, $b_1(X)=3$

e) $q(X)=2$, $p_g(X)=1$, $b_1(X)=4$.

We shall now consider these cases separately, though not in alphabetical order.

Case a) is easily excluded: from $p_g(X)=0$ we find $b^+(X)=0$, and this would (because of Theorem IV.2.6) imply that b_1 is negative.

Case b) is equally easy: using Proposition 3.2 we find from Riemann-Roch that $h^0(\mathscr{K}_X^\vee)\geq 1$. Hence $\mathscr{K}_X \cong \mathscr{O}_X$, and X is a K3-surface.

In case e) we know from Corollary IV.2.10 that there is a 2-dimensional Albanese torus $\mathrm{Alb}(X)$. Let $f\colon X\to \mathrm{Alb}(X)$ be a standard map. Then $f(X)$ can't be a curve, otherwise we would have $\mathrm{a}(X)\geq 1$. So f is surjective. If C is any curve on X, then $f(C)$ must be a point, otherwise we would again obtain by translation an infinity of curves on $\mathrm{Alb}(X)$ and hence on X, which is excluded by Theorem IV.6.2. So f maps a finite number of curves onto points $p_1,\ldots,p_k$, and is outside of these of maximal rank (ramification would again lead to an infinity of curves on $\mathrm{Alb}(X)$ and X). Stein factorisation yields a factorisation $f=h\circ g$, where g is a connected map from X onto some surface X', and $h\colon X'\to \mathrm{Alb}(X)$ is a finite map. The restriction $h|X'\setminus h^{-1}(p_1\cup\ldots\cup p_k)$ is an unramified covering of $\mathrm{Alb}(X)\setminus(p_1\cup\ldots\cup p_k)$. But the embedding of this last subspace into $\mathrm{Alb}(X)$ induces an isomorphism of fundamental groups (both injectivity and surjectivity are elementary exercises), hence X' is a smooth unramified covering of $\mathrm{Alb}(X)$. Now $g\colon X\to X'$ is connected; since X is minimal it must be an isomorphism. Hence X is an unramified covering over a torus and therefore is a torus itself.

Next we exclude case d) by proving the rather subtle

(5.1) **Theorem.** *Every minimal surface X with $q(X)=2$, $p_g(X)=1$ and $b_1(X)=3$ is elliptic.*

As a preliminary result we need

(5.2) **Proposition.** *A minimal surface X with $\mathrm{a}(X)=0$, $q(X)=2$, $p_g(X)=1$ and $b_1(X)=3$ contains no curves.*

Proof. From the adjunction formula, the minimality of X and Proposition V.4.3 we obtain that $(\mathscr{K}_X, C)\geq 0$ for every curve C, and in particular that $(\mathscr{K}_X,\mathscr{K}_X)=0$. If $(\mathscr{K}_X, C)>0$ for some curve C, then $c_1^2(\mathscr{K}_X^{\otimes n}\otimes\mathscr{O}_X(C))$ would be strictly positive for n large, and X would be algebraic (Theorem IV.5.2). So $(\mathscr{K}_X, C)=0$ for every curve C, and the adjunction formula yields that an irreducible curve C on X is either a (-2)-curve, or a curve of genus $g=1$ (smooth elliptic or rational with a node or a cusp). If X would contain a (-2)-curve, then we would obtain by the unbranched covering trick a minimal surface Y with $\mathrm{a}(Y)=0$, containing at least five disjoint (-2)-curves, which are independent over $\mathbb{R}$, hence $b_2(Y)$ would be at least 5. The surface Y would have to belong to one of the classes a)–e) above and it is easy to see that it would belong to class d) again. For such a surface however we have $e(X)=12\chi(X)-K_X^2=0$, and $b_2(X)=e(X)-2+2b_1(X)=4$.

Now suppose that X contains a smooth elliptic curve C. Since $b_1(C)=2$ and $b_1(X)=3$ we would then find unbranched coverings Y of X of arbitrarily high degree d, which are trivial over C, i.e. such that the inverse image of C consists of d disjoint curves isomorphic to C. As before, these surfaces Y would again be in class d). If D is the union of those d smooth elliptic curves, then, considering the exact sequence

$$H^1(\mathcal{O}_Y)\to H^1(\mathcal{O}_D)\to H^2(\mathcal{O}_Y(-D))$$

we would obtain a contradiction: on the one hand $h^1(\mathcal{O}_Y)=q(Y)=2$ and $h^2(\mathcal{O}_Y(-D))=h^0(\mathcal{K}_Y\otimes\mathcal{O}_Y(D))\leq 1$, but on the other hand $h^1(\mathcal{O}_D)=d$. The case that C is rational is similar, but simpler. □

In the sequel, if θ is a real closed 1-form, then we shall denote its Hodge decomposition by $\theta=\theta_0+\bar\theta_0$, with θ_0 of type $(1,0)$.

Proof of Theorem 5.1. We assume $a(X)=0$. From Theorem IV.2.6 we know that $h^{1,0}(X)=1$, in other words there is a holomorphic 1-form $\omega\neq 0$, determined up to a constant. This form can't vanish on a curve, since there are no curves on X by Proposition 5.2. But it also can't vanish in a number of isolated points, for the number of these (each counted with the proper strictly positive multiplicity) equals $c_2(\mathcal{T}_X^\vee)=e(X)=12\chi(X)-c_1^2(X)=0$. There is also a holomorphic 2-form on X which (again by Proposition 5.2) nowhere vanishes. From $e(X)=0$ and $b_1(X)=3$ we find $b_2(X)=4$, and from $c_1^2(X)=e(X)=0$ we find that the index $\tau(X)=0$, hence $b^+(X)=b^-(X)=2$.

The real 2-form $\sqrt{-1}(\omega\wedge\bar\omega)$ is exact, for if $c\in H^2(X,\mathbb{R})$ is its de Rham class, then $c^2=0$ hence $c=0$ by Theorem IV.2.13, and there is a real 1-form ρ with $d\rho=\sqrt{-1}(\omega\wedge\bar\omega)$.

Now we come to the first crucial point of the proof, namely the construction of a real closed 1-form σ, which has the following properties:

(i) $\omega+\bar\omega$, $\sqrt{-1}(\omega-\bar\omega)$ and σ form a de Rham base of $H^1(X,\mathbb{R})$;
(ii) $d\sigma_0=\omega\wedge\bar\omega$.

Since $b_1(X)=3$, there is a real closed 1-form τ, such that $\omega+\bar\omega$, $\sqrt{-1}(\omega-\bar\omega)$ and τ form a de Rham base of $H^1(X,\mathbb{R})$ and also of $H^1(X,\mathbb{C})$. From the sequence IV, (1)

$$0\to\Gamma(\Omega_X^1)\to H^1(X,\mathbb{C})\to H^1(\mathcal{O}_X)\to 0,$$

the fact that $q(X)=h^{0,1}(X)=2$, and the way the Dolbeault isomorphism is constructed, it follows that $\bar\omega$ and $\bar\tau_0$ form a Dolbeault base for $H^1(\mathcal{O}_X)$. Now $\bar\partial\bar\rho_0=0$, hence $\sqrt{-1}\,\bar\rho_0\sim\lambda\bar\omega+\mu\bar\tau_0$ in the sense of Dolbeault, i.e.

$$\sqrt{-1}\,\bar\rho_0=\lambda\bar\omega+\mu\bar\tau_0+\bar\partial f\quad(\lambda,\mu\in\mathbb{C}).$$

If we define $\sigma_0=-\sqrt{-1}\,\rho+\lambda\bar\omega+\mu\tau+df$, then

$$d\sigma_0=-\sqrt{-1}\,d\rho,\quad\text{and}$$

$$\begin{aligned}\sigma_0 &= -\sqrt{-1}(\rho_0+\bar{\rho}_0)+\lambda\bar{\omega}+\mu(\tau_0+\bar{\tau}_0)+(\partial+\bar{\partial})f\\ &= -\sqrt{-1}\,\rho_0+\mu\tau_0+\partial f-\sqrt{-1}\,\bar{\rho}_0+\lambda\bar{\omega}+\mu\bar{\tau}_0+\bar{\partial}f\\ &= -\sqrt{-1}\,\rho_0+\mu\tau_0+\partial f.\end{aligned}$$

So σ_0 is of type $(1,0)$ and $\sigma=\sigma_0+\bar{\sigma}_0$ is a real closed 1-form, satisfying already (ii). As to (i), it is sufficient to show that ω, $\bar{\omega}$ and σ are independent in $H^1(X,\mathbb{C})=H^1(X,\mathbb{R})\otimes\mathbb{C}$. If this were not the case, then we would have

$$\sigma_0=\lambda\omega+\partial g \qquad (g \text{ real}),$$

and hence $d\sigma_0=\bar{\partial}\partial g=\omega\wedge\bar{\omega}$. If g attains a minimum at $x_0\in X$, we define a holomorphic function F in a neighbourhood of x_0 by requiring $dF=\omega$, with $F(x_0)=0$. Then we have

$$\partial\bar{\partial}(|F|^2+g)=0,$$

i.e. $|F|^2+g$ is a harmonic function, which has a minimum in x_0. Thus F is constant and $\omega=0$, a contradiction.

Now let Y be the universal covering space of X. We shall denote a form on X and its pull-back on Y by the same symbol. On Y there is a holomorphic function w, such that $dw=\omega$. Furthermore, we have $d(\sigma_0+\bar{w}\,\omega)=0$, hence there is a second holomorphic function z on Y with $dz=\sigma_0+\bar{w}\,\omega$. The holomorphic 2-form $dw\wedge dz=dw\wedge\sigma_0$ comes from a holomorphic 2-form on X. We claim that it never vanishes. By the remarks made at the very beginning of the present proof it is sufficient to show that it does not vanish identically. So suppose that $\omega\wedge\sigma_0\equiv 0$. Then $\sigma_0=h\omega$, with h a differentiable function on X, and $\omega\wedge\bar{\omega}=d\sigma_0=dh\wedge\omega$. It follows that $\bar{\omega}+dh=g\,\omega$ and $\bar{\omega}=-\bar{\partial}h$, which is impossible since $\bar{\omega}\neq 0$ in the Dolbeault cohomology.

Now let $p\in X$ and let γ_1, γ_2 and γ_3 be three closed paths, starting and ending at p, which represent a base of $H_1(X,\mathbb{Z})$. The vectors

$$\Big(\int_{\gamma_i}\omega+\bar{\omega},\ \sqrt{-1}\int_{\gamma_i}\omega-\bar{\omega},\ \int_{\gamma_i}\sigma\Big),\qquad i=1,2,3$$

are independent in $\mathbb{R}^3$, and thus define a 3-torus T. As in the case of the Albanese torus we can define a (differentiable) map $f\colon X\to T$ by setting

$$f(x)=\left(\int_p^x\omega+\bar{\omega},\ \sqrt{-1}\int_p^x\omega-\bar{\omega},\ \int_p^x\sigma\right),$$

where the integral is taken along some path from p to x. The map f has maximal rank everywhere, for

$$\begin{aligned}-\tfrac{1}{2}(\omega+\bar{\omega})\wedge(\omega-\bar{\omega})\wedge\sigma &= \omega\wedge\bar{\omega}\wedge\sigma=\omega\wedge\bar{\omega}\wedge(\sigma_0+\bar{\sigma}_0)\\ &= \omega\wedge\bar{\omega}\wedge\sigma_0+\omega\wedge\bar{\omega}\wedge\bar{\sigma}_0.\end{aligned}$$

So (for type reasons) the vanishing of df implies that also $\omega\wedge\bar{\omega}\wedge\sigma_0=0$, which can't happen, since it implies $\omega\wedge\sigma_0=0$. Consequently, f maps X onto T. The fibres of f are connected, since by construction $f_*(\pi_1(X))=\pi_1(T)$. In other words, X is a circle bundle over T. In fact it is even an orientable circle bundle, since X and T are orientable.

As part of the exact homotopy sequence we obtain a short exact sequence

$$0\to\pi_1(S^1)\to\pi_1(X)\to\pi_1(T)\to 0.$$

From this it follows that we can replace, if necessary, γ_1, γ_2 and γ_3 by other loops from and to p, such that their classes g_1, g_2 and g_3, together with a generator g_0 of $\pi_1(S^1)$ generate $\pi_1(X)$. From here on we assume that this has been done. Since $\pi_1(T)$ is abelian, we must have

$$g_i g_j g_i^{-1} g_j^{-1} = g_0^{n_{ij}}, \qquad n_{ij}\in\mathbb{Z},$$

for $i,j=1,2,3$. We claim that not all n_{ij} vanish. Namely, if they did, then $\pi_1(X)$ would be abelian since $g_i g_0 g_i^{-1} g_0^{-1}=1$ in any case, for X is an *oriented* bundle over T. But if $\pi_1(X)$ were abelian, then $b_1(X)$ would be 4. So not all integers n_{ij} vanish (in fact we have three of them, since $n_{ji}=-n_{ij}$).

If we choose an identification of $\pi_1(X)$ with the group of covering transformations of Y over X, then the same relations hold for the corresponding covering transformations.

If we put for $y\in Y$:

$$g_i(w)(y)=w(g_i(y))$$

and similarly for z, then we see from $dw=\omega$ and $dz=\sigma_0+\bar{w}\,\omega$, that

$$g_i(w)=w+c_i,$$
$$g_i(z)=z+\bar{c}_i\,w+\beta_i,$$

with $c_0=0$ and $c_i=\int_{\gamma_i}\omega$, $i=1,2,3$. Substituting this into $g_i g_j g_i^{-1} g_j^{-1}=g_0^{n_{ij}}$ we find, since $\beta_0\neq 0$:

$$n_{23}c_1+n_{31}c_2+n_{12}c_3=0.$$

So c_1, c_2, c_3 generate a lattice L in $\mathbb{C}$ (two of them must be independent over $\mathbb{R}$, otherwise $\int_{\gamma_i}\omega+\bar{\omega}$, $\sqrt{-1}\int_{\gamma_i}\omega-\bar{\omega}$, $\int_{\gamma_i}\sigma$ could not be independent in $\mathbb{R}^3$) and the holomorphic map from X onto $\mathbb{C}/L$, given by ω, shows that there are enough curves on X. So X must be elliptic. We have found that the assumption $\mathrm{a}(X)=0$ leads to a contradiction, hence $\mathrm{a}(X)=1$ and X is elliptic. □

Finally we come to case c). Since $b^+(X)$ has to vanish, we find from Theorem IV.2.6 that $b_1(X)=1$. We shall prove that for a surface X of this type $\mathrm{kod}(X)=-\infty$; in other words it is a surface of class VII by definition. Suppose that $\mathrm{kod}(X)=0$. Then there is an $n_0\geq 2$ such that $P_{n_0}(X)=1$ and either $\mathcal{K}_X^{\otimes n_0}$ is trivial or there exists a positive divisor $\sum d_i D_i\in|n_0 K_X|$. If $\mathcal{K}_X^{\otimes n_0}\cong\mathcal{O}_X$, then by I, Sect. 18 there is an unramified covering X' of X with $\mathcal{K}_{X'}=\mathcal{O}_{X'}$, hence X' is minimal and $p_g(X')=1$. Furthermore $b_1(X')\geq b_1(X)\geq 1$, and $\mathrm{kod}(X')=\mathrm{a}(X')=0$. Consequently X' is either a surface of class d) or a torus. This last possibility is excluded, since then X would be itself a torus. So X' is a surface of class d) and $\mathrm{a}(X')=\mathrm{a}(X)=1$ by Theorem 5.1. Therefore $\mathcal{K}_X^{\otimes n_0}\cong\mathcal{O}_X$ is impossible.

On the other hand, if $|n_0 K_X|$ contains a positive divisor $\sum_{j=1}^{l} d_j D_j$, then it follows in the same way as in the proof of Proposition 5.2 that $KD_i=D_i^2=0$ for all i. Hence D_i is smooth elliptic or rational with a cusp or node. In any case we have $\mathscr{K}_{D_i} = \mathcal{O}_{D_i}$. Using the adjunction formula and Serre duality on D_i we obtain an exact sequence

$$H^1(\mathcal{O}_X(n(K_X+D_i))) \to H^1(\mathcal{O}_{D_i}) \to H^2(\mathcal{O}_X(n(K_X+D_i)-D_i)).$$

Since $H^2(\mathcal{O}_X(n(K_X+D_i)-D_i)) \cong H^0(\mathcal{O}_X((1-n)K_X+(1-n)D_i))=0$ for $n \geq 2$, we conclude that $H^1(\mathcal{O}_X(n(K_X+D_i))) \neq 0$ for $n \geq 2$. By Riemann-Roch it follows that $\dim|n(K_X+D_i)| \geq 0$ for $n \geq 2$. If for all i, $1 \leq i \leq l$, there is an integer n_i with $n_i(K_X+D_i)=0$, then we can find an integer $m \neq 0$ with $mK_X=0$. But this is excluded here since by assumption there is an effective divisor in some $|n_0 K_X|$. So there must be one effective divisor in $|n(K_X+D_i)|$ for all $n \geq 2$. Taking $n=2$ and $n=3$ we see that there is an effective divisor $C \in |K_X+D_i|$. Then $n_0 C \in |n_0(K_X+D_i)|$, i.e. $n_0 C$ is equal to the divisor $\sum d_j D_j + n_0 D_i$. This means that C contains D_i with some strictly positive multiplicity, i.e. $C - D_i \in |K_X|$ is non-negative. Since $p_g(X)=0$, we have obtained a contradiction. So $\operatorname{kod}(X) = -\infty$.

6. The Final Step

Since $\operatorname{kod}(X)$, $P_n(X)$ and $b_1(X)$ are invariant under blowing up, whereas the classes 1)–10) can be distinguished by these invariants, every surface has indeed a minimal model in exactly one of the classes 1)–10). We know from Proposition III.4.6 that this minimal model is unique (up to isomorphisms) if $\operatorname{kod}(X) \geq 0$. So the only thing left is to show that for surfaces of class VII the minimal model is unique. Otherwise there would exist a surface of class VII with two intersecting exceptional curves; blowing down one of them would yield a curve C on a surface X of class VII, such that $(\mathscr{K}_X, C) \leq -2$. Since $C^2 \leq 0$ by Theorem IV.5.2, the adjunction formula yields that C is a smooth rational curve with $C^2=0$. This is impossible by Proposition V.4.3, (i).

Remark. Surfaces in classes 1) and 3) can indeed have several minimal models. A simple example with minimal models in class 1) provides the surface X, obtained by blowing up a point on $\mathbb{P}_1 \times \mathbb{P}_1$. This surface has of course $\mathbb{P}_1 \times \mathbb{P}_1$ as a minimal model, but also $\mathbb{P}_2$, as you can see either directly or using Remark 2.4 by blowing down the proper transforms of the "horizontal and vertical fibres" through p, which have become (-1)-curves. (The process consisting of blowing up a point x in a ruled surface and then blowing down the proper transform of the ruling through p, is known as an *elementary transformation.*)

To complete the picture, we give another example of this type, but now with minimal models in class 3). Let B be any elliptic curve and $X = B \times \mathbb{P}_1$. We claim that the surface $\tilde{X}$, obtained from X by blowing up $x_0 \in X$, has two (even topologically) different minimal models. Indeed, in $\tilde{X}$ we can blow down the proper transform of the fibre through x_0, obtaining another ruled surface Y with base B, which is minimal by Lüroth's theorem. It remains to be shown that X and Y are different. But that is easy: on X all self-intersection numbers are even, whereas on Y there is a section with self-intersection number 1. Such a section is obtained by starting from a section in X which does not pass through x_0, taking its proper transform on $\tilde{X}$ and projecting this into Y.

7. Deformations

In this section we consider arbitrary analytic deformations of surfaces. The results are summed up in the following theorem.

(7.1) **Theorem.** (i) *If X is a surface with minimal model in one of the classes* 1) *to* 10), *then every deformation of X has a minimal model in that same class.*

(ii) *If X is minimal and contained in one of the classes* 3) *to* 10), *then every deformation of X is again minimal.*

(iii) *Every deformation of $\mathbb{P}_2$ is again $\mathbb{P}_2$.*

(iv) *A surface Y is a deformation of Σ_n, if and only if Y is isomorphic to Σ_m with $m \equiv n(2)$.*

Taking into account that every minimal rational surface is either $\mathbb{P}_2$ or Σ_n, $n \neq 1$ (Remark 2.4), and that in class VII it really happens that a minimal surface *can* be deformed into a non-minimal one ([Ka2], p. 61), we see that the above result is satisfactory from the point of view of the Enriques-Kodaira classification.

N.B. We don't prove Iitaka's result that all plurigenera are invariant under deformations. In fact, this result follows easily from our considerations, here and in Chap. VII, except for one case: properly elliptic surfaces.

In order to prove Theorem 7.1 we need the following facts.

(I) If X is a minimal surface of general type, then $c_1^2(X) > 0$ (compare Table 10).

(II) If X is a surface of general type, then $h^1(\mathcal{K}_X^{-1}) = 0$ if and only if X is minimal.

(III) If X is a minimal properly elliptic surface X with $p_g(X) = 0$, $q(X) = 1$, $b_1(X) = 1$, then every small deformation of X is again properly elliptic.

We shall prove (I) and (II) in the next chapter (Theorem VII.2.2 and Proposition VII.5.5), but (III) will not be proved in this book, and we have to refer to [Ko4], p. 685–694. Kodaira proves this result by first showing that all these surfaces can be obtained by applying suitable logarithmic transformations (V, Sect. 17) to a product $E \times \mathbb{P}_1$, where E is an elliptic curve. By varying both this curve and all but three of the points on $\mathbb{P}_1$ in which a logarithmic transformation is performed, he obtaines an analytic family $\mathcal{F}$ of properly elliptic surfaces. Then Kodaira calculates $h^1(\mathcal{T}_X)$ and finds that this dimension equals the dimension of $\mathcal{F}$. The proof is completed by showing that $\mathcal{F}$ is "effectively parametrised", i.e. that $\mathcal{F}$ induces the whole of $H^1(\mathcal{T}_X)$.

We shall also use the amusing

(7.2) **Example.** *Let X be a ruled surface of genus $g \geq 1$. Then every surface which is homeomorphic to X is also ruled of genus g.*

Proof. Since $\tau(X)=0$, the integers $b^+(X)$, $b^-(X)$, $p_g(X)$ and $c_1^2(X)$ are topological invariants of the *non-oriented* underlying manifold of X (compare Theorem IV.2.6). So if $g \geq 2$, the result follows from Table 10, together with the fact that blowing up a point diminishes c_1^2 by one.

Now let $g=1$, and Y a surface, homeomorphic to X. Since $p_g(Y)=p_g(X)=0$, and $b^+(Y)=b^+(X)=1$, the surface Y is projective by Theorem IV.5.2. If we apply Stein factorisation to the Albanese map of Y, then we obtain a connected holomorphic map f: $Y \to E$ onto a smooth elliptic curve E (E can't be rational because of Lüroth and $g(E) \geq 2$ is impossible since otherwise there would be too many holomorphic 1-forms on Y); of course, $E=\mathrm{Alb}(Y)$. From Remark III.11.5 we conclude that either f is everywhere of maximal rank, or the general fibre of f is elliptic and f has no other singular fibres but multiple elliptic curves. In the first of these cases the general fibre of f must be rational, otherwise the universal covering space of Y would be contractible, whereas that of X has the homotopy type of S^2. The second case does not occur, for in this case there exists by stable reduction and Theorem III.15.4 an unramified covering of Y which is a locally trivial elliptic fibre bundle over some curve of genus ≥ 1, so the universal covering of Y is again contractible. Using again $e(Y) = e(X)=0$ we conclude that Y is a ruled surface over an elliptic curve. □

In the coming proof we frequently shall use the upper semi-continuity of $h^i(\mathscr{K}_{X_t}^{\otimes n})$ in a family X_t (Theorem I.8.5, (ii)) and the fact that blowing-up does not change b_1 nor the P_i's (Theorem I.9.1, (iv) and (viii)).

Proof of Theorem 7.1.

Step A) Any deformation of a surface in class i), $4 \leq i \leq 8$, is again a surface in class i).

Since minimal surfaces X with $\mathrm{kod}(X)=0$ are characterised among all surfaces by the fact that there is some $n \in \mathbb{Z}$ with $\mathscr{K}_X^{\otimes n}$ trivial, it follows from Proposition IV.3.4 that any deformation of such a surface has again Kodaira dimension 0. Now the classes 4) to 8) can be distinguished by topological invariants, so every deformation must preserve the class from which you start.

Step B) Any deformation of a surface X with minimal model in class i), $4 \leq i \leq 8$, has again its minimal model in class i).

Since blow-ups of surfaces in different classes 4) to 8) can still be distinguished by topological invariants, it is enough to prove that every deformation of X has again Kodaira dimension 0. Thus we have to establish two things: firstly that any small deformation X' of X has again $\mathrm{kod}(X')=0$, and secondly that if X_t, $0 \leq t < 1$ is a "real subfamily" of a complex-analytic family, such that $\mathrm{kod}(X_t)=0$ for $t \neq 0$, then $\mathrm{kod}(X_0)=0$ too.

The first point can be reduced to step A) since by Proposition IV.3.1 we may assume X to be minimal.

In the second case we may assume X_0 to be minimal for the same reason. If any of the surfaces X_t, $t \neq 0$, is minimal, we are ready. If not, then $c_1^2(X_0) = c_1^2(X_t)$ must be strictly negative. But Table 10 shows that this can't occur for a minimal surface of Kodaira dimension ≥ 0.

Step C) Any deformation of a minimal surface of general type is a minimal surface of general type.

Let X_t be a small family, with X_0 minimal, and of general type. Since $h^0(\mathcal{K}_{X_0}^{\otimes n})\geq 2$ for n large enough, we have $h^0(\mathcal{K}_{X_0}^{\vee})=0$ and there exists by upper semi-continuity a neighbourhood E: $|t|<\varepsilon$ of 0 such that $h^0(\mathcal{K}_{X_t}^{\vee})=0$ for $t\in E$. Now $c_1^2(X_t)=c_1^2(X_0)>0$ by (I), so X_t is either rational or of general type. But if X_t is rational, then Riemann-Roch yields that $h^0(\mathcal{K}_{X_t}^{\vee})\neq 0$. So X_t is of general type for all $t\in E$. And if $|t|$ is small enough, then X_t is minimal by (II) and upper semi-continuity.

Now let X_t, $0\leq t<1$ be a "real family" with X_t minimal and of general type for $0<t<1$. Then by Riemann-Roch there exist constants α, β, with $\alpha\geq 1$ by (I), such that for $n\geq 2$ $P_n(X_t)\geq\alpha\frac{n(n-1)}{2}+\beta$. Hence $P_n(X_0)\geq\alpha\frac{n(n-1)}{2}+\beta$ for $n\geq 2$, and X_0 is of general type. And it is minimal because of Proposition IV.3.1.

Step D) Any deformation of a surface of general type is again of general type.

The case of a small deformation follows from step C) and Proposition IV.3.1, whereas the "limit case" can be proved as in C).

Step E) Any deformation of a surface of class VII is again of class VII.

Let X_t be a small deformation with X_0 of class VII. By upper semi-continuity there exists a neighbourhood E: $|t|<\varepsilon$ of 0 in the base space such that $P_{12}(X_t)=0$ for $t\in E$. Table 10 and step B) show that any surface X_t, $t\in E$, is either properly elliptic with $a(X_t)=1$ or of class VII. But the first possibility is excluded because of Theorem V.18.5.

For the case of a family X_t, $0\leq t<1$, with X_t in class VII for $t\neq 0$, Table 10 shows again that X_0 is either in class VII or properly elliptic. But this last case is excluded by (III).

Step F) Any deformation of a ruled surface of genus $g\geq 1$ is again a ruled surface of genus g.

This follows from Example 7.2.

Step G) Any deformation of a blown-up ruled surface of genus $g\geq 1$ is again a blown-up ruled surface of genus g.

The case of a small deformation follows from step F) and Proposition IV.3.1

As to a family X_t, $0\leq t<1$, with X_t blown-up ruled of genus g for $t\neq 0$, we may again assume that X_0 is minimal. So $c_1^2(X_0)=c_1^2(X_t)\leq 0$, with all X_t minimal and $g=1$ if $c_1^2(X_0)=0$. So if $c_1^2(X_0)<0$ we are ready by Table 1 and if $c_1^2(X_0)=0$ we are back in the preceding case.

Step H) Every deformation of a minimal properly elliptic surface is again minimal properly elliptic.

Combining Table 10 with the preceding steps we see that, as far as small deformations X_t are concerned we only have to show that in a sufficiently small neighbourhood $|t|<\varepsilon$ none of the surfaces X_t is rational. But $c_1^2(X_0)=0$ and $h^0(\mathcal{K}_{X_0}^{\vee})=0$, whereas for a rational surface Y with $c_1^2(Y)=0$ Riemann-Roch yields $h^0(\mathscr{K}_Y^\vee)>0$. So our claim is again a consequence of upper semi-continuity.

As to the case of a "family" X_t, $0\leq t<1$, with X_t minimal properly elliptic for $t\neq 0$, the only case left would be that X_0 is rational. This is impossible by Castelnuovo's criterion 2.1 and upper semi-continuity.

Step I) Every deformation of a properly elliptic surface is properly elliptic.

This follows for a small deformation from step H), combined with Proposition IV.3.1, whereas the case X_t, $0\leq t<1$ is again a consequence of the preceding steps and Castelnuovo's criterion.

Step J) Any deformation of a rational surface is a rational surface.

This follows from Table 10 and steps A)–I).

Step K) Every deformation of $\mathbb{P}_2$ is again $\mathbb{P}_2$. This has been proved earlier (Example V.1.3).

Step L) Any deformation of a Σ_n is isomorphic to Σ_m with $m\equiv n \pmod 2$.

From step J) and the fact that, among rational surfaces, Σ-surfaces are characterised by $b_2=2$, we conclude that any deformation of a Σ_n is another Σ-surface. Since on Σ_m all self-intersection numbers are even if and only if $m\equiv 0 \pmod 2$, we infer that Σ_n can only be deformed into Σ_m if $m=n \pmod 2$.

Step M) It remains to be shown that every pair of Σ-surfaces, Σ_n and Σ_m, can be deformed into each other if $n\equiv m \pmod 2$. This can be done in the following way. Let $(x_0:x_1)$ and $(y_0:y_1:y_2)$ be homogeneous coordinates in $\mathbb{P}_1$ and $\mathbb{P}_2$ respectively. Then, almost by definition, the surface, given in $\mathbb{P}_1\times\mathbb{P}_2$ by $x_0^n y_1 - x_1^n y_2=0$ is isomorphic to Σ_n. If we take on $\mathbb{P}_1\times\mathbb{P}_2\times\mathbb{C}$ the hypersurface $x_0^n y_1 - x_1^n y_2 + t x_0^{n-m} x_1^m y_0=0$, and consider it as a holomorphic family X_t, $t\in\mathbb{C}$, of Σ-surfaces over $\mathbb{C}$, then X_0 is isomorphic to Σ_n, whereas for $t\neq 0$ the surface X_t is isomorphic to Σ_{n-2m} (see [M-K], p. 26). □

Reference: [Ii].

VII. Surfaces of General Type

In this chapter the use of the words "surface" and "curve" is the same as in Chap. VI.

Preliminaries

1. Introduction

The minimal surfaces of general type can be parametrised in a satisfactory way, namely by a countable number of quasi-projective families. More precisely we have Gieseker's theorem ([Gi], p. 236):

There exists a quasi-projective coarse moduli scheme for the minimal surfaces of general type X with fixed Chern numbers $c_1^2(X)$ and $c_2(X)$.

Remark. By Theorem III.4.5 and Proposition III.4.6 a surface of general type has a unique minimal model. For that reason classifying surfaces of general type essentially means classifying minimal surfaces of general type.

The proof of Gieseker's theorem is based on the fact that for $n \geq 5$ an n-canonical map f_n (IV, Sect. 1) is a birational morphism from X onto a normal 2-dimensional subvariety of degree $n^2 c_1^2(X)$, with only rational double points, of a fixed projective space $\mathbb{P}_N$ (where N only depends on $c_1^2(X)$ and $c_2(X)$). The variety $f_n(X)$ determines X up to isomorphisms, but, given X, the map f_n depends on the choice of a base in $\Gamma(X, \mathcal{K}_X^{\otimes n})$. For fixed $c_1^2(X)$ and $c_2(X)$, let $\mathcal{S}$ be the set of all varieties in $\mathbb{P}_N$ thus obtained (for all surfaces X and all choices of a base). Then $PGL(N, \mathbb{C})$ operates on $\mathcal{S}$ in a natural way, and the isomorphy classes of surfaces X are in one-to-one correspondence with the points of $\mathcal{S}/PGL(N, \mathbb{C})$. Now Gieseker shows, using a criterion of Hilbert-Mumford, that for n large enough, this quotient is quasi-projective and indeed a coarse moduli space for the surfaces X.

In this chapter we shall prove the result on the n-canonical maps, but Gieseker's work will not be dealt with.

In general, little is known about the structure of the Gieseker scheme. Even the very first question: for which pairs (c_1^2, c_2) is the Gieseker scheme not empty, has not been completely answered. The known restrictions on c_1^2 and c_2 are expressed by

(1.1) **Theorem.** *If X is any minimal surface of general type, then:*
(i) $c_1^2(X)+c_2(X)\equiv 0(12)$,
(ii) $c_1^2(X)>0$ *and* $c_2(X)>0$,
(iii) $c_1^2(X)\leq 3c_2(X)$,
(iv) $5c_1^2(X)-c_2(X)+36\geq 0$ ($c_1^2(X)$ *even*)
$5c_1^2(X)-c_2(X)+30\geq 0$ ($c_1^2(X)$ *odd*).

Property (i) is of course an immediate consequence of Noether's formula, and actually holds for any almost-complex surface (compare IV, Sect. 7). The inequalities (ii)–(iv) will be proved below. The inequalities (iv) might appear strange, but they become more natural if you look at them from another side, namely as the Noether inequalities (Theorem 3.1). They are more conveniently stated by using c_1^2 and the Todd genus $\chi=\frac{1}{12}(c_1^2+c_2)$ instead of c_1^2 and c_2:

(iv)′ $\chi(X)\leq\frac{1}{2}c_1^2(X)+3$ ($c_1^2(X)$ even)
$\chi(X)\leq\frac{1}{2}c_1^2(X)+\frac{5}{2}$ ($c_1^2(X)$ odd).

Many people prefer the use of c_1^2 and χ instead of c_1^2 and c_2. We shall mainly stick to c_1^2 and c_2, leaving the obvious translation to the reader.

"Most" pairs of integers, satisfying the conditions of Theorem 1.1, can be realised as the Chern numbers of at least one minimal surface of general type. In particular, most pairs on the boundary of the domain in the plane, determined by the theorem, can be realised. However, at this moment it is not known whether all pairs, satisfying the conditions of the theorem, can be represented.

Apart from the existence of at least one surface for most pairs (c_1^2, c_2), satisfying the conditions of Theorem 1.1, some information is available concerning the Gieseker scheme for certain special values of (c_1^2, c_2). For the case that the Noether inequalities are equalities, Horikawa has obtained detailed results. (One or two simple cases were known before.) And, due to the efforts of many, a good deal is known about the Gieseker scheme for some (c_1^2, c_2) with low c_1^2, in particular $c_1^2=1$.

The existence of the Hilbert scheme implies that by desingularisation of 2-dimensional irreducible varieties of given degree in a fixed $\mathbb{P}_N$ you can obtain only finitely many diffeomorphism types.

So the fact that every minimal surface X of general type can be obtained by desingularising a 2-dimensional variety (with rational singularities) of degree d in $\mathbb{P}_N$, with d only depending on c_1^2 and c_2 (Theorem 5.2) means in particular that for given c_1^2 and c_2 there is only a *finite number* of diffeomorphism types for such surfaces (because of Theorem 1.1, (ii) and (iii), and the fact that blowing down a (-1)-curve increases c_1^2 with 1 and decreases c_2 with 1, this statement remains true if we consider *all* surfaces of general type, minimal or not). The topological index theorem expresses their index in c_1^2 and c_2, and the Noether inequalities yield an upper bound for $b_1=2q$ in terms of c_1^2 and c_2, but apart from such rather obvious remarks in general nothing is known about the diffeomorphism types that actually occur. No general bound is known for the number of irreducible components of the Gieseker scheme, but there is a bound on the dimension of these components (compare [Cat 2]). A question remaining completely open is the question of whether there is some "main stream" for surfaces of general type, i.e. whether there exists some construction yielding a component of the Gieseker scheme which is the largest in dimension for most pairs (c_1^2, c_2).

The further content of the chapter is as follows. In Sect. 2 we prove a certain number of basic facts, including Theorem 1.1, (ii). Sections 3 and 4 are respectively devoted to the proof of Theorem 1.1, (iv) and (iii). In Sects. 5–8 we study pluricanonical maps and in the Sects. 9–11 we briefly describe what is known about surfaces with given c_1^2 and c_2.

2. Some General Theorems

(2.1) **Proposition.** *If there exists on an algebraic surface X an algebraic system of effective divisors, of dimension at least* 1, *such that the general member is a (possibly singular) rational or elliptic curve, then* $\mathrm{kod}(X) \leq 1$.

Proof. Without loss of generality we may assume the system to be irreducible and of dimension 1. Then the assumption implies that there exists a smooth compact curve C with the following property. The product $X \times C$ contains an irreducible 2-dimensional subvariety, projecting onto X, such that its desingularisation Y is either a blown-up ruled or an elliptic surface (V, Sect. 4 and Sect. 7). Hence $\mathrm{kod}(Y) \leq 1$ by V, Sect. 4 and Proposition V.12.4, and Theorem I.7.4 yields that also $\mathrm{kod}(X) \leq 1$. □

(2.2) **Theorem.** *If X is a minimal surface of general type, then $K_X^2 = c_1^2(X) > 0$.*

Proof. Let H be a smooth hyperplane section of X. We consider the exact sequence

$$0 \to \mathcal{O}_X(nK_X - H) \to \mathcal{O}_X(nK_X) \to \mathcal{O}_H(nK_X) \to 0,$$

and the corresponding cohomology sequence. By Theorem I.7.2 there is a $c > 0$, such that $h^0(\mathcal{O}_X(nK_X)) > cn^2$ for n large, whereas $h^0(\mathcal{O}_H(nK_X))$ grows linearly with n. So there is an $n_0 > 0$, such that $n_0 K_X - H$ can be represented by an effective divisor R. Since $K_X R \geq 0$ by Proposition III.2.3, we find:

$$n_0^2 K_X^2 = (n_0 K_X)(H + R) \geq (n_0 K_X) H = H^2 + HR \geq H^2 > 0. \quad \square$$

(2.3) **Corollary.** *Let X be a minimal surface of general type and C an irreducible curve on X. Then $K_X C \geq 0$ and $K_X C = 0$ if and only if C is a (-2)-curve.*

Proof. The first statement is already included in Proposition III.2.3.

If C is a (-2)-curve, then $KC=0$ by the adjunction formula. Conversely, if $KC=0$, then by the Algebraic index theorem IV.2.15 and Theorem 2.2 above we must have $C^2<0$, since an effective divisor is never rationally homologous to 0 on an algebraic surface. The adjunction formula now implies that C is a (-2)-curve. □

(2.4) **Proposition.** *If X is any surface of general type, then $c_2(X)>0$.*

Proof. We distinguish between two cases:

a) if X admits a connected holomorphic map onto a curve of genus ≥ 2, then by Proposition 2.1 the general fibre must have genus ≥ 2, hence $c_2(X) = e(X) \geq 4$ by Proposition III.11.4.

b) if X does not admit such a map, then by Proposition IV.4.2 and Corollary IV.4.4 we find that $e(X)>0$, unless $q(X)=1$, $p_g(X)=0$ or $q(X)=2$, $p_g(X)=1$.
So let us assume that there is a surface X with $q(X)>0$, $\chi(X)=0$ and $c_2(X)\leq 0$. By Theorem 2.2 we must have $c_2(X)<0$. Since $b_1(X)=2q(X)\neq 0$, we know from Theorem I.18.1, (i) that X admits unramified coverings of arbitrarily high order. They all are of general type by Theorem I.7.4. We pick one, say Y, with $c_2(Y)<-3$. Now Y can't admit a connected holomorphic map onto a curve of genus ≥ 2, otherwise $c_2(Y)$ would be at least 4 by a). But now we have arrived at a contradiction, for Proposition IV.4.2 would imply $c_2(Y)\geq -3$. So a surface X as described above can't exist, and we are ready. □

(2.5) **Proposition.** *Let X be a minimal surface of general type. Then we have*

(i) *The number of (-2)-curves on X is finite. They are independent over $\mathbb{Q}$ and their number is at most equal to $\rho(X)-1$, where $\rho(X)$ is the Picard number of X.*

(ii) *On the subspace of $H_2(X,\mathbb{Q})$, spanned by the (-2)-curves, the intersection form is negative definite.*

Proof. To prove the first part of (i) it is sufficient to show the following: if $C_1, \ldots, C_l$ are (-2)-curves on X, such that

$$\sum_{i=1}^{k} \lambda_i C_i = \sum_{j=k+1}^{l} \lambda_j C_j \quad \text{in } H_2(X,\mathbb{Q})$$

for some k, $1\leq k\leq l$, and with all λ_i, $\lambda_j \geq 0$, then $\lambda_i=0$ for $i=1,\ldots,l$.

We have

$$\left(\sum_{i=1}^{k} \lambda_i C_i\right)^2 = \left(\sum_{i=1}^{k} \lambda_i C_i\right)\left(\sum_{j=k+1}^{l} \lambda_j C_j\right) \geq 0.$$

If we combine this inequality with $K_X^2>0$ (Theorem 2.2), then the index theorem yields $\sum_{i=1}^{k} \lambda_i C_i=0$ in rational homology, i.e. $\lambda_1=\ldots=\lambda_l=0$.

Since each rational homology class contains at most one (-2)-curve, the number of these curves is at most $\rho(X)$. But $KH>0$ for a hyperplane section H, hence H is not homologous to a sum $\sum \lambda_i C_i$, and the number of (-2)-curves is at most $\rho(X)-1$.

Part (ii) is an immediate consequence of Theorem 2.2 and the index theorem. □

Two Inequalities

3. Noether's Inequality

(3.1) **Theorem** (Noether's inequality). *Let X be a minimal surface of general type. Then*

$$p_g(X) \leq \tfrac{1}{2} c_1^2(X) + 2.$$

Proof. Since $c_1^2(X) \geq 1$ (Theorem 2.2) we may assume that $p_g(X) \geq 3$. Let $|K| = |C| + V$, where V is the fixed part of $|K|$. The linear system $|C|$ is either composed with a pencil or its general member is irreducible (IV, Sect. 1).

In the first case there is a 1-dimensional algebraic system of effective divisors on X, the general member F of which is irreducible, such that K is homologous to $nF+V$ and $p_g = \dim|C|+1 \leq n+1$. We claim that $K^2 \geq 2n$. Since $K^2 = (nF+V)K \geq nFK$ by Corollary 2.3, this will be proved as soon as we have shown that $KF \geq 2$. If $KF \leq 1$ we find from $KF = (nF+V)F \leq 1$ that $F^2 \leq 1$. Since also $F^2 \geq 0$, and $KF \equiv F^2(2)$, we only have to consider the cases $KF = F^2 = 1$ and $KF = F^2 = 0$. The first of these cases leads to $n=1$ and $p_g(X) \leq 2$, whereas we have assumed $p_g(X) \geq 3$. The second case is excluded by Corollary 2.3.

If $|C|$ is not composed with a pencil, then $C^2 > 0$ (IV, Sect. 1). We take an irreducible member C of $|C|$ and consider the standard exact sequence

$$0 \to \mathcal{O}_X(K) \to \mathcal{O}_X(K+C) \to \mathcal{O}_C(K+C) \to 0.$$

Since $H^1(\mathcal{O}_X(K+C)) \cong H^1(\mathcal{O}_X(-C)) = 0$ by Corollary IV.8.2 and also $H^2(\mathcal{O}_X(K+C)) \cong H^0(\mathcal{O}_X(-C)) = 0$, we find from the associated exact cohomology sequence and the Riemann-Roch theorem that

$$h^0(\mathcal{O}_C(K+C)) = h^0(\mathcal{O}_X(K+C)) - p_g(X) + q(X) = \tfrac{1}{2}(C^2+KC)+1.$$

Now if $\mathcal{L}$ and $\mathcal{M}$ are two line bundles on an irreducible variety with

$$h^0(\mathcal{L}) \geq 1, \quad h^0(\mathcal{M}) \geq 1, \quad \text{then } h^0(\mathcal{L} \otimes \mathcal{M}) \geq h^0(\mathcal{L}) + h^0(\mathcal{M}) - 1.$$

So we have:

$$\begin{aligned} h^0(\mathcal{O}_C(K+C)) &= h^0(\mathcal{O}_C(2C+V)) \geq h^0(\mathcal{O}_C(2C)) \geq 2h^0(\mathcal{O}_C(C)) - 1 \\ &\geq 2h^0(\mathcal{O}_X(C)) - 3 = 2h^0(\mathcal{O}_X(C+V)) - 3 = 2p_g(X) - 3. \end{aligned}$$

But $2K^2 - (KC + C^2) = 2KV + CV \geq 0$, hence $p_g(X) \leq \tfrac{1}{2}K_X^2 + 2$. □

Remark. In case $|C|$ is not composed with a pencil, the inequality can be sharpened a little bit by refining the argument in the following way. Firstly, we can replace the lines following "On the other hand we have" by

$$\begin{aligned}\dim r_C(H^0(\mathcal{O}_X(K+C))) &= \dim r_C(H^0(\mathcal{O}_X(2C+V))) \\ &\geq \dim r_C(H^0(\mathcal{O}_X(2C))) \geq 2\dim r_C(H^0(\mathcal{O}_X(C))) - 1 \\ &= 2p_g(X) - 3.\end{aligned}$$

Since $\dim H^0(\mathcal{O}_C(K+C)) = \dim r_C(H^0(\mathcal{O}_X(K+C))) + q(X)$, we thus obtain

$$p_g(X) \leq \tfrac{1}{2}K_X^2 + 2 - \tfrac{1}{2}q(X) - \tfrac{1}{2}K_X V - \tfrac{1}{4}CV.$$

This generalisation can be quite useful in special situations.

(3.2) **Corollary.** *Let X be a minimal surface of general type. Then*

$$5c_1^2(X) - c_2(X) + 36 \geq 0 \quad (c_1^2(X) \textit{ even})$$
$$5c_1^2(X) - c_2(X) + 30 \geq 0 \quad (c_1^2(X) \textit{ odd}).$$

Proof. Let $c_1^2(X)$ be even. Then by Noether's formula we have $\chi(X) \leq p_g(X) + 1 \leq \frac{1}{2}c_1^2(X) + 3$, i.e. $5c_1^2(X) - c_2(X) + 36 \geq 0$.

If $c_1^2(X)$ is odd, then we automatically have $p_g(X) \leq \frac{1}{2}c_1^2(X) + \frac{3}{2}$, and we can proceed as before. □

(3.3) **Corollary.** *If X is a minimal surface of general type with*

$$5c_1^2(X) - c_2(X) + 36 = 0 \; (c_1^2(X) \textit{ even}) \textit{ or}$$
$$5c_1^2(X) - c_2(X) + 30 = 0 \; (c_1^2(X) \textit{ odd}),$$

then $q(X) = 0$,

Remark. In the case that $|K|$ is not composed with a pencil there is an equivalent way of phrasing the proof of Theorem 3.1, based on "counting quadrics". This has been pointed out to us by M. Reid. In rough outline he argues as follows.

1) If $S \subset P = \mathbb{P}_n$ is a 2-dimensional irreducible variety spanning P, H a hyperplane of P, $C = S \cap H$, the restriction map

$$Q_S = \Gamma(\mathcal{I}_{S|P}(2H)) \to Q_C = \Gamma(\mathcal{I}_{C|H}(2H))$$

is an isomorphism for generic H (there pass as many quadrics through S as there pass through a generic hyperplane section).

2) Since $\mathbb{P}(Q_C) \subset \mathbb{P}(\Gamma(\mathcal{O}_H(2H)))$ does not meet the $2(n-1)$-dimensional set of reducible quadrics in H, its codimension is $\geq 2(n-1)$.

3) Combining 1) and 2) we find that the codimension of $\mathbb{P}(Q_S)$ in $\mathbb{P}(\Gamma(\mathcal{O}_P(2H)))$ is $\geq 3n-1$.

4) Applying 3) to $S = f_1(X) \subset \mathbb{P}_{p_g - 1}$ one finds that

$$\operatorname{rank}(\Gamma(\mathcal{O}_P(2H)) \to \Gamma(\mathcal{O}_X(2K_X))) \geq 3(p_g - 1)$$

and in particular, since $P_2 \geq 1 - q + p_g + K^2$ by Riemann-Roch, we find that $K^2 \geq 2p_g - 4$.

This proof has some advantages if you want to obtain more detailed information. For instance, it can be refined to give a proof of Castelnuovo's second inequality (Theorem 8.5 below).

4. The Inequality $c_1^2 \leq 3c_2$

(4.1) **Theorem.** *For every surface of general type X the inequality* $c_1^2(X) \leq 3c_2(X)$ *holds.*

The proof of this theorem will be preceded by a number of auxiliary results.

(4.2) **Proposition.** *If on the algebraic surface X there is a line bundle* $\mathscr{L}$ *with* $h^0(\mathscr{L}^\vee \otimes \Omega_X^1) \neq 0$, *then there is a constant c such that* $h^0(\mathscr{L}^{\otimes k}) \leq ck$ *for all* $k \geq 1$.

Proof. Since otherwise the result is trivial, we assume that $h^0(\mathscr{L}^{\otimes k_0}) \geq 2$ for some $k_0 \geq 1$.

We start with the case $k_0 = 1$. Let $s_1, s_2 \in \Gamma(\mathscr{L})$ be linearly independent, and let h be a homomorphism from $\mathscr{L}$ into Ω_X^1, $h \neq 0$. Then $h(s_1)$ and $h(s_2)$ are linearly independent 1-forms on X with $h(s_1) \wedge h(s_2) \equiv 0$. So we are in a position to apply Proposition IV.4.1. Consequently, there exists a holomorphic, connected map $f\colon X \to Y$ from X onto a smooth curve Y, such that both $h(s_1)$ and $h(s_2)$ are the pull-back of 1-forms on Y. It follows that if s_1 vanishes on a curve, then this curve is contained in the sum of some fibres of f. Hence $\mathscr{L} = \mathcal{O}_X(D)$, where every component of the non-negative divisor D is contained in some fibre of f. Since $(D - nF)A < 0$ for n sufficiently large and A ample, there are no non-negative divisors on X, which are homologous to $k(D - nF)$, where F is a fibre, n sufficiently large and k any natural number. Let the divisor F_k consist of ck general (hence smooth) fibres of f.

From the standard exact sequence

$$0 \to \mathcal{O}_X(kD - F_k) \to \mathcal{O}_X(kD) \to \mathcal{O}_{F_k}(kD) \to 0$$

we find $h^0(\mathscr{L}^{\otimes k}) \leq h^0(\mathcal{O}_{F_k}(kD)) \leq ck$ for all $k \geq 1$.

As to the general case, by Theorem I.18.3 and Theorem IV.5.5 there exists an algebraic surface Y and a holomorphic surjective map $g\colon Y \to X$, such that $g^*(\mathscr{L})$ has two independent sections. Since $h^0(\mathscr{H}om(\mathscr{L}, \Omega_X^1)) \neq 0$ implies $h^0(\mathscr{H}om(g^*(\mathscr{L}), \Omega_Y^1)) \neq 0$, we can apply to Y and $g^*(\mathscr{L})$ the result for $k = 1$. In other words there exist a constant c, such that for all $k \geq 1$ the inequality $h^0((g^*(\mathscr{L}))^{\otimes k}) \leq ck$ holds. But $h^0(\mathscr{L}^{\otimes k}) \leq h^0((g^*(\mathscr{L}))^{\otimes k})$, and the proposition has been proved. □

(4.3) **Proposition.** *Let X be an algebraic surface,* $\mathcal{O}_X(D)$ *a line bundle on X, and* $\mathscr{F}$ *a locally free, rank-two subsheaf of* Ω_X^1, *such that:*

(i) $c_1(\mathscr{F})S \geq 0$ *for every effective divisor S on X,*

(ii) $h^0(\mathscr{H}om(\mathcal{O}_X(D), \mathscr{F})) \neq 0$.

Then $c_1(\mathscr{F})D \leq \max(c_2(\mathscr{F}), 0)$.

Proof. Since $\Gamma(\mathscr{H}om(\mathcal{O}_X(D), \mathscr{F})) \cong \Gamma(\mathscr{F} \otimes \mathcal{O}_X(-D)) \neq 0$, there is a non-negative divisor S on X, such that $\mathscr{F} \otimes \mathcal{O}_X(-D-S)$ admits a section with at most isolated zeros.

Hence

$$c_2(\mathscr{F}\otimes\mathcal{O}_X(-D-S))=(D+S)^2-c_1(\mathscr{F})(D+S)+c_2(\mathscr{F})\geq 0$$

$$c_1(\mathscr{F})D\leq(D+S)^2-c_1(\mathscr{F})S+c_2(\mathscr{F}).$$

Since by assumption $c_1(\mathscr{F})S\geq 0$, the proposition is already clear if $(D+S)^2\leq 0$. On the other hand, if $(D+S)^2>0$, then application of the Riemann-Roch theorem, together with Serre duality gives

$$h^0(\mathcal{O}_X(n(D+S)))+h^0(\mathcal{O}_X(K_X-n(D+S)))>dn^2$$

for some constant $d>0$ and n large enough. So we have either

$$h^0(\mathcal{O}_X(n(D+S)))>\tfrac{1}{2}dn^2$$

or

$$h^0(\mathcal{O}_X(K_X-n(D+S)))>\tfrac{1}{2}dn^2$$

for an infinite number of positive values of n. The first of these possibilities is excluded by Proposition 4.2 (for $h^0(\mathscr{H}om\,(\mathcal{O}_X(D+S),\Omega_X^1))\neq 0$ since $h^0(\mathscr{H}om\,(\mathcal{O}_X(D+S),\mathscr{F}))\neq 0$). In the second case we have $c_1(\mathscr{F})(D+S)\leq\frac{1}{n}(c_1(\mathscr{F})K_X)$ for an infinite number of n's, i.e. $c_1(\mathscr{F})D\leq -c_1(\mathscr{F})S\leq 0$. □

(4.4) **Proposition.** *Let X be an algebraic surface, $\mathcal{O}_X(D)$ a line bundle on X and $\mathscr{F}$ a locally free rank-two subsheaf of Ω_X^1, such that*

(i) $c_1(\mathscr{F})S\geq 0$ *for every effective divisor S on X,*
(ii) $h^0(\mathscr{H}om(\mathcal{O}_X(D),S^n\mathscr{F}))\neq 0$.

Then $c_1(\mathscr{F})D\leq\max(nc_2(\mathscr{F}),0)$.

Proof. Let $Z=\mathbb{P}(\mathscr{F})=\mathbb{P}(\mathscr{F}^\vee)$, and let $p\colon Z\to X$ be the projection. Then Theorem I.5.1 implies the existence of a divisor class $H=H_{\mathscr{F}}$ on Z, such that for every divisor E on X there is a canonical isomorphism between $\Gamma(\mathcal{O}_Z(nH+p^*(E))$ and $\Gamma(\mathcal{O}_X(S^n\mathscr{F}\otimes\mathcal{O}_X(E)))$. So in our case there is an effective divisor G on Z with $\mathcal{O}_Z(G)=\mathcal{O}_Z(nH-p^*(D))$. By the branched covering trick (Theorem I.18.2) there exists an algebraic surface Y, together with a surjective holomorphic map $f\colon Y\to X$ (of degree k, say) such that under the induced bundle map from $\mathbb{P}(f^*(\mathscr{F}))$ onto $\mathbb{P}(\mathscr{F})=Z$ the pull-back of G decomposes into a sum of positive divisors, representing $H_{q^*(\mathscr{F})}-q^*(D_i)$, where $q\colon\mathbb{P}(f^*(\mathscr{F}))\to Y$ denotes the projection. The D_i's need not to be effective, but $\sum D_i=D$. Since $f^*(\mathscr{F})$ is a subsheaf of Ω_Y^1, since by construction $h^0(\mathscr{H}om\,(\mathcal{O}_Y(D_i),\ f^*(\mathscr{F}))\neq 0$, and since $c_1(f^*(\mathscr{F}))\cdot T=c_1(\mathscr{F})\cdot f_*(T)\geq 0$ for every effective divisor T on Y, we can conclude from Proposition 4.3:

$$c_1(f^*(\mathscr{F}))\cdot D_i\leq\max(c_2(f^*(\mathscr{F})),0)$$

$$f^*(c_1(\mathscr{F})\cdot D)\leq\max(nc_2(f^*(\mathscr{F})),0)$$

$$kc_1(\mathscr{F})D\leq k\max(nc_2(\mathscr{F}),0)$$

$$c_1(\mathscr{F})D\leq\max(nc_2(\mathscr{F}),0).\quad\square$$

Now we can prove Theorem 4.1. The idea is quite simple: assuming $c_1^2 > 3c_2$, a contradiction is obtained by showing that for suitable $\lambda \in \mathbb{Q}$, $n \in \mathbb{Z}$ with $n\lambda \in \mathbb{Z}$, the cohomology groups $H^i(S^n\Omega_X^1 \otimes \mathcal{O}_X(n\lambda K_X))$ vanish for $i=0,2$ and n large, whereas on the other hand

$$\chi(S^n\Omega_X^1 \otimes \mathcal{O}_X(n\lambda K_X)) > 0 \quad \text{for } n > n_0.$$

Proof of Theorem 4.1. Since blowing up a point decreases c_1^2 by 1 and increases c_2 by 1, we may assume that X is minimal. Then we have $c_1^2(X) > 0$ and $c_2(X) > 0$ by Theorem 2.2 and Proposition 2.4 respectively.

We shall derive a contradiction from the assumption that

$$\alpha = \frac{c_2(X)}{c_1^2(X)} < \tfrac{1}{3}.$$

Let $\beta = \frac{1}{4}(1-3\alpha)$, and let n be a natural number such that $n(\alpha+\beta) \in \mathbb{Z}$. We consider the vector bundle

$$\mathcal{V}_n = S^n\Omega_X^1 \otimes \mathcal{O}_X(-n(\alpha+\beta)K_X)$$

and claim: $h^0(\mathcal{V}_n) = h^2(\mathcal{V}_n) = 0$, provided that n is large enough. The vanishing of $h^0(\mathcal{V}_n)$ is an immediate consequence of Proposition 4.4: you take $\mathcal{F} = \Omega_X^1$ and $D = n(\alpha+\beta)K_X$, then you use Corollary 2.3 and the assumptions concerning α and β. As to $h^2(\mathcal{V}_n)$, we find, using Serre duality and the fact that $\mathcal{T}_X \cong \Omega_X^1 \otimes \mathcal{O}_X(-K)$:

$$\begin{aligned} h^2(\mathcal{V}_n) &= h^0(S^n\mathcal{T}_X \otimes \mathcal{O}_X((n(\alpha+\beta)+1)K_X)) \\ &= h^0(S^n\Omega_X^1 \otimes \mathcal{O}_X((n(\alpha+\beta-1)+1)K_X)). \end{aligned}$$

If n is large enough we find as before that this dimension vanishes. We conclude that the Euler characteristic

$$\chi(S^n\Omega_X^1 \otimes \mathcal{O}_X(-n(\alpha+\beta)K_X)) \ \text{is} \ \leq 0$$

if n is large enough. On the other hand, by the Riemann-Roch theorem this Euler characteristic can be written as a polynomial of degree 3 in n with strictly positive leading coefficient.

To see this, let $Y = \mathbb{P}(\mathcal{T}_X)$, let $p\colon Y \to X$ be the projection and let $\mathcal{L}^\vee$ be the tautological bundle on Y. Then we derive from Theorem I.5.1 and I, (10):

$$\begin{aligned} &\chi(S^n\Omega_X^1 \otimes \mathcal{O}_X(-n(\alpha+\beta)K_X)) \\ &= \chi(\mathcal{L}^n \otimes p^*(\mathcal{O}_X(-n(\alpha+\beta)K_X)) \\ &= \frac{c_1^3(\mathcal{L} \otimes \mathcal{O}_X(-(\alpha+\beta)K_X))}{6} n^3 + \gamma n^2 + \delta n + \varepsilon. \end{aligned}$$

So what we have to do is to show that $c_1^3(\mathcal{L} \otimes \mathcal{O}_X(-(\alpha+\beta)K_X)) > 0$. Putting $c_1(\mathcal{L}) = c$, we have by I, Sect. 5

$$c^2 + p^*(c_1(X))c + p^*(c_2(X)) = 0.$$

Since $c \cdot p^*(\text{natural generator of } H^4(X,\mathbb{Z})) = \text{natural generator of } H^6(Y,\mathbb{Z})$ we thus find

$$c^3 = c_1^2(X) - c_2(X).$$

We obtain:

$$\begin{aligned}c_1^3(\mathscr{L}\otimes\mathscr{O}_X(-(\alpha+\beta)K_X)) &= (c-(\alpha+\beta)p^*(-c_1(X)))^3\\ &= c^3+3(\alpha+\beta)c^2p^*(c_1(X))+3(\alpha+\beta)^2cp^*(c_1^2(X))\\ &= \frac{c_1^2}{16}(3\alpha^2-22\alpha+7)>0.\end{aligned}$$

So the assumption $\alpha<\frac{1}{3}$ leads to a contradiction, and we find $\alpha\geq\frac{1}{3}$, i.e. $c_1^2(X)\leq 3c_2(X)$.

All the inequalities expressed by Theorem 1.1 are more or less classical, except for (iii). A weak form of this last inequality, namely $c_1^2\leq 8c_2$, was proved in [Ve1]. In 1975 Bogomolov obtained the inequality $c_1^2\leq 4c_2$ (see [Bog] and [Rei1]), and a year later Miyaoka and S.T. Yau independently proved (iii), see [Mi3] and [Y3]. The proof given here is a simplified version of Miyaoka's (compare [Ve3]).

In [Sak] Sakai extends the inequality $c_1^2\leq 3c_2$ to the case of "logarithmic Chern numbers"; this result is used by Miyaoka ([Mi5]) to prove a.o. the following conjecture of Hirzebruch: *if the minimal surface of general type X contains k disjoint* (-2)*-curves, then* $k\leq\frac{2}{9}(3c_2(X)-c_1^2(X))$.

Pluricanonical Maps

5. The Main Results

The best way of formulating results on pluricanonical maps (IV, Sect. 1) is by way of abstract canonical models. To do this, however, you need to know that the canonical ring of a surface of general type is finitely generated. Since we obtain this fact only as a consequence of our results on pluricanonical maps, we shall first state these last results in a more primitive way, returning to the proper highbrow interpretation a little bit later. In this light our Definition 5.1 should be seen for what it is: an ad hoc tool.

The structure of the canonical maps $f_{\mathscr{K}^{\otimes n}}=f_n$ is very simple for $n\geq 5$. This holds also for $n=4$ or 3, except for some low values of K_X^2. These exceptions have been studied intensively during the last years and the structure of f_n is now rather well known for these exceptional classes of surfaces. For f_2 a fairly general result is known, but our knowledge is not yet as detailed as for the case $n\geq 3$. Finally, for $n=1$ only some general facts are known.

We start by describing (and proving) the general case for $n\geq 2$, leaving the exceptional cases for f_2, f_3 and f_4 as well as the whole case f_1 for later.

(5.1) **Definition.** *Let X be a surface and* $\mathscr{L}=\mathscr{O}_X(D)$ *a line bundle on X. We say that* $\mathscr{L}$ *provides a C-isomorphism if:*

(i) $|D|$ *has no fixed components and* $f_{\mathscr{L}}$ *is a holomorphic map;*

(ii) $f_{\mathscr{L}}(X)$ *is a normal 2-dimensional variety with only a finite number of singular points:* $p_1,\dots,p_k$, *which are all rational double points;*

(iii) *if* $C^{(i)}=f_{\mathscr{L}}^{-1}(p_i)$, *then the* $C^{(i)}$*'s are the connected components of the union of all* (-2)*-curves on* X;

(iv) $f_{\mathscr{L}}$ *maps* $X \setminus \bigcup_{i=1}^{k} C^{(i)}$ *isomorphically onto* $f_{\mathscr{L}}(X) \setminus \bigcup_{i=1}^{k} p_i$.

Automatically, $f_{\mathscr{L}}$ then provides a minimal resolution of $p_1, \ldots, p_k$.

The preceding definition enables us to formulate the main facts about pluricanonical maps in the following simple way.

(5.2) **Theorem.** *Let X be a minimal surface of general type. Then:*

(i) f_n *is a C-isomorphism for* $n \geq 5$;

(ii) f_4 *is a C-isomorphism if* $K_X^2 \geq 2$;

(iii) if $K_X^2 \geq 3$, *then* $|3K_X|$ *has no fixed components, and* f_3 *is a birational holomorphic map for* $K_X^2 \geq 3$, *whereas* f_3 *is a C-isomorphism if* $K_X^2 \geq 6$;

(iv) *if* $K_X^2 \geq 10$ *and* $p_g(X) \geq 6$, *then* f_2 *is a birational map if and only if X is not a fibre space with a curve of genus 2 as general fibre.*

As an easy consequence, we indeed obtain

(5.3) **Proposition.** *Let X be a surface of general type. Then its canonical ring* $R(X)$ *is a finitely generated noetherian ring.*

Proof. By Theorem I.9.1, (viii) we may assume that X is minimal.

We have $R(X) = \sum_{i \geq 0} \Gamma(\mathscr{K}_X^{\otimes 5i} \otimes \mathscr{S})$, where

$$\mathscr{S} = \mathscr{O}_X \oplus \mathscr{K}_X^{\otimes 2} \oplus \mathscr{K}_X^{\otimes 3} \oplus \mathscr{K}_X^{\otimes 4}.$$

Since by Theorem 5.2 the 5-canonical map $f_5\colon X \to \mathbb{P}_{k(5)}$ is a C-isomorphism, we find, using Theorem III.3.4 and $f_5^*(\mathscr{O}_{\mathbb{P}_{k(5)}}(1)) = \mathscr{K}_X^{\otimes 5}$, that

$$\sum_{i=0}^{\infty} \Gamma(\mathscr{K}_X^{\otimes 5i} \otimes \mathscr{S}) \simeq \sum_{i=0}^{\infty} \Gamma(\mathscr{T}(i)),$$

with $\mathscr{T} = f_{5*}(\mathscr{S})$. If we put $S = \sum_{j=0}^{\infty} \Gamma(\mathscr{O}_{\mathbb{P}_{k(5)}}(j))$, then by [Se2], § 3 $\sum_{i=0}^{\infty} \Gamma(\mathscr{T}(i))$ is a finitely generated S-module. If we combine this with the fact that S is a finitely generated ring we find that $R(X)$ is also a finitely generated ring. □

Now we are ready for the proper formulation of the results, expressed by Theorem 5.2 in a primitive way.

If we define the abstract canonical model X_{can} as Proj $(R(X))$ (see I, Sect. 7) then X_{can} becomes a 2-dimensional irreducible projective variety (because for a surface of general type $R(X)$ has transcendency degree 2 over $\mathbb{C}$). We set $R^n = \sum_{m=0}^{\infty} R_m{}^n$, where $R_m{}^n$ is the subspace of $\Gamma(X, \mathscr{K}_X^{\otimes mn})$, generated by products of sections in $\Gamma(X, \mathscr{K}_X^{\otimes n})$, and define the n-th canonical model $X_{\text{can}}^{(n)}$ of X as Proj(R^n). Then the inclusion of R^n in $R(X)$ induces a rational map $k_n\colon X_{\text{can}} \to X_{\text{can}}^{(n)}$, and Theorem 5.2 becomes

(5.4) **Theorem.** *Let X be a minimal surface of general type. Then*

(i) k_n *is an isomorphism for* $n \geq 5$;
(ii) k_4 *is an isomorphism if* $K_X^2 \geq 2$;
(iii) k_3 *is a birational morphism if* $K_X^2 \geq 3$ *and an isomorphism if* $K_X^2 \geq 6$;
(iv) *if* $K_X^2 \geq 10$ *and* $p_g(X) \geq 6$, *then* k_2 *is birational if and only if X is not a fibre space with a curve of genus 2 as general fibre.*

The proof of Theorem 5.2 is rather complicated, but the underlying idea is quite simple as we shall now explain.

Let X be a surface, and $\mathcal{L} = \mathcal{O}_X(L)$ a line bundle on X. To show that $f_{\mathcal{L}}$ (compare IV, Sect. 1) is a holomorphic map, it is sufficient to prove the following: given any point $x \in X$, there is a section $s \in \Gamma(X, \mathcal{L})$ with $s(x) \neq 0$. If $p: (\bar{X}, E) \to (X, x)$ denotes the blowing up of X at x, then there is such a section s if and only if there exists a section in $\Gamma(\bar{X}, p^*(\mathcal{L}))$ which does not vanish on E. Since $\mathcal{O}_E(p^*(\mathcal{L})) \cong \mathcal{O}_E$, this existence is assured as soon as $H^1(\bar{X}, p^*(\mathcal{L}) \otimes \mathcal{O}_{\bar{X}}(-E)) = 0$, or, by Serre duality and Theorem I.9.1, (vii), as soon as $H^1(\bar{X}, p^*(\mathcal{K}_X \otimes \mathcal{L}^\vee) \otimes \mathcal{O}_{\bar{X}}(2E)) = 0$. Ramanujam's theorem (Corollary IV.8.2) tells us that this last cohomology group vanishes if there is a 1-connected divisor $\bar{D} \in |p^*(L - K_X) - 2E|$ with $\bar{D}^2 > 0$. Now it is easily seen that if D is a 2-connected divisor in $|L - K_X|$ with $D^2 \geq 5$ which has (at least) a double point at x, then $p^*(D)$ contains $2E$, and $D' = p^*(D) - 2E$ is a 1-connected divisor in $|p^*(L - K_X) - 2E|$ with $(D')^2 > 0$. The existence of a $D \in |L - K_X|$ with a double point at x is assured as soon as $\dim |L - K_X| \geq 3$. So we conclude that $f_{\mathcal{L}}$ is a holomorphic map as soon as three conditions are satisfied: (i) $(L - K_X)^2 \geq 5$, (ii) $\dim |L - K_X| \geq 3$ and (iii) all divisors in $|L - K_X|$ are 2-connected.

In the case $\mathcal{L} = \mathcal{K}_X^{\otimes n}$ $(n \geq 1)$ it follows from general connectivity properties of pluricanonical divisors and Riemann-Roch that these conditions are fulfilled if certain numerical conditions involving n, K_X^2 and $c_2(X)$ are satisfied.

Similarly, to prove that $f_{\mathcal{L}}$ is one-to-one on a subset $U \subset X$ it is sufficient to show that if you blow up two points in U, then there exists on the blown-up surface a 1-connected effective divisor with such and such properties. Again, in the case of f_n, the existence of such a divisor is assured if certain numerical conditions, involving n, K_X^2 and $c_2(X)$ are satisfied.

Remark. The use of the preceding idea is not restricted to pluricanonical maps. For example, it can also be applied to the study of surfaces with a hyperelliptic plane section, compare [Ve5] and [So1].

We mention some more consequences of Theorem 5.2, one of which we have already used in Chap. VI.

(5.5) **Proposition.** *Let X be a surface of general type and $n \in \mathbb{Z}$, $n \neq 0, 1$. Then X is minimal if and only if* $H^1(\mathcal{K}_X^{\otimes n}) = 0$.

Proof. If X is minimal, the result follows from Theorem 5.2, Mumford's vanishing theorem IV.8.5 and Serre duality. Conversely, if $n \geq 2$ and if $\bar{X}$ is obtained from X by blowing up at least once, we have

$$h^1(\mathcal{K}_{\bar{X}}^{\otimes n}) = P_n(\bar{X}) - \frac{n(n-1)}{2} K_{\bar{X}}^2 - \chi(\bar{X}) \quad \text{(Riemann-Roch)}$$

$$> P_n(X) - \frac{n(n-1)}{2} K_X^2 - \chi(X) = h^1(\mathcal{K}_X^{\otimes n}).$$

For $n \leq -1$ the argument is similar, but with $h^2(\mathcal{K}_X^{\otimes n})$, instead of $P_n(X)$.

(5.6) **Corollary.** *If X is a minimal surface of general type, then*

$$P_n(X) = \frac{n(n-1)}{2} K_X^2 + \chi(X)$$

for all $n \geq 2$.

The classical geometers certainly were aware of the fact that, except sometimes for low values of n, the pluricanonical maps f_n are birational ([Enr2]). Mumford proved in [Mu2] that f_n is a C-isomorphism for n large enough and that the canonical ring is always finitely generated. In [Saf] Moishezon showed the birationality of f_n for $n \geq 8$. After important work of Kodaira (see [Ko6]), Bombieri finally proved the precise results expressed by Theorem 5.2 ([Bom2]). We follow him rather closely in this chapter. For more recent work on the exceptional cases left open by Bombieri, we refer to Sect. 8.

An important new development is the birth of a theory of "*logarithmic Kodaira dimension*" and "*logarithmic pluricanonical maps*"; compare for example [Sak] and the references given there.

6. Connectedness Properties of Pluricanonical Divisors

To carry out the sketch for the proof of Theorem 5.2 given in Sect. 5, we start with some connectivity properties of pluricanonical divisors.

(6.1) **Proposition.** *Let X be a minimal surface of general type and $D \in |nK_X|$, $n \geq 1$. Then*

(i) *D is 1-connected;*

(ii) *except for the case $n=2$, $K_X^2=1$, the divisor D is even 2-connected.*

Proof. Let $D = D_1 + D_2$ be a splitting of D; by definition D_1 and D_2 are effective. If, say $KD_1 = 0$, then the index theorem implies that $D_1^2 < 0$. But $D_1^2 \equiv KD_1 (\text{mod}\, 2)$ by the adjunction formula, hence $D_1^2 \leq -2$. So $D_1 D_2 = D_1(nK - D_1) \geq 2$ in this case.

So let from here on $KD_1 \geq 1$ and $KD_2 \geq 1$. We set

$$\frac{KD_1}{K^2} = \lambda \in \mathbb{Q} \quad \text{and} \quad D_1 = \lambda K + E \in H_2(X, \mathbb{Q}).$$

Then the algebraic index theorem yields $E^2 \leq 0$. Furthermore,

$$D_1 D_2 = (\lambda K + E)((n-\lambda)K - E) = \lambda(n-\lambda)K^2 - E^2 \geq \lambda(n-\lambda)K^2 \geq n - \frac{1}{K^2},$$

since $\frac{1}{K^2} \leq \lambda \leq \frac{n-1}{K^2}$. This implies our result for $n \geq 2$. And if $n=1$, we get $D_1 D_2 \geq 1$ as soon as $K^2 \geq 2$. But $D_1 D_2 = 2KD_1 - (D_1^2 + KD_1) \equiv 0 \pmod 2$, hence $D_1 D_2 \geq 2$. Finally, the case $K^2=1$, $n=1$ is excluded here since $KD_1 + KD_2 = 1$ implies that either KD_1 or KD_2 vanishes. □

(6.2) **Proposition.** *Let X be a minimal surface of general type and $D \in |nK_X|$, $n \geq 2$. If $D = D_1 + D_2$ is a splitting of D, with $K_X D_1 \geq 1$, $K_X D_2 \geq 1$, then $D_1 D_2 \geq 3$, except in the cases $n=2$, $K_X^2 = 1$ or 2, and $n=3$, $K_X^2=1$.*

Proof. In the same way as in the proof of Proposition 6.1 we obtain $D_1 D_2 \geq 3$ as soon as $n \geq 3$ and $K_X^2 \geq 2$. So we are left with the case $n=2$. If $KD_1 \geq 2$ and $KD_2 \geq 2$, then $D_1 D_2 \geq 4 - \frac{4}{K^2}$, since $\frac{2}{K^2} \leq \lambda \leq \frac{n-2}{K^2}$, hence $D_1 D_2 \geq 3$ if $K_X^2 \geq 3$. Finally, if, say, $KD_1 = 1$, then by the algebraic index theorem $D_1^2 \leq \frac{1}{K_X^2}$. So (since $K^2 \geq 3$ and $D_1^2 \equiv KD_1 \pmod 2$) $D_1^2 \leq 1$ and $D_1 D_2 = D_1(2K - D_1) \geq 3$. □

(6.3) **Proposition.** *Let X be a minimal surface of general type, $x_0 \in X$ and $p\colon (\bar{X}, E) \to (X, x_0)$ the blowing-up of x_0 in X. Let D be a 2-connected divisor on X. If there exists an effective divisor D' on $\bar{X}$, such that $D' + 2E = p^*(D)$, then D' is 1-connected.*

Proof. Let $D' = D_1' + D_2'$ be any splitting of D'. Then

$$D_1' = p^*(D_1) + aE,$$
$$D_2' = p^*(D_2) - (a+2)E,$$

with D_1, D_2 non-negative, $D = D_1 + D_2$ and $a \in \mathbb{Z}$. If $D_1 \neq 0$, $D_2 \neq 0$, then $D_1' D_2' = p^*(D_1 D_2) + a(a+2) = D_1 D_2 + a(a+2) \geq 1$, since also D_1 and D_2 are effective. And if, say, $D_1 = 0$, then $a > 0$ and $D_1' D_2' = a(a+2) \geq 3$. □

Let C be the union of all (-2)-curves on X (by Proposition 2.5 we know that they are finite in number). The connected components $C^{(i)}$ of C are exceptional by the algebraic index theorem, combined with Proposition 2.5. By Proposition III.2.4 the normal singularity obtained by blowing down $C^{(i)}$ is rational, and we denote by $Z^{(i)} = \sum a_j C_j^{(i)}$, $a_j > 0$, its fundamental cycle (III, Sect. 3).

(6.4) **Proposition.** *Let X be a minimal surface of general type and $n \geq 2$. Suppose that neither $K_X^2 = 1$ or 2, $n=2$ nor $K_X^2=1$, $n=3$. If $Z^{(i)}$ and $Z^{(j)}$ are two not necessarily different fundamental cycles on X, and $D \in |nK_X - Z^{(i)} - Z^{(j)}|$, then D is 1-connected.*

Proof. Firstly we consider the case that $i \neq j$.

Let $D = D_1 + D_2$, with D_1, D_2 effective. We apply Proposition 6.2 to $D_1 + D_2 + Z^{(i)} + Z^{(j)} \in |nK|$. Thus if $KD_1 \geq 1$, $KD_2 \geq 1$ we have

$$D_1(D_2 + Z^{(i)} + Z^{(j)}) \geq 3$$

and similarly

$$D_2(D_1+Z^{(i)}+Z^{(j)})\geq 3.$$

Hence $2D_1D_2-(Z^{(i)})^2-(Z^{(j)})^2\geq 6$, and $D_1D_2\geq 1$. If $KD_1=0$, then D_1 consists of (-2)-curves only. Let $D_1=D_1^{(i)}+R$, with $\operatorname{supp}(D_1^{(i)})\subset\operatorname{supp}(Z^{(i)})$ and $\operatorname{supp}(R)\cap\operatorname{supp}(Z^{(i)})=\emptyset$. Then $D_1Z^{(i)}=D_1^{(i)}Z^{(i)}\leq 0$ by III, Sect. 3. Similarly $D_1Z^{(j)}\leq 0$, and $D_1D_2\geq 1$ by Proposition 6.1.

If $i=j$, and $KD_1\geq 1$, $KD_2\geq 1$, we find in the same way $(D_1+Z)(D_2+Z)\geq 3$, hence $D_1D_2-Z^2\geq 3$ and $D_1D_2\geq 1$. And if, say, $KD_1=0$, we put again $D_1=D_1'+R$, with D_1' consisting of (-2)-curves in Z and $R\cdot Z=0$, hence $D_1'Z\leq 0$ and $D_1(D_2+2Z)\geq 1$ by Proposition 6.1, so $D_1D_2\geq 1$. □

7. Proof of the Main Results

Next we prove a weak form of Theorem 5.2(i), namely

(7.1) **Proposition.** *Let X be a minimal surface of general type. Then f_n has the following properties for $n\geq 5$:*

(i) *f_n is a holomorphic map;*

(ii) *f_n is one-to-one and of maximal rank outside the union C of all (-2)-curves in X.*

Proof. (i) We have to show that in every point $x_0\in X$ there is a section $s\in\Gamma(X,\mathscr{K}_X^{\otimes n})$, with $s(x_0)\neq 0$. Let $p\colon(\bar{X},E)\to(X,x_0)$ be the blowing-up of X at x_0. Since we have a commutative diagram

$$\begin{array}{ccc}
\Gamma(\bar{X},p^*(\mathscr{K}_X^{\otimes n})) & \xrightarrow{\;r\;} & \Gamma(E,p^*(\mathscr{K}_X^{\otimes n})|E)\cong\mathbb{C}\\
\wr\uparrow & & \wr\uparrow\\
\Gamma(X,\mathscr{K}_X^{\otimes n}) & \longrightarrow & \Gamma(x_0,\mathscr{K}_X^{\otimes n}|x_0)\cong\mathbb{C}
\end{array}$$

a section s as required will exist as soon as r is surjective, in particular as soon as $H^1(\bar{X},p^*(\mathscr{K}_X^{\otimes n})\otimes\mathscr{O}_{\bar{X}}(-E))=0$. By Theorem I.9.1, (vii) and Serre duality

$$H^1(\bar{X},p^*(\mathscr{K}_X^{\otimes n})\otimes\mathscr{O}_{\bar{X}}(-E))\cong H^1(\bar{X},p^*(\mathscr{K}_X^{\otimes(-n+1)})\otimes\mathscr{O}_{\bar{X}}(2E)),$$

so by Ramanujam's vanishing theorem it is sufficient to produce an effective divisor $D'\in|p^*(n-1)K_X-2E|$, with $(D')^2>0$, which is 1-connected. Now by Riemann-Roch $\dim|(n-1)K_X|\geq\dfrac{(n-1)(n-2)}{2}K_X^2+\chi(X)-1\geq 6$, hence there exists an effective divisor $D\in|(n-1)K|$, which has a double point at x_0. Consequently $p^*(D)-2E$ is effective, and 1-connected by Propositions 6.1, (ii) and 6.3. Since furthermore $(p^*((n-1)K_X-2E))^2=(n-1)^2K_X^2-4\geq 12$ the proof of (i) is complete.

To show that also (ii) is true we shall have to accomplish two things: we have to show that given any two points x_0, $x_1\in X\setminus C$, there is a section $s\in\Gamma(X,\mathscr{K}_X^{\otimes n})$, vanishing at x_0, but not at x_1, and we have to prove that f_n is of maximal rank in every point $x_0\in X\setminus C$.

So let $x_0, x_1 \in X \setminus C$, and $p: (\bar{X}, E_0 \cup E_1) \to (X, x_0 \cup x_1)$ be the blowing-up of X at $x_0 \cup x_1$. Exactly as in the proof of (i) we see that it is sufficient to show that

$$H^1(\bar{X}, p^*(\mathscr{K}_X^{\otimes n}) \otimes \mathcal{O}_{\bar{X}}(-E_0 - E_1)) = 0,$$

and this again will follow as soon as we have produced a 1-connected divisor

$$D' \in |p^*((n-1)K_X) - 2E_0 - 2E_1| \quad \text{with} \quad (D')^2 \geq 1.$$

But since $\dim|(n-1)K_X| \geq 6$ we still have enough freedom to make a $D \in |(n-1)K_X|$, which has double points at x_0 and x_1, and therefore a $D' \in |p^*((n-1)K_X) - 2E_0 - 2E_1|$ which we claim to be 1-connected. Indeed, if $D' = D'_1 + D'_2$ is a splitting of D', then

$$D'_1 = p^*(D_1) + aE_0 + bE_1$$
$$D'_2 = p^*(D_2) - (a+2)E_0 - (b+2)E_1$$

and we see that $D'_1 D'_2 \geq 1$ if $D_1 D_2 \geq 3$. Now we know by Proposition 6.2 that $D_1 D_2 \geq 3$ as soon as $KD_1 \geq 1$, $KD_2 \geq 1$. If $KD_1 = 0$, then D_1 consists solely of (-2)-curves by Corollary 2.3, i.e. $\operatorname{supp}(D_1) \subset C$, whereas $\operatorname{supp}(D_1 \cap D_2)$ does meet neither x_0 nor x_1. Hence $D'_1 D'_2 = D_1 D_2 \geq 2$ by Proposition 6.1, (ii). Since $(D')^2 = (n-1)^2 K_X^2 + 8 \geq 8$, we have proved the first part of (ii).

Finally, let x_0 again be any point of $X \setminus C$. Now quite generally, if we have a smooth curve Y and a line bundle $\mathscr{L}$ on Y, which is generated by its global sections, then $f_{\mathscr{L}}$ will be of maximal rank at $y_0 \in Y$ as soon as there is a section $s \in \Gamma(Y, \mathscr{L})$, which vanishes simply at y_0, i.e. if

$$\Gamma(Y, \mathscr{L} \otimes \mathcal{O}_Y(-y_0)) \to \Gamma(y_0, \mathscr{L} \otimes \mathcal{O}_{y_0}(-y_0))$$

is surjective. In our 2-dimensional case f_n will be of maximal rank at x_0 if the restriction of f_n to any line through x_0 (with respect to a local coordinate system) will be of maximal rank. By the preceding remark this will certainly be true as soon as

$$\Gamma(\bar{X}, p^*(\mathscr{K}_X^{\otimes n}) \otimes \mathcal{O}_{\bar{X}}(-E)) \to \Gamma(E, p^*(\mathscr{K}_X^{\otimes n}) \otimes \mathcal{O}_E(-E))$$

is surjective, and in particular if $H^1(\bar{X}, p^*(\mathscr{K}_X^{\otimes n}) \otimes \mathcal{O}_{\bar{X}}(-2E))$ vanishes. Thus we are back in an already familiar situation:

$$H^1(\bar{X}, p^*(\mathscr{K}_X^{\otimes n}) \otimes \mathcal{O}_{\bar{X}}(-2E)) \cong H^1(\bar{X}, p^*(\mathscr{K}_X^{-(n-1)}) \otimes \mathcal{O}_{\bar{X}}(3E))$$

and this last cohomology group vanishes if we can find a 1-connected $D' \in |p^*(n-1)K_X - 3E|$. Such a D' is obtained as before, but now starting from a $D \in |(n-1)K_X|$ which has a triple point at x_0.

Thus the proof of Proposition 7.1 is complete. □

Proof of Theorem 5.2. Once we have Proposition 7.1, three things still have to be shown to prove Theorem 5.2, (i), namely

a) given a point $x_0 \notin C$ and a fundamental cycle $Z = Z^{(i)}$ on X, there exists for $n \geq 5$ a divisor $D \in |nK|$, vanishing on $\operatorname{supp}(Z)$, but not at x_0;

b) given two fundamental cycles $Z^{(i)}$ and $Z^{(j)}$, there is a $D \in |nK|$, vanishing on $\operatorname{supp}(Z^{(i)})$, but not on $\operatorname{supp}(Z^{(j)})$;

c) $f_n(Z)$ is a normal point of $f_n(X)$.

As to a) it is sufficient to prove that there is a $D \in |nK - Z|$ which does not pass through x_0. Let as before $p: (\bar{X}, E) \to (X, x_0)$ be the blowing-up of X at x_0. Again it will be sufficient to show that the restriction homomorphism

$$\Gamma(\bar{X}, \mathcal{O}_{\bar{X}}(p^*(nK_X - Z))) \to \Gamma(E, \mathcal{O}_E(p^*(nK_X - Z)))$$

is surjective. Anew this is accomplished by showing that

$$H^1(\bar{X}, \mathcal{O}_{\bar{X}}(p^*(nK_X - Z) - E)) = 0.$$

By Serre duality and Ramanujam's vanishing theorem it will be enough to produce a 1-connected divisor $D' \in |p^*((n-1)K_X - Z) - 2E|$, with $(D')^2 > 0$. We know from Proposition III.3.4 that $O_Z(K_X) \cong \mathcal{O}_Z$ and $h^0(Z, \mathcal{O}_Z) = 1$, so the exact sequence

$$0 \to \Gamma(X, \mathcal{O}_X((n-1)K_X - Z)) \to \Gamma(X, \mathcal{O}_X((n-1)K_X)) \to \Gamma(Z, \mathcal{O}_Z)$$

together with Riemann-Roch, applied to $\mathcal{K}_X^{\otimes(n-1)}$ yield that there is a $D \in |(n-1)K_X - Z|$ which has a double point at x_0. Thus $D' = p^*(D) - 2E$ is effective and $(D')^2 > 0$. Because of Proposition 6.3 we shall be ready as soon as we have shown that D is 2-connected. So suppose $D = D_1 + D_2$, with D_1 and D_2 both effective. Then $D + Z \in |(n-1)K_X|$, and $D = (D_1 + Z) + D_2$ is a splitting of $D + Z$. If $KD_1 \neq 0$, $KD_2 \neq 0$, we have by Proposition 6.2:

$$(D_1 + Z)D_2 \geq 3$$
$$(D_2 + Z)D_1 \geq 3,$$

hence $D_1 D_2 \geq 2$. On the other hand, by Corollary 2.3, if $KD_1 = 0$ then D_1 consists only of (-2)-curves. If we put $D_1 = A_1 + R$, with $\operatorname{supp}(A_1) \subset \operatorname{supp}(Z)$ and $\operatorname{supp}(R) \cap \operatorname{supp}(Z) = \varnothing$, then $D_1 Z = A_1 Z \leq 0$ by III, Sect. 3, so $D_1 D_2 \geq 2$, for $D_1(D_2 + Z) \geq 2$ by Proposition 6.1.

To prove b) we only have to observe that in the exact sequence

$$0 \to \Gamma(\mathcal{O}_X(nK_X - Z^{(i)} - Z^{(j)})) \to \Gamma(\mathcal{O}_X(nK_X)) \xrightarrow{\ r\ }$$
$$\to \Gamma(\mathcal{O}_{Z^{(i)} \cup Z^{(j)}}) \to H^1(\mathcal{O}_X(nK_X - Z^{(i)} - Z^{(j)}))$$

the last group vanishes in our case $(n \geq 5)$, because of Ramanujam's vanishing theorem and Proposition 6.4. So r is surjective, and since

$$\Gamma(\mathcal{O}_{Z^{(i)} \cup Z^{(j)}}) \cong \Gamma(\mathcal{O}_{Z^{(i)}}) \oplus \Gamma(\mathcal{O}_{Z^{(j)}}) \cong \mathbb{C} \oplus \mathbb{C}$$

there is a section $s \in \Gamma(X, \mathcal{O}_X(nK_X))$ vanishing on $\operatorname{supp}(Z^{(i)})$, but not on $\operatorname{supp}(Z^{(j)})$.

Finally, the proof of c). If $v: \widetilde{f_n(X)} \to f_n(X)$ is the normalisation map (which by b) is bijective already), and if $q \in \widetilde{f_n(X)}$ is the point corresponding to $p = f_n(Z)$, we have to show that $v^*: \mathcal{O}_p \to \mathcal{O}_q$ is bijective. Injectivity is obvious, and to prove surjectivity, it suffices to do this for the induced map $\mathfrak{m}_p/\mathfrak{m}_p^2 \to \mathfrak{m}_q/\mathfrak{m}_q^2$ (differential criterion in [G-R3]). Since by Proposition III.3.6 there is a natural isomorphism $\mathfrak{m}_q/\mathfrak{m}_q^2 \to \mathcal{O}_Z(-Z)$, the assertion is equivalent with the surjectivity of

$$\Gamma(\mathcal{O}_X(nK_X - Z)) \to \Gamma(\mathcal{O}_Z(-Z)),$$

and so it follows from $H^1(\mathcal{O}_X(nK_X-2Z))=0$. But this can be shown as in case b) above, when we put $Z^{(i)}=Z^{(j)}$.

Theorem 5.2, (ii) is proved in exactly the same way. The restriction $K_X^2\geq 2$ comes from the fact that you need 3-canonical divisors with certain properties (for example, they should have double points in two prescribed points), and the condition ensures that there is enough freedom to construct them.

And the same holds again for most of Theorem 5.2, (iii). By the same method as in case (i) it can be proved that f_3 is a morphism (onto a surface, since $K_X^2>0$) for $K_X^2\geq 3$ and an isomorphism for $K_X^2\geq 6$. So we still have to prove that f_3 is birational for $3\leq K_X^2\leq 5$.

We distinguish between three cases.

1) $p_g(X)\geq 1$. Suppose that f_3 is not birational. Let $D\in|K|$. It is possible to find two points $x_0, x_1\in X$, such that:

(j) $x_0\notin\mathrm{supp}(D)$, $x_1\notin\mathrm{supp}(D)$,

(jj) $f_3(x_0)=f_3(x_1)$,

(jjj) f_3 is of maximal rank at x_0 and x_1.

Since $P_2(X)=K_X^2+\chi(X)\geq 4$ we can find $E\in|2K|$ which has a double point at x_0. Now every effective divisor in $|3K|$ which has a double point at x_0 has also a double point at x_1, and since $D\cup E\in|3K|$, and $x_1\notin\mathrm{supp}(D)$ we must have that E has a double point at x_1. But then, using E, we can prove as before that $f_3(x_0)\neq f_3(x_1)$, thus obtaining a contradiction.

2) $p_g(X)=0$, $K_X^2=3$. Again we assume that f_3 is not birational. Now $(3K_X)^2=27$, and $P_3(X)=10$, so $f_3(X)$ is a 2-dimensional subvariety Y of $\mathbb{P}_9$ with $\deg(Y)$ equal to a proper divisor of 27. Since Y has to span $\mathbb{P}_9$ we find as only possibility that $\deg(Y)=9$ (it is an elementary exercise to show that every irreducible 2-dimensional variety of degree 3, whether it is singular or not, spans at most a $\mathbb{P}_4$). By [Bom2], p. 217 the variety Y is either the isomorphic image of $\mathbb{P}_2$ by $\mathcal{O}_{\mathbb{P}_2}(3)$, or it contains a 1-dimensional family of lines (genuine lines in $\mathbb{P}_9$, that is). This last possibility can be excluded in the following way. The inverse image of the family consists of a 1-dimensional family of effective divisors D, which at most can have a sum B of (-2)-curves in common, because of Corollary 2.3. We have $(D-B)^2\geq 0$. Since $K(D-B)=KD=1$, this would lead us into conflict with the index theorem: $K(D-B-\frac{1}{3}K)=0$ gives $(D-B)^2<\frac{1}{3}$, and since $1=K(D-B)\equiv(D-B)^2 \pmod 2$, we would find $(D-B)^2<0$.

So Y is isomorphic to $\mathbb{P}_2$. Taking back onto X the images on Y of the lines in $\mathbb{P}_2$, we obtain a linear system $|D|$, without base points, of dimension at least 2 and with $D^2=3$. Since $|3D|=|3K|$, the map f_D can't be birational, hence it must map X onto $\mathbb{P}_2$. It follows that $h^0(X,\mathcal{O}_X(2D))\geq 6$. But since $H^1(X,\mathcal{O}_X(2D))=H^1(X,\mathcal{O}_X(K_X-2D))=0$ by Mumford's vanishing theorem (for $3(2D-K)=3K$), the Riemann-Roch theorem yields $h^0(X,\mathcal{O}_X(2D))=4$. This contradiction implies that f_3 is birational in this case.

3) $p_g(X)=0$, $K_X^2\geq 4$. Again we assume that f_3 is not birational, and then derive a contradiction from this assumption.

Because of Proposition 2.1 and since a Chow variety has only finitely many components, there is only a countable number of rational curves (smooth or not) on X. Thus we can choose points $x_0, x_1 \in X$, such that $f_3(x_0)=f_3(x_1)$, whereas no rational curve passes through x_1. As before we shall reach a contradiction as soon as we can produce an effective 3-canonical divisor, passing through x_0, but not through x_1. As before the existence of such a divisor will follow if

$$H^1(\bar{X}, p^*(\mathcal{O}_X(-2K_X)) \otimes \mathcal{O}_{\bar{X}}(2E_0+2E_1))=0,$$

where $p\colon (\bar{X}, E_0 \cup E_1) \to (X, x_0 \cup x_1)$ is the blowing-up of X at $x_0 \cup x_1$.

Since $P_2(X) \geq 5$ there exists a $D \in |2K|$, which has a double point at x_0 and which passes through x_1. If D happens to have a double point at x_1, then we see in the usual way that

$$H^1(\bar{X}, p^*(\mathcal{O}_X(-2K_X)) \otimes \mathcal{O}_{\bar{X}}(2E_0+2E_1))=0.$$

So let us assume from here on that x_1 is a simple point of D, in other words E_1 is not a component of the divisor $D' = p^*(D) - 2E_0 - E_1$, which is 1-connected by Proposition 6.3. Since $p_g(X)=0$, we find from Theorem 1.1, (ii) that $q(X)=0$, hence $q(\bar{X})=0$, and because of the exact sequence

$$H^1(\mathcal{O}_{\bar{X}}) \to H^1(\mathcal{O}_{\bar{X}}(E_1)) \to H^1(\mathcal{O}_{E_1}(E_1))$$

also that $H^1(\mathcal{O}_{\bar{X}}(E_1))=0$. Since furthermore $\Gamma(\mathcal{O}_{\bar{X}}(-D'+E_1))=0$ we see from the exact sequence

$$0 \to \Gamma(\mathcal{O}_{\bar{X}}(-D'+E_1)) \to \Gamma(\mathcal{O}_{\bar{X}}(E_1)) \to \Gamma(\mathcal{O}_{D'}(E_1)) \to H^1(\mathcal{O}_{\bar{X}}(-D'+E_1)) \to H^1(\mathcal{O}_{\bar{X}}(E_1))$$

that we shall be ready as soon as we have shown that $H^0(\mathcal{O}_{D'}(E_1)) \cong \mathbb{C}$. Because of Lemma II.12.4, this however follows from the way x_0 and x_1 are chosen.

Proof of Theorem 5.2, (*iv*). If $K_X^2 \geq 10$ and $p_g(X) \geq 6$, then f_2 is seen to be a morphism in the usual way. We assume again that f_2 is not birational. Then there is a Zariski-open subset $U \subset X$ such that:

1) $f_2 | U$ is an unramified covering of degree ≥ 2,
2) through no point of U there passes an irreducible curve B with $KB \leq 2$, $B^2 < 0$.

It will be clear that such a U exists as soon as we have proved that there is only a finite number of irreducible curves B on X with $KB \leq 2$, $B^2 < 0$. Because of the second condition it is sufficient to show that there is only a finite number of classes $b \in H_2(X, \mathbb{Q})$ which can be represented by such a curve. Suppose there would be an infinite number of such classes. Then there would be projections $b - \frac{BK}{K^2} k$ into the orthogonal complement of the class k of K with $-\left(b - \frac{BK}{K^2} k\right)^2$ arbitrarily large (algebraic index theorem). But this is impossible, for $B^2 \geq -4$ by the adjunction formula.

Now let $x_0, x_1 \in U$ with $f_2(x_0)=f_2(x_1)$. Since $p_g(X) \geq 6$, and since every canonical divisor passing through x_0 contains x_1, there exists a $D \in |K|$, with double points in x_0 and x_1. Let $p\colon (\bar{X}, E_0 \cup E_1) \to (X, x_0 \cup x_1)$ again be the blowing-up of X at $x_0 \cup x_1$. Then the effective divisor $D' = p^*(D) - 2E_0 - 2E_1$

can't be 1-connected, otherwise we would be able to prove in the usual way that $f_2(x_0) \neq f_2(x_1)$. Hence there is a splitting $D' = D'_1 + D'_2$ with $D'_1 D'_2 \leq 0$. If either D'_1 or D'_2 would be of the form $aE_1 + bE_2$ (with $a \geq 0$, $b \geq 0$ and a and b not both 0), we would have $D'_1 D'_2 \geq 1$. So the splitting $D' = D'_1 + D'_2$ gives by projection rise to a splitting $D = D_1 + D_2$, which must satisfy $D_1 D_2 \geq 2$ by Proposition 6.1, (ii). If we put $\frac{KD_1}{K^2} = \lambda$, we find in exactly the same way as in the proof of Proposition 6.2 that $D_1 D_2 \geq \lambda(1-\lambda) K^2$, hence $D_1 D_2 \geq 3$ if $KD_1 \geq 3$ and $KD_2 \geq 3$. So we find that either $KD_1 \leq 2$ or $KD_2 \leq 2$, say $KD_1 \leq 2$. If D_1 would neither pass through x_0 nor through x_1, then we would have $D'_1 D'_2 = D_1 D_2 \geq 2$. Say D_1 passes through x_0. Through x_0 there does not pass any curve B with $KB \leq 1$, for such a curve automatically satisfies $B^2 < 0$ by the index theorem, and assumption 2) still holds. Hence there is an irreducible component A of D_1 passing through x_0, with $KA = 2$. Then $A^2 \leq 0$ by the index theorem, and hence by assumption 2) we have $A^2 = 0$.

Suppose there would only be a countable number of irreducible curves A on X with $KA = 2$, $A^2 = 0$. Then it would be impossible that for every pair (x_0, x_1) chosen as before there would be such a curve, passing through $x_0 \cup x_1$. So there must be a 1-dimensional family of such curves. It can't have base points and its general member is a smooth curve of genus 2.

Conversely, if X is a fibre space over a curve such that the general fibre F has genus 2, then $\mathcal{K}_X^{\otimes 2} | F \cong \mathcal{K}_F^{\otimes 2}$ by the adjunction formula and Lemma III.8.1, so $f_2 | F$ can't be one-to-one, since the bicanonical map of a smooth curve of genus 2 is two-to-one. It follows that f_2 is not birational.

The proof of Theorem 5.2 is complete. □

8. The Exceptional Cases and the 1-Canonical Map

As far as f_n, $n \geq 3$, is concerned, Theorem 5.2 is the best possible in the sense that f_4 is not always birational if $K_X^2 = 1$, that f_3 is not always birational if $K_X^2 \leq 2$ and that f_3 is not always a morphism if $K_X^2 = 1$.

Let us first consider the question of the birationality of f_3 and f_4.

(8.1) **Proposition.** *Let X be a minimal surface of general type with $K_X^2 = 1$ and $p_g(X) = 2$. Then f_3 and f_4 are not birational.*

Proof. Firstly the case of f_4. As usual, we put $|K| = |C| + V$, where $|C|$ is a linear system of dimension 1 without fixed components, and V the fixed part of $|K|$. Then $K(C+V) = 1$. Since by Proposition 2.2 we have $KD \geq 0$ for every effective D, and $KD = 0$ if and only if D is a sum of (-2)-curves, we must have $KC = 1$, $KV = 0$. Furthermore, the general C must be irreducible. From $K(K-C) = 0$ we find by the index theorem that $C^2 \leq 1$, and since $C^2 \geq 0$ and $KC \equiv C^2 \pmod 2$ we must have $C^2 = 1$. So $V^2 = 0$, and since V consists of (-2)-curves, we derive from Theorem III.2.1 that $V = 0$. So K is a pencil without fixed part and one base point, in which two general members meet with multiplicity 1. Consequently, the general canonical curve is an irreducible smooth curve, which is of genus 2 by the adjunction formula. The restriction of $\mathcal{K}_X^{\otimes 4}$ to such a smooth canonical curve C is $\mathcal{K}_C^{\otimes 2}$, and since $\mathcal{K}_C^{\otimes 2}$ yields a two-to-one map, we can conclude that f_4 is not birational.

As to f_3, $|3K|$ can't have fixed components, since $|K|$ has no fixed component. So if f_3 were birational, then f_4 would be birational, since $p_g(X) \neq 0$.

Minimal surfaces X of general type with $K_X^2=1$ and $p_g(X)=2$ do exist; they are extremal with respect to Noether's inequalities, and as such will appear in Sect. 10 below.

Remark. An alternative way of proving Proposition 8.1 is the following. It is not difficult to show that the canonical ring $R(X)$ of X is generated by four elements x_0, x_1, y, z, with x_0, x_1 spanning $\Gamma(\mathcal{K}_X)$, and x_0^2, x_0x_1, x_1^2, y spanning $\Gamma(\mathcal{K}_X^{\otimes 2})$, whereas $\Gamma(\mathcal{K}_X^{\otimes 5})$ is spanned by x_0^5, $x_0^4x_1$, $x_0^3x_1^2$, $x_0^2x_1^3$, $x_0x_1^4$, x_1^5, x_0^3y, $x_0^2x_1y$, $x_0x_1^2y$, x_1^3y, x_0y^2, x_1y^2 and z. If we give x_0, x_1 degree 1, y degree 2 and z degree 5, then it can be shown that there is exactly one polynomial $f(x_0, x_1, y, z)$ of total degree 10 such that $R(X) \cong \mathbb{C}[x_0, x_1, y, z]/f(x_0, x_1, y, z)$. We must have $R(X) \neq R^2(X)$, $R(X) \neq R^3(X)$, so in particular f_2 and f_3 can't be birational.

(8.2) **Proposition.** *Let X be a minimal surface of general type with $K_X^2=2$ and $p_g(X)=3$. Then*

(i) *$f_n, n \geq 1$, is a holomorphic map, and f_1 maps X generically 2-to-1 onto $\mathbb{P}_2$;*
(ii) *f_3 is a morphism of degree 2.*

We have met surfaces X with $K_X^2=2$, $p_g(X)=3$ before as 2-fold coverings of $\mathbb{P}_2$ ramified over a curve of degree 6 with at most simple singularities. They also form the simplest example of Horikawa surfaces with K_X^2 even (see Sect. 10).

Proof of Proposition 8.2, (i). We start by showing that $|K|$ has no fixed component. So let again $|K|=|C|+V$, where $|C|$ is of dimension 2, without fixed components and where V is the fixed part of $|K|$. Since $K(C+V)=K^2=2$, and $KV \geq 0$, $KC \geq 1$, there are two possibilities: $KC=KV=1$, or $KC=2$, $KV=0$.

Suppose that $KC=KV=1$. Then $K(K-2C)=0$, hence by the index theorem $(K-2C)^2 \leq 0$. Since on the other hand $C^2 \geq 0$, we must have $C^2=0$, which is impossible, for $C^2 \equiv KC \pmod 2$.

Now let us assume that $KC=2$, $KV=0$. From $K(K-C)=0$ and the index theorem we deduce $C^2 \leq 2$. Since $C^2 \geq 0$ and $C^2 \equiv KC \pmod 2$, we must have $C^2=2$ or $C^2=0$. In the first of these cases we have that $K-C$ represents $0 \in H_2(X, \mathbb{Q})$, so V is indeed the zero divisor. It remains to exclude the second possibility, i.e. $C^2=0$. To do this, we start by observing that the general member of $|C|$ can't be irreducible, for otherwise the dimension of $|C|$ would be at most 1. But the general member of $|C|$ can't be reducible either, for then $|C|$ would be composed with a pencil $|D|$, i.e. C would be homologous to aD, $a \geq 2$, and the only possibility left would be that a general member of $|C|$ would consist of two components, say C_1 and C_2, both homologous to D, so that KD would be 1, whereas on the other hand $KD \equiv D^2 \pmod 2$, and $D^2 = \frac{1}{a^2} C^2 = 0$.

So $|K|$ has no fixed part. Furthermore, the general member of $|K|$ is irreducible, since $|K|$ can't be composed with a pencil, for $K^2=2$. For the same reason $|K|$ has at most one base point. But if $|K|$ had one, then f_1 would be a birational map onto $\mathbb{P}_2$, which is impossible. So f_1 has no base points, in other words it is a morphism onto $\mathbb{P}_2$. The fact that $K^2=2$ means that this morphism has degree 2.

(ii) Let s: $\mathbb{P}_2 \xrightarrow{\sim} Q \subset \mathbb{P}_9$ be a Segre embedding of $\mathbb{P}_2$ by $\mathcal{O}_{\mathbb{P}_2}(3)$. Then $(sf_1)^*(\Gamma(\mathcal{O}_Q(1)))$ is a linear subsystem of $\Gamma(\mathcal{K}_X^{\otimes 3})$. But both systems have dimension 10, hence sf_1 is some f_3. □

Remark. Exactly as in the preceding case we can find an alternative proof by showing that

$$R(X) \cong \mathbb{C}[x_0, x_1, x_2, y]/g(x_0, x_1, x_2, y),$$

where degree $x_i=1$, degree $y=2$ and g has total degree 8. Then f_i comes from the inclusion $\mathbb{C}[x_0, x_1, x_2] \subset R(X)$. Since the respective quotient fields yield an extension of degree 2, we find that f_1 is indeed of degree 2. The argument for f_3 is similar.

The two classses of exceptions we just produced are the only ones. In fact we have

(8.3) **Theorem.** *Let X be a minimal surface of general type. If $n \geq 3$, then f_n is birational, except in the following cases:*

(i) $n=4$, $K_X^2=1$, $p_g(X)=2$;

(ii) $n=3$, $K_X^2=2$, $p_g(X)=3$ *and* $K_X^2=1$, $p_g(X)=2$.

Because of Noether's inequality, one has only to consider the cases $p_g(X) \leq 3$ if $K_X^2=2$ and $p_g(X) \leq 2$ if $K_X^2=1$. From these cases all remaining possibilities with $p_g(X)=1$ or 2 were treated by Bombieri in [Bom1], whereas Miyaoka proved the result for the case $K_X^2=1$, $p_g(X)=0$. Finally, in [B-C1] and [B-C2] Bombieri and Catanese settled the cases $K_X^2=1$, $p_g(X)=0$ for f_4 and $K_X^2=2$, $p_g(X)=0$ for f_3.

As to the 2-canonical map, Theorem 5.2, (iv) implies that f_2 is birational except possibly for the members of *finitely many* families of minimal surfaces of general type, for it immediately follows from Gieseker's theorem and Theorem 1.1 that there is only a finite number of families satisfying $K^2 <$ constant. Bombieri has shown that Theorem 5.2 is sharp in the following sense: there exists a surface X with $c_1^2(X)=9$, $p_g(X)=6$ for which f_2 is not birational (see [Bom2], p. 193). We understand that recently Francia has obtained further results on f_2 for the remaining cases.

Finally, a few words about the 1-canonical map. It has been studied systematically for the first time by Beauville in [Be2]. Before this paper appeared, only little was known (in particular Castelnuovo's inequality, see below).

The main results of [Be2] are contained in the following

(8.4) **Theorem.** *Let X be a minimal surface of general type with $p_g(X)\geq 2$. Then there are the following possibilities for the (in general: rational) 1-canonical map f_1.*

A. f_1 is composed with a pencil of curves of genus g. If $\chi(\mathcal{O}_X)\geq 21$, then $2\leq g\leq 5$ and the pencil has no base points.

B. $\dim f_1(X)=2$.

Let Y be a minimal desingularisation of $f_1(X)$

1) *$p_g(Y)=0$. If $\chi(\mathcal{O}_X)\geq 31$, then* $\deg(f_1)\leq 9$.
2) *Y is of general type with $p_g(Y)\geq 4$, and the projection from Y onto $f_1(X)$ is a canonical map for Y. If $\chi(\mathcal{O}_X)\geq 14$, then* $\deg(f_1)\leq 3$.

This theorem (the proof of which rests heavily on the inequality $c_1^2\leq 3c_2$) should be seen in the light of the following facts.

a) Gieseker's theorem and Theorem 1.1 imply that there is only a finite number of families of minimal surfaces of general type, satisfying $\chi(\mathcal{O}_X)<$constant. So A says in particular that the fibre genus g *always* satisfies $2\leq g\leq 5$, except perhaps for the members of finitely many families. A similar remark applies to B.

b) Beauville has constructed minimal surfaces X, belonging to class A, with $g=2$ and 3, and $\chi(\mathcal{O}_X)$ arbitrarily large.

c) There exist surfaces belonging to class B 1) with $\chi(\mathcal{O}_X)$ arbitrarily large, for which f_1 has degree 8 (and similarly for $\deg(f_1)=2$ and 6). Every surface with $p_g=0$ appears at least once as a surface Y in this class.

d) There are few examples known of case B 2). In fact, Beauville gives only one, a double covering of a quintic with twenty nodes in $\mathbb{P}_3$.

Another important result, appearing in [Be2] is

(8.5) **Theorem** (Castelnuovo's second inequality). *Let X be a minimal surface of general type, satisfying $K_X^2<3p_g(X)-7$. Then f_1 is a (rational) map of degree 2 onto a ruled surface.*

Castelnuovo proved $K_X^2\geq 3p_g(X)-7$, assuming that K_X is very ample.

Surfaces with Given Chern Numbers

9. The Geography of Chern Numbers

In IV, Sect. 7 we have seen that, given any ordered pair (n,m) of integers, with $n+m\equiv 0(12)$, there always exists a compact *almost complex* manifold X of real dimension 4 with $c_1^2(X)=n$, $c_2(X)=m$. Combining the results of VI, Sect. 1 with Theorem 1.1, we see that the same is not true if we require X to be a surface. In fact, since blowing up a point increases c_2 by 1 and decreases c_1^2 by 1, we see that no pairs (n,m) (with $n+m\equiv 0(12)$), satisfying $n>\max(2m,3m)$, can be represented by a surface.

We now want to consider the Chern numbers for minimal surfaces of general type more in detail. In the light of Theorem 1.1, our first question is the following one. *If* $D \subset \mathbb{Z}^2(n,m)$ *is given by* $n>0$, $n \leq 3m$ *and* $m \leq 5n+36$, *then for which pairs* $(n,m) \in D$ *does there exist a minimal surface of general type* X *with* $c_1^2(X)=n$, $c_2(X)=m$?

We divide D into two parts D_1 and D_2 given respectively by $n \leq 2m$ and $n>2m$. The simplest examples, like complete intersections or double coverings of $\mathbb{P}_2$, practically always yield a point in D_1. Indeed, for a long time only few examples were known of surfaces X with their Chern pairs $(c_1^2(X), c_2(X))$ in D_2.

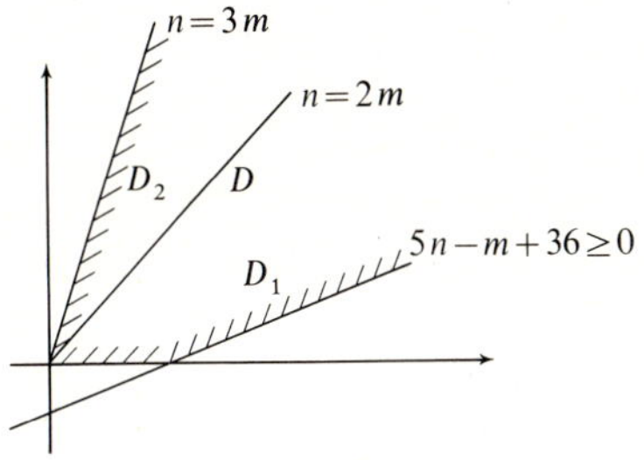

But, trivial as it is to find lots of Chern pairs in D_1, it is a different matter to show that *all* pairs in D_1 can be represented. In fact, this has not been accomplished completely, though the following result of Persson ([Per], Theorem 2) settles the question for most pairs in D_1.

(9.1) **Theorem.** *Given any pair* $(n,m) \in D_1$ *with* $n+m \equiv 0(12)$, *then there exists a minimal surface of general type* X *with* $c_1^2(X)=n$, $c_2(X)=m$, *except maybe if* $n-2m+3k=0$, *where* k *is either* 2, *or* 19, *or odd with* $1 \leq k \leq 15$.

Persson shows even more, namely that in all cases covered by his theorem, it is possible to find a representing surface, admitting a fibration of genus 2, so a very special surface of general type. Nobody doubts that in most cases there are many other types of surfaces representing the point of D_1 in question.

As to D_2, we have already observed that until some years ago only few examples were known, namely quotients of the unit ball (V, Sect. 20) and Kodaira fibrations (V, Sect. 14). In the last few years many more examples have been found, in particular by Mumford (V, Sect. 1), Holzapfel, Livné, Inoue, Miyaoka, Mostow-Siu ([M-Si]), Mostow ([Mo3]), Hirzebruch and Sommese. Holzapfel ([Hol1], [Hol2]) obtains his examples in a way similar to the construction of Hilbert modular surfaces (V, Sect. 21), namely by taking certain quotients of the unit ball by arithmetical groups and compactifying them. Livné ([Liv]) and Inoue ([In4]) use branched coverings of special elliptic surfaces. Miyaoka ([Mi4]) starts from sufficiently general ramified coverings of $\mathbb{P}_2$ and takes the Galois closure. Hirzebruch constructs examples in [Hir7] as minimal models of certain Kummer extensions of the rational function field in two variables, so these examples appear as ramified coverings of blown-ups of

$\mathbb{P}_2$. He e.g. obtains $2\frac{1}{2}$ as an accumulation point for the values c_1^2/c_2 occurring in his examples. And in [Hir8] he constructs more examples along the same lines, but now starting from a special 2-torus instead of $\mathbb{P}_2$. Finally, in [So2] Sommese shows that the set $\left\{\frac{c_1^2(S)}{c_2(S)}\right\}$ is dense in $[\frac{1}{5}, 3]$ if S runs through the minimal surfaces of general type.

Apparently, no example has ever been published of a *simply-connected* minimal surface X of general type with $(c_1^2(X), c_2(X)) \in D_2$, and the question whether or not $\pi_1(X)$ is infinite for every X representing a point in D_2 remains intriguing. (Compare [Mi4], [Hol1], and for the question which points of D_1 can be represented by a simply-connected surface, see also [Per].)

Quite a lot is known about the surfaces on or near the border of D.

As we have already seen in V, Sect. 20, there are infinitely many examples of surfaces of general type X with $c_1^2(X) = 3c_2(X)$, namely quotients of the unit ball $B \in \mathbb{C}^2$. Conversely, it follows from Corollary I.15.5 that a surface of general type X with $c_1^2(X) = 3c_2(X)$ (which is necessarily minimal) is a quotient of B as soon as it admits a Kähler-Einstein metric. According to Theorem I.15.2 this is the case if $\mathcal{K}_X$ is ample, i.e. if X does not contain any (-2)-curves (Theorem 5.2). Now by a recent result of Miyaoka ([Mi5]) surfaces of general type X with $c_1^2(X) = 3c_2(X)$ never contain (-2)-curves; so all these surfaces are quotients of B. Since by [C-V] every small deformation of a compact smooth quotient of B is trivial, this implies that for all $(3c_2, c_2)$ the Gieseker scheme is finite.

Apart from Mumford's surface, some of the examples X constructed by Livné, Inoue and Hirzebruch also satisfy $c_1^2(X) = 3c_2(X)$, so a fortiori their universal covering space is the unit ball.

As to the lower border, i.e. as to those surfaces X for which the Noether inequalities are equalities, they have been classified by Horikawa ([Hor2], [Hor1]). His results form the subject of the next section.

10. Surfaces on the Noether Lines

On a Hirzebruch surface Σ_n (V, Sect. 4) every divisor D is homologous (even linearly equivalent) to $aC_n + bF$ (C_n a section with $C_n^2 = -n$, F a fibre). In the sequel we shall call the ordered pair (a, b) the *type* of D.

In [Hor2] Horikawa has proved

(10.1) **Theorem.** *Let X be a minimal surface of general type with $c_1^2(X)$ even and $p_g(X) = \frac{1}{2}c_1^2(X) + 2$. Then the 1-canonical map f_1 is a holomorphic map of degree 2 onto a 2-dimensional variety Y of degree $p_g - 2$ in $\mathbb{P}_{p_g - 1}$. The minimal resolution Y' of Y is either $\mathbb{P}_2$ or a Hirzebruch surface, and there exists a 2-fold* *branched covering g': $X' \to Y'$, branched along a curve of type (a,b), with simple singularities only, such that X is the minimal resolution of X' and the diagram*

$$\begin{array}{ccc} X & \xrightarrow[\text{res.}]{\text{min.}} & X' \\ {\scriptstyle f_1}\downarrow & & \downarrow{\scriptstyle g'} \\ Y & \xleftarrow[\text{res.}]{\text{min.}} & Y' \end{array}$$

is commutative. For each fixed value of $c_1^2(X)=k$ there is a finite number of possibilities for Y' and a finite number of possibilities for (a,b). Conversely, starting from any of these Y' and any of these (a,b), the minimal resolution X of a 2-fold covering $X' \to Y'$ over a curve of type (a,b) with simple singularities only, is a minimal surface of general type with $c_1^2(X)=k$, $p_g(X)=\frac{1}{2}c_1^2(X)+2$.

All possibilities for Y' and (a,b) are listed in the following table.

Table 11

c_1^2	p_g	Minimal resolution of the canonical model (= image of f_1)	Type of branch locus	Number of deformation classes
2	3	$\mathbb{P}_2$	8	one class
4	4	Σ_0	(6, 5)	one class
		Σ_2	(6, 12)	
6	5	Σ_1	(6, 10)	one class
		Σ_3	(6, 16)	
8	6	$\mathbb{P}_2$	10	two classes
		Σ_4	(6, 20)	
		Σ_0	(6, 8)	
		Σ_2	(6, 14)	
$8k+2$	$4k+3$ $(k \geq 1)$	Σ_{2j+1} $(0 \leq j \leq k+1)$	$(6, 4k+6j+8)$	one class
$8k-4$	$4k$ $(k \geq 2)$	Σ_{2j} $(0 \leq j \leq k)$	$(6, 4k+6j+2)$	one class
$8k-2$	$4k+1$ $(k \geq 2)$	Σ_{2j+1} $(0 \leq j \leq k)$	$(6, 4k+6j+6)$	one class
$8k$	$4k+2$ $(k \geq 2)$	Σ_{2j} $(0 \leq j \leq k)$	$(6, 4k+6j+4)$	two classes
		Σ_{2k+2}	$(6, 16k+16)$	

For surfaces X with $c_1^2(X)$ odd and $p_g(X)=\frac{1}{2}c_1^2(X)+\frac{3}{2}$, Horikawa's results are quite similar. Since however the situation is more complicated for a few low values of $c_1^2(X)$, we don't formulate a theorem, but restrict ourselves to a (slightly less precise) description of the different possibilities, referring to [Hor2] for details.

(i) $c_1^2(X)=1$. Then $p_g(X)=2$, and $|K|$ is a pencil with one base point $x_0 \in X$. If $(\bar{X}, E) \to (X, x_0)$ is the blowing-up of X at x_0, then the rational map induced by f_2 on $\bar{X} \setminus E$ can be extended to a morphism $\bar{f}_2$ of degree 2 from $\bar{X}$ onto a quadratic cone $Y \subset \mathbb{P}_3$. Furthermore, there exists a 2-fold branched covering g: $X' \to \Sigma_2$, branched over a curve of type (6, 10) with simple singularities, such that the diagram

$$\begin{array}{ccc} \bar{X} & \xrightarrow[\text{res.}]{\text{min.}} & X \\ {\scriptstyle \bar{f}_2}\downarrow & & \downarrow{\scriptstyle g} \\ Y & \xleftarrow[\text{res.}]{\text{min.}} & \Sigma_2 \end{array}$$

is commutative. Every curve of type (6, 10) with at most simple singularities occurs in this way.

(ii) $c_1^2(X)=3$. There are two types:

Type I. f_1 is a holomorphic map of degree 3 onto $\mathbb{P}_2$, and X is the minimal resolution of a 3-section in $\mathcal{O}_{\mathbb{P}_2}(2)$. The 3-sections thus occurring can be listed.

Type II. $|K|$ has one base point, and after blowing up this point f_1 extends to a map $\bar{f}_1 \colon \bar{X} \to \mathbb{P}_2$ of degree 2. There is again a commutative diagram

$$\begin{array}{ccc} \bar{X} & \xrightarrow[\text{res.}]{\text{min.}} & \bar{X}' \\ {\scriptstyle \bar{f}_1}\downarrow & \swarrow{\scriptstyle g} & \\ \mathbb{P}_2 & & \end{array}$$

where the 2-fold covering $g \colon \bar{X}' \to \mathbb{P}_2$ is ramified over a curve of degree 10, decomposing into a line L and a curve of degree 9, having three simple triple points on L. Starting from such a curve the process can be reversed to obtain a minimal surface of general type X with $c_1^2(X)=3$ and $p_g(X)=3$.

The types I and II form a single deformation class. Horikawa's proof in [Hor2] can be simplified if we use another way of description. Namely, it follows from his considerations that for a surface X of type I

$$R(X)=\mathbb{C}[x_0, x_1, x_2, y]/f(x_0, x_1, x_2, y)$$

with $\deg(x_i)=1$, $\deg(y)=2$ and $\deg(f)=6$, whereas conversely a surface X with such a canonical ring is of type I, provided f is "generic".

On the other hand, as V. Iliev has shown, for a surface X of type II

$$R(X)=\mathbb{C}[x_0, x_1, x_2, y, z]/(f(x_0, x_1, x_2, y, z), g(x_0, x_1, x_2, y))$$

with $\deg(x_i)=1$, $\deg(y)=2$, $\deg(z)=3$, $\deg(f)=6$, $\deg(g)=3$, while the converse is again true for "generic" f and g.

Since replacing g by $g+tz$, $t \neq 0$, in the above ring yields a ring of the form

$$\mathbb{C}[x_0, x_1, x_2, y]/f(x_0, x_1, x_2, y)$$

this shows that a type-II surface deformes into a surface of type I.

(iii) $c_1^2(X)=5$. Again there are two types, forming one deformation class (see [Hor1]).

Type I. f_1 is a C-morphism (Definition 5.1) onto a quintic in $\mathbb{P}_3$ with at most rational double points. Conversely, the minimal desingularisation of such a quintic is an X with $c_1^2(X)=5$, $p_g(X)=4$.

Type II. $|K|$ has one base point, and after blowing it up, f_1 extends to a holomorphic map of degree 2 onto either a smooth quadric or a quadratic cone in $\mathbb{P}_3$. As before $\bar{X}$ is the minimal resolution of a 2-fold covering over either Σ_0 or Σ_2, the branch locus respectively being a curve of degree $(6,7)$ and $(6,13)$ with only simple singularities. Again all such curves occur.

(iv) $c_1^2(X)\geq 7$. $|K|$ has one base point, and after blowing up this point f_1 extends to a map of degree 2 onto either $\mathbb{P}_2$ or a Σ_d. From this point on the situation is analogous to the one in Theorem 10.1, but for the fact that in a well-determined way some non-simple singularities appear on the branch locus. To be more precise: at most two quadruple points appear. If one appears, after blowing up once, the proper transform has simple singularities only; if two of them appear, the same is to hold, except that the two quadruple points may become "infinitely near", i.e. they may coalesce to an octuple point which splits into two quadruple points after one blowing up. Again, all curves of this type appear. See Table 12 below.

Remark 1. All the surfaces X which we just have described, and which are known as Horikawa-surfaces, are simply-connected ([Hor2], [Hor1]).

Remark 2. The simplest case with $c_1^2(X)$ even, namely $c_1^2(X)=2$, occurred already before, in Proposition 8.2. It was well known long before Horikawa treated the general case ([Saf], Chap. VI, § 3).

Table 12

c_1^2	p_g	Minimal resolution of the canonical model	Type of branch locus	Singularities besides simple ones	Number of deformation types
7	5	Σ_1	$(8,10)$	one quadruple point	two
		Σ_3	$(6,17)$	two quadruple points on same fibre	
		Σ_1	$(6,11)$		
9	6	Σ_2	$(6,14)$	none	two
		Σ_0	$(6,9)$	two quadruple points on same fibre	
		Σ_3	$(6,15)$		
$8k+3$	$4k+3$ $(k\geq 1)$	Σ_{2j+1} $(0\leq j\leq k+1)$	$(6,4k+6j+9)$	two quadruple points on same fibre	one
$8k-3$	$4k$ $(k\geq 2)$	Σ_{2j} $(0\leq j\leq k)$	$(6,4k+6j+3)$		one
$8k-1$	$4k+1$ $(k\geq 2)$	Σ_{2j+1} $(0\leq j\leq k-1)$	$(6,4k+6j+7)$		two
		Σ_{2k+1}	$(6,10k+7)$		
$8k+1$	$4k+2$ $(k\geq 3)$	Σ_{2j} $(0\leq j\leq k)$	$(6,4k+6j+8)$		one

11. Surfaces with $q=p_g=0$

The extremal points on the left hand side of D are the points on the line $c_1^2(X)+c_2(X)=12$, i.e. the points $c_1^2(X)=a$, $c_2(X)=12-a$, with $1\le a\le 9$. All these points can already be represented by surfaces X with $p_g(X)=q(X)=0$ (if $c_1^2(X)=1$, then automatically $p_g(X)=q(X)=0$, since otherwise application of the unbranched covering trick would lead to a violation of Noether's inequality, but if $c_1^2(X)\ge 2$, this is no longer true). Such surfaces of general type have been studied for a long time, for they are very interesting from the point of view of Castelnuovo's criterion (Theorem VI.2.1): if $q(Y)=0$ for the non-rational surface Y, then $P_2(Y)\ge 1$. Nowadays a large number of examples is known; they have been obtained by classical constructions, due to Godeaux, Campedelli, Burniat, as well as by some new methods. The fact that so many examples are known does by no means imply that the moduli scheme is known too. Still, some general results have been proved. If $c_1^2(X)=1$, then $H_1(X,\mathbb{Z})$ is either 0, $\mathbb{Z}_2$, $\mathbb{Z}_3$, $\mathbb{Z}_4$ or $\mathbb{Z}_5$ (see [Rei2], [Mi1]), and if $c_1^2(X)=2$, $p_g(X)=q(X)=0$, then $|H_1(X,\mathbb{Z})|\le 10$ ([Be2]). And there is a complete classification with description of moduli spaces for the cases $c_1^2(X)=1$, $H_1(X,\mathbb{Z})=\mathbb{Z}_3$, $\mathbb{Z}_4$ or $\mathbb{Z}_5$ ([Rei2], [Mi1]) and $c_1^2(X)=2$, $\pi_1(X)=\mathbb{Z}_2+\mathbb{Z}_2+\mathbb{Z}_2$, $\mathbb{Z}_8$, $\mathbb{Z}_2\oplus\mathbb{Z}_4$ or the quaternion group of order 8 (see [Rei3]).

We shall now list several of the aforementioned constructions. For more details we have to refer to the original papers (see [Rei3]) or to Dolgacev's excellent survey [Do], which however covers only results up to 1977.

1. The Godeaux Construction

We have met this construction already in V, Sect. 16. The idea is to start with a minimal surface of general type Y with $\chi(Y)=a$, on which a group G of order a operates without fixed points. Then by Corollary 2.3 the quotient will be a minimal surface of general type with $\chi(X)=1$, and if moreover there are no holomorphic 2-forms on Y, left invariant by G, then $p_g(X)=q(X)=0$. Apart from a smooth quintic in $\mathbb{P}_3$, the construction can also be applied to an intersection of four quadrics in $\mathbb{P}_6$, with G one of the groups $\mathbb{Z}_2+\mathbb{Z}_2+\mathbb{Z}_2$, $\mathbb{Z}_2+\mathbb{Z}_4$, $\mathbb{Z}_8$ or the quaternion group. The quotient X has $c_1^2(X)=2$, and $p_g(X)=q(X)=0$. And there exist also more complicated examples, where G is $\mathbb{Z}_3$ or $\mathbb{Z}_4$ and $c_1^2(X)=1$, $p_g(X)=q(X)=0$ (see [Rei2] and [Mi1]).

2. The Campedelli Construction

The imposition of non-simple singularities on the branch curve of a 2-fold covering diminishes the values of c_1^2 and c_2 (compare V, Sect. 22). Starting from any reduced curve of degree 10 in $\mathbb{P}_2$, having six triple points with infinitely near ordinary triple points (i.e. after blowing up once the triple points become ordinary triple points) one can thus construct a surface X with $c_1^2(X)=2$, $p_g(X)=q(X)=0$. Campedelli ([Cam]) gave examples of such curves; his double planes are special Godeaux-type surfaces ([Pet1]). More surfaces with $c_1^2=2$ or 1 arise if certain other singularities occur on the branch locus, see for instance [O-P].

3. The Burniat Construction

Starting point is a configuration, formed by nine different lines A_0, A_1, A_2, B_0, B_1, B_2, C_0, C_1, C_2 in $\mathbb{P}_2$.

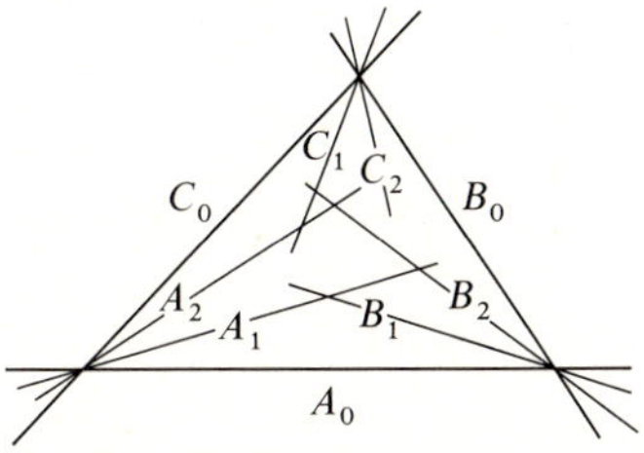

The lines A_0, B_0, C_0 form a non-degenerate triangle Δ, whereas A_1, A_2 both pass through a same vertex of Δ; B_1, B_2 both pass through a second vertex and C_1, C_2 both pass through the third vertex of Δ. The surface in question X is obtained by first taking the double covering X_1 of $\mathbb{P}_2$, branched over $A_0 + A_1 + A_2 + B_0 + B_1 + B_2$, then taking the double covering X_2 of X_1, branched over the inverse image of $C_0 + C_1 + C_2$ (of course, one has to verify that it exists), and finally desingularising X_2 in a minimal way. If m is the number of points through which there passes a line A_i, a line B_j and a line C_k, $1 \le i, j, k \le 2$ (so $0 \le m \le 4$), then $c_1^2(X) = 6 - m$, whereas $p_g(X) = q(X) = 0$. For further details we refer to [Bu] and [Pet2].

4. The Inoue Construction

Let $Z = E_1 \times E_2 \times E_3$ be the product of three 1-dimensional abelian varieties $E_i = \mathbb{C}/\mathbb{Z}1 + \mathbb{Z}\tau_i$, $i = 1, 2, 3$, and Y an irreducible hypersurface on X with the following properties:

(i) Y is of type (2, 2, 2), i.e. if $\rho_i: Z \to E_i$ is the projection, then $\mathcal{O}_Z(Y) \cong \prod_{i=1}^{3} \rho_i^*(\mathcal{L}_i)$, where $\mathcal{L}_i$ is a line bundle of degree 2;

(ii) $\mathcal{O}_Z(Y)$ is very ample;

(iii) as singularities Y has only n nodes, each of them in one of the 2-division points of Z;

(iv) Y is left invariant by the group G of automorphisms of Z, generated by

$$(z_1, z_2, z_3) \to (-z_1 + \tfrac{1}{2}, z_2 + \tfrac{1}{2}, z_3)$$
$$(z_1, z_2, z_3) \to (z_1, -z_2 + \tfrac{1}{2}, z_3 + \tfrac{1}{2})$$
$$(z_1, z_2, z_3) \to (z_1 + \tfrac{1}{2}, z_2, -z_3 + \tfrac{1}{2});$$

(v) G acts without fixed points on Y', the minimal desingularisation of Y.

These requirements imply that $n \equiv 0 \pmod 4$. The quotient $X = Y'/G$ has the invariants $p_g(X) = q(X) = 0$, $c_1^2(X) = 6 - \frac{n}{4}$. In [In3] Inoue constructs a family Y_c, depending on a parameter $c \in \mathbb{C}$, satisfying (i)–(v), such that $n = 0$ for general c, but $n=4$, 8, 12 or 16 for special values of c. Upon replacing one of the curves E_i by a curve of genus 5, embedded in the product of two elliptic curves, he furthermore constructs a surface on which $\overset{5}{\bigoplus}\mathbb{Z}_2$ acts freely; the quotient X has $c_1^2(X)=7$ and $p_g(X)=q(X)=0$. While the first four examples can be shown to be Burniat surfaces, the last oen is a new surface.

5. The Beauville Construction

Let C_1, C_2 be two smooth connected compact curves of genus p_1, p_2 respectively, with $p_1, p_2 \geq 2$. Let $Y=C_1 \times C_2$ and G a finite group of order $(p_1-1)(p_2-1)$, operating on both C_1 and C_2, such that, but for the unit element, no element of G has a fixpoint on both C_1 and C_2. This implies that the induced action of G on Y is fixpoint-free. So if $X=Y/G$, then $c_1^2(X)=8$, and if moreover $C_i/G \simeq \mathbb{P}_1$, then $p_g(X)=q(X)=0$. We refer to [Be1], p. 159 and [Do] for explicit examples.

6. The Catanese Construction

Let $Q \subset \mathbb{P}_3$ be a quintic with exactly twenty ordinary double points, such that the following conditions are satisfied:

(i) the group $\mathbb{Z}_5$ acts freely on Q, and hence freely on the minimal desingularisation $\tilde{Q}$ of Q;

(ii) the line bundle corresponding to the sum of the twenty exceptional (-2)-curves on $\tilde{Q}$ is 2-divisible, so that there is a double covering $\tilde{Y}$ of $\tilde{Q}$, branched over the exceptional locus;

(iii) the $\mathbb{Z}_5$-action on $\tilde{Q}$ can be lifted to $\tilde{Y}$.

If Y is obtained from $\tilde{Y}$ by blowing down the (-1)-curves lying over the twenty (-2)-curves, then $\mathbb{Z}_5$ still acts freely on Y. The quotient $X=Y/\mathbb{Z}_5$ turns out to be minimal with $c_1^2(X)=2$, $p_g(X)=q(X)=0$. Catanese has produced in [Cat1] a 4-dimensional family of quintics satisfying the conditions above. His surfaces Y are simply-connected, as follows from the fact that one of them is the minimal model of a Hilbert modular surface (of a type more general then those described in Chap. V).

7. The Barlow Construction

Suppose we are given a simply-connected surface Y with $c_1^2(Y)=10$, $p_g(Y)=4$, admitting an action of the dihedral group $\mathbb{D}_5$, such that the normal subgroup $\mathbb{Z}_5$ acts freely, whereas each of the five conjugate involutions in $\mathbb{D}_5$ has exactly four isolated fixpoints. Then the desingularisation X of the quotient $Y/\mathbb{D}_5$ is simply-connected and satisfies $c_1^2(X)=1$, $p_g(X)=q(X)=0$. In [Bw] Barlow has constructed a 2-dimensional family of such Y's.

In the following table we have collected many examples of surfaces with $p_g=q=0$, giving for each type π_1, H_1 (in as far as they have been determined), at least one method of construction, and one or two recent references. We have added the little information about moduli, known to us in October 1982.

Table 13. Surfaces with $p_g = q = 0$

c_1^2	π_1	H_1	Method of construction	Moduli scheme	Reference
1	0	0	no 7		[Bw]
	$\mathbb{Z}_2$	$\mathbb{Z}_2$	Barlow & Reid		[Bw]
	$\mathbb{Z}_2$	$\mathbb{Z}_2$	quotient of no 4 with $c_1^2 = 5, 3$		[In3]
		$\mathbb{Z}_2$	no 2		[O-P]
		$\mathbb{Z}_3$	no 1	irreducible	[Rei2]
	$\mathbb{Z}_4$	$\mathbb{Z}_4$	no 1	irreducible	[Rei2]
	$\mathbb{Z}_5$	$\mathbb{Z}_5$	no 1	unirational, dim $= 8$	[Rei2] [Mi1]
2	$\mathbb{Z}_8$	$\mathbb{Z}_8$	no 1	unirational, dim $= 6$	[Rei3]
	$\mathbb{Z}_4 \oplus \mathbb{Z}_2$	$\mathbb{Z}_4 \oplus \mathbb{Z}_2$	no 1	unirational, dim $= 6$	[Rei3]
	$\overset{3}{\bigoplus} \mathbb{Z}_2$	$\overset{3}{\bigoplus} \mathbb{Z}_2$	no 1, no 2	unirational, dim $= 6$	[Rei3], [Pet1]
	$\mathbb{H}$†	$\overset{2}{\bigoplus} \mathbb{Z}_2$	no 1	unirational, dim $= 6$	[Rei3]
	$\mathbb{H}$	$\overset{2}{\bigoplus} \mathbb{Z}_2$	no 3, no 4		[In3], [Pet2]
	$\overset{2}{\bigoplus} \mathbb{Z}_2$	$\overset{2}{\bigoplus} \mathbb{Z}_2$	quotient of no 4		[In3]
	$\mathbb{Z}_5$	$\mathbb{Z}_5$	no 6		[Cat1]
			no 2		[Pet1]
3	$\mathbb{H} \oplus \mathbb{Z}_2$	$\overset{3}{\bigoplus} \mathbb{Z}_2$	no 3, no 4		[In3], [Pet2]
4	$\mathbb{H} \oplus \overset{2}{\bigoplus} \mathbb{Z}_2$	$\overset{4}{\bigoplus} \mathbb{Z}_2$	no 3, no 4		[In3], [Pet2]
5	$\mathbb{H} \oplus \overset{3}{\bigoplus} \mathbb{Z}_2$	$\overset{5}{\bigoplus} \mathbb{Z}_2$	no 3, no 4		[In3], [Pet2]
6	Extension of $\overset{3}{\bigoplus} \mathbb{Z}_2$ by $\mathbb{Z}^6$	$\overset{6}{\bigoplus} \mathbb{Z}_2$	no 3, no 4		[In3], [Pet2]
7			no 4		[In3]
8	infinite	?	number theory (many examples) (Chap. V, Sect. 20)		[Kug]
	infinite	?	no 5 (several examples)		[Be1]
9	infinite	?	p-adic geometry (Chap. V, Sect. 1)		[Mu6]

† $\mathbb{H} = \{\pm 1, \pm i, \pm j, \pm k\}$ with the usual relations (quaternion group of order 8)

VIII. K3-Surfaces and Enriques Surfaces

What is said about the use of "surface" and "curve" (on a surface) at the beginning of Chap. VI also applies in this chapter.

Introduction

1. Notations

Let X be a compact (connected) surface with $b_1(X)$ even, with cupproduct form $(\ ,\ )$ on $H^2(X,\mathbb{C})$. By Theorem IV.2.13 the signature of $(\ ,\ )|H^{1,1}(X,\mathbb{R})$ is $(1,h^{1,1}-1)$, so if $h^{1,1}(X)\geq 2$ the set $\{x\in H^{1,1}(X,\mathbb{R});\ (x,x)>0\}$ consists of two disjoint connected cones, say $\mathscr{C}_X$ and $\mathscr{C}'_X$. Since the Kähler classes, if present, form a convex subcone of $\mathscr{C}_X\cup\mathscr{C}'_X$, they all belong to one of them, say $\mathscr{C}_X$, the *positive cone*.

Another consequence of the Signature Theorem is found upon using Schwarz' inequality:

(0) $x,y\in\overline{\mathscr{C}}_X$ implies $(x,y)\geq 0$ with strict inequality if either x or y is contained in $\mathscr{C}_X$.

Let j: $H^2(X,\mathbb{Z})\to H^2(X,\mathbb{C})$ be induced by the inclusion, and let

$$S_X=H^{1,1}(X,\mathbb{R})\cap \operatorname{Im} j^*(H^2(X,\mathbb{Z}))\ \text{(the \textit{Picard lattice})},$$
$$T_X=S_X^{\perp}\cap \operatorname{Im} j^*(H^2(X,\mathbb{Z}))\ \text{(the \textit{transcendental lattice})}.$$

We shall give names to some special types of elements in $j^{-1}(S_X)$. So let $d\in j^{-1}(S_X)$. We call d *divisorial* if there exists at least one divisor D with $c_1(\mathcal{O}_X(D))=d$. Then d is called *effective* if moreover D can be chosen effective and similarly for *irreducible*. The classes of the (-2)-curves are called the *nodal classes* in S_X. An effective class which is *not* the sum of two other effective classes is called *indecomposable*. N.B. It may very well happen that an irreducible class is decomposable (for example the class of an irreducible conic in $\mathbb{P}_2$).

If X, X' are surfaces, an isomorphism of $\mathbb{Z}$-modules

$$H^2(X,\mathbb{Z})\to H^2(X',\mathbb{Z})$$

is called a *Hodge-isometry* if

i) it preserves the cupproduct (i.e. it is an isometry), and
ii) its $\mathbb{C}$-linear extension $H^2(X,\mathbb{C}) \to H^2(X',\mathbb{C})$ preserves the Hodge-decomposition (IV, Sect. 2).

If X, X' are moreover Kähler surfaces, a Hodge-isometry is called *effective*, if it preserves the positive cones and induces a bijection between the respective sets of effective classes.

For any $d \in j^{-1}(S_X)$ with $(d,d) = -2$ the isomorphism

$$s_d\colon H^2(X,\mathbb{Z}) \to H^2(X,\mathbb{Z})$$

defined by $s_d(x) = x + (x,d)d$ $(x \in H^2(X,\mathbb{Z}))$ is a Hodge-isometry, because $d \in H^{1,1}$ implies that the $\mathbb{C}$-linear extension of s_d is the identity on $H^{2,0} \oplus H^{0,2}$ and preserves $H^{1,1}$. Its $\mathbb{R}$-linear extension to $H^2(X,\mathbb{R})$ is also denoted by s_d. Observe that $s_d(d) = -d$ and that s_d is just the orthogonal reflection in the hyperplane H_d orthogonal to d. We call s_d a *Picard-Lefschetz reflection.* The subgroup of $\mathrm{Aut}(H^{1,1}(X,\mathbb{R}))$ generated by the Picard-Lefschetz reflections is denoted by W_X. In case of a Kähler surface X we define the *Kähler cone* to be the convex subcone of the positive cone consisting of those elements which have positive inner product with any effective class in S_X. The Kähler cone contains all Kähler classes.

We finally introduce some notation in connection with the relevant period domains. We put

$$\begin{aligned} L &= -E_8 \oplus -E_8 \oplus H \oplus H \oplus H && \text{(cf. also I, Sect. 2),}\\ L_{\mathbb{C}} &= L \otimes \mathbb{C} && \text{with } (\ ,\ ) \text{ extended } \mathbb{C}\text{-bilinearly,}\\ L_{\mathbb{R}} &= L \otimes \mathbb{R} && \text{with } (\ ,\ ) \text{ extended } \mathbb{R}\text{-bilinearly.} \end{aligned}$$

For $\omega \in L_{\mathbb{C}}$ we denote by $[\omega] \in \mathbb{P}(L_{\mathbb{C}})$ the corresponding line and set

$$\Omega = \{[\omega] \in \mathbb{P}(L_{\mathbb{C}});\, (\omega,\omega) = 0, (\omega,\bar{\omega}) > 0\}.$$

2. The Results

By definition a *K3-surface* is a surface X with $\mathscr{K}_X$ trivial and $b_1(X) = 0$. In Sect. 8 we shall show that all K3-surfaces are deformations of each other, so they are all diffeomorphic. In particular, since the K3-surface considered by way of an example in V, Sect. 2 is simply-connected, all K3-surfaces are simply-connected.

In Sect. 3 we shall show that $H^2(X,\mathbb{Z})$ is free of rank 22 and isometric to L. Since $p_g(X) = 1$, the choice of an isometry $\phi\colon H^2(X,\mathbb{Z}) \to L$ determines a line in $L_{\mathbb{C}}$ spanned by the $\phi_{\mathbb{C}}$-image of a nowhere vanishing holomorphic 2-form ω_X. The identity $(\omega_X, \omega_X) = 0$ and the inequality $(\omega_X, \bar{\omega}_X) > 0$ imply that this line considered as a point of $\mathbb{P}(L_{\mathbb{C}})$, belongs to Ω (compare also IV, Sect. 3). This point is called the *period point* of the marked K3-surface (X, ϕ).

The first main theorem states that two Kähler K3-surfaces*) are isomorphic if and only if there are markings for them, such that the corresponding

*) We shall show in Sect. 14 that every K3-surface is kählerian.

period points are the same. This is the *Weak Torelli Theorem*, which can also be stated as follows: *two Kähler K3-surfaces X, X' are isomorphic if there exists a Hodge isometry ϕ: $H^2(X',\mathbb{Z}) \to H^2(X,\mathbb{Z})$.* In this form we shall prove it in Sect. 11. It is a consequence of the more refined *Torelli theorem*, which asserts that *if in addition ϕ is assumed to be effective, then it is induced by a unique isomorphism $X \to X'$.*

The second main theorem states that *all points of Ω occur as period points of marked Kähler K3-surfaces* (Theorem 14.2).

An Enriques surface by definition is a surface Y with $\mathscr{K}_Y^{\otimes 2} = \mathcal{O}_Y$, $p_g(Y) = b_1(Y) = 0$ (compare VI, Sect. 1). We recall from V, Sect. 23 that every Enriques surface is projective. In Sect. 16 we shall show that they always admit at least one elliptic fibration over $\mathbb{P}_1$.

All Enriques surfaces turn out to be deformations of each other (Theorem 18.5), so they are all diffeomorphic. Since for an Enriques surface Y we have $p_g(Y) = 0$, we cannot form the period map. Instead we shall pass to the universal covering X of Y, which is easily shown to be a K3-surface doubly covering Y (see Lemma 15.1). We then simply might take any marking for $H^2(X,\mathbb{Z})$ and consider the period point of X as a period point for Y. However, in doing this, we loose the extra information that X comes from an Enriques surface. So we consider instead only those markings ϕ: $H^2(X,\mathbb{Z}) \xrightarrow{\sim} L$ for which $\phi \circ \sigma^* = \rho \circ \phi$, where σ: $X \to X$ is the involution interchanging the sheets of $X \to Y$ and where ρ is a certain involution, to be defined in Sect. 18.

The period points of the special marked K3's then all belong to the intersection Ω^- of Ω with the linear subspace $\mathbb{P}(L^- \otimes \mathbb{C})$ of $\mathbb{P}(L_{\mathbb{C}})$, where $L^- = \{l \in L; \rho(l) = -l\}$. This intersection consists of two connected manifolds of dimension 10, each of which is isomorphic to a bounded domain of type IV. The discrete group Γ of automorphisms of L commuting with ρ acts properly on Ω^- and the quotient D is a quasi-projective variety, the *period domain*. For a fixed Enriques surface Y all of the previously considered special markings differ by elements of Γ, so they give the same point of D, the *period point* of Y.

The counterpart of the first main theorem for K3-surfaces is valid in the following form: *two Enriques surfaces are isomorphic if and only if they have the same period point* (Theorem 21.2).

It is not true that all points of D occur as period points of Enriques surfaces. We have to delete certain hyperplanes from Ω^-, namely the hyperplanes $H_d = \{[\omega] \in \Omega^-; (\omega, d) = 0\}$ for $d \in L$, $(d,d) = -2$, $\rho(d) = -d$. In Sect. 20 we show that their images in D form finitely many algebraic hypersurfaces, and in Sect. 21 we finally prove that all points of D not lying on those, actually occur as period points for Enriques surfaces.

K3-Surfaces

3. Topological and Analytical Invariants

By Theorem IV.2.6, (ii), we have that $b_1(X) = 0$ is equivalent to $q(X) = 0$. So we have $\chi(X) = 2$ and hence by Noethers formula I, (4) and Theorem I.3.1 we obtain:

(3.1) **Proposition.** $c_1(X)=0$, $c_2(X)=24$, $\tau(X)=-16$ *for any K3-surface X.*

(3.2) **Proposition.** *Let X be a K3-surface. Then*
(i) $H^1(X,\mathbb{Z})=H^3(X,\mathbb{Z})=0$, *and*
(ii) $H^2(X,\mathbb{Z})$ *is torsion-free of rank 22 and, when equipped with the cupproduct pairing, isometric to L.*

Proof. To prove (i) and the fact that $H^2(X,\mathbb{Z})$ is torsion-free it suffices to show that $H_1(X,\mathbb{Z})$ has no torsion. So suppose that $H_1(X,\mathbb{Z})$ has n-torsion. Then there exists a non-trivial unramified covering $Y \to X$ of degree n. Since $\mathcal{K}_Y \cong \mathcal{O}_Y$, it follows that $p_g(Y)=1$. Noether's formula for Y now reads

$$2-2q(Y)=\chi(Y)=n\,\chi(X)=2n,$$

hence $n=1$ and so $H_1(X,\mathbb{Z})$ is torsion-free. Since $c_1(X)\equiv w_2(X)=0 \pmod 2$, by Wu's formula ([M-S], p. 732) the cupproduct form on $H^2(X,\mathbb{Z})$ is even. It is also unimodular (by Poincaré-duality) and indefinite (by Proposition 3.1), so (ii) follows from Theorem I.2.8, for both L and $H^2(X,\mathbb{Z})$ have rank 22 and index -16. □

(3.3) **Proposition.** *Let X be any K3-surface. Then:*

$$h^{0,1}(X)=h^{1,0}(X)=h^{2,1}(X)=h^{1,2}(X)=0,$$
$$h^{0,2}(X)=h^{2,0}(X)=1,$$
$$h^{1,1}(X)=20.$$

Proof. Since $b_1(X)=0$, we have $h^{0,1}(X)=h^{1,0}(X)=0$ by Theorem IV.2.6. This and Serre duality imply that $h^{2,1}(X)=h^{1,2}(X)=0$. Secondly, $h^{2,0}(X)=h^{0,2}(X)=1$, hence $h^{1,1}(X)=20$ in view of the Hodge decomposition (Theorem IV.2.9) and the fact that $b_2(X)=22$. □

(3.4) **Corollary.** *For a K3-surface X we have*

$$h^0(\mathcal{T}_X)=h^2(\mathcal{T}_X)=0 \quad \textit{and} \quad h^1(\mathcal{T}_X)=20.$$

In fact, the isomorphism $\mathcal{K}_X \simeq \mathcal{O}_X$ *defines (via the exterior product) an isomorphism* $\mathcal{T}_X \cong \Omega_X^1$.

Next, we study the Picard lattice of a K3-surface.

(3.5) **Proposition.** *The map* c_1: $\mathrm{Pic}(X) \to H^2(X,\mathbb{Z})$ *is injective, hence maps* Pic(X) *isomorphically onto the Picard lattice. In particular, an effective divisor is never homologous to zero.*

Proof. This follows from the cohomology exponential sequence (I, Sect. 6) since $H^1(\mathcal{O}_X)=0$ by Proposition 3.3.

(3.6) **Proposition.**
(i) If $d \in S_X$, $d \neq 0$, *and* $(d,d) \geq -2$, *then* d *or* $-d$ *is effective.**)
(ii) If d *is irreducible, then* $(d,d) \geq -2$ *and equality holds precisely when* d *is a nodal class.*
(iii) An indecomposable class is irreducible.
(iv) A nodal class is represented by only one effective divisor (the (-2)*-curve) and in particular it is indecomposable.*

Proof. (i) Let $d = c_1(\mathscr{L})$, $\mathscr{L} \in \mathrm{Pic}(X)$. Riemann-Roch yields the inequality

$$h^0(\mathscr{L}) + h^0(\mathscr{L}^\vee) \geq 2 + \tfrac{1}{2}(d,d) \tag{1}$$

and hence $\mathscr{L}$ or $\mathscr{L}^\vee$ has a non-trivial section.

(ii) Let D be an irreducible divisor. Since $\mathscr{K}_X \cong \mathscr{O}_X$ the adjunction formula implies that $D^2 \geq -2$, where equality holds if and only if D is a (-2)-curve.

(iii) By definition.

(iv) Let $d = c_1(\mathscr{O}_X(D)) = c_1(\mathscr{O}_X(D'))$ with D a (-2)-curve and D' effective. Since $DD' = -2$, we have that D is a component of D', hence $D' - D \geq 0$. Now by Proposition 3.5 an effective divisor cannot be homologous to 0 on X, so $D' = D$. □

(3.7) **Proposition.** *The set of effective classes on a Kähler K3-surface is the semigroup generated by the nodal classes and the integral points in the closure of the positive cone.*

Proof. Let d be an irreducible class. Since (,) is even by Proposition 3.2, (ii) we see that Proposition 3.6, (ii) implies that either d is nodal or $(d,d) \geq 0$. In the last case $d \in \bar{\mathscr{C}}_X \cup \bar{\mathscr{C}}'_X$, so $d \in \pm \bar{\mathscr{C}}_X$. Since $(d,\kappa) > 0$ for every Kähler class κ (Lemma I.13.1) and since by (0), for every pair $x, y \in \bar{\mathscr{C}}_X$ we have that $(x,y) \geq 0$, it follows that $d \in \bar{\mathscr{C}}_X$ *(and not* $-d \in \bar{\mathscr{C}}_X$*).*

It remains to show that any class $d \in \bar{\mathscr{C}}_X \cap H^2(X,\mathbb{Z})$ is effective. Such a d is contained in the Picard lattice and by Proposition 3.6, (i) either d or $-d$ is effective. The second possibility is excluded as before. □

Next we set $\Delta = \{d \in S_X;\ (d,d) = -2 \text{ and } d \text{ effective}\}$ and for every $d \in \Delta$ we let H_d be the hyperplane of fixpoints of the Picard-Lefschetz reflection s_d corresponding to d. The connected components of $\mathscr{C}_X \setminus \bigcup_{d \in \Delta} H_d$ we call the *chambers* of $\mathscr{C}_X$. If X is Kähler, then the Kähler class is contained in the following chamber

$$\mathscr{C}_X^+ = \{y \in \mathscr{C}_X;\ (y,d) > 0 \text{ for all } d \in \Delta\}.$$

(3.8) **Corollary.** *For a Kähler K3-surface* $\mathscr{C}_X^+$ *is the Kähler cone.*

Proof. By definition the Kähler cone consists of all those $x \in \mathscr{C}_X$ such that $(x,d) > 0$ for all effective classes d. By Proposition 3.7 it is sufficient to test this for nodal classes and for $d \in \bar{\mathscr{C}}_X^+$, but for the latter we automatically have $(x,d) > 0$ by (0). □

---

*) Since X is simply-connected, we can identify $H^2(X,\mathbb{Z})$ with its j-image.

?

(3.9) **Proposition.** *The Picard-Lefschetz reflections of a Kähler K3-surface X leave invariant the positive cone, and the group W_X generated by them operates on this cone in a properly discontinuous fashion. The closure of the Kähler cone in the positive cone is a fundamental domain for W_X in the sense that any W_X-orbit in $\mathscr{C}_X$ meets it in exactly one point.*

Proof. Let $d \in \Delta$ and s_d, H_d as before. Since s_d preserves $(\ ,\ )$, it preserves $\mathscr{C}_X \cup \mathscr{C}'_X$ so either preserves $\mathscr{C}_X$ or interchanges $\mathscr{C}_X$ and $\mathscr{C}'_X$. Since $s_d | H_d = \mathrm{id}$ and H_d meets necessarily $\mathscr{C}_X$ (because the cupproduct has signature (1, 18) on H_d) the last possibility is excluded.

The subgroup G of $\mathrm{Aut}(H^{1,1}(X, \mathbb{R}), (\ ,\ ))$ preserving $\mathscr{C}_X$ acts transitively on the set of half-lines in $\mathscr{C}_X$ making this set into a homogeneous space, in fact, a Lobatchevski space. The isotropy group of a point being isomorphic to 0(19), hence compact, the group G operates properly on $\mathscr{C}_X$. Since W_X is a subgroup of $\mathrm{Aut}(L)$, it is discrete in G and therefore acts properly and discontinuously on $\mathscr{C}_X$. It is a general fact that the fundamental domain of a discrete group generated by reflections in hyperplanes of a Lobatchevski space, is the closure of a chamber (compare [Bou3], Exercise to Chap. V §4, and [Vin]). Since by Corollary 3.8 the Kähler cone is a chamber, the result follows. □

(3.10) **Proposition.** *Let X and X′ be Kähler K3-surfaces and $\phi: H^2(X, \mathbb{Z}) \to H^2(X', \mathbb{Z})$ a Hodge-isometry. The following properties are equivalent:*

(i) *ϕ is effective,*

(ii) *ϕ maps the Kähler cone of X to the one of X′,*

(iii) *ϕ maps an element of the Kähler cone of X into the Kähler cone of X′.*

Proof. The Kähler cone being defined in terms of effective classes and intersection properties only, the implication (i) ⇒ (ii) is trivial. Also (ii) ⇒ (iii) is obvious. As to the implication (iii) ⇒ (i), we observe first of all that $\phi(\mathscr{C}_X) = \mathscr{C}_{X'}$, so by Proposition 3.7 we only have to show that if $d \in S_X$ is nodal, then $d' = \phi(d)$ is effective. But if $x \in \mathscr{C}_X^+$ with $\phi(x) \in \mathscr{C}_{X'}^+$, then $(\phi(x), d') = (x, d) > 0$. So $-d'$ cannot be effective, and by Proposition 3.6, (i) the class d' will be effective. □

(3.11) **Corollary.** *If X is a projective K3-surface, X′ any Kähler K3-surface, and $\phi: H^2(X, \mathbb{Z}) \xrightarrow{\sim} H^2(X', \mathbb{Z})$ an effective Hodge-isometry, then X′ is projective and ϕ maps the class of an ample divisor on X to a similar class on X′.*

Conversely, if X and X′ are projective and ϕ is a Hodge-isometry sending the class of an ample divisor on X to the class of an ample divisor on X′, then ϕ is effective.

Proof. Integral points in the Kähler cone are just the classes of divisors $d \in S_X$ for which $(d, d) > 0$ and $(d, c) > 0$ for all effective $c \in S_X$. So by Corollary IV.5.4 these classes are classes of ample divisors. If ϕ is effective, it preserves such classes; in particular X' is projective. Conversely, if X and X' are projective, then the set of integral points in the respective Kähler cones is non-empty and is preserved by ϕ, so Proposition 3.10 applies. □

4. Digression on Affine Geometry over $\mathbb{F}_2$

We start by recalling some elementary facts from affine geometry, for which we refer to [G-W]. By definition an *affine space* V over a field k is a set on which the additive group of a vector space $T(V)$ acts transitively and effectively. A choice of $v \in V$ determines the bijection ("choice of origin")

$$f_v\colon T(V) \to V \qquad (t \to t(v)).$$

If V and V' are affine spaces over k an *affine-linear map* is a map $\phi\colon V \to V'$ such that for some $v \in V$ the map $T(\phi)\colon T(V) \to T(V')$ defined by $T(\phi) = f_{\phi(v)}^{-1} \circ \phi \circ f_v$, is a k-linear vector space morphism. Then this is true for all $v \in V$ and $T(\phi)$ is independent of $v \in V$. It is called *the linear map induced by* ϕ. A map $A\colon W \to W'$ between k-vector spaces is called *semi-linear* if there exists an automorphism σ of k such that $A(\lambda w) = \sigma(\lambda) A(w)$ for all $\lambda \in k$, $w \in W$ and $A(w_1 + w_2) = A(w_1) + A(w_2)$ for all $w_1, w_2 \in W$. If in the definition for "affine-linear" we replace "k-linear" by "semi-linear" we obtain the definition of a *semi-affine map*. Its importance stems from the *fundamental theorem of affine geometry* which reads as follows:

Let V, V' be affine spaces over k of dimension ≥ 3, and let $\phi\colon V \to V'$ be a bijection having the property that W is an affine subspace of V if and only if $\phi(W)$ is an affine subspace of V'. Then ϕ is semi-affine.

In the sequel we need a special version of it for $k = \mathbb{F}_2$.

(4.1) **Lemma.** *Let V, V' be affine spaces over $\mathbb{F}_2$ and $\phi\colon V \to V'$ a bijection inducing a bijection between the respective sets of affine-linear functions on V and V'. Then ϕ is affine.*

Proof. If $\dim V = \dim V' \leq 2$, any bijection is easily checked to be affine. So it suffices to consider the case $\dim V = \dim V' \geq 3$, where we may apply the fundamental theorem of affine geometry. The assumptions of the lemma imply that ϕ maps hyperplanes to hyperplanes, so for any $W \subset V$ the image $\phi(W)$ is affine if and only if W is affine. It follows that ϕ is semi-affine, hence affine, since $\mathbb{F}_2$ does not have any automorphisms except the identity. $\square$

(4.2) **Lemma.** *Let V be an affine space of finite dimension ≥ 2 over $\mathbb{F}_2$ and let U be a linear subspace of the vector space $\mathbb{F}_2^V$ of $\mathbb{F}_2$-valued functions on V. Suppose that U is invariant under the group* $\mathrm{Aut}(V)$ *of affine transformations of V. Then we have the following possibilities for U:*

a) U consists of the two constant functions (i.e. the polynomial functions of degree 0),

b) U consists of the affine-linear functions (i.e. the polynomial functions of degree ≤ 1),

c) U contains for any codim-2 subspace $W \subset V$ its characteristic function χ_W (they generate the polynomial functions of degree ≤ 2).

Proof. If U contains non-constant affine-linear functions, it contains all of them, since they form an $\mathrm{Aut}(V)$-orbit. But then U also contains the constants, for they are the sum of two non-constant affine-linear functions.

The non-constant affine-linear functions are exactly the characteristic functions for the affine hyperplanes: if f is affine-linear with zero-set the hyperplane W, then f is the characteristic function of the set $V \setminus W$, which is a hyperplane (we are working over $\mathbb{F}_2$) - and conversely $\chi_W = 1 + f$ where f vanishes exactly on W.

Now suppose that U contains a function g which is not affine-linear. We prove by induction on $n = \dim V \geq 2$ that U contains the characteristic functions of all codimension-2 subspaces.

For $n=2$, we argue as follows. *Either* g is the characteristic function of one point and U contains all of those, *or* g is the characteristic function of three points and U contains all of those. In the first case we are ready. In the second case we observe that U also contains the characteristic functions of *two* points (they are sums of characteristic functions of three points) and therefore also the characteristic functions of *one* point (they are sums of characteristic functions of two and three points).

For $n \geq 3$ we first observe that V contains a hyperplane V' such that $g|V'$ is not affine-linear. Indeed, for at least one hyperplane V'' we have $g|V'' \not\equiv 0$ or 1, since g is not linear. Hence either $g|V''$ is not affine-linear and we take $V' = V''$ *or* the zero set of $g|V''$ is a codimension-2 subspace $W \subset V''$. In the last case *either* $g|V \setminus W \equiv 1$ *or* $g|V'$ is not affine-linear for at least one hyperplane V' in the pencil through W. If $g|V \setminus W \equiv 1$ we may take for V' any hyperplane, not in the pencil through W, which has at least one point in common with W.

So we can apply the induction-hypothesis to $U' = \{f|V'; f \in U\}$. It follows that there exists a $g' \in U$, such that $g'|V' = \chi_{W'}$, where W' is an $(n-3)$-dimensional subspace of V'. Let $a \in W'$, $b \in V' \setminus W'$. The set $V \setminus V'$ is a hyperplane and therefore there exists a unique affine transformation t leaving $V \setminus V'$ pointwise fixed and sending a to b. Then t induces a translation by $b-a$ in V', carrying W' to an $(n-3)$-dimensional subspace $t(W')$ of V' which is disjoint from W'. So $g = (g' \circ t) + g' \in U$ is the characteristic function of $W = W' \cup t(W')$. Since W is the $(n-2)$-dimensional subspace of V generated by W' and b, the space U contains at least one characteristic function of a codimension-2 subspace. Since $\mathrm{Aut}(V)$ acts transitively on these subspaces, U contains all of these characteristic functions. □

(4.3) **Lemma.** *Let V be a finite-dimensional affine space over $\mathbb{F}_2$, $T(V)$ its vector space of affine translations and U_k the vector space of polynomial functions of degree $\leq k$. For fixed $v \in V$ and $g \in U_2$ the bilinear form $Q(g)$ on $T(V)$ defined by*

$$Q(g)(s,t) = g(f_v(s+t)) + g(v) + g(f_v(s)) + g(f_v(t))$$

is symplectic. It is independent of the choice of $v \in V$, and $Q(g) = 0$ if and only if $g \in U_1$. The resulting $\mathbb{F}_2$-linear map

$$U_2/U_1 \to \Lambda^2 \mathrm{Hom}(T(V), \mathbb{F}_2) \qquad (g \bmod U_1 \to Q(g))$$

is an isomorphism.

Proof. Recall that a form Q on a vector space W is symplectic if $Q(x,x)=0$ for all $x \in W$. By [Bou2], III, §7, no. 4, Prop. 7, p. 80, these *forms* form a vector space isomorphic to $\bigwedge^2 \mathrm{Hom}(W, \mathbb{F}_2)$. Since $Q(g)$ is obviously symplectic and independent of $v \in V$, the assignment $g \to Q(g)$ defines a map $U_2 \to \bigwedge^2 \mathrm{Hom}(T(V), \mathbb{F}_2)$. The proof of the remaining assertions is straightforward. □

Remark. The choice of an origin $0 \in V$ identifies V with $T(V)$. If Z' is a codim-2 affine subspace, we may write $Z' = z + Z$, with $z \in V$ and Z a linear subspace of $T(V)$. One verifies directly that in the above isomorphism $\chi_{Z'} \in U_2$ corresponds to the decomposable vector $f_1 \wedge f_2 \in \bigwedge^2 \mathrm{Hom}(T(V), \mathbb{F}_2)$ for which $(f_1=0) \cap (f_2=0) = Z$. □

5. The Picard Lattice of Kummer Surfaces

In this section Y will be a complex analytic 2-torus and X its Kummer surface. Using the notation of V, Sect. 16 we have a commutative diagram

$$\begin{array}{ccc} \overline{Y} & \xrightarrow{\bar{p}} & X \\ {\scriptstyle\sigma}\downarrow & & \downarrow \\ Y & \xrightarrow{p} & Y/\langle 1, i \rangle. \end{array}$$

Let $\alpha = \bar{p}_! \circ \sigma^*\colon H^2(Y, \mathbb{Z}) \to H^2(X, \mathbb{Z})$.

(5.1) **Proposition.** *For all $x, y \in H^2(Y, \mathbb{Z})$ we have $(\alpha(x), \alpha(y)) = 2(x, y)$. In particular α is a monomorphism.*

Proof. Since every element $z \in H^2(\overline{Y}, \mathbb{Z})$ is invariant under the covering involution for $\overline{Y} \to X$ we can write $z = \bar{p}^*(w)$, $w \in H^2(X, \mathbb{Z})$, and the projection formula I, (1) shows that $\bar{p}^*(\bar{p}_! z) = 2z$. So $(\bar{p}_! z_1, \bar{p}_! z_2) = 2(z_1, z_2)$ for any pair $z_1, z_2 \in H^2(\overline{Y}, \mathbb{Z})$ and therefore $\bar{p}_!$ and α multiply the intersection numbers by 2. □

(5.2) **Proposition.** *For the complexification $\alpha_{\mathbb{C}}$ of α we have $\alpha_{\mathbb{C}}(H^{2,0}(Y)) = H^{2,0}(X)$.*

Proof. Since $h^{2,0}(X) = h^{2,0}(Y)$ it suffices to observe that any nowhere-zero holomorphic 2-form on Y is i-invariant. □

We proceed by introducing some more notation. We set

V: the set of points of order 2 on Y, equipped with its natural structure of 4-dimensional affine space over $\mathbb{F}_2$.

e_v: the class of $\bar{p}(\sigma^{-1}(v))$ in $H^2(X, \mathbb{Z})$, the *distinguished nodal class* of $v \in V$.

W: $\{e_v;\ v \in V\}$.

M: the smallest primitive sublattice of $H^2(X, \mathbb{Z})$ containing W.

$M^\vee$: the dual of M, i.e. $M^\vee = \{m' \in M \otimes_{\mathbb{Z}} \mathbb{Q};\ (m', m) \in \mathbb{Z} \text{ for all } m \in M\}$ (see also I, Sect. 2).

Since $(w,w)=-2$ for $w\in W$ and $(w,w')=0$ for $w'\neq w$ in W we have inclusions

$$\mathbb{Z}^W\subset M\subset M^\vee\subset(\tfrac{1}{2}\mathbb{Z})^W.$$

Let $(\frac{1}{2}\mathbb{Z})^W\to(\frac{1}{2}\mathbb{Z}/\mathbb{Z})^W=\mathbb{F}_2^W$ be the natural projection. So there results an epimorphism

$$r\colon(\tfrac{1}{2}\mathbb{Z})^W\to\mathbb{F}_2^V.$$

Explicitly, if $x=\sum_{v\in V}x_v e_v$ with $x_v\in\frac{1}{2}\mathbb{Z}$ then $r(x)(v)=2x_v \bmod 2$. With these observations the next result is obvious.

(5.3) **Proposition.** *For any subset $V'\subset V$ the characteristic function $\chi_{V'}\in\mathbb{F}_2^V$ is precisely $r(\sum_{v\in V'}\frac{1}{2}e_v)$.*

Ultimately we want to show that $\operatorname{Im}\alpha=(\mathbb{Z}^W)^\perp$. In other words the lattice spanned by the distinguished nodal classes determines $\operatorname{Im}\alpha$ as its orthogonal complement. As a first step we observe:

(5.4) **Proposition.** $\operatorname{Im}\alpha$ *is of finite index in* $(\mathbb{Z}^W)^\perp$.

Proof. By construction $\operatorname{Im}\alpha\subset\operatorname{Im}(H^2(X,\bar{p}(\sigma^{-1}(V))\,\mathbb{Z})\to H^2(X,\mathbb{Z}))$, so $\operatorname{Im}\alpha\subset(\mathbb{Z}^W)^\perp$. Since $\operatorname{Im}\alpha$ and $(\mathbb{Z}^W)^\perp$ are both free $\mathbb{Z}$-modules of rank 6 (Proposition 5.1) the index $((\mathbb{Z}^W)^\perp\colon\operatorname{Im}\alpha)$ is finite. □

We compute this index, noting that by Lemma I.2.1, (i) its square equals

$$d(\operatorname{Im}\alpha)\cdot(d(\mathbb{Z}^W)^\perp)^{-1},$$

where d denotes the determinant introduced in I, Sect. 2. Since

$$d(\operatorname{Im}\alpha)=2^6\quad\text{and}\quad d(M^\perp)=d(M)$$

by Lemma I.2.5, it suffices to compute $d(M)$. But $M\subset(\frac{1}{2}\mathbb{Z})^W$ as observed before and since $r(M)\cong M/\mathbb{Z}^W$ it follows that $d(M)=d(\mathbb{Z}^W)\cdot u^{-2}$ where u is the cardinality of $r(M)$. This shows the importance of the subspace $r(M)$. In fact, we shall presently show

(5.5) **Proposition.** *The subspace $U=r(M)\subset\mathbb{F}_2^V$ consists precisely of the affine-linear functions on V.*

As a corollary we infer that in the above computations $u=2^5$ and so $[(\mathbb{Z}^W)^\perp\colon\operatorname{Im}\alpha]^2=2^6\cdot 2^{-16}\cdot 2^{10}=1$, i.e.

(5.6) **Corollary.** $\operatorname{Im}\alpha=(\mathbb{Z}^W)^\perp$.

For the proof of Proposition 5.5 we need the following very geometric

(5.7) **Lemma.** *Let $t\colon V\to V$ be an affine transformation. Then there exists an isometry $\tau\colon H^2(X,\mathbb{Z})\to H^2(X,\mathbb{Z})$ such that $\tau(e_v)=e_{t(v)}$ $(v\in V)$.*

Proof. Suppose $T\colon Y\to Y$ is a homeomorphism with the following properties

1) $T|V=t$,
2) T and i commute,
3) there exists a translation-invariant metric on Y such that on a suitably small ball centred at $v\in V$ the translation by $t(v)-v$ coincides with T.

Such a T lifts to a homeomorphism of $\bar{Y}$ onto itself commuting with the involution of $\bar{Y}$, and – mapping $\sigma^{-1}(v)$ onto $\sigma^{-1}(t(v))$ in an orientation-preserving manner – it descends to a homeomorphism of X onto itself, inducing an isometry of $H^2(X,\mathbb{Z})$ with the required properties.

So it suffices to construct such a T. This will be done by way of a homeomorphism of the universal covering $\mathbb{C}^2$ of Y onto itself, preserving the lattice $\Gamma=\bigoplus_{i=1}^{4}\mathbb{Z}e_i$ for which $\mathbb{C}^2/\Gamma\cong Y$. In case t is fixed point-free, it is a translation and for T we take the corresponding translation. If $v_0\in V$ is a fixed point for t we may assume that $0\in\mathbb{C}^2$ maps to v_0. Then $v_0\in V$ serves as origin of the $\mathbb{F}_2$-space V and t is a linear map, given by

$$t(\tfrac{1}{2}e_i)=\sum_{j=1}^{4}t_{i,j}(\tfrac{1}{2}e_j)\qquad(t_{i,j}\in\mathbb{F}_2).$$

Clearly t admits an $\mathbb{R}$-linear orientation-preserving extension T_0 to $\mathbb{C}^2$. Moreover, since t and hence also T_0 is of finite order, we may replace the euclidean metric on $\mathbb{C}^2$ by an equivalent T_0-invariant one, so that $T_0\in SO(4,\mathbb{R})$. Let $\varepsilon>0$ be so small that for any two distinct $x,y\in\frac{1}{2}\Gamma$ the balls $B(x,\varepsilon)$ and $B(y,\varepsilon)$ are disjoint. The map T_0 takes $B(x,\varepsilon)$ orientation- and length-preserving onto $B(t(x),\varepsilon)$, so T_0 and the translation by $t(x)-x$ are homotopic, when restricted to $B(x,\varepsilon)$. This makes it possible to replace T_0 by $t(x)-x$ on every $B(x,\frac{1}{2}\varepsilon)$, while smoothly adjusting it to T_0 in the spherical shells $B(x,\varepsilon)\setminus B(x,\frac{1}{2}\varepsilon)$ $(x\in\frac{1}{2}\Gamma)$. Moreover, we can assume that the resulting homeomorphism of $\mathbb{C}^2$ onto itself commutes with $-\mathrm{id}$. Since it also respects Γ, it descends to a homeomorphism $T\colon Y\to Y$ with the required properties. □

Now we give the proof of Proposition 5.5.

We first need an estimate on the cardinality of U. Since $\operatorname{Im}\alpha\subset M^\perp$ we have $2^6=d(\operatorname{Im}\alpha)\geq d(M^\perp)=d(M)$ (the last identity follows from Lemma I.2.5). Since $U\cong M/\mathbb{Z}^W$, the square of the cardinality of U equals $d(\mathbb{Z}^W)\,d(M)^{-1}\geq 2^{16}\cdot 2^{-6}=2^{10}$, so U contains at least 2^5 elements.

By Lemma 5.7 the subspace U is invariant under the affine group of V. If U would contain a function which is not affine-linear, according to Lemma 4.2 the space U would contain the characteristic functions of *all* codim-2 subspaces of V. Let V', V'' be two such subspaces having exactly one point in common. Since $\chi_{V'},\chi_{V''}\in U$, it follows from Proposition 5.3 that $\sum_{v\in V'}\frac{1}{2}e_v$ and $\sum_{v\in V''}\frac{1}{2}e_v$ both belong to M. But their intersection product is $\frac{1}{2}$, so this is impossible. Again by Lemma 4.2, either U consists of the two constant functions or U contains all the affine-linear functions. The previous estimate rules out the first alternative, so the proposition is proved. □

Next, we give various interpretations of $H^2(Y, \mathbb{F}_2)$ in terms of the lattice M generated by the distinguished nodal classes of X. There is an isomorphism of $\mathbb{Z}$-modules

$$\beta\colon \mathrm{Hom}_{\mathbb{Z}}(\Gamma, \mathbb{Z}) \to H^1(Y, \mathbb{Z}),$$

whose $\mathbb{F}_2$-reduction is an isomorphism of $\mathbb{F}_2$-vector spaces

$$\beta_2\colon \mathrm{Hom}(T(V), \mathbb{F}_2) \to H^1(Y, \mathbb{F}_2)$$

(we use that $T(V) = \Gamma/2\Gamma$ as an $\mathbb{F}_2$-vector space).

According to Proposition 5.1 the map α multiplies (,) by 2. Hence $(\mathrm{Im}\,\alpha)^\vee = \frac{1}{2}\mathrm{Im}\,\alpha$. Applying Corollary 5.6 we find:

$$H^2(Y, \mathbb{F}_2) \cong H^2(Y, \mathbb{Z})/2H^2(Y, \mathbb{Z}) = \mathrm{Im}\,\alpha/2\,\mathrm{Im}\,\alpha \xrightarrow{\sim} \tfrac{1}{2}\mathrm{Im}\,\alpha/\mathrm{Im}\,\alpha$$
$$= (\mathrm{Im}\,\alpha)^\vee/\mathrm{Im}\,\alpha = (M^\perp)^\vee/M^\perp.$$

Since by I, (2) there is a natural isomorphism $(M^\perp)^\vee/M^\perp \to M^\vee/M$ and since $M^\vee/M \xrightarrow{\sim} r(M^\vee)/r(M)$ we obtain an isomorphism

$$\gamma\colon H^2(Y, \mathbb{F}_2) \to r(M^\vee)/r(M).$$

For the proof of the next proposition we need an explicit representation for the image under $\gamma \circ (\beta_2 \wedge \beta_2)$ of a decomposable element. So let $z_i \in \mathrm{Hom}(\Gamma, \mathbb{Z})$ $(i = 1, 2)$ and $z \in \bigwedge^2 \mathrm{Hom}(T(V), \mathbb{F}_2)$ be the mod-2 reduction of $z_1 \wedge z_2$. Then Z, the mod-2 reduction of $\{z_1 = 0\} \cap \{z_2 = 0\}$, is a 2-codimensional linear subspace of $T(V)$ and we claim

$$\gamma(\beta_2 \wedge \beta_2)(z) = r\Big(\tfrac{1}{2}\sum_{v \in Z} e_v\Big) \mod r(M). \tag{2}$$

To establish (2), we first observe that the definition of α implies that the Poincaré-dual of $\alpha(\beta(z_1) \wedge \beta(z_2)) - \sum_{v \in Z} e_v$ is the class of the $\bar{p}$-image of the proper transform on $\bar{Y}$ of the cycle which on Y is given by $\{z_1 = 0\} \cap \{z_2 = 0\}$. This cycle is invariant under the covering involution, so the class of its image is 2-divisible in $H^2(X, \mathbb{Z})$, i.e.

$$\tfrac{1}{2}\Big[\alpha(\beta(z_1) \wedge \beta(z_2)) - \sum_{v \in Z} e_v\Big] \in H^2(X, \mathbb{Z}).$$

This means that $\frac{1}{2}\alpha(\beta(z_1) \wedge \beta(z_2))$ and $\frac{1}{2}\sum_{v \in Z} e_v$ correspond under the isomorphism of I, (2), thereby proving (2).

(5.8) **Proposiiton.** *The subspace $r(M^\vee) \subset \mathbb{F}_2^V$ consists of all polynomial functions of degree ≤ 2. Hence the isomorphism of Lemma 4.3 gives a natural isomorphism of $\mathbb{F}_2$-vector spaces*

$$\delta\colon r(M^\vee)/r(M) \to \mathrm{Hom}(T(V), \mathbb{F}_2).$$

This isomorphism is the inverse of $\gamma \circ (\beta \wedge \beta)$.

Proof. The subspace $r(M^\vee)$ contains $|r(M)| \cdot [r(M^\vee) : r(M)] = 2^{11}$ elements. On the other hand it is preserved by the affine group, hence by Lemma 4.2

contains the subspace U_2 of $\mathbb{F}_2^V$ of polynomial functions of degree ≤ 2 on V, which itself already consists of 2^{11} elements. So $U_2 = r(M^\vee)$ and we may apply Lemma 4.3.

The last assertion need only be checked for decomposable elements. By (2) and Proposition 5.3 we have that $\gamma \circ (\beta \wedge \beta)(z) = \chi_Z$, where z is the mod-2 reduction of $z_1 \wedge z_2 \in \bigwedge^2 \mathrm{Hom}(\Gamma, \mathbb{Z})$. On the other hand, the Remark following Lemma 4.3 shows that $\delta(\chi_Z)$ is the mod-2 reduction of $z_1 \wedge z_2$. So indeed $\gamma \circ (\beta \wedge \beta)(z) = \delta^{-1}(z)$. $\square$

Finally, we shall compare the Hodge structure on two Kummer surfaces $X = Km(Y)$, resp. $X' = Km(Y')$. For objects on Y', dashed symbols are used having similar meaning as the corresponding undashed ones for Y. Moreover, in addition to $U = r(M)$ we set

$$^\circ U = r(M^\vee).$$

We recall from V, Sect. 16 that every Kummer surface is a Kähler surface, so we can speak of an effective Hodge isometry between the cohomology groups of two Kummer surfaces.

(5.9) **Proposition.** *Let $X = Km(Y)$, $X' = Km(Y')$ and $\phi: H^2(X, \mathbb{Z}) \to H^2(X', \mathbb{Z})$ an isometry inducing a bijection $v: V \to V'$. Assume moreover that $H^2(Y, \mathbb{Z})$ (and hence $H^2(X, \mathbb{Z})$) contains effective classes and that ϕ is an effective Hodge-isometry. Then ϕ is induced by an isomorphism $X' \to X$.*

Proof. Since ϕ is an isometry sending M isometrically to M', the same holds for the orthogonal complements. So, by Proposition 5.1 and Corollary 5.6 we obtain an isometry

$$\psi: H^2(Y, \mathbb{Z}) \to H^2(Y', \mathbb{Z}).$$

This is even a Hodge-isometry, by Proposition 5.2. We are going to show that the assumptions of Theorem V.3.2 are satisfied. So we consider the mod-2 reduction ψ_2 of ψ. Now $U \to {}^\circ U$ is contravariant, so by naturality of the isomorphisms γ and γ' we obtain a commutative diagram

$$\begin{array}{ccc} H^2(Y, \mathbb{F}_2) & \xrightarrow[\psi_2]{} & H^2(Y', \mathbb{F}_2) \\ \downarrow{\scriptstyle\gamma} & & \downarrow{\scriptstyle\gamma'} \\ {}^\circ U/U & \xleftarrow[{}^\circ\phi]{} & {}^\circ U'/U' \end{array} \tag{3}$$

where ${}^\circ\phi$ is induced by ϕ.

On the other hand, v induces a map v^*, making the following diagram commutative

$$\begin{array}{ccc} M & \xrightarrow[\phi|M]{} & M' \\ \downarrow{\scriptstyle r} & & \downarrow{\scriptstyle r'} \\ \mathbb{F}_2^V & \xleftarrow[v^*]{} & \mathbb{F}_2^{V'}. \end{array}$$

Since $r(M)=U$, resp. $r'(M')=U'$ consists of the affine-linear functions of V, resp. V', an application of Lemma 4.1 shows that ν is affine, linear, inducing an isomorphism

$$\nu^{\vee}\colon \operatorname{Hom}(T(V'),\mathbb{F}_2)\to \operatorname{Hom}(T(V),\mathbb{F}_2)$$

which, according to Proposition 5.8, fits into the following commutative diagram

$$\begin{array}{ccc} \bigwedge^2 \operatorname{Hom}(T(V),\mathbb{F}_2) & \xleftarrow[\nu^{\vee}\wedge\nu^{\vee}]{} & \bigwedge^2 \operatorname{Hom}(T(V',\mathbb{F}_2) \\ \uparrow \wr\, \delta & & \uparrow \wr\, \delta \\ {}^{\circ}U/U & \xleftarrow[{}^{\circ}\phi]{} & {}^{\circ}U'/U'. \end{array} \tag{4}$$

By Proposition 5.8 we have $\delta^{-1}=\gamma\circ(\beta_2\wedge\beta_2)$ and similarly for $(\delta')^{-1}$, so (3) and (4) imply that $\psi_2=\chi_2^{-1}\wedge\chi_2^{-1}$ with $\chi_2=(\beta')^{-1}\circ\nu^{\vee}\circ\beta$. The assumptions of Theorem V.3.2 are indeed satisfied, so we have that $\psi_2=\pm f^*$ for some biholomorphic map

$$f\colon Y'\to Y;$$

in fact we have $\psi_2=f^*$, since ϕ and hence ψ takes effective classes into effective classes.

Since composing f with a translation does not affect the induced homomorphisms in cohomology, we may assume that g is a group homomorphism. So f commutes with i. Next, we may compose f with a translation by a point of order 2, if necessary, such that the resulting map (still denoted by f) coincides with ν^{-1} for at least one $v'\in V'$. Now the new f still commutes with i and we obtain a biholomorphic map $g\colon X'\to X$ which by construction has the property that the induced map in 2-cohomology coincides with ϕ on $\operatorname{Im}\alpha$ and on e_v ($v=\nu^{-1}(v')$). Clearly f induces an affine-linear isomorphism $V'\to V$, which differs from ν at most by a translation (they both induce the same isomorphism $\nu^{\vee}\wedge\nu^{\vee}$ on $\bigwedge^2\operatorname{Hom}(T(V),\mathbb{F}_2)$). This means that the map induced by g^* on V and the map ϕ differ at most by a translation. Since $g^*(e_v)=\phi(e_v)$, this translation is the identity, so $g^*|M=\phi|M$ and therefore $g^*=\phi$. $\square$

6. The Torelli Theorem for Kummer Surfaces

The weakness of Proposition 5.9 lies in the fact that one of its assumptions, namely that the set of distinguished nodal classes be preserved, is far too strong. In this section we show how to remove this assumption for projective Kummer surfaces.

(6.1) **Proposition.** *Let X be a K3-surface containing sixteen disjoint (-2)-curves $C_1,\dots,C_{16}$ such that $\mathcal{O}_X\left(\sum_{i=1}^{16} C_i\right)$ is 2-divisible in* Pic(X). *Then X has the structure of a Kummer surface with $\{C_1,\dots,C_{16}\}$ as the set of distinguished (-2)-curves.*

Proof. Since $\mathcal{O}_X\left(\sum_{i=1}^{16} C_i\right)$ is 2-divisible in $\mathrm{Pic}(X)$, by Proposition I.18.1 there exists a double covering $\rho: Z \to X$ whose branch locus is $\sum_{i=1}^{16} C_i$. The curves $D_i = \phi^{-1}(C_i)$ are all (-1)-curves on Z and upon contracting them we get a smooth surface Y. The formulas of Lemma I.7.1, (iii) and Theorem I.8.1, (vii) for the behaviour of the canonical bundle under branched coverings and blowing up imply that $\mathcal{K}_Y \cong \mathcal{O}_Y$. For $e(Y)$ we find $e(Y) = e(Z) - 16 = 2(e(X) - 16) - 16 = 0$ upon applying V, (8) and Proposition 3.1. So Y is minimal, and in fact a torus by the Enriques classification. Moreover the involution on Z interchanging the sheets of ρ descends to an involution i on Y. Since the i-invariant part of $H^1(Y, \mathbb{Q})$ is canonically isomorphic to $H^1(X, \mathbb{Q}) = 0$, i acts as $-\mathrm{id}$ on $H^1(X, \mathbb{Q})$ and so $X = Km(Y)$ by definition. □

Before stating the next theorem we have to make the following remark. If X and X' are K3-surfaces and $\phi: H^2(X, \mathbb{Z}) \to H^2(X', \mathbb{Z})$ is an isometry preserving the Hodge-decomposition, projectivity of X implies projectivity of X'. Indeed, $\phi(S_X) \subset S_{X'}$ and S_X contains an element d with $d^2 > 0$, hence $\phi(d) \in S_{X'}$ has positive norm and X' is projective by Theorem IV.5.2.

It follows that under the above assumptions it makes sense to speak of an effective Hodge-isometry, and we shall do this several times in the sequel.

(6.2) **Theorem** (Torelli theorem for projective Kummer surfaces). *Let X' be a K3-surface and X a projective Kummer surface. Suppose we are given an effective Hodge-isometry $\phi: H^2(X, \mathbb{Z}) \to H^2(X', \mathbb{Z})$. Then $\phi = f^*$, where $f: X' \to X$ is a biholomorphic map.*

Proof. Let $\{c_1, \ldots, c_{16}\}$ be the set of distinguished nodal classes on X. Since ϕ and ϕ^{-1} preserve effective classes, they also preserve indecomposable classes. So, if we put $c'_i = \phi(c_i)$ $(i = 1, \ldots, 16)$ we know that $c'_i = c_1(\mathcal{O}_X(C'_i))$ for a unique (-2)-curve C'_i. Since $\sum_{i=1}^{16} c_i$ is 2-divisible in the Picard lattice of X, also $\sum_{i=1}^{16} c'_i$ and therefore, by Proposition 3.5, $\sum_{i=1}^{16} \mathcal{O}_X(C'_i)$ is 2-divisible in $\mathrm{Pic}(X')$. From Proposition 6.1 we conclude that X' is a Kummer surface having $\{C'_1, \ldots, C'_{16}\}$ as its distinguished set of (-2)-curves. Since X is projective, it contains an effective class orthogonal to $\{c_1, \ldots, c_{16}\}$. Such a class gives an effective class on the torus from which X is constructed. Hence all the hypotheses of Proposition 5.9 are satisfied and the result follows. □

(6.3) **Corollary** (Weak Torelli theorem for projective Kummer surfaces). *Let X' be a K3-surface and X be a projective Kummer surface. Suppose that there exists a Hodge-isometry $H^2(X, \mathbb{Z}) \to H^2(X', \mathbb{Z})$. Then X and X' are biholomorphically equivalent.*

Proof. By composing the Hodge-isometry with $-\mathrm{id}$, if necessary, we may assume that it preserves the positive cones. As observed before, the hypothesis of

Corollary 6.3 implies that X' is projective and a fortiori a Kähler surface. If we compose the given Hodge-isometry with an appropriate element of W_X, it will preserve the Kähler cones (Proposition 3.9), so the resulting Hodge-isometry is effective by Proposition 3.10 and we may apply Theorem 6.2 to obtain the corollary. □

7. The Local Torelli Theorem for K3-Surfaces

Let $\Delta=\{z\in\mathbb{C};\ |z|<\varepsilon\}$ and $\gamma\colon \Delta\to\mathbb{P}(L_{\mathbb{C}})$ a holomorphic map with $\gamma(0)=l$ and tangent $\theta=\gamma'(0)$. Let $\tilde{\gamma}\colon \Delta\to L_{\mathbb{C}}\setminus\{0\}$ be a lifting of γ. Associating to l the class of $\tilde{\gamma}'(0)$ in $L_{\mathbb{C}}/l$ gives a homomorphism $l\to L_{\mathbb{C}}/l$, which is independent of the lifting of γ. In fact it is uniquely determined by θ. We denote it by $h(\theta)$. Clearly $\theta\to h(\theta)$ gives an injective map $h\colon \mathcal{T}_{\mathbb{P}(L_{\mathbb{C}})}(l)\to\operatorname{Hom}(l, L_{\mathbb{C}}/l)$, hence an isomorphism.

(7.1) **Proposition.** *For any $l=[\omega]\in\Omega$ the isomorphism h gives an identification of $\mathcal{T}_{\Omega}(l)$ with* $\operatorname{Hom}(l, l^{\perp}/l)$).

Proof. $\mathcal{T}_{\Omega}(l)$ is the subspace of those θ for which a lifting $\tilde{\gamma}$ satisfies $\tilde{\gamma}(\Delta)\subset\Omega$, i.e. for which $(\tilde{\gamma}(0), \tilde{\gamma}'(0))=0$, or equivalently $h(\theta)\in\operatorname{Hom}(l, l^{\perp}/l)$. □

Suppose that we have a family of K3-surfaces $p\colon Y\to S$ together with a trivialisation $\phi\colon p_{*2}\mathbb{Z}_Y\xrightarrow{\sim} L_S$. Consider its associated period mapping

$$\tau\colon S\to\Omega$$

defined by sending $s\in S$ to the complex line $\phi_{\mathbb{C}}(s)(H^{2,0}(Y_s))$. The map τ is holomorphic by IV, Sect. 3. We want to compute the induced map $(d\tau)_0$ on the tangent space $\mathcal{T}_S(0)$.

(7.2) **Proposition.** *Via the identification of Proposition* 7.1, *we have that*

$$(d\tau)_0\colon \mathcal{T}_S(0)\to\operatorname{Hom}(H^{2,0}(Y_0), H^{1,1}(Y_0))$$

is the composition of the Kodaira-Spencer map and the homomorphism

$$\nabla\colon H^1(\mathcal{T}_{Y_0})\to\operatorname{Hom}(H^{2,0}(Y_0), H^{1,1}(Y_0))$$

obtained by the cupproduct

$$\cup\colon H^1(\mathcal{T}_{Y_0})\otimes H^0(\Omega^2_{Y_0})\to H^1(\Omega^1_{Y_0}).$$

Proof. By IV, Sect. 3 the bundle $\bigcup_S H^0(\Omega^2_{Y_s})$ is a holomorphic subbundle of $f_{*2}\mathbb{C}_X\otimes\mathcal{O}_S$. It admits a never vanishing holomorphic section $\omega(s)$ over S. We may assume that for all $s\in S$ the differentiable manifold underlying Y_s is the same, say Y. We identify Y with Y_0 and consider local holomorphic coordinates $\{z_1, z_2\}$ on Y_0 as differentiable coordinates on Y. Moreover, by [K-N-S] we may find holomorphic coordinates $\{w_1, w_2\}$ on Y_s such that $w_1=w_1(z,s)$, $w_2=w_2(z,s)$ (differentiably in z and holomorphically in s). Let $\Delta=\{t\in\mathbb{C};\ |t|<\varepsilon\}$

and $s\colon \Delta \to S$ a holomorphic map with $s(0)=0$. We identify $(ds)_0(\partial/\partial t)$ with $\partial/\partial t$. If $\omega'(0) \equiv \partial/\partial t\, \omega(s(t))|_{t=0}$ mod(2,0)-forms we have that $(d\tau)_0(\partial/\partial t) \in \mathrm{Hom}(l, l^\perp/l)$ ($l = \mathbb{C}\cdot\tau(0)$) associates to $\omega(0)$ the class $\phi_{\mathbb{C}}(0)\,\omega'(0)$.

Let $\omega(s)$ in local coordinates be given by $\omega(s) = f(z,s)\,dw_1 \wedge dw_2$. Then we have

$$\omega'(0) \equiv f(z,0)\,[\bar\partial(w_1'(z,0)) \wedge dz_2 + \bar\partial(w_2'(z,0)) \wedge dz_1] \quad \mathrm{mod}(2,0)\text{-forms}.$$

By [M-K] a Dolbeault representative for the Kodaira-Spencer class $\rho(\partial/\partial t)$ reads in local coordinates

$$\bar\partial(w_1'(z,0))\,\partial/\partial z_1 + \bar\partial(w_2'(z,0))\,\partial/\partial z_2,$$

and hence

$$\omega'(0) = \rho(\partial/\partial t) \cup \omega(0). \quad \square$$

(7.3) **Theorem** (Local Torelli theorem). *The Kuranishi family for a K3-surface X is universal at all points in a small neighbourhood S of the point in the base corresponding to X. This base is smooth and of dimension 20, and the period map is a local isomorphism at each point of S.*

Proof. The first two assertions follow from Corollary 3.4 and Theorem I.10.5. It also follows that the Kodaira-Spencer map is an isomorphism at $s \in S$, so the local injectivity follows from Proposition 7.2 and the fact that "$\cup$" is an isomorphism by Corollary 3.4. The last statement of the theorem follows, since $\dim S = 20 = \dim \Omega$. $\square$

8. A Density Theorem

We call a surface X *exceptional*, if its Picard number is maximal, i.e. rank $S_X = h^{1,1}(X)$. This is equivalent to saying that $H^{2,0}(X) \oplus \overline{H^{2,0}(X)}$ is defined over $\mathbb{Q}$.

For a K3-surface X with $H^{2,0}(X) = \mathbb{C}\cdot\omega_X$ we therefore have that X is exceptional if and only if the subspace $E(\omega_X)$ of $H^2(X,\mathbb{R})$ spanned by $\{\mathrm{Re}\,\omega_X, \mathrm{Im}\,\omega_X\}$ is defined over the rationals. Then we have (for the notation see Sect. 1):

$$E(\omega_X) = T_X \otimes \mathbb{R},$$

and T_X becomes an oriented euclidean lattice of rank 2 such that $(\ ,\ )|T_X > 0$.

If $X = Km(Y)$, we have by Propositions 5.1 and 5.2 an injection $\alpha\colon H^2(Y,\mathbb{Z}) \to H^2(X,\mathbb{Z})$ preserving the Hodge decomposition, so X is exceptional if and only if Y is.

(8.1) **Proposition.** *Let T be a primitive oriented sublattice of rank 2 of L such that $(\ ,\ )|T > 0$ and $(x,x) \in 4\mathbb{Z}$ for all $x \in T$. Then there exists an exceptional Kummer surface X and an isometry $\phi\colon H^2(X,\mathbb{Z}) \to L$, mapping T_X isometrically onto T.*

Proof. Firstly, we construct a torus Y such that $T_Y \cong \frac{1}{2}T$. Let $\Gamma = \sum_{i=1}^{4} \mathbb{Z} e_i$ be an oriented $\mathbb{Z}$-module, i.e. we are given an isomorphism det: $\wedge^4 \Gamma \to \mathbb{Z}$. Then the lattice

$$\wedge^2 \Gamma = \mathbb{Z} e_1 \wedge e_2 + \mathbb{Z} e_3 \wedge e_4 + \mathbb{Z} e_1 \wedge e_3 + \mathbb{Z} e_4 \wedge e_2 + \mathbb{Z} e_1 \wedge e_4 + \mathbb{Z} e_2 \wedge e_3,$$

when equipped with the bilinear form defined by $(x, y) = \det(x \wedge y)$ for $x, y \in \bigwedge^2 \Gamma$, is clearly isometric to $H \oplus H \oplus H$, so by Theorem I.2.9 we have an isometric primitive embedding of $\frac{1}{2}T$ into $\wedge^2 \Gamma$. Let T' be the image, $\{t_1, t_2\}$ an oriented orthogonal basis for $T' \otimes \mathbb{R}$ and let $\{t_3, t_4, t_5, t_6\}$ be a basis for the orthogonal complement of $T' \otimes \mathbb{R}$ in $\wedge^2 \Gamma \otimes \mathbb{R}$. Furthermore let $\{t_1^\vee, \ldots, t_6^\vee\}$ be the basis of $\mathrm{Hom}(\wedge^2 \Gamma, \mathbb{C})$ dual to $\{t_1, \ldots, t_6\}$. We provide the latter $\mathbb{C}$-vector space with a $\mathbb{C}$-bilinear form in the usual way, by demanding that the 'evaluation'-map $\wedge^2 \Gamma \otimes \mathbb{C} \to \mathrm{Hom}(\wedge^2 \Gamma, \mathbb{C})$, i.e. the map given by $v \to (v, -)$, is an isometry. The form thus obtained is nothing but the one defined by $\alpha \wedge \beta = (\alpha, \beta) \cdot \det$, for $\alpha, \beta \in \mathrm{Hom}(\wedge^2 \Gamma, \mathbb{C})$. If we put $\omega = t_1^\vee + \sqrt{-1}\, t_2^\vee$, it follows that $\omega \wedge \omega = 0$ and $\omega \wedge \bar{\omega} = 2 \det$. The first equation shows that $\omega = \omega_1 \wedge \omega_2$ for $\omega_j \in \mathrm{Hom}(\Gamma, \mathbb{C})$ and the second one implies that if we map Γ into $\mathbb{C}^2$ via $\gamma \to (\omega_1(\gamma), \omega_2(\gamma))$ we obtain a lattice of maximal rank. Identifying Γ with this lattice we obtain a 2-torus $Y = \mathbb{C}^2/\Gamma$ and a canonical isomorphism $\psi: \Gamma \xrightarrow{\sim} H_1(Y, \mathbb{Z})$ of oriented lattices, inducing an isomorphism $\psi^\vee \wedge \psi^\vee: H^2(Y, \mathbb{Z}) \to \mathrm{Hom}(\wedge^2 \Gamma, \mathbb{Z})$ of euclidean lattices. Since ω_1, ω_2 are exactly the coordinate functions on $\mathbb{C}^2$, they are complex-linear and so the complexification of $\psi^\vee$ maps $H^{1,0}(Y)$ to the subspace of $\mathrm{Hom}(\Gamma, \mathbb{C})$ spanned by the complex-linear functionals, i.e. $\mathbb{C}\omega_1 \oplus \mathbb{C}\omega_2$. Then $H^{2,0}(Y)$ corresponds to $\mathbb{C}\omega_1 \wedge \omega_2 = \mathbb{C}\omega$, S_Y to $\mathrm{Ker}\, \omega$ and T_Y to T'.

If $X = Km(Y)$, then by Proposition 5.1 the map α: $H^2(X, \mathbb{Z}) \to H^2(Y, \mathbb{Z})$ multiplies $(\ ,\ )$ by 2 and its complexification sends $H^{2,0}(X)$ to $H^{2,0}(Y)$. It follows that T_X is isometric to T.

Let $\phi': H^2(X, \mathbb{Z}) \xrightarrow{\sim} L$ be any isometry. Since T and $\phi'(T_X)$ are isometric by construction, we may apply Theorem I.2.9, (i) to find an automorphism of L which sends $\phi'(T_X)$ to T in such a way that the ordering of two bases, coming from orientation, correspond. Composing ϕ' with this automorphism we get an isometry ϕ: $H^2(X, \mathbb{Z}) \to L$ as desired. □

(8.2) **Proposition.** *The set of the rationally defined 2-planes $P \subset L_{\mathbb{R}}$ satisfying $(x, x) \in 4\mathbb{Z}$ for all $x \in P \cap L$ is dense in the Grassmannian $Gr_2(L_{\mathbb{R}})$.*

For the proof we need

(8.3) **Lemma.** *Let $m, n \in \mathbb{N}$ and let M be a lattice containing a primitive vector e_0 with $(e_0, e_0) \equiv m (\mathrm{mod}\, n)$. Then the set of lines $l \subset M_{\mathbb{R}}$ spanned by a primitive vector e with $(e, e) \equiv m (\mathrm{mod}\, n)$ is dense in $\mathbb{P}(M_{\mathbb{R}})$.*

Proof. Let U be an open subset of $\mathbb{P}(M_{\mathbb{R}})$, $U \neq \emptyset$. It contains a rationally defined element $l = \mathbb{R} \cdot e$ where $e \in M$ may be chosen primitive. If $e = \pm e_0$ there is nothing to prove. So assume that this is not the case. Then e and e_0 are linearly independent. Let $f \in (\mathbb{R}e + \mathbb{R}e_0) \cap M = M'$ such that $\{e, f\}$ is a basis of M' and let $e_0 = ae + bf$, with $a, b \in \mathbb{Z}$. Since e_0 is primitive, $\gcd(a, b) = 1$. Then, for any $N \in \mathbb{Z}$, we have that $e_N = e_0 + Nbe = (a + Nb)e + bf$ is also primitive. If N is a multiple of n we have $(e_N, e_N) \equiv (e_0, e_0) \equiv m (\mathrm{mod}\, n)$ and if N is large enough $\mathbb{R} e_N = \mathbb{R}\left(e + \frac{1}{Nb} e_0\right)$ is a line in U. So U contains an element with the required properties. □

Proof of Proposition 8.2. Let U be an open subset of $Gr_2(L_{\mathbb{R}})$, $U \neq \emptyset$. Since H contains a primitive vector of norm 4 so does $L \supset H$. By Lemma 8.3 there exists a $P' \in U$ containing a primitive vector e_1 with $(e_1, e_1) \equiv 4 (\mathrm{mod}\, 8)$. Let M be the orthogonal complement of e_1 in L. Since by Theorem 2.9, (ii) there is an automorphism of L sending e_1 to a primitive vector in $H \subset L$ of length (e_1, e_1), we see that M contains a sublattice isometric to $H \oplus H \oplus (-E_8) \oplus (-E_8)$. Hence it contains a primitive vector of norm 64. Again Lemma 8.3 implies that M contains a primitive vector e_2 with $(e_2, e_2) \equiv 0 (\mathrm{mod}\, 64)$ such that the 2-plane spanned by e_1 and e_2 belongs to U.

We claim that for all $f \in P \cap L$ we have $(f, f) \in 4\mathbb{Z}$. Since

$$(e_1, e_1) f - (e_1, f) e_1 \in P \cap L$$

is orthogonal to e_1, it is an integral multiple $a e_2$ of e_2. So $(e_1, e_1) f = (e_1, f) e_1 + a e_2$ and taking norms we get

$$(e_1, e_1)^2 (f, f) = (e_1, f)^2 (e_1, e_1) + a^2 (e_2, e_2).$$

Now (f, f) is even and if it were not 4-divisible the left hand side of the above equality would be divisible by 2^3, but by no larger power of 2. On the other hand the right hand side is either exactly divisible by 2^2 (in case (e_1, f) is odd) or by at least 2^6 (in case (e_1, f) is even). The contradiction shows that $(f, f) \in 4\mathbb{Z}$. □

(8.4) *Remark.* This proof shows that the following holds: given $x \in L$ with $(x, x) \in 4\mathbb{Z}$, the set of $y \in L_{\mathbb{R}}$ such that $(\mathbb{R} x + \mathbb{R} y) \cap L$ is a primitive rank-2 sublattice of L, all of whose vectors have norm in $4\mathbb{Z}$, is dense in $L_{\mathbb{R}}$.

(8.5) **Corollary.** *The period points of marked projective Kummer surfaces are dense in* Ω.

Proof. Let $G_2^+(L)$ be the set of oriented 2-planes of $L_{\mathbb{R}}$ on which $(\, , \,)$ is positive definite, and consider the map $\pi: \Omega \to G_2^+(L)$ sending $l = [\omega]$ to the oriented 2-plane $E(\omega)$ with oriented basis $\{\mathrm{Re}\,\omega, \mathrm{Im}\,\omega\}$. This is a bijection, since if E is an oriented 2-plane on which $(\, , \,)$ is positive, and $\{\omega_1, \omega_2\}$ an oriented orthogonal basis for it, the line $[\omega_1 + \sqrt{-1}\,\omega_2] = [\omega]$ is in Ω and $E(\omega) = \mathbb{R}\omega_1 + \mathbb{R}\omega_2$.

So it suffices to show that the set of oriented 2-planes $E(\omega)$ belonging to marked projective Kummer surfaces is dense in the Grassmannian $Gr_2(L)$, in which $G_2^+(L)$ is Zariski open. By Proposition 8.2 the set of $[\omega] \in \Omega$ such that (i) the 2-plane $E(\omega)$ is rationally defined, say $E(\omega) = P \otimes \mathbb{Q}$, and (ii) $(x, x) \in 4\mathbb{Z}$ for all $x \in P \cap L$, is dense in $Gr_2(L)$. According to Lemma 8.1 any such $[\omega]$ is the image under the period map of a marked exceptional Kummer surface X. Since for such an X the Picard lattice contains elements with positive norm, X is projective. □

(8.6) **Corollary.** *Any two K3-surfaces are diffeomorphic. In particular any K3-surface is simply-connected.*

Proof. Since any two 2-tori are isomorphic as real Lie groups, the corresponding Kummer surfaces are diffeomorphic. If X is a K3-surface, let $p\colon Y \to S$ be its Kuranishi deformation. Any trivialisation of $p_{*2}\mathbb{Z}_Y$ induces a period map which by Theorem 7.3 realises S as an open subset of Ω. By Remark 8.4 this open set contains period points of projective Kummer surfaces. So X is diffeomorphic to a Kummer surface. As was observed before, the last assertion is a consequence of the fact that any smooth surface of degree 4 in $\mathbb{P}_3$ is a simply-connected K3-surface (Theorem V.2.1). □

Remark. In fact, the proof of Corollary 8.6 shows that any two K3-surfaces are deformations of each other.

9. Behaviour of the Kähler Cone Under Deformations

Let us recall some notation from Sect. 8 and at the same time introduce some more.

For $[\omega] \in \Omega$ we let $E(\omega)$ be the oriented 2-plane of $L_{\mathbb{R}}$ spanned by the oriented base $\{\mathrm{Re}\,\omega, \mathrm{Im}\,\omega\}$, viewed as an element of the manifold $G_2^+(L)$ of oriented 2-planes of $L_{\mathbb{R}}$ on which $(\ ,\ )$ is positive definite. Let

$$K\Omega = \{(\kappa, [\omega]) \in L_{\mathbb{R}} \times \Omega; (\kappa, E(\omega)) = 0, (\kappa, \kappa) > 0\}$$

and for $(\kappa, [\omega]) \in K\Omega$ let $E(\kappa, \omega)$ be the oriented 3-space of $L_{\mathbb{R}}$ spanned by the oriented basis $\{\kappa, \mathrm{Re}\,\omega, \mathrm{Im}\,\omega\}$ and viewed as an element of the manifold $G_3^+(L)$ of oriented 3-planes of $L_{\mathbb{R}}$ on which $(\ ,\ )$ is positive definite. So we obtain a map

$$\begin{aligned} \Pi\colon\ & K\Omega \to G_3^+(L) \\ & (\kappa, [\omega]) \to E(\kappa, \omega). \end{aligned}$$

(9.1) **Proposition.** *The map Π exhibits $K\Omega$ as an $SO(3)$-fibre bundle over $G_3^+(L)$ with fibre $\mathbb{R}^3 \setminus \{0\}$. In fact $\Pi^{-1}(E(\kappa, \omega)) \cong E(\kappa, \omega) \setminus \{0\}$. The group* $\mathrm{Aut}(L_{\mathbb{R}}) \cong O(3, 19)$ *acts Π-equivariantly on $K\Omega$. This action is proper.*

Proof. All assertions except the last one are immediately clear. To prove that $\mathrm{Aut}(L_{\mathbb{R}})$ acts properly, we observe that it acts properly on the base space (the stabiliser of a 3-plane is $O(3) \times O(19)$, hence compact) and so it also acts properly on $K\Omega$. □

(9.2) **Corollary.** *The set*

$$(K\Omega)^0 = \{(\kappa, [\omega]) \in K\Omega; (\kappa, d) \neq 0 \textit{ for every } d \in L \textit{ with } (d, d) = -2 \textit{ and } (\omega, d) = 0\}$$

is open in $K\Omega$.

Proof. Aut(L) is discrete in Aut($L_{\mathbb{R}}$), so Proposition 9.1 implies that it acts properly on $K\Omega$. A fortiori this holds for the group W generated by the reflections s_d, for $d \in L$, $(d,d) = -2$, $(d,\omega) = 0$, where $s_d \in \mathrm{Aut}(L)$ is defined as usual by $s_d(x) = x + (x,d)d$, $x \in L$. We obtain $(K\Omega)^0$ from $K\Omega$ by deleting all reflection hyperplanes H_d for s_d. Since W is discrete and acts properly on $K\Omega$, these form a locally finite, hence closed set: every point of $K\Omega$ has an open neighbourhood U such that $U \cap s_d(U) \neq \emptyset$ for only finitely many $s_d(U)$, so a fortiori only finitely many H_d meet U. □

(9.3) **Lemma.** *Let* $p\colon Y \to S$ *be a family of kählerian K3-surfaces. Then* $\bigcup_{s \in S} H^{1,1}(Y_s, \mathbb{R})$ *is a real-analytic subbundle of* $p_{*2}\mathbb{R}_Y$ *in which the union of the Kähler cones is open.*

Proof. The assertion being local we may take for S any sufficiently small open polydisc, and replace p by a locally universal family, i.e. we may suppose that $S \subset \Omega$ and we may identify $H^2(Y_s, \mathbb{Z})$, $s \in S$ with L. Since $H^{1,1}(Y_s, \mathbb{R}) = E(s)^{\perp}$, this space varies in a real analytical way with s and the first assertion follows from [Hi4]. Let $K_s \subset H^{1,1}(Y_s, \mathbb{R})$ be the Kähler cone of Y_s, and let $\kappa \in K_s$. If $\kappa_0 \in K_s$ is a Kähler class, the segment $[\kappa, \kappa_0]$ belongs to K_s. Let C_s be a compact connected neighbourhood of $[\kappa, \kappa_0]$. Since $K_s \times \{s\} \subset (K\Omega)^0$ by compactness and Proposition 9.1, we may find an open neighbourhood V of C_s inside the union $\bigcup_{s \in S} H^{1,1}(Y_s, \mathbb{R}) \subset K\Omega$ such that $V \subset (K\Omega)^0$. Moreover we may assume

(i) if $\kappa_1 \in V \cap H^2(Y_s, \mathbb{R})$, then $(\kappa_1, \kappa_1) > 0$, and
(ii) for every $s_1 \in S$, the set $V \cap H^{1,1}(Y_{s_1}, \mathbb{R})$ is connected.

According to [K-S2], Thm. 15 the set of Kähler classes is open in $\bigcup_{s \in S} H^{1,1}(Y_s, \mathbb{R})$. Since C_s contains a Kähler class, this implies that for some neighbourhood U of s in S the set $V \cap H^{1,1}(Y_{s_1}, \mathbb{R})$ contains a Kähler class. By (i), (ii) and Corollary 3.8 this set is contained in the Kähler cone of Y_{s_1}. The lemma follows. □

(9.4) **Proposition.** *Let* $p\colon Y \to S$ *and* $p'\colon Y' \to S$ *be two families of Kähler K3-surfaces and let* $\Phi\colon p'_{*2}\mathbb{Z}_{Y'} \xrightarrow{\sim} p_{*2}\mathbb{Z}_Y$ *be an isomorphism of local systems such that* $\Phi(s)$ *is a Hodge-isometry for each* $s \in S$*. Then the set of* $s \in S$ *with the property that* $\Phi(s)$ *maps the Kähler cone of* Y'_s *to the Kähler cone of* Y_s*, is open in* S*.*

Proof. The conditions imposed on Φ imply that Φ induces bundle isomorphisms between the real-analytic bundles $\bigcup_{s \in S} H^{1,1}(Y'_s, \mathbb{R})$ and $\bigcup_{s \in S} H^{1,1}(Y_s, \mathbb{R})$. By Lemma 9.3 the set $s \in S$ for which an element in the Kähler cone of Y'_s is mapped into the Kähler cone of Y_s is open in S. But Proposition 3.10 teaches us that for such s the entire Kähler cone of Y'_s is mapped onto the Kähler cone of Y_s. □

10. Degenerations of Isomorphisms Between Kähler K3-Surfaces

(10.1) **Proposition.** *Let* X *be a Kähler K3-surface and let* $\{C_i, D_i\}_{i \in I}$ *be a finite collection of curves on* X*, with* $c_i = c_1(\mathcal{O}_X(C_i))$ *and* $d_i = c_1(\mathcal{O}_X(D_i))$.

If the map $\phi: H^2(X,\mathbb{Z}) \to H^2(X,\mathbb{Z})$ *defined by* $\phi(x) = x + \sum_{i \in I} (c_i, x) d_i$, $x \in H^2(X,\mathbb{Z})$, *is an effective Hodge-isometry, then* $I = \emptyset$, i.e. $\phi = \mathrm{id}$.

Proof. Let us take $x \in \mathscr{C}_X^+$, the Kähler cone of X. Then the numbers $r_i = (c_i, x)$ are positive and since ϕ is an isometry, we have

$$0 = (\phi(x), \phi(x)) - (x,x) = (\phi(x) - x, \phi(x) + x)$$
$$= (\sum_{i \in I} r_i d_i, \phi(x) + x).$$

By Proposition 3.10, (iii) we have $\phi(x) \in \mathscr{C}_X^+$, hence $x + \phi(x) \in \mathscr{C}_X^+$ and so $(x + \phi(x), d_i) > 0$ for all $i \in I$, and the preceding equation shows that $I = \emptyset$. $\square$

We now come to a crucial result, expressed by Theorem 10.6 below. It says that, under certain circumstances, isomorphisms between corresponding members of two sequences of K3-surfaces, both converging to a limit K3-surface, extend to these limiting surfaces.

Our point of departure is the following.

(10.2) **Data**

i) Two families $p: Y \to S$, $p': Y' \to S$ of Kähler K3-surfaces over a polydisc S, centred at 0.
ii) A sequence $\{s_n\}$ in S converging to 0.
iii) An isomorphism $\Phi: p'_{*2}\mathbb{Z}_{Y'} \to p_{*2}\mathbb{Z}_Y$ of local systems.
iv) Isomorphisms $f_n: Y_{s_n} \to Y'_{s_n}$ such that the induced map in 2-cohomology coincides with $\Phi(s_n)$ for all $n \in \mathbb{N}$.

We consider the graphs Γ_n of f_n in $Y_{s_n} \times Y'_{s_n}$ as compactly supported cycles of $Y \times_S Y'$ and ask whether there exists a limit cycle Γ_0 in the topology introduced by Barlet in [Ba1]. In fact Barlet gives the set of compactly supported cycles on a fixed reduced analytic space Z the structure of a reduced analytic space $H(Z)$. Moreover, over $H(Z)$ there exists a universal family of compactly supported cycles of Z. Hence, if two points of $H(Z)$ are in the same connected component of $H(Z)$ they are cohomologous. A subset of $H(Z)$ is compact whenever the corresponding cycles all lie in a compact subset of Z and if moreover all volumes are uniformly bounded in a suitable but fixed metric on Z ([Ba2]). We apply this to the set consisting of the Γ_n.

(10.3) **Lemma.** *There exists a subsequence of* $\{s_n\}$ *such that the corresponding* Γ_n *converge to a limit cycle* Γ_0 *in Barlet's topology.*

Proof. We may assume that S is taken so small that

i) $Y \times_S Y'$ is differentiably trivial,
ii) each Y_s (resp. Y'_s) carries a Kähler metric whose (1,1)-form θ_s (resp. θ'_s) depends differentiably on s (see [K-S2]).

Let us take a metric on $Y \times_S Y'$ whose (1,1)-form θ on each fibre $Y_s \times Y'_s$ coincides with $p_1^*(\theta_s) + p_2^*(\theta'_s)$, where p_1 (resp. p_2) is the projection onto Y_s (resp. Y'_s).

If $[\Gamma_n] \in H^4(Y \times_S Y')$ is the dual cohomology class of the cycle Γ_n, then

(5) $$\mathrm{Vol}(\Gamma_n) = \tfrac{1}{2} \int_{\Gamma} (\theta \cup \theta) = \tfrac{1}{2} [\Gamma_n] \cup (\theta_{s_n} \cup \theta_{s_n}).$$

Since $Y \times_S Y'$ is differentiably trivial, the given isometry Φ in 10.2, (iii) is completely determined by its value at one point, i.e. if we choose a trivialisation $f\colon Y \times_S Y' \to Y_0 \times Y_0' \times S$ and if $f(s)$ is the induced diffeomorphism $Y_s \times Y_s' \to Y_0 \times Y_0'$ we have that $\Phi(s) = (f(s)^*)^{-1} \Phi(0) f(s)^*$. Here we view $\Phi(s)$ as an element of $H^4(Y_s \times Y_s', \mathbb{Z})$. In particular, $[\Gamma_n] = (f(s_n)^*)^{-1} \Phi(0) f(s_n)^*$, in view of the assumption 10.2, (iv). Therefore and because of (5) twice the volume of Γ_n is the value at s_n of the continuous function $(f(s)^{*-1} \Phi(0) f^*(s)) \cup \theta_s \cup \theta_s$, hence is uniformly bounded on any compact neighbourhood of $0 \in S$. □

(10.4) *Remark.* Two compactly supported cycles of $Y \times_S Y'$ in the same component of $H(Y \times_S Y')$ are homologous. In particular $[\Gamma_0] = \lim_{n \to \infty} [\Gamma_n]$ is the same as $\Phi(0) \in H^4(Y_0 \times Y_0', \mathbb{Z})$.

Before we proceed to the main Theorem 10.6 we make some general remarks. Any purely 2-dimensional cycle C on a product of two compact surfaces X and Y induces linear maps in homology

$$[C]_*\colon H_k(X) \to H_k(Y) \qquad (k = 0, \ldots, 4)$$

as follows: if $p_2\colon X \times Y \to Y$ denotes projection onto the second factor, $[C]_* \xi = p_{2*}([\xi \times Y] \cdot C)$ for all $\xi \in H_k(X)$.

By Poincaré-duality we obtain similar maps in cohomology, also denoted by $[C]_*\colon H^k(X) \to H^k(Y)$. We apply this to $\Gamma_n \subset Y_{s_n} \times Y_{s_n}'$ and $\Gamma_0 \subset Y_0 \times Y_0'$. Of course $\lim_{n \to \infty} [\Gamma_n]_* = [\Gamma_0]_*$ in all dimensions. In particular $[\Gamma_0]_*$ is an isomorphism. This motivates the hypotheses in the following

(10.5) **Proposition.** *Let X and Y be compact surfaces with $p_g(Y) \neq 0$. Let C be a purely 2-dimensional effective cycle on $X \times Y$, such that $[C]_*$ is an isomorphism in all dimensions. Then $C = \Delta + \sum_{i=1}^{N} C_i \times D_i$, where Δ gives a bimeromorphic correspondence and C_i (resp. D_i) are curves on X (resp. Y).*

Proof. Since $[C]_*$ is an isomorphism in 4-homology, the degree of the projection $C \to Y$ must be one. So

$$C = \Delta + C',$$

where Δ projects bimeromorphically onto Y and C' maps to lower dimensional cycles. Similarly, *either* Δ projects onto a lower dimensional subvariety of X and $C' \to X$ has degree one *or* $\Delta \to X$ projects birationally. The first alternative is impossible, since in this case the image of $H^2(X, \mathbb{Z})$ in $H^2(Y, \mathbb{Z})$ via $[C]_*$ would entirely be contained in the Picard lattice. But $p_g(Y) \neq 0$, so this lattice is strictly smaller than $H^2(Y, \mathbb{Z})$ while $[C]_*$ is an isomorphism. It follows that Δ establishes a bimeromorphic correspondence $X \to Y$ and that $C' = \sum C_i \times D_i$ for some curves $C_i \subset X$ and $D_i \subset Y$. □

(10.6) **Theorem.** *Given two families $p: Y \to S$, $p': Y' \to S$ of Kähler K3-surfaces over a polydisc S. Then the existence of a sequence and isomorphisms as in* 10.2, (ii)–(iv) *imply that Y_0 and Y_0' are isomorphic. If moreover $\Phi(0)$ is an effective Hodge-isometry, then a subsequence of $\{f_n\}$ converges uniformly in the Barlet topology to an isomorphism $f_0: Y_0 \to Y_0'$ with the property that $(f_0)^* = \Phi(0)$.*

Proof. The first assertion is a consequence of Lemma 10.3 and Proposition 10.5. Indeed, passing to a subsequence if necessary, $\lim_{n\to\infty} \Gamma_n = \Gamma_0$ exists as a purely 2-dimensional cycle of $Y_0 \times Y_0'$ and by Proposition 10.5 the surfaces Y_0 and Y_0' are bimeromorphically equivalent, hence isomorphic, since both are minimal and non-ruled. (Apply Proposition III.4.6.) More precisely $\Gamma_0 = \Delta_0 + \sum_{i\in I} C_i \times D_i$, where Δ_0 is the graph of an isomorphism f_0 and the C_i's (resp. the D_i's) are curves on Y_0 (resp. Y_0'). By Remark 10.4 we have that $[\Gamma_0] = \Phi(0) \in H^4(Y_0 \times Y_0', \mathbb{Z})$. Let us for a moment identify Y_0 and Y_0' via f_0 and let us translate the identity $[\Gamma_0] = \Phi(0)$ as an identity between automorphisms of $H^2(Y_0, \mathbb{Z})$:

$$\Phi(0)\,x = x + \sum_{i\in I} ([C_i], x)\, D_i, \qquad x \in H^2(Y_0, \mathbb{Z}).$$

So, if $\Phi(0)$ is effective, an application of Proposition 10.1 gives that $I = \emptyset$, i.e. $\Phi(0) = f_0^*$ as desired. □

(10.7) **Corollary.** *The group of automorphisms of a Kähler K3-surface X inducing the identity in $H^2(X, \mathbb{Z})$ is finite.*

Proof. Since $H^0(\mathcal{T}_X) = 0$ by Corollary 3.4, we have that the group of automorphisms G is discrete. It is also compact: let $g_n \in G$, $n \in \mathbb{N}$, and apply Theorem 10.6 to the trivial family $p = p'$, $\Phi^* = \mathrm{id}$, $\{s_n\}$ a sequence converging to $0 \in S$ and $f_n = g_n$; then a subsequence of g_n converges within G. □

11. The Torelli Theorems for Kähler K3-Surfaces

(11.1) **Theorem** (Torelli theorem). *Let X and X′ be Kähler K3-surfaces and suppose that there exists an effective Hodge-isometry $\phi: H^2(X', \mathbb{Z}) \to H^2(X, \mathbb{Z})$. Then $\phi = f^*$, with $f: X \to X'$ biholomorphic.*

Proof. Let $p: Y \to S$, resp. $p': Y' \to S'$ be the Kuranishi families for $X = Y_0$, resp. $X' = Y_0'$, and let S and S' be small polydiscs about 0. Then first of all we can choose a marking $\alpha: p_{*2}\mathbb{Z}_Y \xrightarrow{\sim} L_S$ and secondly we can extend ϕ to an isomorphism of local systems $\Phi: p'_{*2}\mathbb{Z}_{Y'} \to p_{*2}\mathbb{Z}_Y$, thereby obtaining a marking $\alpha \circ \Phi$ for the local system $p_{*2}\mathbb{Z}_Y$. The two markings determine period maps $\tau: S \to \Omega$, resp. $\tau': S' \to \Omega$ and, by construction, we have $\tau(0) = \tau'(0)$. By the local Torelli theorem 7.3 we may shrink S and S' if necessary in such a way that there is a biholomorphic map $q: S \to S'$ such that $\tau' \circ q = q$. The family which p' via q induces over S is of course still locally universal, so that we may as well assume that $S = S'$, $q = \mathrm{id}$ and (hence) $\tau = \tau'$.

Now by construction $\Phi(s)$ is a Hodge-isometry for every $s \in S$ and by Proposition 9.4 we may replace S by a smaller neighbourhood of $0 \in S$ such that not only $\Phi(0)=\phi$ but such that moreover $\Phi(s)$ maps the Kähler cone of Y'_s to the Kähler cone of Y_s for all $s \in S$. By Corollary 8.5 the period points of projective Kummer surfaces are dense in Ω, so we can find a sequence $\{s_n\}$ in S converging to 0 such that $\tau(s_n)$ is the period point of a projective Kummer surface. The Torelli theorem for projective Kummer surfaces 6.2 implies that $\Phi(s_n)=f_n^*$ for a biholomorphic map $f_n\colon Y_{s_n} \to Y'_{s_n}$. It now follows from Theorem 10.6 that $\Phi(0) = \phi$ is induced by an isomorphism $f\colon X \to X'$. □

(11.2) **Corollary** (Weak Torelli theorem). *Let X and X' be Kähler K3-surfaces and suppose that there exists a Hodge-isometry $H^2(X',\mathbb{Z}) \to H^2(X,\mathbb{Z})$. Then X and X' are isomorphic.*

Proof. This is an immediate consequence of Theorem 11.1 and Proposition 3.9. □

(11.3) **Proposition.** *Let X be a kählerian K3-surface. Any automorphism inducing the identity on $H^2(X,\mathbb{Z})$ is itself the identity.*

Proof. By Corollary 10.7 any such automorphism g has finite order, which we may assume to be prime, say l. We shall show that the assumption $l \neq 1$ leads to a contradiction. Since g has finite order, locally around a fixed point $x \in X$ we may linearise the action ([Car]). Let (n_x, m_x) be the weights (III, Sect. 5). Since $g^*|H^2(X,\mathbb{C})=\mathrm{id}$, for any nowhere vanishing holomorphic 2-form ω we have that $g^*(\omega)=\omega$, hence $\det(dg)_x=1$, i.e. $n_x=-m_x$ and in particular $n_x \neq 0$, $m_x \neq 0$. This means that x is an isolated fixed point and we may apply Lefschetz' fixed point formula

$$\sum(-1)^k \operatorname{Tr} g^*|H^k(X,\mathbb{R})=\mu,$$

where μ is the number of fixed points. So $\mu=e(X)=24$. Secondly, since $\det(1-dg)_x \neq 0$ we can apply the holomorphic fixed point formula ([A-S], p. 567) to g:

$$2=\sum_k(-1)^k \operatorname{Tr}(g^*|H^{0,k}(X))=\sum_x \det(1-(dg)_x)^{-1},$$

where the summation is over the μ fixed points x. Since

$$\det(1-(dg)_x)=(1-\lambda)(1-\lambda^{-1})=4\sin\theta/2 \leq 4$$

$(\lambda=\mathrm{e}(\theta))$ it follows that $\mu \leq 8$, contradicting $\mu=24$. □

(11.4) **Corollary.** *In Theorem 11.1 the isomorphism f is the unique isomorphism with $f^*=\phi$.*

12. Construction of Moduli Spaces

Let X_0 be any Kähler K3-surface and let $p\colon X \to U$ be the Kuranishi family of X_0. We assume that U is taken small enough in order that (i) all fibres X_s,

$s \in U$ are Kähler (possible by [K-S]), (ii) p is the Kuranishi family for all $s \in U$ (see Theorem 7.3), (iii) U is contractible. By (iii) we can find a marking $\phi: p_{*2}\mathbb{Z}_Y \xrightarrow{\sim} L_U$. Let $\tau: U \to \Omega$ be the period map determined by this marking. We assume that U is so small that (iv) τ is an embedding. This is again possible by Theorem 7.3. The last condition guarantees that no two marked K3-surfaces $(X_s, \phi(s))$, $(X_{s'}, \phi(s'))$, $s, s' \in U$ are isomorphic. We consider all pairs $(p: X \to U, \phi)$ of marked Kuranishi families of Kähler K3-surfaces, with base U satisfying the above conditions (i) up to (iv). Let us glue the U's by identifying two points for which the corresponding marked K3-surfaces are isomorphic. In this manner we obtain an analytic space M_1 such that every point has a neighbourhood isomorphic to some U. In particular M_1 is smooth (but possibly non-Hausdorff). The remark after I.10.6 implies that if, in glueing the U's, we glue the universal families over it, no two points in the same fibre come together so that over M_1 we obtain a marked family of Kähler K3-surfaces. By construction it is a *universal* such family. Except for the fact that the base is non-Hausdorff, this proves the following theorem.

(12.1) **Theorem.** *There exists a universal marked family of Kähler K3-surfaces. The base space is a non-Hausdorff "smooth analytic space" of dimension* 20.

As to the fact that the base isn't Hausdorff we make the following remark.

(12.2) *Remark.* The following example, due to Atiyah ([At2]), shows that the base space M_1 is not a Hausdorff space. Consider the family of quartics $\{X_t\}$, $t \in \Delta = \{t \in \mathbb{C};\, |t| < \varepsilon\}$, which in affine coordinates is given by

$$x^2(x^2-2)+y^2(y^2-2)+z^2(z^2-2)=2t^2.$$

For $t \neq 0$ the surface X_t is smooth and X_0 has a unique double point at $x_0 = (0,0,0)$ which is also the unique singular point of the total space X. As shown in V, Sect. 2 the surfaces X_t $(t \neq 0)$ are K3-surfaces. The tangent cone at x_0 is $x^2+y^2+z^2+t^2=0$, and one σ-process suffices to desingularise X, say $\sigma: \bar{X} \to X$. Let $E = \sigma^{-1}(x_0)$ be the exceptional manifold. The proper transform $\bar{X}_0$ of X_0 is a smooth K3-surface meeting E $(\cong \mathbb{P}_1 \times \mathbb{P}_1)$ transversally in a (-2)-curve on $\bar{X}_0$ (a curve of bidegree $(1,1)$ on E). Each of the rulings of E defines a contraction of E inside $\bar{X}$ onto this curve, and by an explicit calculation one shows that the resulting threefolds X_1 and X_2 are smooth. So we obtain two families $p_1: X_1 \to \Delta$, $p_2: X_2 \to \Delta$ of Kähler K3-surfaces, identical over $\Delta \setminus \{0\}$. These families are not isomorphic, since otherwise there would be an automorphism of X extending the identity map on $X \setminus X_0$ but acting nontrivially on the tangent cone at x_0. Choose a marking for p_1. This induces a marking for p_2, since over $\Delta \setminus \{0\}$ the families p_1 and p_2 coincide. By construction the resulting period maps coincide over $\Delta \setminus \{0\}$, but the images of $\{0\}$ are different. This is only possible if M_1 is non-Hausdorff.

Associated to the universal family $p: Y \to M_1$ of Theorem 12.1, we have the period map (see IV, Sect. 3) $\tau_1: M_1 \to \Omega$. The point $\tau_1(x)$ is called the *period point of* x.

We next define the notion of '*marked pairs*' consisting of a K3-surface and a Kähler class on it. In the fibre bundle $K\Omega \to \Omega$ from Sect. 9 we take for every $[\omega] \in \Omega$ the cone $C_\omega = \{x \in L_{\mathbb{R}}; (x, \omega) = 0, (x, x) > 0\}$. Since by [He], Chap. IX, 4.3 $O(a,b)$ has the homotopy type of $O(a) \times O(b)$, the exact homotopy sequence for the fibration $SO(2) \times O(1,19) \to O(3,19) \to \Omega$ easily shows that Ω is connected and simply connected. So we may choose a connected component of C_ω, say C_ω^+ in such a way that it varies continuously with $[\omega] \in \Omega$. If now X is a Kähler K3-surface and $\kappa \in H^{1,1}(X, \mathbb{R})$ any Kähler class, we say that (X, κ) is *a marked pair* if first of all X is marked, say $\phi \colon H^2(X, \mathbb{Z}) \xrightarrow{\sim} L$ and secondly $\phi_{\mathbb{C}}(\kappa) \in C_\omega^+$, where $[\omega] \in \Omega$ is the point $\tau_1(X, \phi)$.

Let $p \colon Y \to S$, $\phi \colon p_{*2} \mathbb{Z}_Y \to L_S$ be a marked family of K3-surfaces and let $\kappa \in \Gamma(S, p_{*2} \mathbb{R}_Y \otimes_{\mathbb{R}} \mathcal{D}_S)$ have the property that for every $s \in S$ the value $\kappa(s)$ is a Kähler class such that $\phi(s)$ gives $(Y_s, \kappa(s))$ the structure of a marked pair. In this situation we speak of a *family of marked pairs*.

Let M_2' be the total space of the real-analytic vector bundle $\bigcup_{t \in M_1} H^{1,1}(Y_t)$ (see Lemma 9.3) and let $M_2 \subset M_2'$ be the subset of Kähler classes in it. Then M_2 is an open subset of M_2' by [K-S 2], Thm. 15, hence M_2 is a real-analytic manifold of dimension 60. It is not difficult to see that M_2 is a universal object for marked pairs (it is even a "fine moduli space" for such pairs) and recalling Corollary 9.2 we have a real-analytic map

$$\tau_2 \colon M_2 \to (K\Omega)^0,$$

defined as follows: if $\kappa \in H^{1,1}(Y_t)(t \in M_1)$, we let $\tau_2(\kappa) = (\phi_{\mathbb{C}}(\kappa), \tau_1(t))$ – where ϕ is the marking for Y_t. This map we call the *refined period map*.

In this way the period map τ_1 and the refined period map τ_2, together with the forgetful maps

$$M_2 \to M_1, \qquad (K\Omega)^0 \to \Omega$$

make the self-explanatory diagram

$$\begin{array}{ccc} M_2 & \xrightarrow{\tau_2} & (K\Omega)^0 \\ \downarrow & & \downarrow \\ M_1 & \xrightarrow{\tau_1} & \Omega \end{array}$$

commutative.

In view of this, the Torelli theorems (11.1) and (11.3) can be reformulated as follows.

(12.3) **Theorem.** *The map τ_2 is injective (in particular M_2 is Hausdorff).*

In the light of this result it is natural to ask for the image of τ_2, a question we deal with in Sect. 14.

13. Digression on Quaternionic Structures

Let X be a differentiable manifold equipped with a Riemannian metric g and an almost-complex structure I. The metric is hermitian if $g(IY, IZ) = g(Y, Z)$ for all $Y, Z \in \mathcal{T}_X(p)$, $p \in X$. A hermitian metric g is called a *Kähler metric*, if I is parallel with respect to the Levi-Cività connection. The existence of a Kähler metric implies that I is integrable ([K-N], Chap. IX, §4) and the Kähler metric is a Kähler metric in the usual sense.

Let X be any differentiable manifold. An *almost-quaternionic structure* on X consists of two almost-complex structures I and J on X such that $I \circ J + J \circ I = 0$. Any almost-quaternionic structure $\{I, J\}$ defines an action of the algebra of quaternions on each tangent space $\mathcal{T}_X(p)$, $p \in X$, by setting $K = I \circ J$ and letting each quaternion $q = a + bi + cj + dk$ act as $Q = a + bI + cJ + dK$. Now Q is an almost-complex structure if and only if $a = 0$ and $b^2 + c^2 + d^2 = 1$. So to every almost-quaternionic structure $\{I, J\}$ there is associated a 2-sphere $S(I, J)$ of almost-complex structures. This sphere we call the associated *sphere of almost-complex structure* $S(X)$.

If (X, g) is a Riemannian manifold, an almost-quaternionic structure $\{I, J\}$ on X is called *quaternionic*, if g is Kähler with respect to I and J. It is easily verified that then g is Kähler with respect to any of the almost-complex structures from $S(I, J)$. In particular all of those are integrable and we can speak of the *associated sphere of complex structures.*

(13.1) **Lemma.** *Let X be a simply-connected complex manifold admitting a metric g, which is Kähler with respect to the underlying complex structure I.*

(i) *If the Ricci-tensor of g vanishes, the canonical bundle can be trivialised by a flat (hence holomorphic) section ω_I.*

(ii) *If moreover $\dim_{\mathbb{R}} X = 4$, then X can be given a quaternionic structure in the following way. We define $J_1 \colon \mathcal{T}_X(p) \to \mathcal{T}_X(p)$, $p \in X$ by $g(J_1 Y, Z) = \operatorname{Re} \omega_I(Y, Z)$, $Y, Z \in \mathcal{T}_X(p)$. Then for suitable $\lambda \in \mathbb{R}^*$ the operator $J = \lambda J_1$ is an almost-complex structure such that $\{g, I, J\}$ is a quaternionic structure on X.*

Proof (communicated to us by Hitchin). (i) The curvature of $\mathcal{K}_X$ is a multiple of the Ricci tensor ([K-N], Chap. IX, §4, §10). It follows that $\mathcal{K}_X$ is flat and since X is simply-connected it admits a flat section.

(ii) Let $\omega = \omega_I$. Since

$$g(I \circ J_1 Y, Z) = -g(J_1 Y, IZ) = -\operatorname{Re} \omega(Y, IZ) = -\operatorname{Re} \omega(IY, Z) = -g(J_1 \circ IY, Z)$$

(ω is of type $(2,0)$), it follows that $I \circ J_1 = -J_1 \circ I$. Now since J_1 is skew-adjoint, J_1^2 is self-adjoint with non-positive eigenvalues. Since $J_1 \neq 0$ we may find a non-zero eigenvalue, say $-\lambda^{-2}$ and we let V be its eigenspace. It we put $J = \lambda J_1$, then $J^2 = -\mathrm{id}$ on V. Since J^2 and I commute, I preserves V and $\{I, J, I \circ J\}$ gives V a quaternionic structure. So $\dim V \geq 4$ and hence $V = \mathcal{T}_X(p)$. Since $g(JY, JZ) = -g(Y, J^2 Z) = g(Y, Z)$ and since J is parallel, it follows that $\{g, I, J\}$ is a quaternionic structure on X. $\square$

Next, let X be a surface, $Y \in \mathcal{T}_X(p)$, $Y \neq 0$. Since $g(Y, HY) = 0$ for any $H \in S(X)$, it follows immediately that $\{Y, IY, JY, KY\}$ is an orthogonal basis for the tangent space at p and in this order gives an orientation consistent with the orientation, induced by any of the complex structures $H \in S(X)$. Let $\{\varphi, \varphi_I, \varphi_J, \varphi_K\}$ be the dual basis for $\mathscr{T}_X^\vee(p)$. Recall ([K-N], §4) that the Kähler form $\kappa_H \in \Gamma(\mathscr{D}_X^{1,1})$ of g with respect to $H \in S(X)$ is defined by

$$g(Y, HZ) = \kappa_H(Y, Z).$$

In terms of the above basis for $\mathscr{T}_X(p)$, it is easy to see that for suitable μ we have

$$\kappa_I(p)\, \varphi_I \wedge \varphi + \varphi_K \wedge \varphi_J$$
$$\kappa_J(p) = \varphi_J \wedge \varphi + \varphi_I \wedge \varphi_K = \mu \operatorname{Re} \omega_I(p)$$
$$\kappa_K(p) = \varphi_K \wedge \varphi + \varphi_J \wedge \varphi_I = \mu \operatorname{Im} \omega_I(p).$$

So we obtain:

(13.2) **Proposition.** *The subspace* $E(\kappa_I, \omega_I)$ *is spanned by* κ_I, κ_J *and* κ_K (cf. Sect. 9).

Since $\mathscr{K}_X$ is trivial for a K3-surface X we can apply Theorem I.15.1 to any Kähler K3-surface X. So for a given Kähler class κ there exists a unique Ricci-flat Kähler metric g (i.e. $\operatorname{Ric}(g) = 0$) and we can apply Lemma 13.1 to this g. We find

(13.3) **Theorem.** *Let X be a Kähler K3-surface and κ a Kähler class. There exists a unique Ricci-flat Kähler metric g of class κ. If I denotes the complex structure on X, a second complex structure J on X exists such that $\{g, I, J\}$ is a quaternionic structure.*

The above metric g is called *the canonical Kähler-Einstein metric of* (X, κ). The quaternionic structure itself is not uniquely determined by (X, κ), but an easy calculation shows that if $\{g, I, J'\}$ is another quaternionic structure, then $J' \in S(I, J)$. So the sphere of complex structures is uniquely determined by (X, κ). *We call it the canonical sphere of complex structures.* Application of Proposition 13.2 gives:

(13.4) **Remark.** Let X be a Kähler surface with given Kähler class κ and let H be one of the complex structures of the canonical sphere of complex structures. Let ω_H be a non-zero (2,0)-form, holomorphic with respect to H and let κ_H be the Kähler form with respect to H. Then the orthogonal complement of κ_H in $E(\kappa_I, \omega_I) = E(\kappa_H, \omega_H) = E$ is exactly $E(\omega_H)$ and since $\{\kappa_I, \kappa_J, \kappa_K\}$ is a basis of E the map $H \to \omega_H$ is a surjection of the canonical sphere of complex structures onto the Grassmannian of oriented 2-planes contained in E. So we have

(13.5) **Proposition.** *The image of the refined period map* (Sect. 12) *consists of fibres of*

$$\Pi: K\Omega \to G_3^+(L).$$

14. Surjectivity of the Period Map

In Sect. 12 we have introduced the refined period map $\tau_2 \colon M_2 \to (K\Omega)^0$.

(14.1) **Theorem.** *For Kähler K3-surfaces the refined period map is surjective.*

Proof. The proof consists of three steps.

Step 1: if $(\kappa,\omega)\in(K\Omega)^0$ such that $E(\kappa,\omega)\cap L$ contains a primitive rank-2 lattice M, all of whose vectors have a norm divisible by 4, then (κ,ω) belongs to the image of τ_2.

By Proposition 13.5 we may replace (κ,ω) by any other element in the fibre $\Pi^{-1}E(\kappa,\omega)$. We therefore may assume that $M\subset E(\omega)$. Then Proposition 8.1 implies the existence of a marked exceptional Kummer surface (X,ϕ) such that $\phi_{\mathbb{C}}(H^{2,0}(X))=[\omega]$. By composing ϕ with an element of $\{\pm 1\}\times W_X$ we may suppose that $\phi^{-1}(\kappa)$ is in the Kähler cone of X (cf. Proposition 3.9). We have to show that it is in fact a Kähler class. Theorem IV.5.1 implies that the integral elements of the Kähler cone are classes of ample divisors, hence are Kähler classes. So all rational elements in the Kähler cone are Kähler classes. The fact that X is exceptional implies that $H^{1,1}(X,\mathbb{R})$ is defined over $\mathbb{Q}$ and hence the rational elements in the Kähler cone are dense in it. Since the set of Kähler classes also forms an open convex subcone of the Kähler cone, it follows that this subcone actually coincides with the Kähler cone.

Step 2: if $(\kappa,\omega)\in(K\Omega)^0$ is such that $E(\kappa,\omega)\cap L$ contains a primitive vector x with $(x,x)\in 4\mathbb{N}$, then $(\kappa,\omega)\in\operatorname{Im}(\tau_2)$.

As in Step 1, we may suppose that $x\in E(\omega)$. Let

$$K\subset(K\Omega)^0\cap\{L_R\times[\omega]\}=C^0_\omega$$

be the component of C^0_ω containing κ. Let $\eta\in K$ be chosen such that

$$M_\eta=(\mathbb{R}\cdot x+\mathbb{R}\cdot\eta)\cap L$$

is a rank-2 lattice, all of whose vectors have norm in $4\mathbb{Z}$. By Remark 8.4 those η are dense in K. By Step 1 we then have $(\eta,[\omega])=\tau_2(X_\eta,\kappa_\eta,\phi_\eta)$ for some marked pair (X_η,κ_η). It follows from the injectivity of τ_2 (see Theorem 12.3) that for all different choices of η in K the isomorphism type of X_η is the same. So all of the η thus chosen via the marking ϕ_η, correspond to Kähler classes on the same K3-surface, hence as in Step 1 they form a dense open convex subset in the Kähler cone of X_η. It follows that the entire Kähler cone of X_η consists of Kähler classes, in particular $\phi_\eta^{-1}(\kappa)$ is a Kähler class, i.e. $(\kappa,[\omega])\in\operatorname{Im}\tau_2$.

Step 3: the final argument.

Let $(\kappa,[\omega])\in(K\Omega)^0$ and let K be the connected component of C^0_ω containing κ as in Step 2. If $\eta\in K$ is such that $E(\omega)+\mathbb{R}\cdot\eta$ contains an indivisible vector with norm in $4\mathbb{N}$, then $(\eta,[\omega])\in\operatorname{Im}\tau_2$ by Step 2. Such η are dense in K by Remark 8.4. Since the cone $(K\times[\omega])\cap(\operatorname{Im}\tau_2)$ is convex, it follows as in the proof of the preceding two steps, that $(\kappa,[\omega])\in\operatorname{Im}\tau_2$. □

(14.2) **Corollary.** *Every point of Ω occurs as the period point of a marked Kähler K3-surface.*

Finally, we sketch Beauville's version of Siu's proof (see [Si] and [Be3]) that all K3-surfaces are kählerian. The idea of the proof is this. Let (X,ϕ) be a marked K3-surface. According to Corollary 14.2 (surjectivity of the period map) there exists a marked Kähler K3-surface (X', ϕ') with the same period point as (X, ϕ). One wants to adapt the proof of Theorem 11.1 (Torelli theorem) in order to show that X and X' are isomorphic. The principal difficulty is to extend Theorem 10.6 to the situation where only one of the K3-surfaces Y_0 and Y_0' is assumed to be Kähler. Siu's idea is to use a particular kind of hermitian metric. In Beauville's version a somewhat different type of hermitian metric is employed, introduced by Gauduchon ([Ga]).

Definition. A hermitian metric on a surface is called a *standard* hermitian metric if its 2-form κ satisfies $\partial\bar{\partial}\kappa = 0$.

A fundamental result proved in [Ga] is the following

(14.3) **Proposition.** *Let h be a hermitian metric on the surface X. Then there exists a standard hermitian metric conformally equivalent to h. It is unique up to multiplication with a positive constant.*

The proof of this proposition heavily relies on the theory of second order elliptic partial differential operators on compact manifolds. The result is used in combination with the following

(14.4) **Lemma.** *Let X be a K3-surface, κ the 2-form of a standard hermitian metric on X.*

a) *There exists a closed 2-form θ on X whose $(1,1)$-part is κ;*
b) *the component k in $H^{1,1}(X,\mathbb{R})$ of the class $[\theta] \in H^2(X,\mathbb{R})$ is independent of the choice of θ; it satisfies the two inequalities:*

$k^2 > 0$

$k \cdot d > 0$ *for all classes d of effective divisors D on X;*

c) *if X is Kähler, then $k \in \mathscr{C}_X^+$, the Kähler cone of X.*

Proof. a) Since $H^1(\Omega_X^2) = 0$, the $\bar{\partial}$-closed form $\partial\kappa$ is $\bar{\partial}$-exact, so there exists a $(2,0)$-form α with $\bar{\partial}\alpha + \partial\kappa = 0$. We may take $\theta = \alpha + \kappa + \bar{\alpha}$. Note that α is uniquely determined mod $\mathbb{C} \cdot \omega$, where ω is a holomorphic 2-form. Hence k, the component of the class of θ in $H^{1,1}(X,\mathbb{R})$, is uniquely determined by κ.

b) We only have to prove the inequalities. In view of the previous remark, we may add to α a suitable multiple of ω. We choose it such that the new θ satisfies $\int_X \theta \wedge \omega = 0$, so that $k = [\theta]$. With these choices of α and θ we have

$$k^2 = \int_X \theta \wedge \theta = 2\int_X \alpha \wedge \bar{\alpha} + \int_X \kappa \wedge \kappa > 0$$

$$k \cdot d = \int_D \theta = \int_D \kappa > 0.$$

c) It suffices to show that $k \in \mathscr{C}_X$. Let ξ be a Kähler form, then $\kappa \wedge \xi$ is a positive multiple of the volume form, hence

$$k \cdot [\xi] = \int_X \kappa \wedge \xi > 0$$

and k cannot belong to $-\mathscr{C}_X$ and so belongs to $\mathscr{C}_X$. □

The next step is to prove Theorem 10.6 if only $p'\colon Y' \to S'$ is assumed to be a family of Kähler K3-surfaces. In the proof of this theorem a crucial rôle is played by Lemma 10.3. In fact, if this lemma holds, the proof of Theorem 10.6 goes over verbatim.

To prove Lemma 10.3 we needed to estimate $\mathrm{Vol}(\Gamma_n)$ with respect to a suitable hermitian metric on $Y \times_S Y'$. In the situation at hand, we take on Y' a hermitian metric inducing a Kähler metric on each fibre. We let κ'_s be its associated 2-form. On Y we construct a metric as follows. Let κ_0 be the 2-form of a standard hermitian metric on Y_0 and let θ_0 be a closed 2-form whose $(1,1)$-part is κ_0. Possibly after shrinking S a bit, we can extend θ_0 to a closed 2-form θ on Y with the property that $\theta_s = \theta | Y_s$ has positive definite $(1,1)$-part κ_s. Now we give Y a hermitian metric which on Y_s induces the standard hermitian metric with form κ_s. The two metrics on Y and Y' combine to give one on $Y \times_S Y'$ and we compute volumes with respect to this metric.

$$\mathrm{Vol}(\Gamma_n) = \mathrm{Vol}(Y_{s_n}) + \mathrm{Vol}(Y'_{s_n}) + \int_{Y'_{s_n}} \kappa'_{s_n} \wedge f_n^*(\kappa_{s_n}).$$

The first two terms are obviously bounded near 0, whereas the third equals

$$\int_{Y_{s_n}} \kappa_{s'_n} \wedge f_n^*(\theta_{s_n}) = [\kappa'_{s_n}] \cdot \phi_{s_n}([\theta_{s_n}])$$

which is the value at s_n of a continuous function on S, so this term also remains bounded.

Now let us return to the proof of Theorem 11.1, to show that in our situation it can be modified to yield that X and X' are isomorphic. First we observe that Theorem 14.1 not only provides us with a marked Kähler K3-surface (X', ϕ') with the same period point as (X, ϕ), but also a Kähler class κ' such that $\phi'([\kappa']) = \phi([\kappa])$, where κ is the 2-form of a standard hermitian metric (it is here that we use part b) of the previous lemma; it says that $[\kappa] \in (K\Omega)^0$, cf. Corollary 9.2). Using the notation of the proof of Theorem 11.1 we need to show that for some sequence $s_n \to 0$ the Hodge-isometries $\Phi(s_n)\colon H^2(Y'_{s_n}, \mathbb{Z}) \to H^2(Y_{s_n}, \mathbb{Z})$ are indeed effective, since then the assumptions of Theorem 10.6 hold and as in the proof of Theorem 11.1 we can conclude that $X' = Y'_0$ and $X = Y_0$ are isomorphic. As before we introduce κ'_s, θ'_s, θ_s and κ_s. By construction $\Phi(0)$ maps $[\kappa'_0]$ to $[\kappa_0]$, in other words $\Phi(0)(\kappa'_0) = \kappa_0$. Since θ', resp. θ are closed forms extending θ'_0, resp. θ_0 it follows that $\Phi(s)([\theta'_s]) = [\theta_s]$, and since $\Phi(s)$ is a Hodge-isometry $\Phi(s)[\kappa'_s] = [\kappa_s]$. So, if Y'_s and Y_s are kählerian, $\Phi(s)$ is effective by part c) of the preceding lemma. This concludes the proof that X' and X are isomorphic. So we obtain

(14.5) **Theorem.** *Every K3-surface is kählerian.*

Enriques Surfaces

15. Topological and Analytic Invariants

In this section Y denotes an *Enriques surface*. By definition (V, Sect. 23) this means

$$\mathscr{K}_Y^{\otimes 2}=\mathcal{O}_Y, \quad \text{but } \mathscr{K}_Y \neq \mathcal{O}_Y,$$

and $q(Y)=0$. We have seen (loc. cit.) that Y is always projective. The conditions on $\mathscr{K}_Y$ imply $c_1^2(Y)=0$ and $p_g=0$. So $\chi(\mathcal{O}_Y)=1$ and Noether's formula becomes

$$e(Y)=12.$$

Riemann-Roch for a line bundle $\mathscr{L}$ takes the form

$$\chi(\mathscr{L})=1+\tfrac{1}{2}c_1^2(\mathscr{L}),$$

and for a rank-2 bundle $\mathscr{V}$ we have

$$\chi(\mathscr{V})=2+\tfrac{1}{2}c_1^2(\mathscr{V})-c_2(\mathscr{V}).$$

(15.1) **Lemma.** *Let Y be an Enriques surface. Then*

(i) $h^{1,0}(Y)=h^{0,1}(Y)=h^{2,0}(Y)=h^{0,2}(Y)=0$, $h^{1,1}(Y)=10$.

(ii) *The fundamental group of Y is $\mathbb{Z}_2$ and the universal covering X of Y is a K3-surface.*

(iii) *The intersection form on $H^2(X,\mathbb{Z})_f$ is isometric to $(-E_8)\oplus H$.*

Proof. (i) From the preceding remarks we conclude that

$$h^{1,0}=h^{0,1}=h^{2,0}=h^{0,2}=0, \quad \text{so} \quad h^{1,1}=b_2=e(Y)-2=10.$$

(ii) By I, Sect. 18 there exists an unramified double covering $\pi\colon X\to Y$ such that $\mathscr{K}_X=\pi^*(\mathscr{K}_Y)=\mathcal{O}_X$. Since $e(X)=2e(Y)=24$ and $p_g(X)=1$, Noether's formula for X implies $q(X)=0$. So by definition (VI, Sect. 1) X is a K3-surface and therefore it is simply-connected (see Corollary 8.6).

(iii) Since $p_g=0$, every $d\in H^2(Y,\mathbb{Z})$ is divisorial, say $d=\mathcal{O}_Y(D)$. Hence $d^2=D^2\equiv DK_Y\equiv 0 \bmod 2$. Since $\tau(Y)=\frac{1}{3}(c_1^2(Y)-2c_2(Y))=-8$ by the topological index theorem, this form is indefinite. The assertion then follows from Theorem I.2.8. □

(15.2) **Proposition.** *For every Enriques surface Y the map*

$$\operatorname{Pic}(Y)\xrightarrow[c_1]{} H^2(Y,\mathbb{Z})=\mathbb{Z}^{10}\oplus\mathbb{Z}_2$$

is an isomorphism.

Proof. The isomorphism is a consequence of the exponential sequence and of the vanishing of $H^i(\mathcal{O}_Y)$, $i=1,2$. □

When considering deformations of Y we need

(15.3) **Proposition.** *For every Enriques surface Y*

$$h^0(\mathcal{T}_Y)=h^2(\mathcal{T}_Y)=0, \qquad h^1(\mathcal{T}_Y)=10.$$

Proof. The universal covering X of Y does not admit holomorphic vector fields by Corollary 3.4, so $h^0(\mathcal{T}_Y)=h^0(\mathcal{T}_X)=0$. Since $\mathcal{K}_Y\otimes\mathcal{T}_Y^\vee=\mathcal{K}_Y^\vee\otimes\Omega_Y^1=\mathcal{T}_Y$, by Serre duality we have $h^2(\mathcal{T}_Y)=h^0(\mathcal{T}_Y)=0$, so $h^1(\mathcal{T}_Y)=10$ by Riemann-Roch. □

16. Divisors on an Enriques Surface Y

In the sequel D will be a divisor on Y and $d\in H^2(Y,\mathbb{Z})$ its class.

(16.1) **Proposition.** *Let D be any divisor on the Enriques surface Y, $D\neq 0, K_Y$.*

(i) *If D is irreducible, then $D^2\geq -2$ and $D^2=-2$ if and only if D is a (-2)-curve.*

(ii) *If $D^2\geq 0$, then either $|D|\neq\emptyset$ or $|-D|\neq\emptyset$, (but not both). If $D^2\geq 0$ and $|D|\neq\emptyset$, then also $|K_Y+D|\neq\emptyset$.*

(iii) *If $D^2\geq 0$, $|D|\neq\emptyset$, and if $|K_Y+D|$ contains a reduced connected divisor, then*

$$\dim|D|=\tfrac{1}{2}D^2.$$

Proof. (i) By the adjunction formula $D^2=\deg(\omega_D)\geq -2$, and $D^2=-2$ implies that D is smooth rational.

(ii) The Riemann-Roch inequality implies

$$h^0(\mathcal{O}_Y(D))+h^0(\mathcal{O}_Y(K_Y-D))\geq\tfrac{1}{2}D^2+1. \tag{6}$$

So, if $D^2\geq 0$, either $|D|\neq\emptyset$ or $|K_Y-D|\neq\emptyset$, but not both at the same time. Since $D^2=(K_Y+D)^2$, the same argument shows that $|K_Y+D|\neq\emptyset$ or $|-D|\neq\emptyset$. So if $|D|\neq\emptyset$ the first alternative must hold.

(iii) The assumptions imply $h^0(\mathcal{O}_{D'})=1$ for some divisor $D'\in|K_Y+D|$. Since $h^1(\mathcal{O}_Y)=0$ by definition, from the cohomology sequence for

$$0\to\mathcal{O}_Y(-D')\to\mathcal{O}_Y\to\mathcal{O}_{D'}\to 0$$

we find $h^1(\mathcal{O}_Y(-D'))=0$. Then $h^1(\mathcal{O}_Y(D))=0$ by Serre duality and in (6) we have equality. □

As the intersection form on $H^2(Y,\mathbb{R})$ has signature $(1,9)$, the set of classes $d\in H^2(Y,\mathbb{R})$ with $d^2>0$ is the union of two convex open cones, say $\mathscr{C}$ and $\mathscr{C}'$. Since

$$d_1\cdot d_2>0 \quad \text{if both } d_1,d_2\in\mathscr{C},$$
$$d_1\cdot d_2<0 \quad \text{if } d_1\in\mathscr{C},\ d_2\in\mathscr{C}',$$

only one of them, say $\mathscr{C}$, contains ample classes. Let $\tilde{\mathscr{C}}\in H^2(Y,\mathbb{Z})$ be the semigroup of classes $d\neq 0$ mapping into the closure $\bar{\mathscr{C}}$.

(16.2) **Proposition.** *For a divisor D on Y of non-torsion class $d \in H^2(Y, \mathbb{Z})$ the following statements are equivalent:*

a) $d \in \tilde{\mathscr{C}}$,

b) $D^2 \geq 0$ and $|D| \neq \varnothing$.

Proof. If $d \in \tilde{\mathscr{C}}$, then $D^2 \geq 0$. If $D^2 \geq 0$, then either d or $-d$ are effective (Proposition 16.1, (ii)). But a class in $-\tilde{\mathscr{C}}$ cannot be effective. □

Next, let $c \in H^2(Y, \mathbb{Z})$ be a class with $c^2 = -2$. Just as on a K3-surface such a class defines a Picard-Lefschetz reflection

$$s_c\colon H^2(Y, \mathbb{Z}) \to H^2(Y, \mathbb{Z}), \qquad s_c(d) = d + (d, c)c.$$

(16.3) **Proposition.** *The reflection s_c leaves $\tilde{\mathscr{C}}$ invariant. In particular, if $d \in H^2(Y, \mathbb{Z})$ is effective with $d^2 \geq 0$, then so is $s_c(d)$.*

Proof. The proof is the same as that of the corresponding statement in Proposition 3.9. □

We consider $H^2(X, \mathbb{Z})_f$ as a euclidean lattice. The decomposition $H^2(Y, \mathbb{Z})_f = -E_8 \oplus H$ is not unique of course. So $H^2(Y, \mathbb{Z})_f$ contains many hyperbolic planes H. Such a plane is generated by the images of classes $d_1, d_2 \in H^2(Y, \mathbb{Z})$ satisfying

$$d_1^2 = d_2^2 = 0, \qquad d_1 \cdot d_2 = 1. \tag{7}$$

(16.4) **Proposition.** *If Y is an Enriques surface and $d_1, d_2 \in H^2(Y, \mathbb{Z})$ satisfy* (7), *then either both, d_1 and d_2, or both, $-d_1$ and $-d_2$, are effective.*

Proof. Both, d_1 and d_2, belong to $\tilde{\mathscr{C}} \cup (-\tilde{\mathscr{C}})$. Since $d_1 \cdot d_2 > 0$, the classes d_1 and d_2 cannot belong to different components of this double cone. The assertion now follows from Proposition 16.2. □

A pair of effective classes d_1, $d_2 \in H^2(Y, \mathbb{Z})$ satisfying (7) will be called an *effective pair of generators of a hyperbolic plane.*

17. Elliptic Pencils

In this section we show that every Enriques surface Y admits an elliptic fibration over $\mathbb{P}_1$. This means that there exists a holomorphic map $f\colon Y \to \mathbb{P}_1$, whose generic fibre is a smooth elliptic curve, or, equivalently, there exists a base point free rational pencil on Y, whose generic member is smooth elliptic. Such pencils will be called elliptic pencils in the sequel. Before proving the main result (Theorem 17.7) we make some simple observations concerning elliptic pencils on Enriques surfaces.

(17.1) **Lemma.** *Every elliptic fibration $f\colon Y \to \mathbb{P}_1$ of an Enriques surface Y has exactly two multiple fibres $2F$ and $2F'$, and*

$$\mathscr{K}_Y = \mathscr{O}_Y(F' - F) = \mathscr{O}_Y(F - F').$$

Proof. The canonical bundle formula for an elliptic fibration (in fact Corollary V.12.3) yields

$$\mathcal{K}_Y = f^*\mathcal{O}_{\mathbb{P}_1}(-1) \otimes \mathcal{O}_Y\left(\sum_{i=1}^{k} (r_i - 1)F_i\right),$$

where $r_i F_i$, $i = 1, \ldots, k$, are the multiple fibres of f. So

$$\mathcal{O}_Y = \mathcal{K}_Y^{\otimes 2} = f^*\mathcal{O}_{\mathbb{P}_1}(-2) \otimes \mathcal{O}_Y\left(\sum_{i=1}^{k} (2r_i - 2)F_i\right).$$

Restricting to F_i and using that $\mathcal{O}_{F_i}(F_i)$ has order exactly r_i in $\mathrm{Pic}(\mathcal{O}_{F_i})$ (Lemma III.8.3), we find that all r_i equal 2. From the preceding formula we infer $k=2$. We put $F_1 = F$, $F_2 = F'$, whereupon the canonical bundle formula becomes:

$$\mathcal{K}_Y = f^*\mathcal{O}_{\mathbb{P}_1}(-1) \otimes \mathcal{O}_Y(F + F').$$

The result follows, because $\mathcal{O}_Y(2F) = \mathcal{O}_Y(2F') = f^*\mathcal{O}_{\mathbb{P}_1}(1)$. □

The curves F, F' in Lemma 17.1 are called the *halfpencils*. By an *elliptic configuration* on Y we mean a curve F appearing in Kodaira's table (V, Table 3) of singular fibres in elliptic fibrations, but not a multiple of such a fibre.

(17.2) **Lemma.** *Let $F \subset Y$ be a connected curve with $F^2 = 0$ and $F \cdot C \geq 0$ for all (-2)-curves C on Y. Then $F = mF_0$, $0 < m \in \mathbb{N}$, where F_0 is an elliptic configuration.*

Proof. By Proposition 16.2 we have $F \cdot C \geq 0$ for all effective curves C satisfying $C^2 \geq 0$. So Proposition 16.1, (i) and the assumption show $F \cdot C \geq 0$ for all effective curves C on Y. If $F = A + B$ is an effective decomposition, then $F \cdot A + F \cdot B = F^2 = 0$. This proves that $F \cdot C = 0$ for each irreducible component $C \subset F$. Now we apply Lemma I.2.10 (as in the proof of Zariski's lemma) to prove $D^2 \leq 0$ for all divisors D consisting of components of F, with $D^2 = 0$ only if $D = rF$, $r \in \mathbb{Q}$. In particular, if F contains more than one component, all its irreducible components are (-2)-curves. The assertion now follows just as in V, Sect. 7. □

(17.3) **Lemma.** *Let F be an elliptic configuration on Y. Then either $\mathcal{O}_F(F) = \mathcal{O}_F$ and $|F|$ is an elliptic pencil, or $\mathcal{O}_F(F) \in \mathrm{Pic}(F)$ is a nontrivial 2-torsion element and $|2F|$ is an elliptic pencil of which F is a halfpencil.*

Proof. By the adjunction formula $\deg(\omega_F | C) = 0$ for all irreducible components $C \subset F$. Since F has arithmetical genus 1, Riemann-Roch on F shows $h^0(\omega_F) = 1$. Then ω_F is trivial by Lemma II.12.2. Since $\omega_F = \mathcal{K}_Y \otimes \mathcal{O}_F(F)$, and $\mathcal{K}_Y^{\otimes 2} = \mathcal{O}_Y$, we have $\mathcal{O}_F(2F) = \mathcal{O}_F$.

If $\mathcal{O}_F(F)$ is already trivial, the sequence

$$0 \to \mathcal{O}_Y \to \mathcal{O}_Y(F) \to \mathcal{O}_F(F) \to 0$$

shows $h^0(\mathcal{O}_Y(F)) = 2$, and $|F|$ is an elliptic pencil.

If $\mathcal{O}_F(F)$ is non-trivial, then $h^0(\mathcal{O}_F(F))=h^1(\mathcal{O}_F(F))=0$ and the above sequence shows $h^0(\mathcal{O}_Y(F))=1$ and $h^1(\mathcal{O}_Y(F))=0$. From the sequence

$$0\to\mathcal{O}_Y(F)\to\mathcal{O}_Y(2F)\to\mathcal{O}_F(2F)\to 0$$

we then infer $h^0(\mathcal{O}_Y(2F))=2$ and $|2F|$ is an elliptic pencil. $\square$

Remark. The two cases distinguished in Lemma 17.3 also differ with respect to the behaviour of F under the universal covering $\pi\colon X\to Y$. Since $\mathcal{O}_F(F)=\mathcal{K}_Y|F$, the inverse image $\pi^*(F)\subset X$ is connected if $\mathcal{O}_F(F)\neq\mathcal{O}_F$, and it decomposes into two connected components mapped isomorphically onto F, if $\mathcal{O}_F(F)=\mathcal{O}_F$. If $|2F_1|$ is an elliptic pencil on Y with halfpencil F_1, then $|\pi^*(F_1)|$ is an elliptic pencil on the K3-surface X. The corresponding fibrations $g\colon X\to\mathbb{P}_1$, $f\colon Y\to\mathbb{P}_1$ form a commutative diagram

$$\begin{array}{ccc} X & \xrightarrow{\ \pi\ } & Y \\ {\scriptstyle g}\downarrow & & \downarrow{\scriptstyle f} \\ \mathbb{P}_1 & \longrightarrow & \mathbb{P}_1, \end{array}$$

where the induced map $\mathbb{P}_1\to\mathbb{P}_1$ is a double covering, ramified over the image points $f(F_1)$, $f(F_2)$ of the half-pencils F_1, F_2 in $|2F_1|$. Let $\sigma\colon X\to X$ be the covering involution for π. This σ lifts to an involution σ^*: $\mathcal{O}_X(\pi^*(F_1))\to\mathcal{O}_X(\pi^*(F_1))$ on the line bundle level. If $g_1,g_2\in H^0(\mathcal{O}_X(\pi^*(F_1)))$ are "equations" for F_1, F_2, then it can be assumed that $\sigma^*(g_1)=g_1$. But this implies $\sigma^*(g_2)=-g_2$.

(17.4) **Lemma.** *Let E be an effective divisor on Y with $E^2=0$ and such that its class $e\in H^2(Y,\mathbb{Z})_f$ is primitive (i.e., $e=me'$, $m\in\mathbb{Z}$, $e'\in H^2(Y,\mathbb{Z})_f$ implies $m=\pm 1$). Then there are finitely many (-2)-curves $C_1,\ldots,C_k$ on Y with classes $c_1,\ldots,c_k\in H^2(Y,\mathbb{Z})$ such that $(s_{c_k}\circ\ldots\circ s_{c_1})(e)$ is the class of a halfpencil in some elliptic fibration of Y.*

Proof. If there is a (-2)-curve C_1 on Y with $E\cdot C_1=-n<0$, we consider $e_1=s_{c_1}(e)=e-nc_1$. By Proposition 16.3, there is an effective divisor E_1 with class e_1. If $E_1\cdot C_2<0$ for some (-2)-curve C_2, we repeat this procedure and so on. With each reflection s_{c_i} the degree of E_{i-1} (w.r.t. an arbitrary projective embedding of Y) decreases by one at least. So after finitely many reflections this process will terminate. We arrive at a divisor F with class $(s_{c_k}\circ\ldots\circ s_{c_1})(e)$ satisfying $F\cdot C\geq 0$ for all (-2)-curves C on Y.

All the connected components F_i of F share the property $F_i^2=0$ and $F_i\cdot C\geq 0$. So Lemma 17.2 shows that $F=\sum m_iF_i$, where the F_i are elliptic configurations on Y. By Lemma 17.3 their classes $f_i\in H^2(Y,\mathbb{Z})$ are all proportional. Then primitivity shows that there is only one connected component F_1, which itself is a halfpencil, and $m_1=1$. $\square$

An obvious consequence of Lemma 17.4 is

(17.5) **Theorem.** *Every Enriques surface admits an elliptic fibration over the projective line.*

We shall need more:

(17.6) **Proposition.** *Given an elliptic fibration of the Enriques surface Y over $\mathbb{P}_1$, there is a 2-section G for this fibration with $G^2 \leq 0$.* (Here, a 2-section means an irreducible curve G with $F \cdot G = 2$ for all the fibres F.)

Proof. Let D_1 be one of the half-pencils for the elliptic fibration and let $d_1 \in H^2(Y, \mathbb{Z})_f$ be its class. To show that d_1 is primitive, we assume $d_1 = md'$, $0 < m \in \mathbb{Z}$, $d' \in H^2(Y, \mathbb{Z})_f$. This d' is effective by Proposition 16.1, (ii). If D' is an effective divisor representing d', from Lemma 17.3 we infer that $h^0(\mathcal{O}_Y(D_1)) = 1$, so we have $D_1 = mD'$ with $m = 1$, because $2D_1$ is a fibre of multiplicity 2.

Since d_1 is primitive, the intersection product with d_1 defines a map of the unimodular lattice $-E_8 \oplus H$ onto $\mathbb{Z}$. So there is some d with $d \cdot d_1 = 1$. Since d^2 is an even integer, we may form $d_2 = d - \frac{d^2}{2} d_1$. Then $d_2^2 = 0$ and $d_1 \cdot d_2 = 1$. Now Proposition 16.4 implies the existence of some effective divisor D_2 with $D_2^2 = 0$ and $D_1 \cdot D_2 = 1$. Writing $D_2 = G + D$ with D consisting of all components contained in fibres, we see that there is some 2-section G. It remains to find such a 2-section with $G^2 \leq 0$.

So assume $G^2 \geq 2$. By Riemann-Roch $h^0(\mathcal{O}_Y(G)) \geq 2$. Since $\mathcal{O}_{D_1}(G)$ is an invertible sheaf of degree 1 on a reduced curve of genus 1, we have $h^0(\mathcal{O}_{D_1}(G)) = 1$. This implies that there is some divisor in $|G|$ containing D_1. Let this divisor be $D_1 + G' + R$ with G' a 2-section and R consisting of fibre components. Then

$$(G')^2 = (G - D_1 - R)^2 = G^2 - 2 - 2G \cdot R + (D_1 + R)^2 < G^2,$$

because $(D_1 + R)^2 \leq 0$ by Zariski's lemma. After repeating this procedure sufficiently many times, we obtain a 2-section with self-intersection ≤ 0. □

In view of Lemma 16.1, (i), there are two possibilities for the 2-section G which occurs in Proposition 17.6:

(i) G is a (-2)-curve, or

(ii) $G^2 = 0$, and G is an elliptic configuration by Lemma 17.2. Since $G \cdot F_1 = 1$ for a halfpencil F_1 in the elliptic fibration given, the class of G in $H^2(Y, \mathbb{Z})_f$ is primitive. By Lemmas 17.3 and 17.1 it follows that $|2G|$ itself is an elliptic pencil, one of whose halfpencils is G.

To formulate the final result of this section, we use the following

Definition. An Enriques surface is called *special*, if it carries an elliptic pencil together with a 2-section which is a (-2)-curve.

Then we have proved:

(17.7) **Theorem.** *Every Enriques surface admits an elliptic fibration. On a non-special Enriques surface there are two effective divisors D_1 and D_2 with $D_1^2 = D_2^2 = 0$, $D_1 \cdot D_2 = 1$, occurring as halfpencils in two distinct elliptic pencils.*

18. Double Coverings of Quadrics

In this section we use elliptic pencils on the Enriques surface Y, as constructed in the preceding section, to represent Y in terms of a double covering of a quadric. Following Horikawa in [Hor4] we use this representation to prove e.g. the connectedness of the space of all Enriques surfaces.

So we assume first, that on Y we have two halfpencils D_1, D_2 with $D_1 \cdot D_2 = 1$, as they exist on every *non-special* Y by Theorem 17.7. We denote by C_i, $i=1,2$, their inverse image $\pi^*(D_i) \subset X$. Both linear systems $|C_1|$ and $|C_2|$ are free from base points and fixed components. The same holds for the system $|C_1+C_2|$ and the map $f_{C_1+C_2} = f_{\mathcal{O}_X(C_1+C_2)}$ is well-defined. The dimension $h^0(\mathcal{O}_X(C_1+C_2))$ is computed easily: $|C_1+C_2|$ contains reduced connected curves, so as in the proof of Proposition 16.1, (iii) we conclude

$$h^1(\mathcal{O}_X(C_1+C_2))=0 \quad \text{and} \quad h^0(\mathcal{O}_X(C_1+C_2))=2+\tfrac{1}{2}(C_1+C_2)^2=4,$$

because $C_1 \cdot C_2 = 2$. Hence we have $f_{C_1+C_2}\colon X \to \mathbb{P}_3$. The canonical map

$$H^0(\mathcal{O}_X(C_1)) \otimes H^0(\mathcal{O}_X(C_2)) \to H^0(\mathcal{O}_X(C_1+C_2))$$

is injective, and even an isomorphism. This shows that the image of X in $\mathbb{P}_3$ is a non-singular quadric $Q = \mathbb{P}_1 \times \mathbb{P}_1$, where the two rulings of Q on X induce the elliptic fibrations defined by $|C_1|$ and $|C_2|$.

As $C_1 \cdot C_2 = 2$, the map $X \to Q$ is generically 2-to-1. The general elliptic curve in $|C_1|$ or $|C_2|$ is mapped 2-to-1 onto a projective line, so it is branched over four points. This shows that there is a branch curve $B \subset Q$ of bidegree (4, 4). Over $Q \setminus B$ the map $f_{C_1+C_2}$ is a non-ramified double covering, and the inverse images of the finitely many singularities of B are exceptional curves in X contracted to points under $f_{C_1+C_2}$. Also, all the singularities of B must be simple. Indeed, if not, then by Theorem III.7.2 a canonical divisor on X would be strictly negative, leading to the contradiction $p_g(X)=0$ for the K3-surface X.

Finally, the covering involution σ for $\pi\colon X \to Y$ acts on the whole situation. To be specific, we introduce bihomogeneous coordinates $(x_0 : x_1)$, $(y_0 : y_1)$ on Q as follows. Let D_1', $D_2' \subset Y$ be the halfpencils adjoint with D_1, D_2 and let $C_1' = \pi^*(D_1')$, $C_2' = \pi^*(D_2')$. Let $g_1, g_1' \in H^0(\mathcal{O}_X(C_1))$ be equations for C_1, C_1' and g_2, $g_2' \in H^0(\mathcal{O}_X(C_2))$ equations for C_2 and C_2'. Then homogeneous coordinates $(x_0 : x_1)$, resp. $(y_0 : y_1)$ on $\mathbb{P}_1$ can be chosen in such a way that the fibration defined by $|C_1|$ has equation $g_1 : g_1' = x_0 : x_1$, and that similarly $g_2 : g_2' = y_0 : y_1$ defines the fibration of the pencil $|C_2|$.

We introduce coordinates $(z_0 : z_1 : z_2 : z_3)$ on $\mathbb{P}_3$ such that Q is embedded by

$$z_0 = x_0 y_0, \quad z_1 = x_1 y_1, \quad z_2 = x_0 y_1, \quad z_3 = x_1 y_0.$$

Then $f_{C_1+C_2}$ is given by

$$z_0 = g_1 g_2, \quad z_1 = g_1' g_2', \quad z_2 = g_1 g_2', \quad z_3 = g_1' g_2.$$

Using the remark after Lemma 17.3 we lift σ to an involution σ^* on $\mathcal{O}_X(C_1)$ with $\sigma^*(g_1) = g_1$, $\sigma^*(g_1') = -g_1$ and on $\mathcal{O}_X(C_2)$ such that $\sigma^*(g_2) = g_2$, $\sigma^*(g_2') = -g_2$.

So if we define the involution τ on $\mathbb{P}_3$ by

$$\tau(z_0:z_1:z_2:z_3)=(z_0:z_1:-z_2:-z_3),$$

then Q is τ-invariant with τ acting on Q by

$$\tau((x_0:x_1)(y_0:y_1))=(x_0:-x_1)(y_0:-y_1).$$

If we let the generator of $\mathbb{Z}_2$ act by σ on X and by τ on $\mathbb{P}_3$, then also $f_{C_1+C_2}$ is $\mathbb{Z}_2$-equivariant. On Q the involution τ has the four fixed points

$$(x_0:x_1)(y_0:y_1)=(1:0)(1:0),\ (1:0)(0:1),\ (0:1)(1:0),\ (0:1)(0:1).$$

The branch curve B does not contain any of them. Indeed, if one of them would be a smooth point of B, the involution σ would leave invariant the corresponding point on X. This is impossible because σ has no fixed points. And if B would have a simple curve singularity at a point x with $\tau(x)=x$, then σ would act as an involution on the A-D-E configuration in X lying over x. But any such involution has a fixed point, yielding again a contradiction. The polynomial of bidegree $(4,4)$ defining B is either invariant or anti-invariant under τ. The absence of fixed points on B shows that it must be invariant.

Altogether this proves:

(18.1) **Theorem** (Horikawa's representation of non-special Enriques surfaces). *Let Y be an Enriques surface and $D_1, D_2 \subset Y$ be two halfpencils with $D_1 \cdot D_2=1$. If τ and σ are defined as above, then there is a τ-invariant bihomogeneous polynomial of bidegree $(4,4)$ in $(x_0:x_1)$, $(y_0:y_1)$, with zero-set B on the smooth quadric Q, such that the universal covering X of Y is the minimal resolution of the double covering of Q ramified over B. The curve B is reduced with at worst simple singularities and does not contain any fixed point of τ. The involution σ on X is induced by the involution τ on Q. The two rulings of Q define the two elliptic pencils $|\pi^*(D_1)|$, $|\pi^*(D_2)|$ on X.*

Theorem 18.1 has a converse: given a τ-invariant curve as in the theorem, the K3-surface X and an Enriques surface $Y=X/\sigma$ can be constructed from it. This has already been shown in V, Sect. 23.

To obtain a representation of all Enriques surfaces Y, we still have to treat the *special* case. So we now assume that on Y we are given a halfpencil D_1 and a (-2)-curve D_2 with $D_1 \cdot D_2=1$.

Again we put $C_1=\pi^*(D_1)$. The elliptic pencil $|C_1|$ on X is free of base points and fixed components. The curve $\pi^*(D_2)$ decomposes into two disjoint (-2)-curves, say E_1 and E_2. On X we consider the linear system $|\pi^*(2D_1+D_2)| =|2C_1+E_1+E_2|$. It contains reduced connected curves, so as in the non-special case we conclude

$$h^1(\mathcal{O}_X(2C_1+E_1+E_2))=0$$

and

$$h^0(\mathcal{O}_X(2C_1+E_1+E_2))=2+\tfrac{1}{2}(2C_1+E_1+E_2)^2=4.$$

Since $\mathcal{O}_{E_i}(2C_1+E_1+E_2)=\mathcal{O}_{E_i}$, we have an exact sequence

$$0\to H^0(\mathcal{O}_X(2C_1))\to H^0(\mathcal{O}_X(2C_1+E_1+E_2))\to H^0(\mathcal{O}_{E_1+E_2}).$$

Together with $h^0(\mathcal{O}_X(2C_1))=3<h^0(\mathcal{O}_X(2C_1+E_1+E_2))$ this implies that $|2C_1+E_1+E_2|$ has no base points. So $\varphi=f_{2C_1+E_1+E_2}: X\to\mathbb{P}_3$ is well-defined.

Next we choose sections $g_1, g_1'\in H^0(\mathcal{O}_X(C_1))$ as in the non-special case and a generator $e\in H^0(\mathcal{O}_X(E_1+E_2))$. Then $eg_1^2, eg_1g_1', eg_1'^2\in H^0(\mathcal{O}_X(2C_1+E_1+E_2))$ can be complemented to a basis of this space by some section f. On $\mathbb{P}_3$ coordinates $(z_0:z_1:z_2:z_3)$ are introduced such that φ is given by

$$(z_0:z_1:z_2:z_3)=(eg_1^2:eg_1'^2:eg_1g_1':f).$$

So $\varphi(X)$ is contained in the quadratic cone Q_0 with equation $z_0z_1=z_2^2$. Both (-2)-curves E_1 and E_2 are contracted by φ, and their image is the vertex $(0:0:0:1)$ of the cone Q_0. On the general curve in $|C_1|$ there are two zeros of f, so φ is generically 2-to-1, mapping the members of $|C_1|$ onto the generators of the cone Q_0. Again there is a branch curve $B\subset Q_0$, not containing the vertex, which intersects the general line on Q_0 in four points. So B is the intersection of Q_0 with a surface of degree 4. As in the non-special case, B can have at most simple singularities.

Again we trace the action of the covering involution σ. As before it lifts to $\mathcal{O}_X(C_1)$ and hence to some σ^* on $\mathcal{O}_X(2C_1)$ such that $\sigma^*(g_1^2)=g_1^2$, $\sigma^*(g_1'^2)=g_1'^2$, $\sigma^*(g_1g_1')=-g_1g_1'$. We can lift σ to some σ^* on $\mathcal{O}_X(E_1+E_2)$ with $\sigma^*(e)=e$. So we have an induced involution, again denoted by σ^*, on the line bundle $\mathcal{O}_X(2C_1+E_1+E_2)$. Now f can be chosen as above, but additionally as an eigenvector for σ^*. Then

$$\sigma^*(eg_1^2)=eg_1^2,\quad \sigma^*(eg_1'^2)=eg_1'^2,\ \sigma^*(eg_1g_1')=-eg_1g_1',\ \sigma^*(f)=\pm f.$$

To determine the last sign, we first define τ on $\mathbb{P}_3$ by

$$\tau(z_0:z_1:z_2:z_3)=(z_0:z_1:-z_2,\ \pm z_3),$$

such that φ is $\mathbb{Z}_2$-equivariant again. If τ would leave z_3 invariant, the plane $z_2=0$, consisting of fixed points under τ only, would intersect the branch curve B. As above we would find a fixed point for σ, that is, a contradiction. So we have the minus-sign, i.e., τ is the same involution as in the non-special case. Also B does not pass through any of the three fixed points of τ on Q_0: the vertex, and the two points $(1:0:0:0)$, $(0:1:0:0)$. And again we find, that B is defined by some τ-invariant polynomial of degree 4.

All this proves the special-version of Theorem 18.1:

(18.2) **Theorem** (Horikawa's representation of special Enriques surfaces). *Let Y be an Enriques surface, $D_1\subset Y$ a halfpencil and $D_2\subset Y$ a (-2)-curve with $D_1\cdot D_2=1$. Then there is a τ-invariant homogeneous polynomial of degree 4 in z_0,z_1,z_2,z_3, vanishing on the quadric cone $Q_0=\{z_0z_1=z_2^2\}$ exactly in the points of the curve B, such that the universal covering X of Y is the minimal resolution of the double covering of Q_0 ramified over B. The curve B is reduced with at most simple singularities and does not contain a fixed point of τ. The involution σ on X is induced by τ on Q_0. The system of generators on Q_0 defines the elliptic pencil $|\pi^*(2C_1)|$ on X, and $\pi^*(D_2)$ consists of the two (-2)-curves over the vertex of Q_0.*

Also Theorem 18.2 has a converse: given B, the surfaces X and Y can be constructed.

To apply the representations given above, we use the following notion:

(18.3) **Definition.** Any Enriques surface Y, which can be represented as in the construction of Theorem 18.1 is called *general*, if B is smooth.

This use of the word general is justified by the obvious fact, that almost every non-special surface is general, and by the following, less obvious observation.

(18.4) **Theorem.** *Every special Enriques surface Y is a deformation of general ones.*

Proof. Let Y be represented as in Theorem 18.2. Let $f(z_0, z_1, z_2, z_3)$ be some τ-invariant polynomial defining B. This f is not uniquely determined by B, the same curve B is defined by any

$$g = f + q_0 \cdot q,$$

where $q_0 = z_0 z_1 - z_2^2$ is an equation for the cone Q_0 and q some τ-invariant quadratic polynomial. I.e., q is a linear combination of $z_0^2, z_0 z_1, z_1^2, z_2^2, z_2 z_3, z_3^2$. These six polynomials have no zero in common, so by Bertini's theorem we can choose q such that the quartic surface $g=0$ is non-singular away from Q_0. We can deform Q_0 into $Q_t = \{z_0 z_1 - z_2(z_2 + t z_3) = 0\}$. Then Q_t is τ-invariant and non-singular for $t \neq 0$, and $B_t = Q_t \cap \{g=0\}$ is smooth for general t. For small t, the curve B_t will not contain a fixed point of $\tau | Q$. Let $\mathscr{X}$ be the double covering of $\mathbb{P}_3$ branched over $\{g=0\}$. The involution τ lifts to an involution σ on $\mathscr{X}$. Let X_t, $t \neq 0$, be the part of $\mathscr{X}$ over Q_t and X_0 the minimal resolution of the part of $\mathscr{X}$ over Q_0. By Brieskorn's theorem ([Bri]), X_0 is a deformation of the smooth surfaces X_t, $t \neq 0$ and small. The involution σ can be chosen without fixed points over X_0, hence over X_t for t small. Then $Y_t = X_t/\sigma$ is a deformation of $Y = Y_0$ and Y_t is general for small $t \neq 0$. □

The following is the main result of this section:

(18.5) **Theorem.** *Any two Enriques surfaces are deformations of each other.*

The proof follows from Theorem 18.4 and the next proposition.

(18.6) **Proposition.** *Any two general Enriques surfaces are deformations of each other.*

Proof. Let $(x_0 : x_1)(y_0 : y_1)$ be bihomogeneous coordinates on Q. The vector space of τ-invariant polynomials of bidegree $(4,4)$ is spanned by the thirteen polynomials.

$$(x_0^k x_1^{2-k})^2 \cdot (y_0^l y_1^{2-l})^2 \qquad k, l = 0, 1, 2,$$

$$x_0 x_1 y_0 y_1 \cdot (x_0^k x_1^{1-k})^2 (y_0^l y_1^{1-l})^2 \quad k, l = 0, 1.$$

Denote by P the corresponding projective space, considered as space of curves on Q, and by $P^0 \subset P$ the Zariski-open subset of smooth curves, not passing through any of the four fixpoints of τ on Q. (From Bertini's theorem it follows immediately that $P^0 \neq \emptyset$.) Let $\Gamma \subset P^0 \times Q$ be the hypersurface $\{(F,x) \in P^0 \times Q;\ x \in F\}$, and $q\colon P^0 \times Q \to Q$ the projection. We can cover P^0 by open sets $\mathcal{U}$ such that $\Gamma \cap (\mathcal{U} \times Q)$ is determined by a section in $p^*(\mathcal{O}_Q(4,4)) = p^*(\mathcal{O}_Q(2,2)^{\otimes 2})$. So there is a double covering $\mathcal{X}_{\mathcal{U}}$ of $\mathcal{U} \times Q$ branched over the hypersurface $\Gamma \cap (\mathcal{U} \times Q)$. All the fibres X_t of $\mathcal{X}_{\mathcal{U}} \to \mathcal{U}$ are K3-surfaces, double coverings of $\{t\} \times Q$, branched over the smooth curve $\Gamma \cap (\{t\} \times Q)$. The involution $\mathrm{id}_{\mathcal{U}} \times \tau$ lifts to an involution $\sigma_{\mathcal{U}}$ on $\mathcal{X}_{\mathcal{U}}$ without fixed points. Forming the quotient $\mathcal{X}_{\mathcal{U}}/\sigma_{\mathcal{U}}$ we obtain a family over $\mathcal{U}$ containing all the Enriques surfaces $X_t/(\sigma_{\mathcal{U}}|X_t)$. Covering P^0 by such open sets $\mathcal{U}$, we prove the assertion. □

19. The Period Map

First we introduce the following notation. As in Sect. 1–14 we put

$$L = -E_8 \oplus -E_8 \oplus H \oplus H \oplus H,$$

the cohomology lattice of a $K3$-surface. We define an involution $\rho\colon L \to L$ by

$$\rho(x \oplus y \oplus z_1 \oplus z_2 \oplus z_3) = (y \oplus x \oplus -z_1 \oplus z_3 \oplus z_2).$$

Its $\mathbb{C}$-linear extension to $L_{\mathbb{C}}$ is denoted by $\rho_{\mathbb{C}}$. The ρ-(anti-)invariant sublattices of L are

$$L^+ = \{l \in L;\ \rho(l) = l\}, \qquad L^- = \{l \in L;\ \rho(l) = -l\}.$$

Then L^+ is isometric to $-2E_8 \oplus 2H$ and in particular

$$l \in L^+ \quad \text{implies } (l,l) \equiv 0 \mod 4.$$

The unimodular lattice $\frac{1}{2}L^+ = L^0$ is isometric to $-E_8 \oplus H$, the cohomology lattice of an Enriques surface. We moreover put

$$\Omega^- = \mathbb{P}(L^- \otimes \mathbb{C}) \cap \Omega$$

$$\Gamma = \mathrm{restr}_{L^-} \{g \in \mathrm{Aut}(L);\ g\,\rho = \rho\, g\}$$

$$D = \Omega^-/\Gamma.$$

(19.1) **Lemma.** *Let $\pi\colon X \to Y$ be the universal covering of an Enriques surface Y, and let $\sigma\colon X \to X$ be the covering involution. Then there exists an isometry*

$$\phi\colon H^2(X,\mathbb{Z}) \to L$$

such that

$$\phi \circ \sigma^* = \rho \circ \phi.$$

Proof. Since by Proposition 18.6 all Enriques surfaces have the same deformation type, to prove the lemma it suffices to consider the particular Enriques surface Y of V, Sect. 23. The K3-surface X doubly covering it contains two disjoint E_8-configurations, interchanged by the covering involution $\sigma\colon X \to X$. They generate a sublattice of $H^2(X,\mathbb{Z})$, and we take an identification of L with

$H^2(X,\mathbb{Z})$, such that this sublattice becomes a direct summand $(-E_8)\oplus(-E_8)$ of L. Then σ^* and ρ coincide on it. Both σ^* and ρ operate on the orthogonal complement $H\oplus H\oplus H$. In general they are different there, but both $(+1)$-eigenspaces are isometric to $2H$. For σ^* this follows since $\pi^*(H^2(Y,\mathbb{Z}))=2H\oplus -2E_8$ is the $(+1)$-eigenspace, and for ρ it can be checked easily. Now by Theorem I.2.9 we may compose σ^* with an automorphism of $H\oplus H\oplus H$ such that these two eigenspaces coincide. But then the (-1)-eigenspaces coincide, since for any involution of a euclidean lattice the (-1)-eigenspace is the orthogonal complement of the $(+1)$-eigenspace. It follows that σ^* and ρ coincide on a lattice of finite index in $H^2(X,\mathbb{Z})$ and hence on all of $H^2(X,\mathbb{Z})$. □

(19.2) **Definition.** A *marked Enriques surface* is a pair (Y,ϕ) with Y an Enriques surface and $\phi\colon H^2(X,\mathbb{Z})\to L$ an isometry satisfying $\phi\circ\sigma^*=\rho\cdot\phi$, as in Lemma 19.1.

Similarly we may speak of *a marked family of Enriques surfaces*. If we have a family $p\colon Y\to S$ of Enriques surfaces over a contractible base S, we may form the universal covering $q\colon X\to S$, and an isometry $\phi(s)\colon H^2(X_s,\mathbb{Z})\to L$ extends to a unique marking $\phi\colon q_{*2}\mathbb{Z}_X \xrightarrow{\sim} L_S$. If, moreover $\phi(s)\circ\sigma_s^*=\rho\circ\phi(s)$ (where σ_s is the covering involution of $X_s\to Y_s$) then this relation holds for all points of S (two markings coinciding at $s\in S$ coincide over all of S).

A marked family defines a period map. In fact the resulting marked family of K3-surfaces determines a period map

$$\tilde{\tau}\colon S\to\Omega.$$

The extra relation $\phi(s)\circ\sigma_s^*=\rho\circ\phi(s)$ implies that the image of $\tilde{\tau}$ belongs to $\Omega^- =\{[\omega]\in\Omega;\ \rho_{\mathbb{C}}(\omega)=-\omega\}$. Indeed, if ω_s is a holomorphic 2-form on X_s we have $\sigma_s^*(\omega_s)=-\omega_s$ (since on Y_s there is no holomorphic 2-form), which translates into $\rho_{\mathbb{C}}(\omega_s')=-\omega_s'$, where $\omega_s'=\phi_{\mathbb{C}}(\omega_s)$. We let

$$\tau\colon S\to\Omega^-$$

be the resulting period map.

(19.3) **Theorem** (Local Torelli theorem). *The Kuranishi family for an Enriques surface Y_0 is universal at all points in a small neighbourhood U about the point corresponding to Y_0. This base is smooth and has dimension* 10. *The period map is a local isomorphism at each point of U.*

Proof. Since $h^0(\mathscr{T}_Y)=h^2(\mathscr{T}_Y)=0$ and $h^1(\mathscr{T}_Y)=10$, the first assertions follow from Theorem I.10.5. To prove that the period map is a local isomorphism, we may assume the base U to be contractible. Let us denote by $p\colon Y\to U$ the universal family and by $q\colon X\to U$ its universal covering. We have a commutative diagram

$$\begin{array}{ccccccccc}
0 & \longrightarrow & q^*\mathscr{T}_u|X_0 & \longrightarrow & f^*\mathscr{T}_Y|X_0 & \longrightarrow & f^*\mathscr{T}_{Y_0} & \longrightarrow & 0\\
 & & \| & & \downarrow & & \downarrow & & \\
0 & \longrightarrow & q^*\mathscr{T}_u|X_0 & \longrightarrow & \mathscr{T}_X|X_0 & \longrightarrow & \mathscr{T}_{X_0} & \longrightarrow & 0
\end{array}$$

where $f: X \to Y$ is the covering map. By definition of the Kodaira-Spencer map (I, Sect. 10) this diagram yields a commutative diagram

$$\begin{array}{ccc} & \overset{\rho_p}{\nearrow} & H^1(\mathcal{T}_{Y_0}) \\ \mathcal{T}_u(0) & & \downarrow \\ & \underset{\rho_q}{\searrow} & H^1(\mathcal{T}_{X_0}). \end{array}$$

The vertical arrow embeds $H^1(\mathcal{T}_{Y_0})$ into $H^1(\mathcal{T}_{X_0})$ as the subspace of its σ_0-invariants, so ρ_q is injective and we may view q as a subfamily of the Kuranishi deformation of X_0. Marking the family of Enriques surfaces we obtain a commutative diagram of period maps

$$\begin{array}{ccc} & \overset{\tau}{\nearrow} & \Omega^- \\ U & & \downarrow \\ & \underset{\tilde{\tau}}{\searrow} & \Omega \end{array}$$

and since $\tilde{\tau}$ is locally injective by Theorem 7.3, it follows that τ is locally injective, hence locally bijective since $\dim \Omega^- = 10$.

20. The Period Domain for Enriques Surfaces

A bounded domain of type IV in $\mathbb{C}^n$ is given by $\{z \in \mathbb{C}^n;\ |(z,z)|^2 + 1 - 2(z,\bar{z}) > 0,\ |(z,z)| < 1\}$, where $(z,z') = \sum_{i=1}^{n} z_i z'_i$.

(20.1) **Lemma.** *Let V be a real vector space of dimension $n+2$ equipped with a symmetric bilinear form $(\ ,\ )$ of signature $(2,n)$. The set*

$$\Omega(V) = \{[\omega] \in \mathbb{P}(V \otimes \mathbb{C});\ (\omega,\omega) = 0,\ (\omega,\bar{\omega}) > 0\}$$

is biholomorphically equivalent to a disjoint union of two copies of a bounded domain of type IV in $\mathbb{C}^n$. In fact, if a basis of V is taken such that $(\ ,\ )$ has diagonal form $\mathbb{1}_2 \oplus -\mathbb{1}_n$, the two connected components are distinguished by the sign of $\operatorname{Im}(\omega_0/\omega_1)$, where $\{\omega_0, \ldots, \omega_n\}$ are the coordinate functions on $V \otimes \mathbb{C}$ with respect to this basis.

For a proof we refer to [Pi].

Putting $n = 10$ we find that the period domain Ω^- consists of two connected components, each of which is a bounded domain of type IV in $\mathbb{C}^{10}$.

(20.2) **Proposition.** *The involution $\lambda = \mathbb{1} \oplus \mathbb{1} \oplus -\mathbb{1} \oplus \mathbb{1} \oplus \mathbb{1}$ of L commutes with ρ and the induced involution of Ω^- interchanges the two connected components.*

Proof. Clearly $\lambda \circ \rho = \rho \circ \lambda$. To prove the second statement we use the following basis for L^- (which incidentally shows that $L^- = -2E_8 \oplus H \oplus 2H$). If $\{e_1, \ldots, e_8\}$, resp. $\{e'_1, \ldots, e'_8\}$ are bases for the first, resp. the second copy of $-E_8$ and

$\{f_i, g_i\}$ $(i=1,2,3)$ for the i-th copy of H we take

(8) $$\{e_1-e_1', e_2-e_2', \ldots, e_8-e_8', f_1, g_1, f_2-f_3, g_2-g_3\}$$

and observe that in this basis $\lambda(f_1)=-f_1$, $\lambda(g_1)=-g_1$ and $\lambda=\mathbb{1}$ on the orthogonal complement of $\mathbb{Z}f_1+\mathbb{Z}g_1$. Of course this basis cannot be used to apply Lemma 20.1. However, with respect to the basis (8) the sign in question is nothing but the sign of $\mathrm{Im}(\omega_8+\omega_9/\omega_{10}+\omega_{11})$, if $[\omega]=(\omega_0:\ldots:\omega_{11})$. Then λ obviously changes this sign. □

(20.3) **Lemma.** *The group $\Gamma=\{g|L^-;\ g\in\mathrm{Aut}(L),\ g\circ\rho=\rho\circ g\}$ is of finite index in* $\mathrm{Aut}(L^-)$, *hence Γ is an arithmetic subgroup of* $\mathrm{Aut}(L^-\otimes\mathbb{R})$.

Proof. Let $g\in\mathrm{Aut}(L^-)$ and suppose $g\equiv\mathrm{id} \bmod 2L$. Using the basis (8), it is easy to check that an extension $h\in\mathrm{Aut}(L)$ exists with $h|L^+=1$. By construction $h\circ\rho=\rho\circ h$, hence $g\in\Gamma$. So Γ contains a congruence subgroup of $\mathrm{Aut}(L^-)$, and hence Γ is of finite index in $\mathrm{Aut}(L^-)$. □

(20.4) **Corollary.** *The group Γ acts properly discontinuously on Ω^- and the complex space $D=\Omega^-/\Gamma$ admits the structure of a quasi-projective variety.*

Proof. The first assertion follows from [Bou1], Chap. III.4.2. and the second from [B-B]. □

(20.5) **Proposition.** *If $d\in L^-$, $(d,d)=-2$, then no point of*

$$H_d=\{[\omega]\in\Omega^-;\ (\omega,d)=0\}$$

can be the period point of a marked Enriques surface.

Proof. Suppose (Y,ϕ) is a marked Enriques surface such that its period point belongs to H_d. Then the class $\delta=\phi^{-1}(d)$ is in the Picard lattice of X, the universal covering of Y. So, by Proposition 3.7, (i) the class $\pm\delta$ is effective. But no effective class can be anti-invariant as is $\pm\delta$. □

(20.6) **Proposition.** *There are only finitely many Γ-equivalence classes of elements $d\in L^-$ with $(d,d)=-2$.*

Proof. The group Γ is of finite index in $\mathrm{Aut}(L^-)$, so we may replace Γ by $\mathrm{Aut}(L^-)$. Let Γ_1 be the subgroup of $\mathrm{Aut}(L')$ with $L'=-2E_8\oplus H\oplus H\supset L^-$ which preserves L^-. Since restriction defines an embedding $\Gamma_1\hookrightarrow\mathrm{Aut}(L^-)$, we may replace $\mathrm{Aut}(L^-)$ by Γ_1. This latter group contains the congruence subgroup $\{g\in\mathrm{Aut}(L');\ g\equiv\mathbb{1} \bmod 2L'\}$ of $\mathrm{Aut}(L')$, and we may replace Γ_1 by $\mathrm{Aut}(L')$. We claim that all elements $d'\in L'$, satisfying $(d',d')=-2$ are conjugate to each other under the action of $\mathrm{Aut}(L')$. If L' would be unimodular, this would follow from Theorem I.2.9. The proof of this theorem, as presented in [L-P], p. 156 can easily be modified to cover our case (for details, see [Hor4], II, p. 223). □

(20.7) **Corollary.** *The union* $\bigcup_d H_d/\Gamma$ *for* $d \in L^-$, $(d,d) = -2$ *consists of finitely many irreducible algebraic hypersurfaces in* $D = \Omega^-/\Gamma$, *so*

$$D^0 = D \setminus (\bigcup_d H_d)/\Gamma$$

is quasi-projective.

Proof. Each H_d consists of two connected components isomorphic to a bounded domain of type IV. Since Γ operates properly and discontinuously, the transforms of H_d form a locally finite collection of connected parts of hyperplanes because these hyperplanes are fixed points of reflections belonging to Γ (cf. the proof of Corollary 9.2). In virtue of Proposition 20.6 the same holds for the union of all H_d. So its image in D consists of finitely many irreducible (analytic) hypersurfaces. The Baily-Borel compactification D^* has the property that $\dim D^* \setminus D = 1$ ([Pi], §4, Lemme 1). A theorem of Remmert and Stein ([R-S], Satz 13) then implies that all of the above hypersurfaces extend to hypersurfaces in D^*. These are algebraic, by Chow's theorem I.19.2, since D^* is projective ([B-B]). So D^0, like D, is quasi-projective. □

21. Global Properties of the Period Map

Firstly, we shall prove a Torelli theorem for Enriques surfaces. As a preliminary we prove:

(21.1) **Proposition.** *If X is the universal covering of an Enriques surface, σ the corresponding involution and $l \in S_X$ with $\sigma^*(l) = l$, $(l,l) > 0$ and $(l,d) \neq 0$ for all $d \in S_X$ with $(d,d) = -2$, then there exists $w \in W_X$ commuting with σ^*, such that $\pm w(l)$ is the class of an ample divisor.*

Proof. By replacing l by $-l$, if necessary, we may assume that l belongs to the positive cone. In particular l and all of its W_X-images are effective. If l itself is the class of an ample divisor we take $w = \mathrm{id}$. If not, we proceed as follows. Let $\pi\colon X \to Y = X/\{1, \sigma\}$ be the quotient map and let $m = \pi_*(l)$. If l is not the class of an ample divisor, this also holds for m, since l is invariant. So there exists an irreducible class $e_1 = c_1(\mathcal{O}_Y(E_1))$ with $(m, e_1) \leq 0$. The curve $\pi^{-1}(E_1)$ cannot be irreducible, since Proposition 3.6, (i) implies that it would be nodal, whereas σ-invariant curves by (6) have self-intersection divisible by 4. So $\pi^{-1}(E_1)$ consists of two irreducible curves D_1 and $\sigma(D_1)$. These are (-2)-curves, again by Proposition 3.6, (i). If these curves meet, then $(D_1 + \sigma(D_1))^2 \geq 0$. So the class of $D_1 + \sigma(D_1)$ belongs to $\bar{C}_X$, and has strictly positive intersection with any element of C_X, such as l. But by construction this is not the case. It follows that D_1 and $\sigma(D_1)$ are disjoint, so the product of their classes d_1 and $d_2 = \sigma^*(d_1)$ is zero, and the product of the corresponding Picard-Lefschetz reflections commutes with σ^*. Moreover $(l, d_2) = (l, \sigma^*(d_1)) = (\sigma^*(l), d_1) = (l, d_1) \leq 0$, hence < 0, by assumption. By induction we obtain a finite ordered set of nodal classes $d_1, \ldots, d_{2k-1}, d_{2k} = \sigma^*(d_{2k-1})$ such that the product $s_{d_{2k}} \circ s_{d_{2k-1}} \circ \ldots \circ s_{d_1}$ of the corresponding reflections commutes with σ^*. If $w_j = s_{d_j} \circ s_{d_{j-1}} \circ \ldots \circ s_{d_1}$, we moreover have

$$(w_{r-1}(l), d_r) = -\alpha_r < 0.$$

We shall show that this process terminates after finitely many steps, namely as soon as

$$l_r = w_r(l)$$

is the class of an ample divisor. To this end observe that

$$l_r = s_r(l_{r-1}) = l_{r-1} - \alpha_r d_r \qquad (\alpha_r > 0).$$

So, if L_k is a divisor representing l_k, we find

$$\dim |L_k| \leq \dim |L_{k-1}|.$$

If the process would not terminate, $\dim |L_k|$ would stabilise from $k = N$ on and then $\sum_{k>N} \alpha_k D_k$ would be contained in $|L_N|$ as fixed part. This clearly is impossible. So the process stops at some stage r, when l_r is the class of an ample divisor. □

(21.2) **Theorem** (Global Torelli theorem for Enriques surfaces). *The isomorphism class of an Enriques surface is uniquely determined by its period point.*

Proof. Let Y, resp. Y' be two Enriques surfaces, X, resp. X' their universal coverings and σ, resp. σ' the corresponding involutions. Choose markings ϕ, resp. ϕ' for X, resp. X' such that $\rho \circ \phi = \phi \circ \sigma^*$ and similarly for ϕ'. If the period points of (Y, ϕ) and (Y', ϕ') are the same, the isometry

$$\psi = \phi^{-1} \circ \phi' \colon H^2(X', \mathbb{Z}) \to H^2(X, \mathbb{Z})$$

is a Hodge-isometry with $\psi \circ (\sigma')^* = \sigma^* \circ \psi$. If $l' \in H^2(X', \mathbb{Z})$ is the class of an ample divisor, invariant under σ', we may apply Proposition 21.1 to $l = \psi(l')$ and replace ψ by $\psi_1 = \pm w \circ \psi$ such that still $\psi_1 \circ (\sigma')^* = \sigma^* \circ \psi_1$, but now with $\psi_1(l)$ ample. By Proposition 3.10 ψ_1 is an effective Hodge-isometry and by the Torelli theorem 11.1, it is induced by an isomorphism $g \colon X' \to X$. Since $g \circ \sigma' \circ g^{-1} \circ \sigma$ induces the identity on $H^2(X', \mathbb{Z})$, by Proposition 11.3 we have that it is itself the identity, i.e. $g \circ \sigma' = \sigma \circ g$ and hence g induces an isomorphism between Y and Y'. □

Next, we shall show that all points of D^0 correspond to marked Enriques surfaces. The next lemma plays a central role in this proof.

(21.3) **Lemma.** *Let X be a projective K3-surface and let $\phi \colon H^2(X, \mathbb{Z}) \to L$ be a marking such that the period point of (X, ϕ) belongs to Ω^-.*

(i) *The involution $j = \phi^{-1} \circ \rho \circ \phi$ is a Hodge-isometry.*

(ii) *Either for some $d \in S_X$, $(d, d) = -2$, $w \in W_X$ we have*

$$w^{-1} \circ j \circ w(d) = -d$$

or, there exists $w \in W_X$ such that

$$w^{-1} \circ j \circ w = w(j)$$

is effective.

Proof: (i) Obvious.

(ii) Upon replacing j by $-j$, if necessary, we may assume that j preserves the positive cone. Choose a class l of an ample divisor on X. Then $j(l)$ as well as $w(j)(l)$ is effective for all $w \in W_X$. If $j(l)$ is already the class of an ample divisor we take $w = \mathrm{id}$ and we are ready. If not, there exists an irreducible class d with $(j(l), d) \leq 0$, which then necessarily is nodal by Proposition 3.7. We also observe that $d' = -j(d)$ is effective. Indeed, if not, then Proposition 3.6, (i) implies that $j(d)$ would also be effective and the inequality $(l, j(d)) > 0$ would contradict $(l, j(d)) = (j(l), d) \leq 0$.

We let s, resp. s' be the Picard-Lefschetz reflections in d, resp. d' and observe that

$$s \circ j \circ s\,(l) = j(l) - (l, d)\, d' - (l, s(d'))\, d$$
$$s' \circ j \circ s'(l) = j(l) - (l, d')\, d - (l, s'(d))\, d'.$$

So, if $s(d')$ is effective the coefficients of both d and d' in $s(j)(l)$ are strictly negative and we can take $d_1 = d$. If $s'(d)$ is effective, the same holds for the coefficients of d and d' in $s'(j)(l)$ and we set $d_1 = d'$. If neither $s(d')$ nor $s'(d)$ is effective, then $(d, d') < 0$. Since Proposition 3.6, (i) implies that $-s(d') = -d' - (d, d')d$ is effective, it follows that $d = d'$, i.e. $j(d) = -d$ and the first alternative holds.

Next we put $l_1 = s_1(j)(l)$. If l_1 is the class of an ample divisor we are ready. If not, we proceed as before with $j_1 = s_1(j)$ instead of l_1. Inductively we find effective (-2)-classes $d_1, \ldots, d_r$, such that, if we let

$$w_r = s_r \circ s_{r-1} \circ \ldots \circ s_1$$

be the product of the corresponding Picard-Lefschetz reflections and

$$w_r(j) = j_r$$
$$j_r(l) = l_r$$
$$d'_r = -j_r(d_r) \quad \text{(an effective class!)},$$

then *either*

$$d'_r = d_r \qquad \text{for some } r,$$

or

$$l_{r+1} = l_r - \beta_r d_r - \gamma_r d'_r, \qquad \beta_r > 0,\ \gamma_r > 0 \ \text{ for all } r \geq 1.$$

As in the proof of Proposition 21.1 the inequalities $\beta_r > 0$ and $\gamma_r > 0$ force the process to stop at some class l_r of an ample divisor. Then j_r is an effective Hodge-isometry by Corollary 3.11. □

(21.4) **Theorem.** *All the points of the variety D^0, introduced in Corollary* 20.7, *occur as period points of Enriques surfaces.*

Proof. The surjectivity of the period map for Kähler K3-surfaces (Theorem 14.1) implies that for a given $[\omega]$ with image in D^0, there is a Kähler K3-surface X and a marking

$$\phi \colon H^2(X, \mathbb{Z}) \to L, \quad \text{with} \quad \phi_{\mathbb{C}}(H^{2,0}(X)) = [\omega].$$

The surface is projective by Theorem IV.5.2, since for any $l' \in L^+$ with $(l', l') > 0$ the class $l = \phi^{-1}(l')$ is the class of a divisor with positive self-intersection. We apply Lemma 21.3 to change the marking. So let $\psi = w \circ \phi$ be a new marking with w as in Lemma 21.3.

Since $[\omega] \in D^0$ the first alternative in Lemma 21.3 cannot occur (Proposition 20.5). So

$$j = \pm \psi^{-1} \circ \rho \circ \psi$$

is an effective Hodge-isometry. Then Theorem 11.1 implies that there exists an automorphism

$$\sigma: X \to X, \quad \sigma^* = j.$$

Since $j^2 = \mathrm{id}$, Proposition 11.3 implies that σ is an involution. Let us compute its *Lefschetz number*, i.e.

$$L(\sigma) = \sum (-1)^j \operatorname{Tr}(\sigma^* | H^j(X, \mathbb{R})).$$

Since on $H^2(X, \mathbb{R})$, $\operatorname{Tr} \sigma^* = \operatorname{Tr} \rho$ up to sign, we find:

$$L(\sigma) = 2 \mp 2. \tag{9}$$

Similarly, for the *holomorphic Lefschetz number*

$$L_{\mathrm{hol}}(\sigma) = 1 + \operatorname{Tr}(\sigma^* | H^{2,0}(X))$$

we find

$$L_{\mathrm{hol}}(\sigma) = 1 \mp 1.$$

Applying the holomorphic Lefschetz formula for an involution ([A-S], Prop. 4.8)

$$L_{\mathrm{hol}}(\sigma) = \tfrac{1}{4} \# \text{ (isolated fixed points)},$$

we find that the number of isolated fixed points of μ equals 0 or 8.

On the other hand, the usual Lefschetz fixed point formula reads ([Ue2], Lemma 1.6):

$$L(\sigma) = \mu + \sum_{j=1}^{t} e(F_j),$$

where $F_1, \ldots, F_t$ are the fixed curves of σ, so combining this with (9) we find

$$\sum_{j=1}^{t} e(F_j) = -4 \text{ (if } \mu = 8) \quad \text{ or } 0 \text{ (if } \mu = 0).$$

If $\mu = 8$ we blow up X at the eight fixed points of σ, and obtain an involution $\tilde{\sigma}$ on the blown-up $\tilde{X}$. Let $Z = \tilde{X}/\{1, \tilde{\sigma}\}$. The fixed locus of $\tilde{\sigma}$ consists of the exceptional locus E and a curve F, the proper transform of $\bigcup F_j$. The canonical bundle formula for double coverings (IV, Sect. 22) and blow-ups (Theorem I.9.1, (viii)) show that $\mathscr{K}_Z$ lifts on X to $\mathcal{O}_X(-F)$, and so $p_g(Z) = 0$. Since $\sigma^* | H^{2,0}(X) = \mathrm{id}$, there is at least one non-zero holomorphic 2-form on Z. This contradiction shows that $\mu = 0$. So the $(+)$-sign holds in the definition of j and hence

$$\sigma^* \circ \psi = \psi \circ \rho. \tag{10}$$

Since by (6) all σ^*-invariant curves have self-intersection divisible by 4, the quotient surface $Z = X/\{1, \sigma\}$ must be minimal (any (-1)-curve would lift to a σ^*-invariant (-2)-curve). Suppose that the fixed point set of σ is non-empty. Then $P_2(Z) = 0$, since $\mathscr{K}_Z$ lifts on X to $\mathcal{O}_X(-F)$. Moreover $q(Z) \leq q(X) = 0$.

Applying Castelnuovo's criterion VI.2.1, we conclude that Z is rational. Since $e(Z)=\frac{1}{2}e(X)=12$ and for a minimal rational surface $e=3$ or 4 (VI, Theorem 1.1), we obtain a contradiction. So $F=0$, σ is a fixpoint-free involution and Z is an Enriques surface. Because of (10) the map ψ is a marking for Z in the sense of Definition 19.2, and so the image of $[\omega]$ in D^0 is the period point of Z. □

The somewhat mysterious name "K3" has been explained by A. Weil in the comment on his final report on contract AF 18(603)-57: "ainsi nommées en l'honneur de Kummer, Kähler, Kodaira et de la belle montagne K2 au Cachemire" (cf. [Wei2], p. 546).

The classical results on Kummer surfaces, which go back to the beginning of this century, are presented in the final chapter of [G-H].

The idea of studying deformations of K3's via the periods of their holomorphic 2-forms stems from A. Andreotti and A. Weil. The local Torelli-property is due to them (unpublished). Proofs have been given by Kodaira ([Ko4], part I, Theorem 17) for arbitrary K3's and by G.N. Tjurina ([Saf], Chap. IX, Theorem 2) who needed the Kähler assumption.

The famous conjectures:

(i) all K3-surfaces constitute one connected family,
(ii) all K3-surfaces are kählerian,
(iii) the period map is surjective,
(iv) (a form of) the global Torelli theorem holds,

were made by Andreotti and Weil (see [Wei2]).

It should be said that Weil's definition of a K3-surface differs from ours in that he calls any surface "K3" if it carries the differentiable structure of a smooth quartic surface in $\mathbb{P}_3$. Adopting our definition, (i) was proved independently by Kodaira ([Ko4], part I, Theorem 19) and – under the Kähler assumption – by G.N. Tjurina ([Saf], Chap. IX, Theorem 7).

Conjecture (ii) was recently proved by Siu (Sect. 14). Conjecture (iv) was solved in the affirmative for projective K3's by Piatečkii-Shapiro and Šafarevič in [Pi-S], but their proof, although in principle correct, contained several gaps and errors, some of which are rather serious. These were later corrected by M. Rapoport and independently by T. Shioda. The results have not been published in full (see however [Shi]). A detailed and corrected account of the original proof can be found in [L-P], in which paper also a simplified version is presented of the work of D. Burns and M. Rapoport on the period map for Kähler K3's ([B-R]). Our exposition is a partly rewritten version of this last work. Conjecture (iii) was first solved for special classes of algebraic K3-surfaces by J. Shah ([Sha1], [Sha2], [Sha3]) and independently by E. Horikawa ([Hor3]). Then V. Kulikov ([Kul]) gave a proof for projective K3's (without restriction), but his proof needed clarification at several points, subsequently provided by U. Persson and H. Pinkham in [P-P]. Relying on these results and making essential use of the Atiyah-Hitchin-Yau results – as presented in Sect. 13 – A. Todorov gave a proof of the surjectivity for the period map for Kähler K3's ([To]). The proof we give does not use the surjectivity for projective K3's and is due to E. Looijenga ([Lo]).

Enriques surfaces bear the name of their inventor. He constructed these surfaces to give examples of non-rational surfaces with $q=p_g=0$ ([Enr2]). Many of his assertions were proved rigorously (in all characteristics except 2) by M. Artin in his Harvard thesis (not published). Several of his ideas have been used freely in Sect. 17. The idea to use double coverings of quadrics to show that any two Enriques surfaces are deformations of each other, seems to be new (Artin, following Enriques, uses instead the fact that all Enriques surfaces can be represented as sixth-degree surfaces in $\mathbb{P}_3$ passing doubly through the edges of a tetrahedron). But it remains nevertheless true that our treatment of Enriques surfaces stays close to Horikawa's.

Results on the period map were first obtained by E. Horikawa in [Hor4]. Our proof of the injectivity is basically the same as his, except for some simplifications. The surjectivity we prove however in a considerably different and shorter way, by making use of the corresponding statement for K3-surfaces. (Our proof does not make use of degenerations, so stays entirely within the realm of non-singular surfaces.)

Bibliography

[Ad] Andreotti, A.: On the complex structure of a class of simply connected manifolds, in *Algebraic Geometry and Topology*, Princeton Univ. Press, Princeton (1957), 58–77

[Ae] Aeppli, A.: Modifikation von reellen und komplexen Mannigfaltigkeiten, Comment. Math. Helv. *32* (1957), 217–301

[A-K] Altmann, A., Kleiman, S.: *Introduction to Grothendieck duality theory*, Lect. Notes Math. *146*, Springer, Heidelberg (1970)

[An1] Artin, M.: *On Enriques surfaces*, thesis Harvard (1960)

[An2] Artin, M.: Some numerical criteria for contractibility of curves on algebraic surfaces, Am. J. Math. *84* (1962), 485–496

[An3] Artin, M.: On isolated rational singularities of surfaces, Am. J. Math. *88* (1966), 129–136

[Ar] Arakĕlov, S.Ju.: Families of algebraic curves with fixed degeneracies, Math. USSR Izv. *5* (1971), 1277–1302

[A-S] Atiyah, M.F., Singer, I.: The index of elliptic operators III, Ann. Math. *87* (1968), 546–604

[At1] Atiyah, M.F.: Vector bundles over an elliptic curve, Proc. Lond. Math. Soc. *7* (1957), 414–452

[At2] Atiyah, M.F.: On analytic surfaces with double points, Proc. R. Soc. Lond., Ser A *245* (1958), 237–244

[At3] Atiyah, M.F.: The signature of fibre bundles, in *Global Analysis*, Univ. Tokyo Press, Tokyo (1969), 73–84

[Ba1] Barlet, D.: Espace analytique reduit des cycles analytiques complexes compacts, Sém. Norguet, Vol. I, Lect. Notes Math. *482*, Springer, Heidelberg (1975), 1–158

[Ba2] Barlet, D.: Convexité de l'espace des cycles, Bull. Soc. Math. France, *106* (1978), 373–397

[Bad] Bădescu, L.: *Suprafeţe algebriche*. Edit. Acad. Rep. Soc. Romania, Bucarest (1981)

[B-B] Baily, W.L., Jr., Borel, A.: Compactification of arithmetic quotients of bounded symmetric domains, Ann. Math. *84* (1966), 422–528

[Bc] Borcea, C.: Some remarks on deformations of Hopf manifolds, Rev. Roum. Math. Pures Appl. *26* (1981), 1287–1294

[B-C1] Bombieri, E., Catanese, F.: The tricanonical map of a surface with $K^2=2$, $p_g=0$, in *C.P. Ramanujam, a tribute*, Springer Verlag, Heidelberg (1978), 279–290

[B-C2] Bombieri, E., Catanese, F.: Birationality of the quadricanonical map for a numerical Godeaux surface. To appear

[Be1] Beauville, A.: *Surfaces algébriques complexes*, Astérisque *54*, Soc. Math. France, Paris (1978)

[Be2] Beauville, A.: L'application canonique pour les surfaces de type général, Invent. Math. *55* (1979), 121–140

[Be3] Beauville, A.: Surfaces K3, Sém. Bourbaki *609* (1982/83)

[B-F] Bagnera, G., Franchis, M. de: Le superficie algebriche, de quali ammettono una rappresentazione parametrica mediante funzione iperellitiche di due argomenti, Mem. Soc. Ital. delle Sci. III Ser. *15* (1908), 251–343

[B-Ha] Borel, A., Haefliger, A.: La classe d'homologie fondamentale d'un espace analytique, Bull. Soc. Math. Fr. *89* (1961), 461–513

[B-Hu] Bombieri, E., Husemoller, D.: Classification and embeddings of surfaces, in *Algebraic Geometry, Arcata 1974*, A.M.S. Proc. Symp. Pure Math. *29* (1975), 329–420

[B-J] Bröcker, Th., Jänich, K.: *Einführung in die Differentialtopologie*, Heidelberger Taschenbücher 143. Springer, Heidelberg (1973)

[Bl] Blanchard, A.: Sur les variétés analytiques complexes, Ann. Sci. Éc. Norm. Sup. *73* (1958), 157–202

[B-M1] Bombieri, E., Mumford, D.: Enriques' classification in char. p II, in *Complex Analysis and Algebraic Geometry*, Iwanami-Shoten, Tokyo (1977), 23–42

[B-M2] Bombieri, E., Mumford, D.: ibid., III, Invent. Math. *35* (1976), 197–232

[Bog] Bogomolov, F.: Holomorphic tensors and vector bundles on projective varieties, Math. USSR, Izv. *13* (1979), 499–555

[Bom1] Bombieri, E.: The pluricanonical map of a complex surface, in *Several Complex variables I, Maryland 1970. Lect. Notes Math. 155*, Springer, Heidelberg (1971)

[Bom2] Bombieri, E.: Canonical models of surfaces of general type, Publ. Math. Inst. Hautes Etud. Sci. *42* (1973), 171–219

[Bor] Borel, A.: Compact Clifford-Klein forms of symmetric spaces, Topology *2* (1963), 111–222

[Bou1] Bourbaki, N.: *Topologie générale, Chap. 1–4.* Hermann, Paris (1971)

[Bou2] Bourbaki, N.: *Algèbre, Chap. 1–3.* Hermann, Paris (1970). *Chap. 9*, ibid. (1959)

[Bou3] Bourbaki, N.: *Groupes et algèbres de Lie, Chap. 4, 5, 6.* Hermann, Paris (1968)

[Br] Berger, M.: *Géometrie 1. Action des groupes, espaces affines et projectifs.* Cedic/Nathan, Paris (1977)

[B-R] Burns, D., Rapoport, M.: On the Torelli problems for Kählerian K3-surfaces, Ann. Sci. Ec. Norm. Super. IV Ser. *8* (1975), 235–274

[Bra] Brand, R.: *Parallelizability of compact complex surfaces*, thesis Leiden (1980)

[Bre] Bredon, G.: *Sheaf theory*, McGraw-Hill, New York (1967)

[Bri] Brieskorn, E.: Singular elements of semi-simple algebraic groups, in *Actes Congr. Int. Math. Nice 2 (1970)*, 279–284

[Bro] Brotherton, N.: Some parallelizable manifolds not admitting a complex structure, Bull. Lond. Math. Soc. *10* (1978), 303–304

[B-S] Bănică, C., Stănăşilă, O.: *Algebraic methods in the global theory of complex spaces.* John Wiley & Sons, New York (1976)

[Bu] Burniat, P.: Sur les surfaces de genre $P_{12}>0$, Ann. Pura Appl., IV Ser. *71* (1966), 1–24

[Bw] Barlow, R.N.: *Some new surfaces with* $p_g=0$, thesis Warwick (1982)

[Cam] Campedelli, L.: Sopra alcuni piani doppi notevoli con curve di diramazione del decimo ordine. Atti Acad. Naz. Lincei *15* (1932), 536–542

[Car] Cartan, H.: Quotient d'un espace analytique par un groupe d'automorphismes, in *Algebraic Geometry and Topology*, Princeton Univ. Press, Princeton (1957), 90–102

[Cas1] Castelnuovo, G.: Sulle superficie di genere zero, Mem. Soc. Ital. Sci., II Ser. *10* (1898) 103–126 = *Mem. Scelte*, Zanichelli, Bologna (1937), 307–334

[Cas2] Castelnuovo, G.: Sulle superficie aventi il genere aritmetico negativo, Rend. Circ. Math. Palermo *20* (1905), = Mem. Scelte, Zanichelli, Bologna (1937), 501–507

[Cat1] Catanese, F.: Babbage's conjecture, contact of surfaces, symmetrical determinantal varieties and applications, Invent. Math. *63* (1981), 433–466

[Cat2] Catanese, F.: On the moduli spaces of surfaces of general type, preprint (1982)

[C-E] Castelnuovo, G., Enriques, F.: Die algebraischen Flächen vom Gesichtspunkt der birationalen Transformationen aus, *Enz. Math. Wissensch.* III_2 (1914), 677–768

[Ci1] Ciliberto, C.: Canonical surfaces with $p_g=p_a=4$ and $K^2=5,\ldots,10$, Duke Math. J. *48* (1981), 1–37

[Ci2] Ciliberto, C.: Canonical surfaces with $p_g=p_a=5$ and $K^2=10$, Ann. Sc. Norm. Super. Pisa, IV Ser. *9* (1982), 282–336

[C-K] Chow, W.L., Kodaira, K.: On analytic surfaces with two independent meromorphic functions, Proc. Natl. Acad. Sci. USA *38* (1952), 319–325

[Cl] Clemens, H., et al.: *Seminar on degeneration of algebraic varieties.* Inst. Adv. Study, Princeton (1969)

[C-V] Calabi, E., Vesentini, E.: On compact locally symmetric Kähler manifolds, Ann. Math. *71* (1960), 472–507

[Da] Dabrowski, K.: Moduli spaces for Hopf surfaces, Math. Ann. *259* (1982), 201–226

[Dd] Dold, A.: *Lectures on algebraic topology*, Grundl. Math. Wiss. *200*, Springer, Heidelberg (1972)

[Del] Deligne, P.: Théorème de Lefschetz et critères de dégénérescence de suites spectrales, Publ. Math. Inst. Hautes Etud. Sci. *35* (1968), 107–126

[Dem] Demazure, M.: A, B, C, D, E, F etc. in *Sémin. sur les singularités de surfaces*, Lect. Notes Math. 777, Springer, Heidelberg (1980), 222–227

[D-M] Deligne, P., Mumford, D.: The irreducibility of the space of curves of given genus, Publ. Math. Inst. Hautes Etud. Sci. *36* (1969), 75–109

[Do] Dolgachev, I.: Algebraic surfaces with $p_g = q = 0$, in *Algebraic Surfaces*, Liguori, Napoli (1981)

[Dy] Douady, A.: Le problème des modules pour les variétés analytiques complexes, Sém. Bourbaki, exp. *277* (1964/'65)

[En] Enoki, I.: Surfaces of class VII_0 with curves, Tôhoku Math. J. *33* (1981), 453–492

[Enr1] Enriques, F.: Sulla classificazione delle superficie algebriche e particolarmente sulle superficie die genere $p^1 = 1$ (2 Notes) Atti Accad. Naz. Lincei, V Ser. 23^1 (1914)

[Enr2] Enriques, F.: *Le superficie algebriche*. Zanichelli, Bologna (1949)

[Es] Esnault, H.: Classification des variétés de dimension 3 et plus (d'après T. Fujita, S. Iitaka, Y. Kawamata, K. Ueno, E. Viehweg), Sém. Bourbaki, exp. *568* (1980/'81), Lect. Notes Math. 901, Springer, Heidelberg (1981), 111–131

[F] Fischer, G.: *Complex analytic geometry*, Lect. Notes Math. *538*, Springer, Heidelberg (1976)

[F-F] Fischer, G., Forster, O.: Ein Endlichkeitssatz für Hyperflächen auf kompakten komplexen Räumen. J. Reine Angew. Math. *306* (1979), 88–93

[F-G] Fischer, W., Grauert, H.: Lokal triviale Familien kompakter komplexer Mannigfaltigkeiten, Nachr. Akad. Wiss. Göttingen, II. Math. Phys. Kl. (1965), 89–94

[F-K] Forster, O., Knorr, K.: Ein Beweis des Grauertschen Bildgarbensatzes nach Ideen von B. Malgrange, Manuscr. Math. *5* (1971), 19–44

[F-L] Fulton, W., Lazarsfeld, R.: Connectivity and its applications to algebraic geometry, in *Algebraic Geometry*, Lect. Notes Math. *862*, Springer, Heidelberg (1981), 26–92

[Fu] Fujiki, A.: Kählerien normal complex spaces. Tôhoku Math. J., 2nd series *35* (1983), 101–118

[Ga] Gauduchon, P.: Le théorème de l'excentricité nulle. C.R. Acad. Sci. Paris *285* (1977), 387–390

[Ge] Geppert, H.: Die Klassifikation algebraischer Flächen, Jahresb. Dtsch. Math.-Ver. *41* (1932), 18–39

[Gf] Griffiths, Ph.: Periods of integrals on algebraic manifolds I, II, Am. J. Math. *90* (1968), 568–626, 805–865

[G-H] Griffiths, Ph., Harris, J.: *Principles of algebraic geometry*, John Wiley & Sons, New York (1978)

[Gi] Gieseker, D.: Global moduli for surfaces of general type, Invent. Math. *43* (1977), 233–282

[Gk1] Grothendieck, A.: Sur la classification des fibrés holomorphes sur la sphère de Riemann, Am. J. Math. *79* (1957), 121–138

[Gk2] Grothendieck, A.: Sur quelques points d'algèbre homologique, Tôhoku Math. J. *9* (1957), 119–221

[Go] Godement, R.: *Théorie des faisceaux*, Act. Sci. Ind. *1252*, Hermann, Paris (1958)

[Gr1] Grauert, H.: Ein Theorem der analytischen Garbentheorie und die Modulräume komplexer Strukturen, Publ. Math. Inst. Hautes Etud. Sci. *5* (1960)

[Gr2] Grauert, H.: Über Modifikationen und exzeptionelle analytische Mengen, Math. Ann. *146* (1962), 331–368

[G-R1] Grauert, H., Remmert, R.: Zur Theorie der Modifikationen I, Stetige und eigentliche Modifikationen komplexer Räume, Math. Ann. *129* (1955), 274–296

[G-R2] Grauert, H., Remmert, R.: Komplexe Räume, Math. Ann. *136* (1958), 245–318

[G-R3] Grauert, H., Remmert, R.: *Analytische Stellenalgebren*, Grundl. Math. Wiss. *176*, Springer, Heidelberg (1971)

[G-R4] Grauert, H., Remmert, R.: *Theorie der Steinschen Räume*, Grundl. Math. Wiss. *227*, Springer, Heidelberg (1977)

[G-V] Geer, G. van der, Ven, A. Van de: On the minimality of certain Hilbert modular surfaces, in *Complex Analysis and Algebraic Geometry*, Iwanami Shoten, Tokyo (1977)

[G-W] Gruenberg, K., Weir, A.J.: *Linear Geometry*, Graduate Texts 49, Springer, Heidelberg (1977)

[Gx1] Godeaux, L.: Sur une surface algébrique de genre zero et de bigenre deux, Atti Acad. Naz. Lincei *14* (1931), 479–481

[Gx2] Godeaux, L.: Sur la construction de surfaces non rationelles de genres zero, Bull. Acad. R. Belg. *45* (1949), 688–693

[Ha1] Hartshorne, R.: Ample Vector Bundles, Publ. Math. Inst. Hautes Etud. Sci. *29* (1966), 319–394

[Ha2] Hartshorne, R.: *Algebraic Geometry*, Graduate Texts *49*, Springer, Heidelberg (1977)

[He] Helgason, S.: *Differential geometry and symmetric spaces.* Acad Press, New York/London (1962)

[Hik1] Hironaka, H.: Resolution of singularities of an algebraic variety over a field of characteristic zero I, II, Ann. Math. *79* (1964), 109–326

[Hik2] Hironaka, H.: Bimeromorphic smoothing of complex analytic spaces, preprint Math. Inst. Warwick (1971)

[Hir1] Hirzebruch, F.: Über vierdimensionale Riemannsche Flächen mehrdeutiger analytischer Funktionen von zwei Veränderlichen, Math. Ann. *126* (1953), 1–22

[Hir2] Hirzebruch, F.: Automorphe Formen und der Satz von Riemann-Roch, in *Symp. Int. Top. Alg.*, México, Univ. México (1958), 129–144

[Hir3] Hirzebruch, F.: Komplexe Mannigfaltigkeiten, in *Proc. Int. Congr. Math.* 1958, Cambr. Univ. Press, Cambridge (1960)

[Hir4] Hirzebruch, F.: *Topological methods in Algebraic Geometry*, Grundl. Math. Wiss. *131*, Springer, Heidelberg (1966)

[Hir5] Hirzebruch, F.: Hilbert modular surfaces, Enseign. Math. *19* (1973), 183–281

[Hir6] Hirzebruch, F.: Modulflächen und Modulkurven zur symmetrischen Hilbertschen Modulgruppe, Ann. Scient. Éc. Norm. Sup., IV Ser. *11* (1978), 101–165

[Hir7] Hirzebruch, F.: Arrangements of lines and algebraic surfaces, preprint Max-Planck-Institut f. Mathematik, Bonn (1982)

[Hir8] Hirzebruch, F.: Chern numbers of algebraic surfaces, an example, preprint Max-Planck-Institut f. Mathematik, Bonn (1983)

[H-K] Hirzebruch, F., Kodaira, K.: On the complex projective spaces, J. Math. Pures Appl. *36* (1957), 201–216

[Hol1] Holzapfel, R.-P.: A class of minimal surfaces in the unknown region of surface geography, Math. Nachr. *98* (1980), 221–232

[Hol2] Holzapfel, R.-P.: Invariants of arithmetic ball quotient surfaces, Math. Nachr. *103* (1981), 117–153

[Hop1] Hopf, H.: Zur Topologie der komplexen Mannigfaltigkeiten, in *Studies and Essays presented to R. Courant*, Interscience, New York (1948), 167–185

[Hop2] Hopf, H.: Schlichte Abbildungen und lokale Modifikationen 4-dimensionaler komplexer Mannigfaltigkeiten, Comment. Math. Helv. *29* (1954), 132–156

[Hor1] Horikawa, E.: On deformations of quintic surfaces, Invent. Math. *31* (1975), 43–85

[Hor2] Horikawa, E.: Algebraic surfaces of general type with small c_1^2, I, Ann. Math. *104* (1976), 357–387. II, Invent. Math. *37* (1976), 121–155. III, Inv. Math. *47* (1978), 209–248. IV, Inv. Math. *50* (1979), 103–128

[Hor3] Horikawa, E.: Surjectivity of the period map of K3-surfaces of degree 2, Math. Ann. *228* (1977), 113–146

[Hor4] Horikawa, E.: On the periods of Enriques surfaces, I, Math. Ann. *234* (1978), 73–108. II, Math. Ann. *235* (1978), 217–246

[H-V1] Hirzebruch, F., Ven, A. Van de: Hilbert modular surfaces and the classification of algebraic surfaces, Invent. Math. *23* (1974), 1–29

[H-V2] Hirzebruch, F., Ven, A. Van de: Minimal Hilbert modular surfaces with $p_g=3$, $K^2=2$, Am. J. Math. *101* (1979), 132–148

[H-Z] Hirzebruch, F., Zagier, D.: Classification of Hilbert modular surfaces, in *Complex Analysis and Algebraic Geometry*, Iwanami-Shoten, Tokyo (1977), 43–77

[Ii] Iitaka, S.: Deformations of compact complex surfaces II, J. Math. Soc. Japan *22* (1970) 247–261. III, ibid. *23* (1970), 692–705

[In1] Inoue, M.: On surfaces of class VII_0, Invent. Math. *24* (1974), 269–310

[In2] Inoue, M.: New surfaces with no meromorphic functions, Proc. Int. Congr. Vancouver 1974, 423–426, ibid. II, in *Complex Analysis and Algebraic Geometry.* Iwanami-Shoten, Tokyo (1977), 91–106

[In3] Inoue, M.: Some surfaces of general type with $p_g = q = 0$, preprint (1979)
[In4] Inoue, M.: Some surfaces of general type with positive indices, preprint (1981)
[J] James, I.D.: On Witt's theorem for unimodular quadratic forms, Pac. J. Math. *26* (1968), 303–316
[J-Y] Jost, J., Yau, S.-T.: Harmonic mappings and Kähler manifolds. Math. Ann. *262* (1983), 145–166
[Ka1] Kato, M.: Topology of Hopf surfaces, J. Math. Soc. Japan *27* (1975), 222–238
[Ka2] Kato, M.: Compact complex manifolds containing 'global' spherical shells I, in *Int. Symp. Alg. Geometry, Kyoto 1977*, Kinokuniya, Tokyo (1978), 45–84
[Ki-V] Kiehl, R., Verdier, J.-L.: Ein einfacher Beweis des Kohärenzsatzes von Grauert, Math. Ann. *195* (1971/1972), 24–50
[K-N] Kobayashi, S., Nomizu, K.: *Foundations of Differential Geometry I*, John Wiley & Sons, New York (1963). *Ibid. II* (1969)
[K-N-S] Kodaira, K., Nirenberg, L., Spencer, D.: On the existence of deformations of complex structures, Ann. Math. *68* (1958), 450–459
[Ko1] Kodaira, K.: On Kähler varieties of restricted type (an intrinsic characterisation of algebraic varieties), Ann. Math. *60* (1954), 28–48
[Ko2] Kodaira, K.: On compact complex analytic surfaces I, Ann. Math. *71* (1960), 111–152. II, Ann. Math. *77* (1963), 563–626. III, Ann. Math. *78* (1963), 1–40
[Ko3] Kodaira, K.: On stability of compact submanifolds of complex manifolds, Ann. Math. *85* (1963), 79–94
[Ko4] Kodaira, K.: On the structure of compact complex analytic surfaces I, Am. J. Math. *86* (1964), 751–798. II, Am. J. Math. *88* (1966), 682–721. III, Am. J. Math. *90* (1969), 55–83. IV, ibid., 1048–1066
[Ko5] Kodaira, K.: A certain type of irregular algebraic surfaces, J. Anal. Math. *19* (1967), 207–215
[Ko6] Kodaira, K.: Pluricanonical systems on algebraic surfaces of general type, J. Math. Soc. Japan *20* (1968), 170–192
[Ko7] Kodaira, K.: On homotopy K3-surfaces, in *Essays on Topology and Related Topics*, Springer, Heidelberg (1970), 58–69
[Kr] Krasnov, V.: Compact complex surfaces without meromorphic functions, Math. Zametki *17* (1975), 119–122
[Kug] Kuga, M.: FAFA-Note, 1975
[Kul] Kulikov, V.: Degenerations of K3-surfaces and Enriques surfaces, Izv. Akad. Nauk, SSSR, Ser. Math. *41* (1977), 1008–1042
[Kur] Kurke, H.: *Vorlesungen über algebraische Flächen*, Teubner-Texte zur Math., *43*, Teubner, Leipzig (1982)
[Ks] Kas, A.: On deformations of a certain type of irregular algebraic surface, Am. J. Math. *90* (1968), 789–804
[K-S1] Kodaira, K., Spencer, D.: A theorem of completeness for complex analytic fibre spaces, Acta Math. *100* (1958), 281–294
[K-S2] Kodaira, K., Spencer, D.: On deformations of complex analytic structures III, Stability theorems for complex analytic structures, Ann. Math. *75* (1962), 536–577
[La] Lamotke, K.: Die Homologie isolierter Singularitäten, Math. Z. *143* (1975), 27–44
[Le] Lefschetz, S.: *L'Analyse Situs et la Géometrie Algébrique*, Gauthier-Villars, Paris (1924)
[Li] Lipman, J.: Introduction to resolution of singularities, *Proc. Symp. Pure Math. 24, Algebraic Geometry, Arcata 1974*, A.M.S., Providence, R.I.: (1975), 187–230
[Liv] Livné, R.A.: On certain covers of the universal elliptic curve, thesis, Harvard (1981)
[Lj] Łojasiewicz, S.: Triangulation of semianalytic sets, Ann. Sc. Norm. Super. Pisa, III Ser. *18* (1964), 449–474
[Lo] Looijenga, E.: A Torelli theorem for Kähler-Einstein K3-surfaces, in *Geometry Sympos. Utrecht 1980*, Lect. Notes Math. *894*, Springer, Heidelberg (1981), 107–112
[L-P] Looijenga, E., Peters, C.: Torelli theorems for Kähler K3-surfaces, Compos. Math. *42* (1981), 145–186
[Maa] Maaß, H.: *Siegel's modular forms and Dirichlet series*, Lect. Notes Math. *216*, Springer, Heidelberg (1971)
[Mar] Martens, H.: A new proof of Torelli's theorem, Ann. Math. *78* (1963), 107–111
[May] Mayer, A.: Families of K3-surfaces, Nagoya Math. J. *48* (1972), 1–17

[M-H] Milnor, J., Husemoller, D.: *Symmetric bilinear forms*, Erg. d. Math. *73*, Springer, Heidelberg (1973)

[Mi1] Miyaoka, Y.: Tricanonical maps of numerical Godeaux surfaces, Invent. Math. *34* (1976), 99–111

[Mi2] Miyaoka, Y.: On numerically Campedelli surfaces, in *Complex Analysis and Algebraic Geometry*, Iwanami-Shoten, Tokyo (1977), 112–118

[Mi3] Miyaoka, Y.: On the Chern numbers of surfaces of general type, Invent. Math. *42* (1977), 225–237

[Mi4] Miyaoka, Y.: On algebraic surfaces with positive index, preprint 1980

[Mi5] Miyaoka, Y.: On the Chern numbers of surfaces of general type II, preprint Max-Planck-Institut f. Mathematik, Bonn (1983)

[Mil1] Milnor, J.: On simply connected 4-manifolds, in *Symp. Int. Top. Alg. México*, Univ. México (1958), 122–128

[Mil2] Milnor, J.: *Morse Theory*, Ann. Math. Studies 76, Princeton Univ. Press, Princeton (1974)

[M-K] Morrow, J., Kodaira, K.: *Complex manifolds*, Holt-Rinehart & Winston, New York (1971)

[Mo1] Mostow, G.D.: Existence of a non-arithmetic lattice in $SU(2,1)$, Proc. Natl. Acad. Sci. USA *75* (1978), 3029–3033

[Mo2] Mostow, G.D.: On a remarkable class of polyhedra in complex hyperbolic space, Pac. J. Math. *86* (1980), 171–276

[Mo3] Mostow, G.D.: Complex reflection groups and non-arithmetic monodromy, To appear in Proc. Natl. Acad. Sci. USA

[M-S] Milnor, J., Stasheff, J.: *Characteristic classes*, Ann. Math. Studies 76, Princeton Univ. Press, Princeton (1974)

[M-Si] Mostow, G.D., Siu, Y.-T.: A compact Kähler surface of negative curvature not covered by the ball, Ann. Math. *112* (1980), 321–360

[Mu1] Mumford, D.: The topology of normal singularities of an algebraic surface and a criterion for simplicity, Publ. Math. I.H.E.S. *9* (1961)

[Mu2] Mumford, D.: The canonical ring of an algebraic surface, Ann. Math. *76* (1962), 612–615

[Mu3] Mumford, D.: *Lectures on curves on an algebraic surface*, Ann. Math. Studies *59*, Princeton Univ. Press, Princeton (1966)

[Mu4] Mumford, D.: Pathologies III, Am. J. Math. *89* (1967), 94–104

[Mu5] Mumford, D.: Enriques' classification of surfaces in char. p I, in *Global Analysis*, Princeton Univ. Press, Princeton (1969) 325–339

[Mu6] Mumford, D.: An algebraic surface with K ample, $K^2=9$, $p_g=q=0$, Am. J. Math. *101* (1979), 233–244

[N] Nikulin, V.: On Kummer surfaces, Izv. Akad. Nauk. SSSR, Ser. Math. *39* (1975), 278–293

[O-P] Oort, F., Peters, C.: A Campedelli surface with torsion group $\mathbb{Z}/2$, Indagationes Math. *43* (1981), 399–407

[Pa] Paršin, A.N.: Algebraic curves over function fields I, Math. USSR Izv. *2* (1968), 1145–1170

[Per] Persson, U.: On Chern invariants of surfaces of general type, Compos. Math. *43* (1981), 3–58

[Pet1] Peters, C.: On two types of surfaces of general type with vanishing geometric genus, Invent. Math. *32* (1976), 33–47

[Pet2] Peters, C.: On certain examples of surfaces with $p_g=0$ due to Burniat, Nagoya Math. J. *66* (1977), 109–119

[Pi] Piatečkii-Shapiro, I.I.: *Geometry of Classical Domains and Automorphic Functions* (Russian), Fizmatgiz, Moscow (1961). French transl., Dunod, Paris (1966)

[Pi-S] Piatečkii-Shapiro, I.I., Shafarevič, I.R.: A Torelli theorem for algebraic surfaces of type K-3, Izv. Akad. Nauk. SSSR, Ser. Math. *35* (1971), 503–572

[P-P] Persson, U., Pinkham, H.: Degeneration of surfaces with trivial canonical bundles, Ann. Math. *113* (1981), 45–66

[P-Si] Picard, E., Simart, G.: *Théorie des fonctions algébriques de deux variables indépendents, Vol. I, II*, Gauthiers-Villars, Paris (1897, 1906)

[Ra] Raynaud, M.: Familles de fibrés vectoriels sur une surface de Riemann, Sém. Bourbaki, exp. *316* (1966)

[Ram] Ramanujam, C.P.: Remarks on the Kodaira vanishing theorem, J. Indian Math. Soc. *36* (1972), 41–51

[Rei1] Reid, M.: Bogomolov's theorem $c_1^2 \leq 4c_2$, in *Proc. Int. Symp. Alg. Geom., Kyoto 1977*, Kinokuniya, Tokyo (1977), 623–642

[Rei2] Reid, M.: Surfaces with $p_g=0$, $K^2=1$, J. Fac. Sci., Univ. Tokyo, Sect. IA, 25 (1978), 75–92

[Rei3] Reid, M.: Surfaces with $p_g=0$, $K^2=2$, preprint 1979

[Rem1] Remmert, R.: Meromorphe Funktionen in kompakten komplexen Räumen, Math. Ann. *133* (1956), 277–288

[Rem2] Remmert, R.: Holomorphe und meromorphe Abbildungen komplexer Räume, Math. Ann. *133* (1957), 328–370

[Rh] Rham, G. de: *Variétés différentiables*, Act. Sci Ind. 1222, Hermann, Paris (1954)

[R-R] Ramis, J.P., Ruget, G.: Complexe dualisant et théorèmes de dualité en géométrie analytique complexe, Publ. Math. Inst. Hautes Etud. Sci. *38* (1971), 77–91

[R-R-V] Ramis, J.P., Ruget, G., Verdier, J.-L.: Dualité relative en géométrie analytique complexe, Invent. Math. *13* (1971), 261–283

[R-S] Remmert, R., Stein, K.: Über die wesentlichen Singularitäten analytischer Mengen. Math. Ann. *126* (1953), 263–306

[R-V1] Remmert, R., Ven, T. Van de: Zwei Sätze über die komplex-projektive Ebene. Nieuw. Arch. Wisk. *(3) VIII* (1960), 147–157

[R-V2] Remmert, R., Ven, A. Van de: Zur Funktionentheorie homogener komplexer Mannigfaltigkeiten, Topology *2* (1963), 137–157

[R-W] Reeb, G., Wu Wen Tsun: *Sur les espaces fibrés et les variétés feuilletés*, Act. Sci. Ind. 1183, Hermann, Paris (1952)

[Saf] Šafarevič, I.R., et al.: *Algebraic surfaces*, Proc. Steklov Inst. Math. *75* (1965), A.M.S. Translations Providence, R.I. (1967)

[Sai] Saint-Donat, B.: Projective models of K3-surfaces, Am. J. Math. *96* (1974), 602–639

[Sak] Sakai, F.: Semi-stable curves on algebraic surfaces and logarithmic pluricanonical maps. Math. Ann. *254* (1980), 89–120

[Sem.P.] Séminaire Palaiseau 1978: *Première classe de Chern et courbure de Ricci: preuve de la conjecture de Calabi*, Astérisque *58*, Soc. Math. France, Paris (1978)

[Se1] Serre, J-P.: Un théoreme de dualité, Comment. Math. Helv. *29* (1955), 9–26

[Se2] Serre, J-P.: Faisceaux algébriques cohérents, Ann. Math. *61* (1955), 197–278

[Se3] Serre, J-P.: Sur la dimension homologique des anneaux et des modules noethériens, in *Proc. Int. Symp. Alg. Number Theory*, Tokyo and Nikko 1955, Kasai. Tokyo (1956), 175–189

[Se4] Serre, J-P.: Géométrie algébrique et géométrie analytique, Ann. Inst. Fourier *6* (1956), 1–42

[Se5] Serre, J-P.: *Groupes algébriques et corps de classes*, Act. Sci. Ind. 1264, Hermann, Paris (1959)

[Se6] Serre, J-P.: *Cours d'arithmétique*, Presses Univ. de France, Paris (1970)

[Sev] Severi, F.: Some remarks on the topological classification of surfaces, in Studies presented to R. von Mises, Acad. Press, New York (1954)

[Sha1] Shah, J.: Surjectivity of the period map in the case of quartic surfaces and sextic double planes, Bull. Am. Math. Soc. *82* (1976), 716–718

[Sha2] Shah, J.: A complete moduli space for K3-surfaces of degree 2, Ann. Math. *112* (1980), 485–510

[Sha3] Shah, J.: Degeneration of K3-surfaces of degree 4, Trans. Am. Math. Soc. *263* (1981), 271–308

[Shi] Shioda, T.: The period map of abelian surfaces, J. Fac. Sci. Univ. Tokyo, Sect. IA, *25* (1978), 47–59

[Si] Siu, Y.T.: Every K3-surface is Kähler. Invent. Math. *73* (1983), 139–150

[S-I] Shioda, T., Inose, H.: On singular K3-surfaces, in *Complex Analysis and Algebraic Geometry*, Iwanami-Shoten, Tokyo (1977), 117–136

[S-M] Shioda, T., Mitani, N.: Singular abelian surfaces and binary quadratic forms, in *Classification of algebraic varieties and compact complex manifolds*, Lect. Notes Math. *412*, Springer, Heidelberg (1974), 259–287

[So1] Sommese, A.: Hyperplane sections of projective surfaces I – The adjunction mapping, Duke Math. J. *46* (1979), 377–401

[So2] Sommese, A.: Theorem A. The set ... Manuscript Max-Planck-Institut für Mathematik, Bonn (1983)

[Sp] Spanier, E.: *Algebraic Topology*, Mc. Graw-Hill, New-York (1966)

[St] Stein, K.: Analytische Zerlegungen komplexer Räume, Math. Ann. *132* (1956), 63–93

[S-T1] Singer, I., Thorpe, J.: *Lecture Notes on Elementary Topology and Geometry*, Scott-Foresman and Co, Glenview (1967)

[S-T2] Singer, I., Thorpe, J.: The curvature of 4-dimensional Einstein spaces, in *Global Analysis*, Princeton Univ. Press, Princeton (1969), 355–365

[Su] Suwa, T.: Ruled surfaces of genus 1, J. Math. Soc. Japan *21* (1969), 291–311

[Sv] Svarcman, O.: Simple-connectedness of the factor spaces of the Hilbert-modular group, Func. Anal. Appl. *8* (1974), 188–189

[Tj1] Tjurin, A.N.: On the classification of two-dimensional fibre bundles over an algebraic curve of arbitrary genus, Izv. Akad. Nauk. SSSR, Ser. Math. *28* (1964), 21–52

[Tj2] Tjurin, A.N.: Classification of vector bundles over an algebraic curve of arbitrary genus, Izv. Akad. Nauk. SSSR, Ser. Math. *29* (1965), 657–688

[To] Todorov, A.: Applications of the Kähler-Einstein-Calabi-Yau metric to moduli of K3-surfaces, Invent. Math. *81* (1980), 251–266

[Ue1] Ueno, K.: *Classification theory of algebraic varieties and compact complex spaces*, Lect. Notes Math. *439*, Springer, Heidelberg (1975)

[Ue2] Ueno, K.: A remark on automorphisms of Enriques surfaces, J. Fac. Sci. Univ. Tokyo, Sect. IA, *23* (1976), 149–165

[Ue3] Ueno, K.: Kodaira dimensions for certain fibre spaces, in *Complex Analysis and Algebraic Geometry*, Iwanami-Shoten, Tokyo (1977), 279–292

[Ue4] Ueno, K.: Classification of algebraic manifolds, Proc. Int. Congr. Math. Helsinki 1978, 549–556

[Ve1] Ven, A. Van de: On the Chern numbers of certain complex and almost-complex manifolds, Proc. Natl. Acad. Sci. USA *55* (1966), 1624–1627

[Ve2] Ven, A. Van de: On the Chern numbers of surfaces of general type, Invent. Math. *36* (1976), 285–293

[Ve3] Ven, A. Van de: Some recent results on surfaces of general type, Sém. Bourbaki, exp. 500 (1977), Lect. Notes Math. *677*, Springer, Heidelberg (1978), 155–166

[Ve4] Ven, A. Van de: On the Enriques classification of algebraic surfaces, Sém. Bourbaki, exp. 506 (1977), Lect. Notes Math. *677*, Springer, Heidelberg (1978), 237–251

[Ve5] Ven, A. Van de: On the 2-connectedness of very ample divisors on a surface, Duke Math. J. *46* (1979), 403–407

[Vie] Viehweg, E.: Canonical divisors and the additivity of the Kodaira dimension for morphisms of relative dimension one, Compos. Math. *35* (1977), 197–223

[Vin] Vinberg, E.: Discrete groups generated by reflections, Izv. Akad. Nauk. SSSR, Ser. Math. *35* (1971), 1083–1119

[Wal] Walker, R.: *Algebraic curves*, Princeton Univ. Press, Princeton (1950)

[Wav] Wavrik, J.: Obstructions to the existence of a space of moduli, in *Global Analysis*, Princeton Univ. Press, Princeton (1969), 403–413

[Weh] Wehler, J.: Versal deformations of Hopf surfaces, J. Reine Angew. Math. *328* (1981), 22–32

[Wei1] Weil, A.: *Variétés Kählériennes*, Act. Sci. Ind. 1267, Hermann, Paris (1958)

[Wei2] Weil, A.: *Œuvres Scientifiques (Collected Papers)* I, II, III. Springer, Heidelberg (1980)

[Y1] Yau, S.-T.: Parallelizable manifolds without complex structures, Topology *15* (1976), 51–54

[Y2] Yau, S.-T.: Calabi's conjecture and some new results in algebraic geometry, Proc. Natl. Acad. Sci. USA *74* (1977), 1789–1799

[Y3] Yau, S.-T.: On the Ricci-curvature of a complex Kähler manifold and the complex Monge-Ampère equation, Comment. Pure Appl. Math. *31* (1978), 339–411

[Za1] Zariski, O.: *Introduction to the problem of minimal models in the theory of algebraic surfaces*, Publ. Math. Soc. Japan *4* (1958)

[Za2] Zariski, O.: On Castelnuovo's criterion $p_a = P_2 = 0$. Ill. J. Math. *2* (1958), 303–315

[Za3] Zariski, O.: *Algebraic surfaces*, 2nd ed., Erg. Math. 61, Springer, Heidelberg (1971)

Notations

(see also p. 9)

G_X, the constant sheaf on X with stalk G 9
$h^i(X,\mathcal{S})=h^i(\mathcal{S})=\dim_{\mathbb{C}} H^i(X,\mathcal{S})$ 10
ab, $a.b$, (a,b), cupproduct of $a,b\in H^*(X,G)$ 10
$f_{*i}(\mathcal{S})=f_{*i}\mathcal{S}$, i-th direct image of $\mathcal{S}$ by f 10
$f_*(\mathcal{S})=f_*\mathcal{S}$ direct image of $\mathcal{S}$ by f 10
$f^{-1}(\mathcal{S})$, inverse image of $\mathcal{S}$ by f 10
$\mathcal{P}_X$, Poincaré-duality isomorphism 11
$f^!$, $f_!$ 11
$T^i(X)$, torsion subgroup of $H^i(X,\mathbb{Z})$ 11
$H^i(X,\mathbb{Z})_f=H^i(X,\mathbb{Z})/T^i(X)$ 11
$\langle\ ,\ \rangle$, bilinear form of a lattice 12
$d(L)$, discriminant of a lattice 12
$\pm\mathbb{1}$, the 1-dimensional unimodular lattices 14
H, the hyperbolic plane 14
E_8, the root lattice 14
$Q(\Gamma)$, the quadratic form associated to the graph Γ 16
A_n, B_n, E_n, certain graphs 16, curve singularities 64, surface singularities 87
$\tilde{A}_n$, $\tilde{B}_n$, $\tilde{E}_n$, certain graphs 16
$\mathbb{P}(\mathcal{V})$, the projective bundle associated to a vectorbundle $\mathcal{V}$ 17
$\mathcal{V}(x)$, the fibre at x of a vector bundle $\mathcal{V}$ 17
$w_i(\mathcal{V})$, $w(\mathcal{V})$, the Stiefel-Whitney classes of $\mathcal{V}$ 17
$p_i(\mathcal{V})$, $p(\mathcal{V})$, the Pontrjagin classes of $\mathcal{V}$ 17
$c_i(\mathcal{V})$, $c(\mathcal{V})$, the Chern classes of $\mathcal{V}$ 17
$L(\mathcal{V})$, the L-class of $\mathcal{V}$ 17
$\mathrm{Todd}(\mathcal{V})$, the Todd class of $\mathcal{V}$ 17
$\tau(X)$, the index of X 18
$\mathrm{ch}(\mathcal{V})$, the Chern character of $\mathcal{V}$ 18
b^+, b^- 18
$\mathcal{T}_X$, the holomorphic tangent bundle of X 18
$c_i(X)$, the i-th Chern class of X 18
$e(X)$, the Euler number of X 18
Ω^i_X, the sheaf of germs of holomorphic i-forms 18
$\mathcal{K}_X$, the canonical bundle of X 18
$\mathcal{N}_{Y|X}$, the normal bundle of Y in X 18
$\chi(X,\mathcal{S})$, the Euler characteristic of $\mathcal{S}$ 19
$\chi(X)=\chi(\mathcal{O}_X)$ 19
$p_a(X)$, the arithmetic(al) genus of X 19
$q(X)$, the irregularity of X 19
$p_g(X)$, the geometric genus of X 19
$\deg(\mathcal{L})$, the degree of a line bundle $\mathcal{L}$ on a curve 20
$\mathrm{Pic}(X)$, the Picard group of X 21
$\mathrm{Pic}^0(X)$ 21
$\mathcal{O}_X(D)=\mathcal{O}(D)$, the line bundle associated to a divisor D 22
$|D|$, the linear system associated to D 22
$K_X=K$, any canonical divisor on X 22
$\mathcal{S}|Y$, the analytic restriction of $\mathcal{S}$ to Y 22
$R(X)$, the canonical ring of X 23
$\mathrm{kod}(X)$, the Kodaira dimension of X 23
$P_m(X)$, the m-th plurigenus of X 23
$a(X)$, the algebraic dimension of X 23
$f^*(\mathcal{S})$, the analytic inverse image of $\mathcal{S}$ 24
$\mathcal{I}_{Y|X}=\mathcal{I}_Y$, the ideal sheaf of Y in X 24
X_{red}, the reduction of X 24
X_{norm}, the normalisation of X 25
$\mathcal{S}_y$, stalk of $\mathcal{S}$ at y 25 (except in III.8.5)
X_y, the analytic fibre over y 26
$(f_{*q}(\mathcal{S}))^\wedge_y$, the formal completion of $f_{*q}(\mathcal{S})$ at y 26
$\mathcal{X}=(X,p,S)$, a family of complex manifolds 29
$\rho_{\mathcal{X}}$, the Kodaira-Spencer map of $\mathcal{X}$ 31
H^p_{DR}, the p-th de Rham group 33
$H^{p,q}(X)$, $h^{p,q}(X)$ 33
$F^p(H^k)$ 33
$\mathrm{Alb}(X)$, the Albanes torus of X 35
$\mathrm{Alb}(H_{\mathbb{C}})$ 37
$\mathfrak{H}_g$, the Siegel upper half space 38
$Sp(g,\mathbb{Z})$, the symplectic group (operating on $\mathbb{Z}^{2g}$) 39
Γ_g, the modular group 39
$D_g=\mathfrak{H}_g/\Gamma_g$ 39
ω_C, the dualising sheaf of C 48
$\deg(\mathcal{F})$, the degree of a locally free sheaf $\mathcal{F}$ on a curve 51
res, the residue map 54
tr_D, the trace map of the curve D 53
Tr_Y, the trace map of the surface Y 54
$i_x(C,D)$, the intersection index at x of C and D 65
$DE=(D,E)$, the intersection number of the divisors D and E 66
$(\mathcal{L},D)$, the intersection number of a line bundle $\mathcal{L}$ and a divisor D 66

$(\mathscr{L}, \mathscr{M})$, the intersection number of line bundles $\mathscr{L}$ and $\mathscr{M}$ 66
$g(C)$, the arithmetic genus of a curve C on a surface 68
$X_{\min}$, the minimal model of X 79
Δ, the unit disc 92
$\Omega_{X|S}$, the sheaf of relative differentials 98
$\omega_{X|S}$, the dualising sheaf 98
Tr, the relative trace map 98
$f_{\mathscr{L}}$, the meromorphic map associated to the line bundle $\mathscr{L}$ 113
$f_m = f_{\mathscr{K}_X^{\otimes m}}$ 114
$\rho(X)$, the Picard number of X 120
$\mathscr{GL}(n+1, \mathbb{C})$, $\mathscr{PGL}(n+1, \mathbb{C})$ 140
Σ_n, the n-th Hirzebruch surface 140
$\mathscr{A}_B$, $\mathscr{E}_B$ 143
I_n, II, III, IV, I_n^*, II*, III*, IV*, ${}_m\mathrm{I}_0$, ${}_m\mathrm{I}_1$, ${}_m\mathrm{I}_n$, Kodaira's terminology for the singular elliptic fibres 150
Jac(f), the jacobian fibration of f 153
$X^{\#}$ 155
1_G, the unit element of a group G 170
Km(T), the Kummer surface of the torus T 170
$\mathscr{C}_X$, the positive cone 238
S_X, the Picard lattice of X 238
T_X, the transcendental lattice of X 238
W_X 239
s_d 239
L, $L_{\mathbb{C}}$, $L_{\mathbb{R}}$ 239
Ω 239
L^- 240
Ω^- 240
$\mathscr{C}_X^+$, the Kähler cone 242
$T(V)$, the vector space of translations of the affine space V 244
$E(\omega_X)$ 254
$K\Omega$, $E(\kappa, \omega)$, $(K\Omega)^0$ 257
$S(I, J)$ 265
s_c 272
Γ 280
D 280
H_d 283

We draw special attention to the following conventions.

- Line bundles and vector bundles are denoted by script capitals, so e.g. $\mathscr{K}_X$ denotes the canonical bundle on X. Divisors are denoted by roman capitals, so K_X or K will denote a(ny) canonical divisor. $\mathscr{K}$ and K are not used completely at random; as long as it is not sure that there are any canonical divisors, we only use $\mathscr{K}$, whereas both $\mathscr{K}$ and K are used as soon as the existence of a canonical divisor is assured.
- The symbol ϕ as used in Chapter VIII refers to an isometry, the corresponding capital Φ refers to an isomorphism of local systems.

Index